T0181750

MULTIPLE SCLEROSIS
REHABILITATION

From Impairment to Participation

REHABILITATION SCIENCE IN PRACTICE SERIES

Series Editors

Marcia J. Scherer, Ph.D.

President
Institute for Matching Person and Technology

Professor
Orthopaedics and Rehabilitation
University of Rochester Medical Center

Dave Muller, Ph.D.

Executive
Suffolk New College

Editor-in-Chief
Disability and Rehabilitation

Founding Editor
Aphasiology

Published Titles

Assistive Technology Assessment Handbook,
edited by Stefano Federici and Marcia J. Scherer

Multiple Sclerosis Rehabilitation: From Impairment to Participation,
edited by Marcia Finlayson

Paediatric Rehabilitation Engineering: From Disability to Possibility,
edited by Tom Chau and Jillian Fairley

Quality of Life Technology, *Richard Schultz*

Forthcoming Titles

Ambient Assisted Living, *edited by Nuno M. Garcia, Joel Jose P. C. Rodrigues,*
Dirk Christian Elias, Miguel Sales Dias

Assistive Technology for the Visually Impaired/Blind,
Roberto Manduchi and Sri Kurniawan

Computer Systems Experiences of Users with and without Disabilities:
An Evaluation Guide for Professionals,
Simone Borsci, Masaaki Kurosu, Stefano Federici, Maria Laura Mele

Neuroprosthetics: Principles and Applications, *Justin C. Sanchez*

Rehabilitation Goal Setting: Theory, Practice and Evidence,
edited by Richard Siegert and William Levack

MULTIPLE SCLEROSIS REHABILITATION

From Impairment to Participation

edited by

Marcia Finlayson

CRC Press
Taylor & Francis Group
Boca Raton London New York

CRC Press is an imprint of the
Taylor & Francis Group, an **informa** business

CRC Press
Taylor & Francis Group
6000 Broken Sound Parkway NW, Suite 300
Boca Raton, FL 33487-2742

First issued in paperback 2017

© 2013 by Taylor & Francis Group, LLC
CRC Press is an imprint of Taylor & Francis Group, an Informa business

No claim to original U.S. Government works

Version Date: 20120518

ISBN 13: 978-1-138-07256-5 (pbk)
ISBN 13: 978-1-4398-2884-7 (hbk)

Library of Congress Cataloging-in-Publication Data

Multiple sclerosis rehabilitation : from impairment to participation / [edited by] Marcia Finlayson.
     p. ; cm. -- (Rehabilitation science in practice series)
  Includes bibliographical references and index.
  ISBN 978-1-4398-2884-7 (hardback : alk. paper)
  I. Finlayson, Marcia. II. Series: Rehabilitation science in practice series.
  [DNLM: 1. Multiple Sclerosis--rehabilitation. 2. Activities of Daily Living. 3. Adaptation, Psychological.
4. Quality of Life. WL 360]

616.8'3406--dc23   2012017893

Visit the Taylor & Francis Web site at
http://www.taylorandfrancis.com

and the CRC Press Web site at
http://www.crcpress.com

To the individuals and families living with multiple sclerosis who have challenged my thinking,

encouraged me to seek answers to the questions that are important to their health and functioning,

and inspired my commitment to improve the quality of rehabilitation services that they receive.

Marcia Finlayson

Contents

Section V   Moving into the Future

Acknowledgments

While *Multiple Sclerosis Rehabilitation: From Impairment to Participation* reflects my own vision for a comprehensive text that can serve rehabilitation professionals and students from a range of disciplines, the final product represents the efforts and commitment of many people. Without them, my vision could not have been realized.

I extend my sincere thanks to each of my colleagues who agreed to author or coauthor chapters in this book. The time they committed to this project and the level of expertise they shared are greatly appreciated. The quality and depth of the contents of this book are a direct result of their contributions.

I also extend my thanks and appreciation to the students, practicing clinicians, and researchers who reviewed the chapters and provided feedback. Their comments enabled the authors and me to make revisions and additions to the content and flow so that chapters can meet the needs of the widest audience possible.

Writing, reviewing, and revising are the "big picture" tasks of preparing a book such as this one. So many other tasks go on behind the scenes: tracking down references; formatting reference lists, text, and tables; acquiring permissions; proofreading; keeping track of authors and reviewers; and so on. Over the course of this project, three graduate students from the Department of Occupational Therapy at the University of Illinois at Chicago (UIC) provided the support I needed to get all of this work done: Meghan Ginter, Teresa Morreale, and Emily Frank. I cannot thank them enough for their hard work, patience, and commitment to this project. Other critical people working behind the scenes on this book included Eileen Fitzsimons (copyediting and proofreading), Michael Slaughter and Jill Jurgensen (from Taylor & Francis), and my last minute proofreaders, Staci Molinar and Rebecca Nixon. Their efforts also contributed to the quality of our final product.

I am also indebted to my colleague and friend, Dr. Susan Coote of the Department of Physical Therapy at the University of Limerick, Ireland. She was at UIC for sabbatical during the last 10 weeks of work on this book. Her willingness to review the final chapters, help me frame critical passages, and support me and my team cannot be measured. Thank you does not feel like enough.

Finally, special thanks to my husband Greg; my parents, Frank and Darlene Ptosnick; and my sister, Myra, for their love, support, and patience throughout this project. Without them, this project would never have come to fruition.

Editor

Marcia Finlayson has a bachelor's in medical rehabilitation (occupational therapy), and a master's degree and a PhD in community health sciences from the University of Manitoba in Canada. She is credentialed as an occupational therapist in Canada (OT (C)) and the United States (OTR/L). She is also credentialed as a multiple sclerosis clinical specialist (MSCS). During the preparation of this book, Dr. Finlayson was a full tenured professor in the Department of Occupational Therapy at the University of Illinois at Chicago, Illinois. As of September 2012, she is the Vice Dean (Health Sciences) and Director of the School of Rehabilitation Therapy at Queen's University in Kingston, Canada. Dr. Finlayson also holds faculty appointments in the Department of Community Health Sciences, University of Manitoba, Canada and the School of Occupational Therapy and Social Work, Curtin University, Australia.

Dr. Finlayson has been actively engaged in multiple sclerosis (MS) rehabilitation service delivery, program development and evaluation, advocacy, and research for over 20 years. As a clinician, researcher, and volunteer, she works consistently to enable people with MS and their family members to fully participate in everyday life despite the challenges that can accompany living with and managing this disease. Dr. Finlayson's current efforts focus on building clinically relevant knowledge that can inform the development, implementation and evaluation of self-management programs, rehabilitation services, and comprehensive MS care. Her research program addresses topics centering on aging with MS, fatigue management, falls prevention and management, caregiver support, and patterns and predictors of rehabilitation service use among people with MS. She has published over 80 peer-reviewed papers and delivered over 110 conference presentations. In addition, she has been invited to deliver lectures and workshops to both professional and consumer audiences locally, nationally, and internationally.

This book represents Dr. Finlayson's ongoing commitment to people who are affected by MS.

Contributors

Eynat Ben Ari
Department of Occupational Therapy
Tel Aviv University
Tel Aviv, Israel

Fabricio E. Balcazar
Department of Disability and Human
  Development
University of Illinois at Chicago
Chicago, Illinois

Denisa Baxter
Allied Health
Alberta Health Services
Foothills Medical Centre
Calgary, Alberta, Canada

Carolyn R. Baylor
Department of Rehabilitation Medicine
University of Washington
Seattle, Washington

Ralph H. B. Benedict
Department of Neurology
University at Buffalo
Buffalo, New York

Susan E. Bennett
Department of Rehabilitation Science
and
Department of Neurology
University at Buffalo
Buffalo, New York

Francois Bethoux
Mellen Center for Multiple Sclerosis
  Treatment and Research
The Cleveland Clinic Foundation
Cleveland, Ohio

Michelle Cameron
Department of Neurology
Oregon Health and Science University
Portland, Oregon

Lauren S. Caruso
Clinical Neuropsychologist
New York City, New York

Davide Cattaneo
LaRiCE Lab
Don Gnocchi Foundation
Milan, Italy

Tanuja Chitnis
Pediatric Neurology
Massachusetts General Hospital
Boston, Massachusetts

Dawn M. Ehde
Department of Rehabilitation
  Medicine
University of Washington
Seattle, Washington

Marcia Finlayson
Department of Occupational Therapy
University of Illinois at Chicago
Chicago, Illinois

Susan Forwell
Department of Occupational Science and
  Occupational Therapy
and
Department of Medicine
University of British Columbia
British Columbia, Vancouver, Canada

Robert T. Fraser
Department of Rehabilitation
  Medicine
University of Washington
Seattle, Washington

Neera Garga
Allied Health
Alberta Health Services
Foothills Medical Centre
Calgary, Alberta, Canada

Erin Gervais
Allied Health
Alberta Health Services
Foothills Medical Centre
Calgary, Alberta, Canada

Myrna Harden
Allied Health
Alberta Health Services
Foothills Medical Centre
Calgary, Alberta, Canada

Elizabeth A. Hartman
Department of Neurology and
    Rehabilitation
University of Illinois at Chicago
Chicago, Illinois

Jutta Hinrichs
OPTIMUS Program
Alberta Health Services
Foothills Medical Centre
Calgary, Alberta, Canada

Lisa I. Iezzoni
Harvard Medical School
and
Mongan Institute for Health Policy
Massachusetts General Hospital
Boston, Massachusetts

Mark P. Jensen
Department of Rehabilitation Medicine
University of Washington
Seattle, Washington

Risha Joffe
Allied Health
Alberta Health Services
Foothills Medical Centre
Calgary, Alberta, Canada

Sverker Johansson
Department of Neurobiology, Care
    Sciences and Society
Karolinska Institutet
and
Department of Physical Therapy
Karolinska University Hospital
Stockholm, Sweden

Kurt L. Johnson
Department of Rehabilitation
    Medicine
University of Washington
Seattle, Washington

Johanna Jonsdottir
Department of
    Neurorehabilitation
Don Gnocchi Foundation
Milan, Italy

Jürg Kesselring
Department of Neurology and
    Neurorehabilitation
Rehabilitation Centre
Valens, Switzerland

Daphne Kos
Department of Health Care—
    Occupational Therapy
Artesis University College
    Antwerp
Antwerp, Belgium

Anna L. Kratz
Department of Physical Medicine and
    Rehabilitation
University of Michigan
Ann Arbor, Michigan

Stephen Krieger
Department of Neurology
Mount Sinai School of Medicine
New York City, New York

Janice Lake
Neurosciences
Alberta Health Services
Foothills Medical Centre
Calgary, Alberta, Canada

Nicholas G. LaRocca
Research Programs Department
National Multiple Sclerosis
    Society
New York City, New York

Eva Månsson Lexell
Department of Rehabilitation Medicine
Skåne University Hospital
and
Department of Health Sciences
Lund University
Lund, Sweden

Kevin Lindland
Allied Health
Alberta Health Services
Foothills Medical Centre
Calgary, Alberta, Canada

Nancy Lowenstein
Department of Occupational Therapy
Boston University College of Health
    and Rehabilitation Sciences:
    Sargent College
Boston, Massachusetts

Ruth Ann Marrie
Department of Internal Medicine
    (Neurology)
University of Manitoba
Winnipeg, Manitoba, Canada

David C. Mohr
Department of Preventive
    Medicine
Northwestern University
Chicago, Illinois

Robert W. Motl
Department of Kinesiology and
    Community Health
University of Illinois at
    Urbana-Champaign
Urbana, Illinois

Julie Ann Nastasi
Department of Occupational Therapy
The University of Scranton
Scranton, Pennsylvania

Kenneth I. Pakenham
School of Psychology
The University of Queensland
Brisbane, Australia

Matthew Plow
Department of Biomedical
    Engineering
and
Department of Physical Medicine
    and Rehabilitation
Cleveland Clinic Lerner Research
    Institute
Cleveland, Ohio

James P. Robinson
Department of Rehabilitation
    Medicine
University of Washington
Seattle, Washington

Amy Perrin Ross
Department of Neurosciences
Loyola University Chicago
Chicago, Illinois

Janet C. Rucker
Department of Neurology and
    Ophthalmology
The Mount Sinai Medical
    Center
New York City, New York

Dory Sabata
Occupational Therapy Education
University of Kansas Medical
    Center
Kansas City, Kansas

Yolanda Suarez-Balcazar
Department of Occupational
    Therapy
University of Illinois at Chicago
Chicago, Illinois

Matthew H. Sutliff
Mellen Center for Multiple Sclerosis
    Treatment and Research
Neurological Institute
Cleveland Clinic
Cleveland, Ohio

Celestine Willis
Department of Disability and Human
    Development
University of Illinois at Chicago
Chicago, Illinois

Kathryn M. Yorkston
Department of Rehabilitation
    Medicine
University of Washington
Seattle, Washington

Reviewers

Rebecca Anderson
Fairview MS Achievement
    Center
St. Paul, Minnesota

Tanvi Bhatt
Department of Physical Therapy
University of Illinois at Chicago
Chicago, Illinois

Elissa C. Held Bradford
Department of Physical Therapy and
    Athletic Training
Saint Louis University
St. Louis, Missouri

Cara Brown
Department of Occupational
    Therapy
University of Manitoba
Winnipeg, Canada

Emily Cade
Andrew C. Carlos Multiple Sclerosis
    Institute at Shepherd Center
Atlanta, Georgia

Tracy Carrasco
Multiple Sclerosis Comprehensive Care
    Center of Central Florida
Orlando Health
Orlando, Florida

Susan Coote
Department of Physiotherapy
University of Limerick
Limerick, Ireland

Margarita Corry
School of Nursing and
    Midwifery
Trinity College Dublin
Dublin, Ireland

Kelly Crossley
Rehabilitation Institute of Chicago
Chicago, Illinois

Clint Douglas
School of Nursing and Midwifery
Queensland University of Technology
Kelvin Grove, Queensland, Australia

Peter Feys
Department of Rehabilitation Sciences
    and Physiotherapy
University of Hasselt
Hasselt, Belgium

Emily Frank
Department of Occupational Therapy
University of Illinois at Chicago
Chicago, Illinois

Setareh Ghahari
Department of Occupational Science and
    Occupational Therapy
University of British Columbia
Vancouver, British Columbia, Canada

Meghan Ginter
Marianjoy Rehabilitation Hospital
Wheaton, Illinois

Yael Goverover
Department of Occupational Therapy
New York University
New York City, New York

Lena Hartelius
Institute of Neuroscience and Physiology
The University of Gothenburg
Gothenburg, Sweden

Peggy Jasien
Department of Flex Staff
Rehabilitation Institute of Chicago
Chicago, Illinois

Rosalind Kalb
Professional Resource Center
National Multiple Sclerosis Society
New York City, New York

Stephen Kanter
International Multiple Sclerosis
    Management Practice
New York City, New York

Neera Kapoor
Department of Clinical Sciences
SUNY—State College of Optometry
New York City, New York

Herb Karpatkin
Department of Physical Therapy
Hunter College
New York City, New York

Eva Månsson Lexell
Department of Rehabilitation Medicine
Skåne University Hospital
and
Department of Health Sciences
Lund University
Lund, Sweden

Christine A. Loveland
Department of Sociology-Anthropology
Shippensburg University
Shippensburg, Pennsylvania

Victor W. Mark
Department of Physical Medicine and
    Rehabilitation
University of Alabama at
    Birmingham
Birmingham, Alabama

Teresa Morreale
University of Illinois at Chicago
Chicago, Illinois

Letha J. Mosley
Department of Occupational Therapy
University of Central Arkansas
Conway, Arkansas

Suzanne K. O'Neal
Outpatient Therapy Services
Scottsdale Healthcare—Shea Campus
Scottsdale, Arizona

Michelle Ploughman
Department of Rehabilitation Research
Eastern Health Authority
Memorial University of
    Newfoundland
St. John's, Newfoundland, Canada

Katharine Preissner
Department of Occupational Therapy
University of Illinois at Chicago
Chicago, Illinois

Jennifer Smith
Aspire Children's Services
Hillside, Illinois

Erin Snook
Department of Kinesiology
University of Massachusetts Amherst
Amherst, Massachusetts

Chuck Stapinski
Occupational Therapy Department
Center for Pain Management
Rehabilitation Institute of Chicago
Chicago, Illinois

Don Straube
Department of Physical Therapy
University of Illinois at Chicago
Chicago, Illinois

Lauren B. Strober
Department of Neuropsychology and
    Neuroscience Laboratory
Kessler Foundation Research Center
and
Department of Physical Medicine and
    Rehabilitation
New Jersey Medical School
University of Medicine and Dentistry of
    New Jersey
West Orange, New Jersey

Laura VanPuymbrouck
LinktoStrength, Inc.
and
Flex Staff Occupational Therapist
Rehabilitation Institute of Chicago
Chicago, Illinois

Irene Ward
Rusk Institute of Rehabilitation Medicine
New York University Medical Center
New York City, New York

Roberta Winter
Rehabilitation Institute of Chicago
Chicago, Illinois

Kathleen M. Zackowski
Physical Medicine and
    Rehabilitation
Kennedy Krieger Institute
Johns Hopkins School of Medicine
Baltimore, Maryland

1

Introduction

Marcia Finlayson

CONTENTS

Living with multiple sclerosis (MS) is challenging and multidimensional—it pervades all aspects of life—one's body becomes unpredictable and unreliable, one's identity and sense of self are tested, and relationships with others often change. "MS is always in the back of your mind. If there is something you want to do, you always wonder if the MS will allow you do to it" (Darlene, living with MS for 22 years). MS symptoms emerge and remit; limitations evolve and progress. At each point in the disease course, some degree of restriction in the activities of everyday life can be expected.[1,2] Restrictions may occur as a consequence of an exacerbation and fully reverse after appropriate treatment; on the other hand, restrictions may develop and progress over time as damage to the central nervous system accumulates.

Regardless of progression and prognosis, people with MS worldwide pursue what everyone else pursues: active engagement in the habits and routines of everyday life, performance of the activities that are necessary and important to them, and a balanced and fulfilling life.[3,4] Throughout the chapters of this book, internationally recognized experts draw on the best available evidence to explain and support the various ways in which rehabilitation professionals can contribute to enabling people with MS to achieve these goals. MS rehabilitation is an active, person-centered, and goal-oriented process embedded within a respectful and collaborative partnership between the person with MS and the members of his or her rehabilitation treatment team.[5,6] This book was prepared to support this process and build these partnerships.

Using the *International Classification of Functioning, Disability and Health* (ICF)[7] as a guiding framework, the primary aim of this book is to provide a comprehensive and evidence-based resource to inform and guide clinical reasoning and decision making during each phase of the MS rehabilitation process, from initial referral to postdischarge follow-up. The specific objectives of the book are to:

1. Increase readers' knowledge about the nature and impact of specific impairments, activity limitations, and participation restrictions experienced by people with MS,

2. Increase readers' awareness of and ability to select valid, reliable, and relevant assessment tools for gathering data to inform the development of rehabilitation goals and to evaluate outcomes,

3. Increase readers' knowledge of evidence-based rehabilitation intervention methods available to people with MS, and

4. Increase readers' awareness of the gaps in knowledge and the research needed to strengthen the evidence base of MS rehabilitation.

To achieve these objectives, the authors contributing to this book provide information about the nature and impact of MS on the daily lives of people living with the disease, explain and provide resources for different assessment methods and instruments, and summarize current knowledge to guide goal setting and treatment planning. Several chapters include detailed case studies that illustrate the entire MS rehabilitation process for a specific issue or challenge, from initial presenting problems to assessment findings, goals, and treatment plans.

Given the complexity of MS and its effects on everyday life, this book is intended for a multidisciplinary audience providing service or conducting research across the range of rehabilitation settings (e.g., inpatient, outpatient, homecare, long-term care). In particular, this book targets students, practitioners, educators, and researchers from disciplines such as occupational and physical therapy, psychology, medicine (neurology, physiatry, primary care), speech-language pathology, rehabilitation counseling, social work, and nursing. Each chapter in this book was reviewed by a rehabilitation student, a practicing clinician, and a researcher to ensure that the contents were relevant and accessible to these different audiences.

1.1 Guiding Framework

Currently, there is no known cure for MS. It is chronic and multidimensional; in other words, it affects multiple aspects of physical, cognitive, social, and emotional health and functioning over the course of a person's lifetime.[8] MS alters the structure and function of the central nervous system, which can affect the efforts required and the strategies needed to perform basic activities such as walking, reaching, carrying on a conversation, and so on.[1] These alterations in functioning can limit choice and control in the activities of daily life (self-care, work, leisure) and influence a person's ability to fulfill his or her roles and responsibilities effectively, efficiently, and safely.[9,10] Depending on the context in which the person lives, the physical and social world may further influence his or her capacity, performance, and satisfaction during participation in meaningful and valued activities.[11]

Addressing the multidimensional nature and impact of MS requires diverse knowledge and expertise that no one discipline can provide. Together with the expertise of the person with MS and his or her family or caregivers, contributions from multiple professionals are often required to effectively minimize or eliminate the challenges associated with living with MS. Effective teamwork requires clear and consistent communication, which can be facilitated through the use of common language. Therefore, this book has been organized using the ICF.[7]

The ICF is both a classification system and a conceptual framework for describing and measuring functioning in society.[7] Both components of the ICF acknowledge that functioning occurs at three levels: the level of body parts (body structure and function), of the whole person (activity), and of the person in society (participation). According to the ICF, disability can occur as a consequence of dysfunction at any or all these levels. Dysfunction in body structure and function is called *impairment*, whereas dysfunction at the other levels is referred to as *activity limitation* (the whole person) or *participation restriction* (the person in society).[7] The ICF also acknowledges and emphasizes that functioning and disability are a consequence of the interaction between a person's body and the physical, social, and cultural world in which he or she lives. In other words, functioning and disability are biopsychosocial phenomena and, therefore, interventions to enhance functioning and reduce disability can occur at multiple levels—person, activity, participation, or context.

The literature supporting the use of the ICF to comprehensively describe the health and functioning of people with MS has been growing in recent years.[1,2,8,11–15] Most recently, the ICF Research Branch[16] released the *Comprehensive ICF Core Set for MS* and the *Brief ICF Core Set for MS*. These tools were developed to enable clinicians and researchers to comprehensively describe health and functioning in people with MS. The development process involved several components and culminated in a multidisciplinary, international consensus conference in 2008.[12,13,15,17] Work to validate the core sets is ongoing and to date is showing promise for use in both practice and research. Definitions and examples of the key ICF terms in relation to MS are provided in Table 1.1.

TABLE 1.1

Description and Exemplars of ICF Levels of Functioning Relevant to MS

Level of Functioning	Description	Exemplars Relevant to MS Based on Comprehensive ICF Core Set
Body structure and function/impairment	Captures the functioning or impairment (deviation or loss) of anatomical parts of the body, and physiological and psychological functions	Structure of the brain and spinal cord Energy level Cognitive functions Vision Proprioception Gait pattern
Activity/activity limitation	Captures an individual's ability or difficulty in executing a task or action	Solving problems Changing basic body position Lifting and carrying objects Walking Washing oneself
Participation/participation restriction	Captures an individual's involvement in a life situation or the problems he or she may experience in these situations	Carrying out a daily routine Using transportation Working Engaging in family life Engaging in leisure and recreation
Environmental factors	Captures the physical, social, and attitudinal context in which people carry out their daily lives	Temperature Health professionals Attitudes of immediate family members Mobility products Accessibility of buildings Work accommodations
Personal factors	Captures internal issues that influence how an individual experiences functioning and disability (e.g., age, gender, background, coping, behavior patterns)	

Source: Adapted from Development of ICF Core Sets for Multiples Sclerosis (MS) [Internet]; c2010. Available from: http://www.icf-research-branch.org/icf-core-sets-projects/neurological-conditions/development-of-icf-core-sets-for-multiple-sclerosis-ms.html.

1.2 Application of the ICF to This Book

The first section of this book provides the basics and foundations for the remainder of the chapters. In addition to this introduction, Section I contains three chapters addressing MS basics (epidemiology, etiology, diagnostic strategies, types of MS), the MS rehabilitation process, and the importance of developing the art of MS rehabilitation practice by eliciting stories from people living with the disease.

Section II includes seven chapters describing rehabilitation for specific MS-related impairments of body structures and functions. Chapters address rehabilitation for fatigue, balance disorders, muscle functions (strength, tone, coordination), cognition, pain, depression, and vision. Each chapter summarizes current knowledge about the nature and extent of the particular impairment within the MS population, how the impairment influences activity and participation in daily life, and evidence-based methods for assessment and intervention. Additional topics that are addressed by most chapters include variations in the way the particular impairment is addressed across different rehabilitation settings, gaps in knowledge, and directions for the future.

Section III includes six chapters describing rehabilitation to enhance activity and participation among people with MS. The topics that are addressed across these chapters include communication, mobility, self-care, domestic life, employment, and physical activity and leisure participation. Like the chapters in the previous section, these chapters also address the nature and extent of the particular restriction within the MS population, how the restriction affects the daily lives of people with MS, and evidence-based methods for assessment and intervention. The ways in which particular restrictions are addressed in different rehabilitation settings, gaps in knowledge, and directions for the future are also addressed in most chapters.

Section IV contains six chapters addressing contextual factors that may influence the course or progression of MS or MS rehabilitation efforts. Issues related to personal contexts are addressed in three chapters—one on age and sex influences on MS, one on coping with MS, and one on other health concerns among people with MS. Issues related to the environmental contexts are addressed in the final three chapters—one on the physical environment and home modifications, one on caregiving, and one on cultural considerations in MS rehabilitation.

The final section of the book includes a single chapter examining current transitions in health care delivery and their implications for MS rehabilitation practice and research. Figure 1.1 summarizes the contents of the book in terms of the ICF model.

1.3 Language and Terminology

There are many challenges in producing a book for a multidisciplinary and international audience, one of which is selection of terminology to identify the population of interest and the individuals providing services. Some disciplines involved in the rehabilitation process prefer to use the term "patient" while others prefer to use the term "client." Discussion about which term is more appropriate has been a source of debate in medicine and the allied health professions for at least 30 years.[18–22] Because people with MS are seen across the continuum of care (inpatient to community) and through the full range of

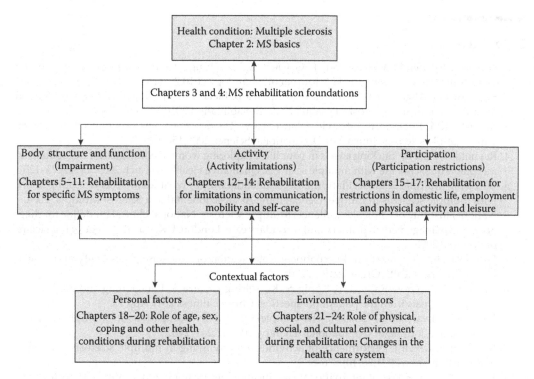

FIGURE 1.1
Application of the ICF model to MS rehabilitation: From impairment to participation.

health and functional ability, both "patient" and "client" are used in this book. Similarly, different terms are used throughout this book to refer to the individuals providing services—professionals, providers, and therapists. In the effort to recognize variations in language, these terms are used interchangeably throughout the text. Chapter authors and reviewers provide a perspective on differences in practice, assessment, and treatment methods, and use of terminology internationally.

1.4 Summary

The multidimensional nature of MS and its impact require a biopsychosocial approach to inform and guide rehabilitation service delivery. By using the ICF as a guiding framework and having internationally recognized leaders in MS rehabilitation as chapter authors, this book contributes to an important and evolving area of practice and research. The chapters provide information and insights to support active, person-centered, collaborative partnerships between people with MS and the rehabilitation team and to identify critical areas for future research efforts. Thoughtful application of the knowledge contained in this book will inform and guide rehabilitation providers to work collaboratively with people with MS and enable them to achieve the goals they have for participation in everyday life.

References

1. Holper L, Coenen M, Weise A, Stucki G, Cieza A, Kesselring J. Characterization of functioning in multiple sclerosis using the ICF. Journal of Neurology 2010;257(1):103–113.
2. Svestkova O, Angerova Y, Sladkova P, Keclikova B, Bickenbach JE, Raggi A. Functioning and disability in multiple sclerosis. Disability and Rehabilitation 2010;32(1):S59–S67.
3. Matuska KM, Erickson B. Lifestyle balance: How it is described and experienced by women with multiple sclerosis. Journal of Occupational Science 2008;15(1):20–26.
4. Reynolds F, Prior S. "Sticking jewels in your life": Exploring women's strategies for negotiating an acceptable quality of life with multiple sclerosis. Qualitative Health Research 2003;13(9):1225–1251.
5. European Multiple Sclerosis Platform. Recommendations for rehabilitation services for persons with multiple sclerosis in Europe. Genoa, Italy: Associazione Italiana Sclerosi Multipla; 2004.
6. National Institute of Clinical Excellence. Multiple sclerosis: National clinical guideline for diagnosis and management in primary and secondary care. London, UK: Royal College of Physicians of London; 2003.
7. World Health Organization. International classification of functioning, disability and health. Geneva: World Health Organization; 2001.
8. Coenen M, Basedow-Rajwich B, Konig N, Kesselring J, Cieza A. Functioning and disability in multiple sclerosis from the patient perspective. Chronic Illness 2011 Aug 12.
9. Lexell EM, Iwarsson S, Lexell J. The complexity of daily occupations in multiple sclerosis. Scandinavian Journal of Occupational Therapy 2006;13:241–248.
10. Mansson E, Lexell J. Performance of activities of daily living in multiple sclerosis. Disability and Rehabilitation 2004;26(10):576–585.
11. Khan F, Pallant JF. Use of international classification of functioning, disability and health (ICF) to describe patient-reported disability in multiple sclerosis and identification of relevant environmental factors. Journal of Rehabilitation Medicine 2007;39(1):63–70.
12. Coenen M, Cieza A, Freeman J, Khan F, Miller D, Weise A, Kesselring J, The members of the Consensus Conference. The development of the ICF core sets for multiple sclerosis: Results of the International Consensus Conference. Journal of Neurology 2011;8(8):1477–1488.
13. Kesselring J, Coenen M, Cieza A, Thompson A, Kostanjsek N, Stucki G. Developing the ICF core sets for multiple sclerosis to specify functioning. Multiple Sclerosis 2008;14(2):252–254.
14. Khan F, Pallant JF. Use of the International Classification of Functioning, Disability, and Health (ICF) to identify preliminary comprehensive and brief core sets for multiple sclerosis. Disability and Rehabilitation 2007;29(3):205–213.
15. Kesselring J. The International Classification of Functioning, Disability, and Health (ICF)—A way to specify functioning in multiple sclerosis. Actuelle Neurologie 2007;34(s2).
16. Development of ICF Core Sets for Multiples Sclerosis (MS) [Internet]; c2010. Available from: http://www.icf-research-branch.org/icf-core-sets-projects/neurological-conditions/development-of-icf-core-sets-for-multiple-sclerosis-ms.html.
17. Kesselring J, Beer S. Rehabilitation in multiple sclerosis. Advances in Clinical Neuroscience and Rehabilitation 2002;2(5):6–8.
18. Ratnapalan S. Shades of grey: Patient versus client. Canadian Medical Association Journal 2009;180(4):472.
19. Shevell MI. What do we call "them"? The "patient" versus "client" dichotomy. Developmental Medicine & Child Neurology 2009;51:770–772.
20. Sharrott GW. Promises to keep: Implications of the referent "patient" versus "client" for those served by occupational therapy. The American Journal of Occupational Therapy 1985;39(6):401–405.
21. Imrie DD. "Client" versus "patient." Canadian Medical Association Journal 1994;151(3):267c.
22. Herzberg SR. Client or patient: Which term is more appropriate for use in occupational therapy? American Journal of Occupational Therapy 1990;44(6):561–564.

Section I

Foundations

2

Multiple Sclerosis Basics

Michelle Cameron, Marcia Finlayson, and Jürg Kesselring

CONTENTS

High-quality rehabilitation for people with multiple sclerosis (MS) draws on broad knowledge about health, disability, and functioning. This knowledge includes a firm foundation in the neuropathological basis of MS, the changes that occur in the body structures and functions of people who have MS, and the typical processes of diagnosis and medical management. The purpose of this chapter is to provide an overview of this information and to prepare readers for the more specific rehabilitation content that follows in later sections of this book. After reading this chapter, you will be able to:

1. Provide a brief historical and epidemiological overview of MS,

2. Provide an overview of the typical clinical presentation and progression of MS,

3. Review basic neurological structures affected by MS and the basic pathophysiology of MS,

4. Explain how MS is diagnosed, and

5. Describe general approaches to the medical management of MS.

2.1 Historical Overview of MS and MS Rehabilitation

The disease that we now know as *multiple sclerosis* has been referred to as *insular sclerosis*, *disseminated sclerosis*, and *sclérose en plaque* in the historical medical literature.[1] Potential cases of MS date as far back as 1421, when Jan van Berieren commented on the illness of Lidwina van Schiedam (1380–1433), who was later canonized by the Catholic Church. At age 16, Lidwina fell when ice skating and broke some ribs. She later developed difficulty walking, pain in her teeth, paralysis in her right arm, loss of sensation, and visual changes. While it is unclear if she actually had MS,[1] her case is frequently noted as the earliest description of the disease.

Histories of MS also include other examples of early cases of MS, including Sir Augustus d'Esté (grandson of George III) (1794–1848), the poet Heinrich Heine (1797–1857), and William Brown, a Hudson Bay trader at York Factory in Canada in the early 1800s.[1] Across each of these cases and others, the major signs and symptoms of what we now recognize as MS can be seen: periods of exacerbation and remission, slow progression, problems with gait and balance, weakness, numbness and other sensory changes, heat sensitivity, pain, and changes in bladder function, vision, and affect.[1–4]

Despite clinical and pathological descriptions and illustrations from several key individuals (e.g., Charles Ollivier, Robert Carswell, Jen Cruveilhier, Friedrich Theodor von Frerichs, Edmé Vulpian), Jean Martin Charcot (1825–1893) is often credited with providing the first comprehensive description of MS incorporating both clinical and pathological information and clearly defining the features of the disease (see Figure 2.1).[1] Charcot's descriptions were part of a series of lectures delivered before the Societé de Biologie in 1868 and later published. Across these lectures, Charcot described different lesions (spinal, bulbar, cerebrospinal), MS tremor, the impact of the disease on the legs, visual changes, difficulty with speech, and intellectual and emotional changes.[1]

Within the historical descriptions of MS, one also finds information about general approaches that have been used to treat the disease over time. For example, the diaries of d'Esté illustrate the wide range of treatments that were attempted for his symptoms, including visiting spas to drink and bathe in "steel-water"; eating beef steaks; drinking sherry and wine; getting massages with alcohol, oils, and opium; brushing the legs; wrapping the body in hot bandages; and riding a horse.[1] Other treatments that are also noted in the histories of MS include sulfur baths, leeches, enemas, diets, arsenic, and music. None of these treatments were scientifically based, but rather selected based on conjecture and prevailing opinion.

Looking back, scientifically based treatments for MS only started to emerge over the past half century, both in medicine and in rehabilitation. The first valid scientific trial for the medical treatment of MS, which identified adrenocorticotropic hormone (ACTH) as having value in the treatment of acute MS attacks, was carried out in 1969.[5] This work eventually led to development and use of intravenous corticosteroids, which are still

CHARCOT
1825 — 1893
Professeur de Clinique des maladies nerveuses à la Faculté de Médecine de Paris
Membre de l'Académie des Sciences et de l'Académie de Médecine
Médecin de l'Hospice de la Salpétrière

FIGURE 2.1
Jean Martin Charcot (1825–1893) is often credited with providing the first comprehensive description of MS. (Courtesy of the National Library of Medicine.)

commonly used today for the treatment of acute MS attacks. The first medication to scientifically demonstrate the ability to alter the course of MS was beta interferonβ-1b (Betaseron®), which was FDA approved in 1993. Subsequent medications were intramuscular beta interferonβ-1a (Avonex®), 1996; glatiramer acetate (Copaxone®), 1996; and subcutaneous beta interferonβ-1a (Rebif®), 2002. Additional pharmaceutical treatments have been introduced over the past 10 years and more are in development. These medications are referred to as "disease-modifying therapies (DMTs)" because they reduce the rate of accumulation of MS lesions in the brain and spinal cord and reduce relapse rate. Further details on DMTs, MS lesions, and the course of the disease are provided in later sections of this chapter.

In terms of rehabilitation, the earliest article in the National Library of Medicine's PubMed database that used both "multiple sclerosis" and "rehabilitation" as index terms was published in *Northwest Medicine* in 1948.[6] Although the article focused primarily on pharmaceutical options, it does acknowledge the role of physiotherapy to improve gait through the treatment of spasticity. The first article on multiple sclerosis in an occupational therapy journal was published in 1950 and focused on the importance of using activities in therapy that "arouse the patient's interest and stimulates his desire for exercise" and that have "emotional appeal"[7] in order to strengthen muscles, develop endurance, and prevent disuse atrophy. The first article on multiple sclerosis in a physical therapy journal appeared a few years later and described a protocol for walking exercises that used two canes and targeted improvements in balance and coordination.[8] Overall, much of the early MS rehabilitation literature describes the features of the disease, the challenges of rehabilitation, and the role of different rehabilitation providers. One author explained that the role of the therapist is "to assist the patient to compensate for his [*sic*] physical difficulties by using all her [*sic*] ingenuity in keeping the patient in action psychologically and physically in a pleasant, constructive way."[9] Even in these early years, the importance of getting to know the client, motivating him or her to engage in therapy, and ensuring that treatment strategies could be carried out at home were emphasized.[7]

Despite these early descriptions of MS rehabilitation, the scientific literature to support a rehabilitation approach has only really emerged in the past 15–20 years. The first randomized controlled trial of a comprehensive inpatient rehabilitation program for people with progressive MS was published in 1997.[10] The researchers found that people with progressive MS who participated in a short period of inpatient rehabilitation (an average of 25 days) significantly improved their level of disability (Functional Independence Measure) and handicap (London Handicap Scale) compared to people who were in a wait-list control group. Since this initial report, many other important studies have documented the benefits of rehabilitation for people with MS. This book describes the rehabilitation assessments and interventions recommended for people with MS and evaluates the evidence examining their use.

2.2 Epidemiological Overview of MS

Epidemiologists ask questions about patterns of health and illness in the population and their findings aid in understanding risk factors for disease, potential causal pathways, and possible directions for treatment protocols. Epidemiologists are particularly interested in describing the incidence (the number of new cases in a specified time period) and prevalence (the number of people with the disease at a specific point in time). Incidence and prevalence studies provide information about the distribution of MS within populations and across geographical areas.[11] Through the efforts of epidemiologists we have a good understanding of how many people have MS, who gets MS and how their disease progresses over time, and what factors are associated with diagnosis and disease progression.

2.2.1 How Many People Have MS?

It is believed that between 2 and 2.5 million people worldwide are living with MS, although the distribution is uneven.[12–14] Prevalence rates vary from as low as 0.77/100,000 in Hong Kong[15] to as high as 240/100,000 (on average) in Canada.[16]

2.2.2 Who Gets MS?

Several different genetic, sociodemographic, and environmental factors have been explored to determine who is more or less likely to get MS. Epidemiologic evidence supports that there are both genetic and environmental risk factors for MS. Current evidence suggests that certain individuals have a genetic susceptibility to MS and they can develop MS if they are exposed to one or more environmental agents that trigger the MS pathological process. At this point, neither genetic nor environmental aspects, nor the way in which they interact, are fully understood.

2.2.2.1 Genetic Factors

Several factors point to a genetic susceptibility to MS: differential rates of MS between men and women, increased risk of MS among people who have a first-degree relative with the disease, and differences in the prevalence of MS across racial groups.

Research has consistently shown that MS is two to three times more common in women than in men.[17,18] Furthermore, MS onset is typically 5 years earlier in women than in men.[19] However, on average, women with MS survive longer (43 years, 95% CI: 39.2–46.8) than men (36 years, 95% CI: 29.0–42.9) after diagnosis.[20]

Genetic susceptibility to MS is also supported by research that indicates that having a first-degree relative with MS (e.g., parent–child or siblings) increases the risk of developing MS by 20–40 times, from approximately 1 per 1000 to 1 per 25 to 50.[21] Although there are some discrepancies in the literature, there may be differing patterns of paternal and maternal inheritance of MS risk.[22,23]

Research on siblings with MS further supports the interaction between genetics and environment in the development of the disease. For example, among nontwin siblings, researchers have reported similarities in the age of symptom onset, course of MS, and disease severity.[24-26] In addition, a Danish research team reported that if a monozygotic twin has MS, the other twin has a 24% risk of developing MS. In comparison, the concordance risk for dizygotic twins is 3%.[8] The fact that monozygotic twins do not have 100% risk of MS supports the role of the environment in the onset of MS.

Genetic susceptibility to MS is also supported by differences in the incidence of MS in different racial groups. Based on an extensive review, Rosati[27] reported that the prevalence of MS in African or African-descendant countries ranges from 3 to 10 per 100,000. Rates in Japan varied from 1 to 4 per 100,000, depending on the location in the country. Rates in China, Hong Kong, and Taiwan were reported as 1 to 2 per 100,000. In other work, Noonan et al.[28] examined MS prevalence in three U.S. communities. Based on this work, researchers concluded that MS is more common among Caucasians compared to Blacks and Hispanics.[28]

Hypotheses for the racial variability of MS include differences in vitamin D metabolism secondary to skin pigmentation and racial differences in the alleles of the human leukocyte antigen (HLA), a group of genes on chromosome 6 that play a major role in immune function.[29]

2.2.2.2 Sociodemographic Factors: Age

MS can start at any age, but the most common age of onset is between 20 and 40 years of age. MS is rarely diagnosed in persons under the age of 15 (3–5%) or over the age of 50 (3–12%).[30-35] In general, there is a reduced life expectancy of about 3–7 years depending on the age group,[36] yet a considerable impact on quality of life.[37,38] For example, one recent study of survival time in MS reported that the median survival time from onset of the disease was 41 years (95% CI: 38.1–43.9) for their MS population, compared to a median survival time of 49 years for the corresponding population without MS.[20] Furthermore, this same study also reported a statistically significantly longer median survival time from onset for relapsing–remitting MS (i.e., 43 years, 95% CI: 41.1–44.9) in comparison with primary progressive MS (i.e., 26 years, 95% CI: 23.2–28.8).[20]

2.2.2.3 Environmental Factors: Latitude and Migration

MS is a disease with a higher prevalence among people living further from the equator. Canada has one of the highest prevalence rates of MS in the world, with an average of 240 cases per 100,000 (95% CI: 210–280).[39] The prevalence rates in the United States have been reported to be as high as 177 per 100,000.[40] In comparison, the prevalence rates in countries closer to the equator (e.g., Japan, China, India, South Africa) are significantly lower, ranging from 1 to 10 per 100,000.[11,41]

The complexity of the association between latitude and MS is illustrated by migration studies. This work indicates that the risk of developing MS appears to decrease among persons who migrate from a higher-risk area to a lower-risk area before the age of 15. However, those who migrate from a lower-risk to a higher-risk area retain their low risk.[42–44]

While genetic factors are likely to play a role in the geographic variability of MS, several other hypotheses have also been put forward. One is that differences in the amount of sunlight, and consequently sun exposure and vitamin D levels, may play a role in the development and distribution of MS across the latitudes. The northern latitudes have fewer hours of sunlight over the course of a year, and this may be a factor in higher MS rates in these regions of the world.[45] Additional information about the role of vitamin D in MS is provided later in this section. It has also been suggested that variability in the types, patterns, and timing of infections during the life course across different regions of the world may play a role.[46]

2.2.2.4 Environmental Factors: Smoking

Smoking has been found to be a risk factor for developing MS.[47,48] Potential explanations for this association include an elevation of peripheral blood leukocytes and inflammatory markers and changes in microvascular blood flow and blood–brain barrier permeability.[46] In addition, there appears to be a dose–response relationship in that heavy smokers (over 20 cigarettes per day) have twice the risk of developing MS (OR = 1.9, 95% CI: 1.2–3.2; RR = 2.2, 95% CI: 1.5–3.2) compared to individuals who have never smoked. Smoking has also been reported to be associated with a more aggressive course of MS (i.e., needing assistance to walk within 5 years).[49]

2.2.2.5 Environmental Factors: Vitamin D

The association between vitamin D levels and the development and severity of MS has been drawing increasing interest and attention over the past several years. Vitamin D supports calcium metabolism, acts as a hormone, and has anti-inflammatory, immunomodulatory, antiproliferative, and neurotransmitter effects.[50] Several factors influence a person's vitamin D levels, including exposure to sunlight, dietary intake, and genetics.[50] Individuals with limited sunlight exposure, due to either geography or lifestyle, are more likely to be vitamin D deficient. Many persons with MS have low serum levels of vitamin D.[51] Vitamin D deficiency has been associated with increased MS prevalence, is reported more among individuals experiencing MS exacerbations, and correlates with the severity of the disease.[52–55] Although a three-decade follow-up study that found a decreased incidence of MS was associated with increased vitamin D intake[56] suggests prevention possibilities, it is still unclear as to whether dietary supplementation of vitamin D is effective or how much is necessary or safe.[50]

2.3 Clinical Aspects

This section of the chapter discusses the course of MS and the signs and symptoms associated with the disease. These influence medical treatment options, including the selection of disease-modifying and symptomatic medications, and they influence the focus and direction of rehabilitation, including which team members become involved and when, and what focus treatment takes (e.g., remedial, compensatory, consultative, educational, preventive). Rehabilitation providers need a thorough knowledge of the clinical aspects of MS to ensure an informed approach to clinical reasoning as they develop treatment goals and plans with their clients.

2.3.1 Course of MS

There are four recognized courses of MS: relapsing–remitting (RR), secondary progressive (SP), primary progressive (PP), and progressive relapsing (PR).[57] As shown in Table 2.1, all forms of MS, with the exception of PPMS, are initially characterized by a pattern of relapses and remissions over time. A relapse is defined as the appearance of new symptoms or the worsening or reactivation of previously present symptoms. The appearance or change in symptoms must last at least 24 h, be separated from a previous relapse by at least 30 days, and not be the consequence of a change in body temperature or an infection. Other terms that are commonly used to describe a relapse include exacerbation, attack, or flare-up. The specific course of MS that a person has will influence disease progression.

MS relapses are highly variable between individuals as well as in a given individual and over time. A person may experience a very mild relapse (e.g., focal numbness) that allows him or her to continue to perform most daily activities. Another time, that same person may experience a relapse involving multiple new or worsening symptoms (e.g., extreme fatigue, imbalance, vision loss, pain) simultaneously that are severe enough to significantly impair function, require hospitalization, and necessitate pharmaceutical treatment.

TABLE 2.1

MS Disease Course Types

Type	Characteristics	Frequency at Disease Onset
RRMS	Clearly defined relapses followed by full or nearly full recovery of function. The disease does not progress in between attacks.	~85%
SPMS	Initial course of relapsing–remitting that is followed by progression, with or without occasional relapses or plateaus in function. The disease progresses over time.	~50% of people with RRMS will transition to SPMS within 15 years from diagnosis
PPMS	Gradual but continual worsening over time; some fluctuations but no distinct relapses.	~10%
PRMS	Progressive course marked by distinct relapses, with or without recovery. The disease continues to progress between attacks.	5%

2.3.2 Progression of MS

Although the Kurtzke Expanded Disability Status Scale (EDSS)[58,59] has come under significant critique in recent years,[60,61] the severity and progression of MS has historically been, and continues to generally be, evaluated by this measure. The EDSS is called a "disability" scale, but from the perspective of the *International Classification of Functioning, Disability and Health* (ICF), it is more accurately described as a tool that evaluates aspects of impairment (i.e., changes across eight body functions) and one aspect of activity (i.e., changes in walking ability). A newer measure, the Multiple Sclerosis Functional Composite (MSFC),[62] has better reliability than the EDSS,[63] is quicker to administer, and is more equally weighted across upper- and lower-extremity function and cognitive function. Nevertheless, the MSFC is also limited by considering only impairment (i.e., Paced Auditory Serial Addition Test [PASAT], Nine-Hole Peg Test) and activity (i.e., 25-foot walk). While the MSFC is commonly used as a research tool, in clinical settings, the EDSS remains the common standard for identifying the severity of MS at a point in time.

The EDSS score is based on examination of eight body functions (called functional systems): pyramidal, cerebellar, brainstem, sensory, bowel and bladder, visual, cerebral, and walking (see Figure 2.2).

Typically, the EDSS score is determined by a neurologist's clinical examination. The resulting score is an ordinal-level measure that increases in half-point increments and extends from 0.0 (normal neurological exam) to 10.0 (death from MS). Individuals who score 1 have no disability and minimal neurological signs in only one functional system whereas individuals with an EDSS of 4.0 are fully ambulatory without aid, are able to manage everyday activities despite relatively severe impairment (e.g., can conduct a full day of activities of ~12 h). People with an EDSS of 4.0 can walk for approximately 500 m without aid or rest. Starting at 5.0, limitations in daily activities are usually

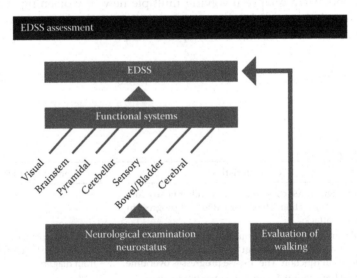

FIGURE 2.2
Factors contributing to the EDSS evaluation and scoring. The functional systems evaluated in the EDSS. (Courtesy of www.neurostatus.net, Ludwig Kappos, MD, Professor and Head, Neurology, University Hospital Basel, Switzerland.)

observed. At 6.0, the individual requires intermittent or constant assistance (cane, crutch, brace) to walk 100 m, with or without resting. At 7.0, the person is essentially restricted to a wheelchair but continues to be able to perform transfers independently (see Figure 2.3).

A score of 6.0 on the EDSS has been often used as an indicator of irreversible disability in MS research.[30,64,65] The time to reach this score depends on several factors, including current age, age of MS onset, MS course, gender, and the symptoms at disease onset. With respect to age, Trojano and colleagues[65] found that EDSS scores increased significantly with increasing age of MS onset and that disability progresses faster in patients diagnosed at an older age than in those diagnosed at a younger age. However, this research team also found that current age, which is associated with disease duration, had a larger effect on disease severity than age at disease onset. Trojano's work is supported by Confavreux and Vukusic's[30] findings that people reached EDSS scores of 4.0, 6.0, and 7.0 at median ages of 44.3 years (CI = 43.3–45.2), 54.7 years (CI = 53.5–55.8), and 63.1 years (CI = 61.0–65.1), respectively.

Other factors associated with faster progression of MS include having a primary progressive disease course, being male, experiencing cerebellar or lower-extremity motor symptoms at onset (versus visual or sensory symptoms), experiencing an incomplete recovery after the first relapse, and experiencing only a short interval between the first two relapses.[64,66–69] Although these factors are generally considered negative prognostic indicators, there are inconsistencies in the literature and not everyone follows this pattern.[70,71] From the perspective of rehabilitation, research suggests that 50% of people with MS can expect to have activity limitations within 15 years of diagnosis, and approximately 5% may have such significant impairment and mobility restriction within a few years of diagnosis that the full-time use of a wheelchair will be required.[72,73] A more recent study reported a longer period before reaching a sustained "moderate disability."[64]

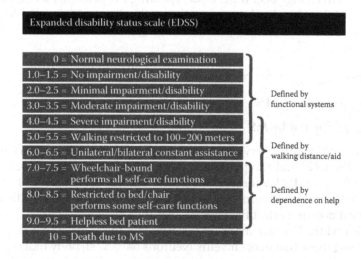

FIGURE 2.3
EDSS Scoring from NEUROSTATUS. Basic scoring of the EDSS. (Courtesy of www.neurostatus.net, Ludwig Kappos, MD, Professor and Head, Neurology, University Hospital Basel, Switzerland.)

2.3.3 Signs and Symptoms

The way in which a person with MS presents clinically to medical and rehabilitation providers can be summarized in several ways. First, there are the *signs* of the disease, that is, what a health care professional observes when the client comes into the office (e.g., ataxic gait) or finds during the clinical examination (e.g., dysarthric speech, spasticity). Next, there are the *symptoms* of the disease, which are the characteristics of the condition that the client experiences and describes to the practitioner (e.g., numbness, fatigue, visual changes). In some cases, signs and symptoms overlap (e.g., problems with balance). Recently, El-Moslimany and Lublin[74] identified over 62 different signs and symptoms associated with MS, although some of them are very rare (e.g., seizures, hearing loss).

While rehabilitation providers often think about the signs and symptoms of MS from the perspective of functional outputs or performance (e.g., problems with gait), neurologists organize signs and symptoms based on the location in the nervous system thought to control them (e.g., sensory cortex, motor cortex, basal ganglia) and whether or not they are considered *positive* or *negative*. *Positive* signs and symptoms are those that produce extra or additional movements or behaviors that are not typically observed or experienced, for example, muscle spasms, and are thought to result from overactive nerve firing. *Negative* signs and symptoms are those reflecting functions that are limited or missing, for example, weakness or absent reflexes, and are thought to result from slowed or blocked nerve conduction. Table 2.2 summarizes the key signs and symptoms of MS that most often lead to a referral for rehabilitation using the framework of control mechanisms (based on EDSS functional systems) and positive and negative signs and symptoms. Further details about the epidemiology, impact, rehabilitation assessment methods, and treatment options for all signs and symptoms marked with an asterisk in this table are provided in later chapters of this book.

It is clear from Table 2.2 that MS is a complex and variable disease. No two people will experience exactly the same signs and symptoms, and, for each of them, the experience with the disease will change over time. Understanding the pathophysiological basis of MS provides some insight into why the presentation of the disease is so variable.

2.4 Pathophysiology of MS

2.4.1 Structures Affected by MS

MS is a disease of the central nervous system (CNS). The nervous system can be divided into two parts, the CNS and the peripheral nervous system (PNS). The CNS consists of the brain, cranial nerves I (olfactory) and II (optic), and the spinal cord. The PNS consists of the nerves outside of the brain and spinal cord and cranial nerves III–XII. MS only affects the CNS; it does not affect the PNS (see Figure 2.4).

Both the CNS and the PNS are made up of nerve cells (neurons) and a range of supporting cells, including those that make myelin. Neurons have a cell body that contains the cell nucleus and a number of processes emanating from the cell body. One of these processes will become much longer than the others and is known as the axon. The other shorter processes are known as the dendrites. The dendrites and the cell body can receive information from other neurons. The axon transmits information from the neuron. Axons may

TABLE 2.2

Signs and Symptoms of MS by Functional System

Functional System (FS)	FS Assessment Areas within the EDSS	Positive Signs and Symptoms	Negative Signs and Symptoms
Visual	Visual acuity Visual fields	Phosphenes	Loss of or changes in vision (e.g., blurred; loss of areas in visual field)
Brainstem	Extraocular movements Hearing Facial sensation and movement Swallowing Speech	Trigeminal neuralgia Nystagmus	Double vision Facial numbness Facial weakness Dysphagia[a] Dysarthria[a] Vertigo
Pyramidal	Upper- and lower-extremity reflexes Strength Tone Babinski	Muscle spasms and cramps Babinski response	Muscle weakness[a]
Cerebellar	Coordination of trunk and limbs, specifically: Gait Fine finger movements Rapid alternating movements of the upper- and lower-extremities Finger to nose and heel to shin	Intention tremor[a] Ataxia of gait[a]	Fine motor dysfunction Incoordination[a] Dysmetria Loss of balance[a]
Sensory	Light touch Vibration Proprioception of the upper- and lower-extremities	Tingling, prickling, and burning sensations[a] Pain[a] L'Hermitte's sign	Numbness Absent or reduced sensitivity to cutaneous stimuli Loss of balance[a]
Bowel/ bladder	Constipation or bowel incontinence Bladder urgency, frequency, incontinence, hesitancy, or need for catheterization	Urinary frequency	Urinary hesitancy Constipation Sexual dysfunction
Cerebral	Presence or absence of depression or euphoria Mild/moderate/marked or severe decrease in cognition	Euphoria Pathological laughing or crying	Loss of attention and concentration[a] Executive dysfunction[a] Difficulties with new learning[a] Fatigue/loss of energy[a] Depression[a]
Walking	Distance walked Degree of assistance		Decreased walking capacity[a] Need for assistive device for walking[a]

[a] Further details about the epidemiology, impact, rehabilitation assessment methods, and treatment options for all signs and symptoms marked with an asterisk are provided in later chapters of this book.

be wrapped in myelin to accelerate nerve conduction and protect the nerve. In the PNS, myelin is made by Schwann cells, and each nerve is myelinated by a number of Schwann cells. In the CNS, myelin is made by oligodendrocytes, and each oligodendrocyte myelinates a number of nerve axons (see Figure 2.5).

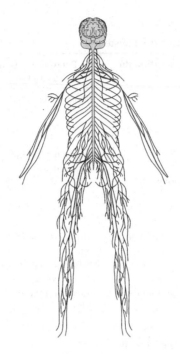

FIGURE 2.4
The CNS consists of the brain, cranial nerves I and II, and the spinal cord. The PNS consists of the nerves outside of the brain and spinal cord and cranial nerves III–XII. MS affects the CNS.

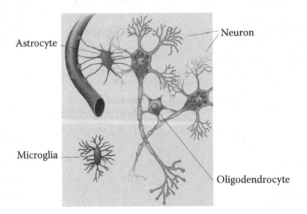

FIGURE 2.5
The cells of the central nervous system.

2.4.2 Pathology

MS is associated with inflammation and damage to the myelin and the axons in the CNS. Early descriptions of MS pathology emphasized the demyelination in MS, but all descriptions, both historic and current, note that although damage to myelin predominates, MS damages both myelin and axons at all stages of the disease.[75,76]

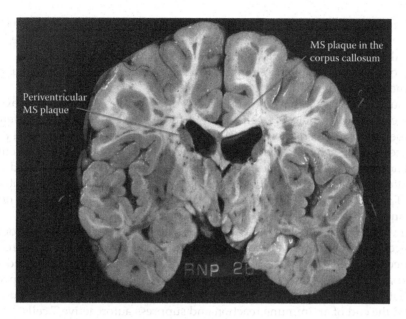

FIGURE 2.6
MS plaques in a cross-section of the brain. (Courtesy of Michele Mass, MD, Portland VA Medical Center and Oregon Health & Sciences University.)

The pathologic hallmark of MS is the formation of sclerotic plaques (see Figure 2.6). These plaques are areas of scarring in the CNS. Sclerotic plaques from MS tend to form in the white matter of the CNS, where axons rather than nerve cell bodies predominate. Plaques can also form in the gray matter of the CNS, and these plaques tend to correlate more with disability than white matter plaques. Plaques form most often around the ventricles, in the corpus callosum, in subcortical areas, in the posterior and lateral spinal cord, and in the optic nerves. MS plaques are not all the same. However, in a single person, all MS lesions tend to have a similar pattern, suggesting that everything we know as MS may not be a single disease but that each individual has only one type of MS.[77]

2.4.3 Pathogenesis

The cause of inflammation and plaque formation in MS is not fully understood. What follows is a simplified and generally accepted description of the current understanding of the pathogenesis of MS. The current favored hypothesis is that MS is an autoimmune disease. The normal healthy immune system identifies, attacks, and eliminates from the body foreign or unhealthy matter such as bacteria, nonbiologic material, or the individual's cells damaged by trauma or malignant malformation. In autoimmune diseases, the immune system turns against healthy parts of the body, and in the case of MS, against the CNS, and damages it.

One source of support for MS being an autoimmune disease comes from the animal model of MS—experimental autoimmune encephalomyelitis (EAE). EAE is an autoimmune demyelinating disorder of the CNS in animals that is similar to MS. EAE is induced in animals by injecting them with myelin proteins. EAE has been induced in various animal species, including rodents and primates, and has been used as an animal model to study MS. EAE and MS have many common features. Both diseases only affect the CNS. Both diseases require lymphocytes, a type of white blood cell that usually fights off infections, to enter the

CNS from the circulation across the blood–brain barrier. Both appear to be primarily mediated by T cells, a subtype of lymphocyte that has T cell receptors on them. The T in T cell stands for thymus because T cells mature in the thymus. MS and EAE also depend on expression of the appropriate HLA antigens, specific T cell receptor-bearing lymphocytes, adhesion molecule expression, and cytokine secretion.

Several subsets of T cells have been discovered, including T helper, cytotoxic T cells, memory T cells, and regulatory T cells. Each of these T-cell subsets has a different function. T helper (T_H) cells, also known as CD4+ T cells because they express CD4 protein on their surface, become activated when presented with antigens by antigen-presenting cells (APCs). When activated they divide rapidly and secrete small proteins called cytokines that regulate immune processes. CD4+ T_H cells can differentiate into different subtypes, including T_H1, T_H2, T_H3, T_H17, and T_{FH}. The names of these subtypes are based on the specific cytokines they produce.[78]

Cytotoxic T cells, also known as CD8+ T cells because they express CD8 glycoprotein on their surface, destroy virally infected cells and tumor cells. Memory T cells are a type of antigen-specific T cell that persists long after an infection has resolved, and they can quickly expand to large numbers of effector T cells when reexposed to their specific antigen. Regulatory T cells, T_{reg} cells, formerly known as suppressor T cells, shut down immunity toward the end of an immune reaction and suppress autoreactive T cells.[78]

CD4+ T_H cells are thought to play an important role in the pathogenesis of MS. T_H1 cell overactivity is traditionally thought to primarily mediate the inflammation associated with the disease, although recently there has also been a growing interest in the role of T_H17 cells in MS pathogenesis.[79] T_H1 cells are primarily proinflammatory, and they also reduce the activity of the primarily anti-inflammatory T_H2 cells. T_H1 cells secrete proinflammatory cytokines, including interleukin-2 (IL-2), interferon gamma (IFNγ), and tumor necrosis factor alpha (TNFα). T_H2 cells secrete IL-4, IL-5, IL-10, and transforming growth factor beta (TGFβ). T_H17 cells secrete IL-17 and IL-22.[78] The roles of these cells and cytokines in MS are supported by studies both in people with MS and in animals with EAE. Many studies investigating the cause of MS and potential treatments for MS focus on the role of these different classes of immune cells and the cytokines they produce.

In EAE and in people with MS, CD4+ cells are activated in the bloodstream outside of the CNS when they are presented with an autoantigen by an APC. These activated T cells then cross the blood–brain barrier, likely through interaction with adhesion molecules, and become activated in the presence of APCs in the CNS bearing the same antigen. When activated, T_H1 cells secrete cytokines, cause breakdown of the blood–brain barrier, attract macrophages, and activate microglia that damage myelin and oligodendrocytes (see Figure 2.7). This T_H1 cell response can be reduced by the cytokines secreted by T_H2 cells. The myelin damage that results from this attack may be followed by some degree of remyelination if oligodendrocytes can extend new processes to cover demyelinated areas of axons, but eventually, with repeated demyelination, remyelination fails. In addition to the demyelination, axonal degeneration also occurs in the CNS, particularly in areas of demyelination. The relationship between the demyelination and the axonal degeneration is uncertain. Axonal damage may be a consequence of the loss of the protective myelin coating, or axonal degeneration may occur at the same time as, or before, the demyelination.

Another source of support for considering MS as an autoimmune disease comes from changes in the immune system seen in people with MS. People with MS have more activated lymphocytes in their circulation and decreased suppressor function of their immune cells. The amount and rate of synthesis of antibodies, specifically immunoglobulin G (IgG), are also often elevated in the cerebrospinal fluid (CSF) of people with MS. In addition, IgG

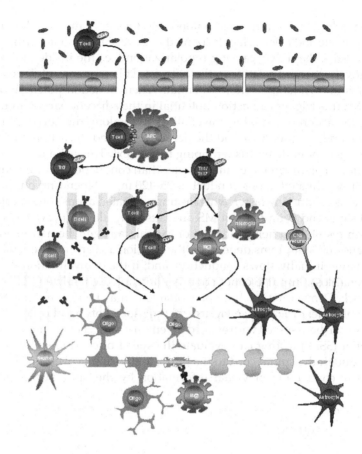

FIGURE 2.7
Myelin damage and axonal degeneration begins when T-cells in the autoimmune system attack the CNS. (Courtesy of Roland Martin, MD, and the Institute for Neuroimmunology and Clinical MS Research (INIMS).)

in the CSF of people with MS is usually oligoclonal in nature, meaning that is produced by a restricted number of cell clones. This suggests that the IgG is being produced against a specific antigen, although the antigen against which this antibody is directed is not known. Because of these common CSF changes in MS, a lumbar puncture is sometimes performed to aid in the diagnosis of MS when the patient history and magnetic resonance imaging (MRI) are not diagnostic. When CSF is obtained to help diagnose MS, the level of IgG, the IgG synthesis rate, the ratio of IgG in the CSF compared to the blood, and the presence of oligoclonal bands are evaluated.

Finally, one of the strongest sources of support for viewing MS as an autoimmune disease is the effectiveness of immunomodulatory medications in improving the course of MS in people with the disease.

2.4.4 Pathophysiology

The CNS has a complex network of neurons, each with a cell body and an axon, to allow for control of all bodily functions. Neurons transmit information with action potentials. At rest, there is an electrical potential across the nerve cell membrane, with the nerve having a negative charge on the inside relative to the outside. With activation, sodium channels on the nerve open, allowing sodium ions to flood into the nerve making the inside positive relative to the outside and briefly depolarizing the nerve. After about a millisecond, potassium ions then leave the nerve, causing repolarization (i.e., the inside of the nerve again becomes negatively charged relative to the outside). This sequence of depolarization and repolarization is known as the action potential. When an action potential occurs in one place on a nerve, this triggers an action potential in the adjacent part of an unmyelinated nerve causing the action potential to travel all the way along the nerve. When a nerve is myelinated, myelin is wrapped around the nerve's axon, and the action potential is transmitted from one gap or node in this wrapping to the next. This is known as saltatory conduction and allows conduction to be much faster than conduction in unmyelinated nerves (see Figure 2.8). Myelinated axons conduct at 25–120 m/s, depending on their thickness, with thicker axons conducting faster, whereas unmyelinated axons conduct at about 1 m/s.

The clinical signs and symptoms of MS are caused by altered nerve conduction in the CNS, which can result from demyelination or axonal damage. The nature and severity of the clinical signs and symptoms depend on the location, nature, and severity of the damage. Demyelination initially slows conduction, and, if the demyelination is more marked or there is axonal loss, impulse conduction is interrupted completely. This is known as conduction block. Demyelination, but not axonal loss, may be restored if a nerve is remyelinated by an oligodendrocyte (see Figure 2.9). The speed of signal conduction in certain areas of the CNS can be assessed with evoked potential testing. For example, somatosensory evoked potential (SSEP) testing can evaluate the speed of conduction of somatosensory information through the spinal cord to the brain, and visual evoked potential (VEP) testing can evaluate conduction of visual information by the optic nerves and the brain.

(a)

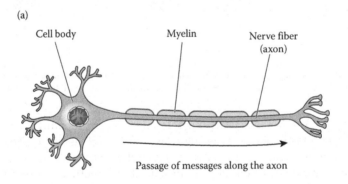

(b)

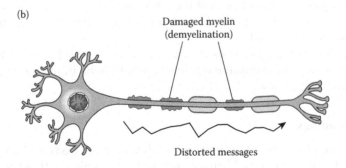

FIGURE 2.8
Comparing myelinated and demyelinated nerves. (a) Normal neuron, (b) demyelination in MS.

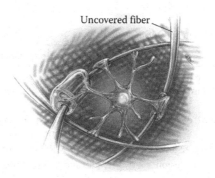

Uncovered fiber

FIGURE 2.9
Oligodendrocytes help restore demyelinated axons. (Courtesy of Wings for Life Spinal Cord Research Foundation.)

Both SSEP and VEP testing are sometimes used to assist in the diagnosis of MS. Slowed or blocked nerve conduction is thought to underlie many of the negative symptoms of MS, such as reduced strength, reduced sensation, and reduced cognitive function.

Temperature can affect the speed of signal conduction in nerves. In healthy nerves, higher temperatures up to about 50°C increase the speed of conduction. Beyond this temperature, conduction block occurs because the action potential in one segment does not last long enough to excite the next segment. In demyelinated nerves, the critical blocking temperature is reduced. If the thickness of the myelin is reduced by a third, even if there are no other changes in the nerve or the myelin, the critical blocking temperature is reduced to about 40°C. In completely demyelinated nerves, an increase of only 0.5°C can be enough to block impulse conduction. As a result, symptoms of MS characteristically become more marked with increases in body temperature. The worsening of MS symptoms with exercise is known as Uhthoff's phenomenon, which classically manifests as temporary visual changes in patients with optic neuritis.[80]

In addition to the CNS demyelination caused by MS reducing nerve conduction and thereby causing negative symptoms, demyelination is also thought to cause paroxysmal and positive symptoms such as pain, paresthesias, and tonic spasms in people with MS. These symptoms are caused by spontaneous excitation of a nerve, excessive discharges in response to mechanical stimulation, or pathological impulse transmission from one nerve fiber to another. In addition, L'Hermitte's sign, which is an electrical sensation in the limbs brought on by flexion of the neck, may be caused by increased activity brought on by the mechanical deformation of the demyelinated cervical cord.

2.5 Diagnosis of MS

The diagnosis of MS is made clinically. There is no single test that can determine if a person does or does not have the disease. MS is a disease defined by neurological symptoms separated (or disseminated) in time and space without a better explanation. Various diagnostic criteria have been used to define who does and who does not have MS.[81–85] The most recently accepted criteria for the diagnosis of MS are the McDonald criteria, 2010 revision, as shown in Table 2.3.

TABLE 2.3

2010 McDonald Diagnostic Criteria for MS

Clinical Presentation	Additional Data Needed for MS Diagnosis
≥2 attacks; objective clinical evidence of ≥2 lesions or objective clinical evidence of 1 lesion with reasonable historical evidence of a prior attack	None
≥2 attacks; objective clinical evidence of 1 lesion	DIS by MRI, or await further clinical attack implicating a different CNS site
1 attack; objective clinical evidence of ≥2 lesions	DIT by MRI, or await second clinical attack
1 attack; objective clinical evidence of 1 lesion (clinically isolated syndrome)	DIS and DIT by MRI
Insidious neurological progression suggestive of MS (PPMS)	1 year of disease progression (retrospectively or prospectively determined) Plus 2 of the following 3: (1) evidence for DIS in the brain based on ≥1 T2 lesions in characteristic area; (2) evidence for DIS in the spinal cord based on ≥2 lesions in the cord; (3) positive CSF (bands or elevated IgG index)

Source: From Polman CH et al. Annals of Neurology 2011;69:292–302. With permission.
Note: CSF = cerebrospinal fluid; DIS = dissemination in space; DIT = dissemination in time; IgG = immunoglobulin G; MS = multiple sclerosis; PPMS = primary progressive MS.

According to the current criteria, MS can be diagnosed with clinical evidence of two attacks at different times (i.e., disseminated in time) and affecting at least two different areas of the CNS (i.e., disseminated in space). A clinical attack of MS is a neurological disturbance of any kind seen in MS that lasts at least 24 hours. The neurological disturbance may include symptoms by subjective report or by objective observation but must not be a pseudoattack, which is an exacerbation of symptoms due to other physiological stresses, such as infections, and must not be a very brief symptom such as a shooting pain or electrical sensation.

This clinical evidence alone, having ruled out alternative explanations, is sufficient to make a diagnosis of MS without any further testing. However, in most circumstances, MRI of the brain is recommended and performed to confirm the diagnosis of MS and rule out alternative explanations for the symptoms, such as stroke, brain tumor, or infection. When there has been only a single clinical attack or there is only clinical evidence of one lesion, MRI can be used to determine if there is dissemination in time or space. According to the 2010 McDonald criteria, there is dissemination in space (DIS) on MRI if there are one or more T2 lesions in at least two of the appropriate areas of the CNS. There is dissemination in time (DIT) by MRI if there is a new T2 and/or gadolinium-enhancing lesion on follow-up MRI compared to a previous baseline MRI, or if asymptomatic gadolinium-enhancing and gadolinium-nonenhancing lesions are present simultaneously on an MRI at any time.

In most people with MS, the brain MRI will show abnormalities consistent with MS, and, in conjunction with the history and physical examination, this will be sufficient to make the diagnosis of MS. It is reported that in more than 95% of people with MS, the brain MRI shows abnormalities, particularly around the ventricle walls, at the border zone between gray and white matter and in the posterior fossa.[86] However, not all abnormalities of the white matter on brain MRI are caused by MS. Particularly in the elderly or in people with an atypical history, a broader range of diagnoses are considered that may include stroke, chronic infections, nutritional deficiencies, other demyelinating disorders such as neuromyelitis optica (NMO) and acute demyelinating encephalomyelitis (ADEM), and rare inherited disorders such as leukodystrophies.[87]

When the clinical history and examination and brain MRI do not clearly indicate if the person has MS, testing for other possible diagnoses, waiting to see if the person has another attack or the MRI changes, and performing other tests that are often abnormal in people with MS, including spinal MRI; CSF evaluation for IgG index, IgG synthesis rate, and IgG oligoclonal bands; VEPs; and SSEPs, may help to determine if someone does or does not have MS. Spinal MRI is helpful because, although a wide range of disorders other than MS can cause lesions on brain MRI, there are very few things other than MS that cause the typical shape and location of spinal cord lesions seen in MS. Similarly, abnormal IgG production in the CSF, including oligoclonal IgG bands, is found in more than 90% of patients with MS, and evoked potential studies in people with MS stereotypically reveal prolonged latencies reflecting the slowing of conduction in the CNS caused by demyelination. Thus, although the diagnosis of MS is usually based on the clinical history and examination and an MRI of the brain alone, some people's diagnosis remains uncertain even after these tests and much additional testing, including various blood tests, MRI of the spine, lumbar puncture to obtain CSF, and evoked potential testing.

The current diagnostic criteria for MS were designed to allow for early diagnosis of MS while minimizing the risk of making an incorrect diagnosis. This allows effective treatments to be introduced early, when they are thought to be most effective, while avoiding the risk of side effects and the adverse consequences of being diagnosed with MS for those who do not actually have the disease.

2.5.1 MRI Basics in MS

MRI of the brain is the primary imaging modality used in the diagnosis of MS and for assessing MS disease progression and activity. MRI imaging of the brain uses a large magnet to produce slice images in various shades of gray. The resolution of the image, and thus the amount of detail visualized, depends largely on the strength of the magnet (typically 1–3 Tesla) and the thickness of the slices (typically 1–5 mm). The shading of the different structures depends on the timing of the MRI's radiofrequency pulses, which are part of the different pulse sequences. T1 sequences show white matter in light gray and gray matter in dark gray and are ideal for visualizing anatomy, overall brain volume, and more severe loss of tissue associated with MS, but they do not show most MS lesions well. T2 sequences show white matter in gray and gray matter and CSF in white. MS lesions show up as white areas within the gray-appearing white matter on T2 sequences. The fluid attenuation inversion recovery (FLAIR) sequence is very similar to the T2 sequence except that the CSF is black. The FLAIR sequence is generally ideal for visualizing MS lesions because many of them form close to the ventricles and are difficult to see on the usual T2 sequence in which the lesion and the CSF are both white (see Figure 2.10).

MRI contrast, usually a gadolinium product, is usually given when imaging to evaluate MS. MRI contrast is given intravenously and penetrates areas of the brain where there is increased blood–brain barrier permeability. This occurs during the inflammation associated with an MS relapse and generally lasts for about 6 weeks. During this period, the lesion will show as a bright area on the postcontrast T1 sequence as well as showing as a bright area on the T2 and FLAIR. This is known as contrast enhancement. As the inflammation resolves, the contrast enhancement appreciable on the postcontrast T1 images also resolves.

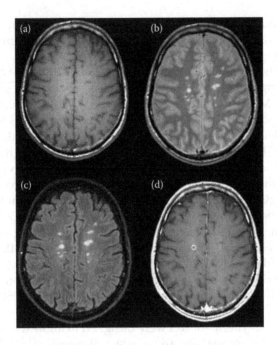

FIGURE 2.10
Brain MRI in MS. (a) T1, (b) T2, (c) FLAIR, and (d) post-contrast T1. (Courtesy of Jack H. Simon, MD, PhD, Portland VA Medical Center and Oregon Health & Sciences University.)

2.6 Medical Management of MS

The medical management of MS is generally divided into three components: management of acute relapses, DMTs, and symptom management.

2.6.1 Management of Acute Relapses

Acute MS relapses are characterized by onset of new signs and symptoms lasting more than 24 h, as described earlier in this chapter. These acute relapses are associated with increased inflammation and edema in the CNS and can be treated with high-dose intravenous steroids if the symptoms interfere sufficiently with the person's quality of life to merit the potential side effects of treatment. Treatment is generally with intravenous methylprednisolone given daily for 3–5 days. This intervention shortens the duration of relapses, likely because of the anti-inflammatory effect, but does not alter the long-term outcome, disease course, or relapse rate.[88] High-dose oral steroids are also occasionally used to manage acute relapses and have been shown to be as effective as intravenous steroids.[89] Plasmapheresis (plasma exchange) is also sometimes used to treat severe relapses that do not respond to high-dose steroids.[90]

2.6.2 Disease-Modifying Therapies

Disease-modifying therapies in MS reduce relapse rate, reduce the rate of accumulation of lesions on MRI, and likely slow progression of disability in people with relapsing MS.

The first disease-modifying therapy to be FDA approved for MS was interferonβ-1b delivered by subcutaneous injection (brand name: Betaseron®), which was approved in 1993. Since then, two other interferon preparations, interferonβ-1a delivered by intramuscular injection (brand name: Avonex®) and interferonβ-1a delivered by subcutaneous injection (brand names: Rebif® and Extavia®) have also been FDA approved as disease-modifying therapies for MS with relapses. In addition, glatiramer acetate (Copaxone®), mitoxantrone (Novantrone®), natalizumab (Tysabri®), and fingolimod (Gilenya®) have all been FDA approved for the treatment of MS with relapses. All these medications are delivered either by self-injection or by infusion except for fingolimod, the first oral disease-modifying therapy for MS, which was FDA approved in September 2010.

How exactly interferonβ modifies MS disease activity is not known. Interferonβ is a polypeptide normally produced by fibroblasts that has antiviral and antiproliferative effects. It is thought to affect MS activity by binding to its receptor on immune cells thereby reducing antigen presentation, T-cell proliferation, and cytokine production and restoring T-cell suppressor function. All the interferons are given by self-injection. The injections vary in frequency from once a week to every other day. The interferons are generally well tolerated. Although 50–75% of patients initially experience flu-like symptoms for up to a day after the injection, this side effect generally resolves over the first 3–6 months of use, may be less severe if the interferon dose is gradually titrated to its full level over the first few weeks, and can often be controlled with acetaminophen or nonsteroidal anti-inflammatory medications such as ibuprofen. Rare side effects of the interferons can also include reduced white blood cell production and elevated liver enzymes, depression, increased spasticity, and injection site reactions. Interferons may also stimulate the production of antibodies, which are likely to reduce their effectiveness.

Glatiramer acetate is a random polymer of four amino acids in myelin basic protein. How exactly glatiramer acetate modifies MS disease activity is also unknown. It may act by shifting the T cells in the individual from Th1 cells, which are proinflammatory, to Th2 cells, which suppress the inflammatory response. Its similarity to myelin basic protein may also divert the autoimmune response from the patient's myelin. Glatiramer acetate requires a once-a-day injection but is otherwise generally well tolerated. It can be associated with inflammatory injection site reactions or permanent indentations at the injection site (known as lipoatrophy) as well as a rare, sporadic, and unpredictable short-term reaction right after injecting the medication that can involve flushing, chest tightness or pain with heart palpitations, anxiety, and trouble breathing. These symptoms can last 30 s to 30 min. This postinjection systemic reaction can be frightening to patients but is not harmful, although its cause is not known.

Mitoxantrone is a chemotherapy agent that disrupts DNA synthesis and repair. Mitoxantrone is given by intravenous infusion every few weeks for a number of months. This drug is rarely used today for the treatment of MS because of its side effects, which include risk of heart damage and heart failure as well as leukemia.

Natalizumab is a humanized monoclonal antibody against the cellular adhesion molecule α4-integrin. The drug is believed to reduce MS disease activity by preventing inflammatory immune cells attaching to and passing through the blood–brain barrier. Natalizumab is given by intravenous infusion once every 4 weeks. This drug is generally well tolerated and very effective, but after it was released to market it was found that its use was associated with the development of progressive multifocal leukoencephalopathy (PML), a rare viral demyelinating disease of the CNS that can be fatal or result in severe disability. Because of this, natalizumab was initially withdrawn from the U.S. market and then rereleased with an enhanced safety and monitoring program (the TOUCH program).

At this time, the use of natalizumab is associated with approximately 1 case of PML per 1000 users worldwide.

The most recent addition to the armamentarium of MS treatment is the first oral MS disease-modifying therapy, fingolimod. Fingolimod is thought to reduce MS disease activity by trapping lymphocytes in the lymph nodes thereby reducing the number of circulating lymphocytes and their migration into the CNS. This drug is taken once a day as a pill and, like the other MS disease-modifying medications, has been shown to reduce relapse rate, MRI markers of MS progression, and progression of disability. Although fingolimod is generally well tolerated, some patients have bradycardia (slowing down of their heart rate) with the first dose and some develop problems with their eyes (macular edema). The studies of fingolimod used two different doses. The higher dose, 1.25 mg/day, was associated with two infection-related deaths. There were no treatment-related deaths with use of the lower dose, 0.5 mg/day, which was released to market in September 2010. Given previous experiences of unexpected serious side effects with natalizumab and mitoxantrone and the prolonged safety track record of interferons and glatiramer acetate in patients with MS, many patients and providers are cautious about using fingolimod until its safety and tolerability in clinical use are clarified over time even though it has been approved as first-line therapy for MS.

2.6.3 Symptom Management

MS is associated with a wide range of symptoms, including cognitive and mood changes, fatigue, muscle weakness and spasticity, sensory changes, balance and coordination changes, bowel- and bladder-related symptoms, and pain (refer to Table 2.2). Management of these symptoms is essential in optimizing participation and quality of life in people with MS. Medications are available to address most of these symptoms to some degree. These medications are generally not directed specifically for people with MS but rather are intended for management of the particular symptom in the general population. For example, antidepressant medications can be helpful to improve mood in people with MS and stool softeners can be helpful to address constipation. Recently, dalfampridine (Ampyra®), approved to improve walking in people with MS, was the first medication specifically FDA approved for symptom management in MS.[91,92] In addition, a wide range of rehabilitation interventions are also often helpful in addressing the symptoms of MS, either alone or in conjunction with medications. These rehabilitation interventions are the focus of the remainder of this book.

References

1. Murray TJ. The history of multiple sclerosis. In: J. S. Burks, K. P. Johnson, editors. Multiple sclerosis: Diagnosis, medical management and rehabilitation. New York: Demos Medical Publishing; 2000.
2. Murray TJ. The history of multiple sclerosis. New York: Demos Medical Publishing; 2005.
3. Compston A. The 150th anniversary of the first depiction of the lesions of multiple sclerosis. Journal of Neurology, Neurosurgery and Psychiatry 1988;51:1249–1252.
4. Kesselring J. History of multiple sclerosis. In: C. S. Raine, H. F. McFarland, R. Hohlfeld, editors. Multiple sclerosis: A comprehensive text. Philadelphia, PA: Elsevier Limited; 2008.
5. Rolak LA. The basic facts: A history of MS. New York: National Multiple Sclerosis Society; 2009.

6. Robson JT. Multiple sclerosis: Diagnosis, therapy and rehabilitation. Northwestern Medicine 1948;47(7):514.
7. Whitaker EW. A suggested treatment in occupational therapy for patients with multiple sclerosis. American Journal of Occupational Therapy 1950;50:247–251.
8. Hansen T, Skytthe A, Stenager E, Petersen HC, Bronnum-Hansen H, Kyvik KO. Concordance for multiple sclerosis in Danish twins: An update of a nationwide study. Multiple Sclerosis 2005;11(5):504–510.
9. Owen T. Multiple sclerosis. Canadian Journal of Occupational Therapy 1957;24:125–129.
10. Freeman JA, Langdon DW, Hobart JC, Thompson AJ. The impact of inpatient rehabiliation on progressive multiple sclerosis. Annals of Neurology 1997;42:236–244.
11. Warren S, Warren KG. Multiple sclerosis. Geneva: World Health Organization; 2001.
12. Confavreux C, Vukusic S, Moreau T, Adeleine P. Relapses and progression of disability in multiple sclerosis. The New England Journal of Medicine 2000;343(20):1430–1438.
13. Compston A, McDonald I, Noseworthy J, Lassmann H, Miller D, Smith K, Wekerle H, Confavreux C. McAlpine's multiple sclerosis. 4th ed. China: Churchill Livingston; 2005.
14. Cook SD, editor. Handbook of multiple sclerosis. New York: Taylor & Francis Group; 2006.
15. Lau KK, Wong LK, Li LS, Chan YW, Li HL, Wong V. Epidemiological study of multiple sclerosis in Hong Kong Chinese: Questionnaire survey. Hong Kong Medical Journal 2002;8(2):77–80.
16. Beck CA, Metz LM, Svenson LW, Patten SB. Regional variation of multiple sclerosis prevalence in Canada. Multiple Sclerosis 2005;11(5):516–519.
17. Orton SM, Herrera BM, Yee IM, Valdar W, Ramagopalan SV, Sadovnick AD. Sex ratio of multiple sclerosis in Canada: A longitudinal study. Lancet Neurology 2006;5(11):932–936.
18. Multiple Sclerosis [Internet]; c2009. Available from: http://www.mssociety.ca/en/information/default.htm.
19. Olek MJ, editor. Multiple sclerosis: Etiology, diagnosis, and new treatment strategies. NJ: Humana Press; 2005.
20. Grytten Torkildsen N, Lie SA, Aarseth JH, Nyland H, Myhr KM. Survival and cause of death in multiple sclerosis: Results from a 50 year follow-up in western Norway. Multiple Sclerosis 2008;14(9):1191–1198.
21. Sadovnick AD, Ebers GC, Dyment DA, Risch NJ. Evidence for genetic basis of multiple sclerosis: The Canadian Collaborative Study Group. Lancet 1996;347(9017):1728–1730.
22. Kantarci OH, Hebrink DD, Achenbach SJ, Pittock SJ, Altintas A, Schaefer-Klein JL. Association of APOE polymorphisms with disease severity in MS is limited to women. Neurology 2004;62(5):811–814.
23. Ebers GC, Sadovnick AD, Dyment DA, Yee IM, Willer CJ, Risch N. Parent-of-origin effect in multiple sclerosis: Observations in half-siblings. Lancet 2004;363(9423):1773–1774.
24. Chataway J, Mander A, Robertson N, Sawcer S, Deans J, Fraser M, Broadley S, Clayton D, Compston A. Multiple sclerosis in sibling pairs: An analysis of 250 families. Journal of Neurology, Neurosurgery & Psychiatry 2001 December 01;71(6):757–761.
25. Hensiek AE, Seaman SR, Barcellos LF, Oturai A, Eraksoi M, Cocco E. Familial effects on the clinical course of multiple sclerosis. Neurology 2007;68(5):376–383.
26. Oturai AB, Ryder LP, Fredrikson S, Myhr KM, Celius EG, Harbo HF. Concordance for disease course and age of onset in Scandinavian multiple sclerosis coaffected sib pairs. Multiple Sclerosis 2004;10(1):5–8.
27. Rosati G. The prevalence of multiple sclerosis in the world: An update. Neurological Sciences 2001;22:117–139.
28. Noonan CW, Williamson DM, Henry JP, Indian R, Lynch SG, Neuberger JS. The prevalence of multiple sclerosis in three US communities. Prevalence of Chronic Diseases 2010;7(1):A12.
29. Giovannoni G, Ebers G. Multiple sclerosis: The environment and causation. Neurology 2007;20:261–268.
30. Confavreux C, Adeleine P. Age at disability milestones in multiple sclerosis. Brain 2006;129:603–605.

31. Duquette P, Murray TJ, Pleines J, Ebers GC, Sadovnick D, Weldon P. Multiple sclerosis in child-hood: Clinical profiles in 125 patients. The Journal of Pediatrics 1987;111(3):359–363.
32. Ghezzi A, Deplano V, Faroni J, Grasso MG, Liguori M, Marrosu G. Multiple sclerosis in child-hood: Clinical features of 149 cases. Multiple Sclerosis 1997;3(1):43–46.
33. Polliack ML, Barak Y, Achiron A. Late-onset multiple sclerosis. Journal of the American Geriatrics Society 2001;49(2):168–171.
34. Pittock SJ, McClelland RJ, Mayr WT, Rodriguez M, Matsumoto JY. Prevalence of tremor in mul-tiple sclerosis and associated disability in the Olmsted County population. Movement Disorders: Official Journal of the Movement Disorder Society 2004;19(12):1482–1485.
35. Sindern E, Haas J, Stark E, Wurster U. Early onset MS under the age of 16: Clinical and para-clinical features. Acta Neurologica Scandinavica 1992;86(3):280–284.
36. Hurwitz BJ. Analysis of current multiple sclerosis registries. Neurology 2011;76:7–13.
37. Patwardhan MB, Matchar DB, Samsa GP, McCrory DC, Williams RG, Li TT. Cost of multiple sclerosis by level of disability: A review of literature. Multiple Sclerosis 2005;11(2):232–239.
38. Kesselring J, Beer S. Rehabilitation in multiple sclerosis. Advances in Clinical Neuroscience and Rehabilitation 2002;2(5):6–8.
39. Kurtzke JF, Heltberg A. Multiple sclerosis in the Faroe Islands: An epitome. Journal of Clinical Epidemiology 2001;54(1):1–22.
40. Mayr WT, Pittock SJ, McClelland RL, Jorgensen NW, Noseworthy JH, Rodriguez M. Incidence and prevalence of multiple sclerosis in Olmsted County, Minnesota, 1985–2000. Neurology 2003;61(10):1373–1377.
41. Pugliatti M, Sotgiu S, Rosati G. The worldwide prevalence of multiple sclerosis. Clinical Neurology and Neurosurgery 2002;104(3):182–191.
42. Kurtzke JF, Lux WE. In defense of death data: An example with multiple sclerosis. Neurology 1985;35(12):1787–1790.
43. Dean G. Annual incidence, prevalence, and mortality of multiple sclerosis in white South-African-born and in white immigrants to South Africa. British Medical Journal 1967;2(5554):724–730.
44. Gale CR, Martyn CN. Migrant studies in multiple sclerosis. Progress in Neurobiology 1995;47:425–448.
45. Handel AE, Handunnetthi L, Giovannoni G, Ebers GC, Ramagopalan SV. Genetic and environ-mental factors and the distribution of multiple sclerosis in Europe. European Journal of Neurology 2010;17(9):1210–1214.
46. Lauer K. Environmental risk factors in multiple sclerosis. Expert Review of Neurotherapeutics 2010;10(3):421–440.
47. Ghadirian P, Dadgostar B, Azani R, Maisonneuve P. A case-control study of the association between socio-demographic, lifestyle, and medical history factors and multiple sclerosis. Canadian Journal of Public Health 2001;92(4):281–285.
48. Hernan MA, Olek MJ, Ascherio A. Cigarette smoking and incidence of multiple sclerosis. American Journal of Epidemiology 2001;15(1):69–74.
49. Gholipour T, Healy B, Baruch NF, Weiner HL, Chitnis T. Demographic and clinical characteris-tics of malignant multiple sclerosis. Neurology 1996;76(23):1996–2001.
50. Pierrot-Deseilligny C. Clinical implications of a possible role of vitamin D in multiple sclerosis. Journal of Neurology 2010;256(9):1468–1479.
51. Ascherio A, Munger KL, Simon KC. Vitamin D and multiple sclerosis. The Lancet Neurology 2010;9(6):599–612.
52. Correale J, Ysrraelit MC, Gaitan MI. Immunomodulatory effects of vitamin D in multiple scle-rosis. Brain 2009;132(5):1146–1160.
53. Ozgocmen S, Bulut S, Ilhan N, Bulkesen A, Ardicoglu O, Ozkan Y. Vitamin D deficiency and reduced bone mineral density in multiple sclerosis: Effect of ambulatory status and functional capacity. Journal of Bone and Mineral Metabolism 2005;23(4):309–313.
54. Soilu-Hanninen M, Laaksonen M, Laitinen I, Eralinna JP, Lilius EM, Mononen I. A longitudinal study of serum 25-hydroxyvitamin D and intact parathyroid hormone levels indicate the

importance of vitamin D and calcium homeostasis regulation in multiple sclerosis. Journal of Neurology, Neurosurgery, & Psychiatry 2008;79(2):152–157.

55. Smolders J, Menheere P, Kessels A, Damoiseaux J, Hupperts R. Association of vitamin D metabolite levels with relapse rate and disability in multiple sclerosis. Multiple Sclerosis 2008;14(9):1220–1224.

56. Munger KL, Zhang SM, O'Reilly E, Hernan MA, Olek MJ, Willett WC. Vitamin D intake and incidence of multiple sclerosis. Neurology 2004;62(1):60–65.

57. Lublin FD, Reingold SC. Defining the clinical course of multiple sclerosis: Results of an international survey. Neurology 1996;46(4):907–911.

58. Kurtzke JF. A new scale for evaluating disability in multiple sclerosis. Neurology 1955;5:580–583.

59. Kurtzke JF. Rating neurologic impairment in multiple sclerosis: An expanded disability status scale (EDSS). Neurology 1983;33:1444–1452.

60. Hobart J, Freeman J, Thompson A. Kurtzke scales revisited: The application of psychometric methods to clinical intuition. Brain 2000;123:1027–1040.

61. Hobart J, Kalkers N, Barkhof F, Uitdehaag B, Polman C, Thompson A. Outcome measures for multiple sclerosis clinical trials: Relative measurement precision of the expanded disability status scale and multiple sclerosis functional composite. Multiple Sclerosis 2004;10:41–46.

62. Fischer JS, Jak AJ, Kniker JE, Rudick RA, Cutter G. Multiple sclerosis functional composite administration and scoring manual. New York: National Multiple Sclerosis Society; 2001.

63. Cohan JA, Fischer JS, Bolibrush DM, Jak AJ, Kniker JE, Mertz LA. Intrarater and interrater reliability of the MS functional composite outcome measure. Neurology 2000;54:802.

64. Tremlett H, Paty D, Devonshire V. Disability progression in multiple sclerosis is slower than previously reported. Neurology 2006;66(2):172–177.

65. Trojano M, Liguori M, Zinatore GB, Bugarini R, Avolio C, Paolicelli D. Age-related disability in multiple sclerosis. Annals of Neurology 2002;51:475–480.

66. Confavreux C, Vukusic S, Adeleine P. Early clinical predictors and progression of irreversible disability in multiple sclerosis: An amnesic process. Brain 2003;126(4):770–782.

67. Vukusic S, Confavreux C. Natural history of multiple sclerosis: Risk factors and prognostic indicators. Current Opinion in Neurology 2007;20(3):269–274.

68. Weinshenker BG. Natural history of multiple sclerosis. Annals of Neurology 1994;36:S6–S11.

69. Renoux C. Natural history of multiple sclerosis: Long-term prognostic factors. Neurologic Clinics 2011;29(2):293–308.

70. Koch M, Mostert J, Heersema D, De KJ. Progression in multiple sclerosis: Further evidence of an age dependent process. Journal of Neurological Sciences 2007;255:35–41.

71. Marrie RA. Demographic, genetic, and environmental factors that modify disease course. Neurologic Clinics 2011;29:323–342.

72. Weinshenker BG, Bass B, Rice GP, Noseworthy J, Carriere W, Baskerville J. The natural history of multiple sclerosis: A geographically based study, part 1, clinical course and disability. Brain: A Journal of Neurology 1989;112(1):133–146.

73. Weinshenker BG, Bass B, Rice GPA, Noseworthy J, Carriere W, Baskerville J, Ebers GC. The natural history of multiple sclerosis: A geographically based study, part 2, predictive value of the early clinical course. Brain 1989;112(6):1419–1428.

74. El-Moslimany H, Lublin F. Clinical features in multiple sclerosis. In: C. Raine, H. Mcfarland, R. Hohlfeld, editors. Multiple sclerosis: A comprehensive text. London: Saunders Elsevier; 2008.

75. Charcot J. Leçons sur les maladies du systèm nerveuz faites a la Salpêtrière. Paris: Cert Etfils; 1880.

76. Trapp BD, Peterson J, Ransohoff RM, Ruddick R, Mork S, Bo L. Axonal transection in the lesions of multiple sclerosis. New England Journal of Medicine 1998;338:278–285.

77. Lucchinetti CF, Bruck W, Parisi J, Schethauer B, Rodriguez M, Lassman H. Heterogeneity of multiple sclerosis lesions: Implications for the pathogenesis of demyelination. Annals of Neurology 2000;47:707–717.

78. Murphy KM, Travers P, Walport M, Shlomik MJ. Janeway's immunobiology. 7th ed. New York: Garland Science; 2007.
79. Tzartos JS, Friese MA, Craner MJ. Interleukin-17 production in central nervous system—A longitudinal study of serum 25-hydroxyvitamin D and intact parathyroid hormone levels indicate the importance of vitamin D and calcium homeostasis regulation in multiple sclerosis. American Journal of Pathology 2008;172(1):146–155.
80. Uhthoff W. Untersuchungen über die bei der multiplen Herdsklerose vorkommenden Augenstörungen. Archiv für Psychiatrie und Nervenkrankheiten 1890;21:303–410.
81. Schumacher GA, Beebe G, Kebler RF. Problems of experimental trials of therapy in multiple sclerosis. Annals of the New York Academy of Science; 1965.
82. Poser CM, Paty DW, Scheinberg L. New diagnostic criteria for multiple sclerosis: Guidelines for research protocols. Annals of Neurology 1983;13(3):227–231.
83. Polman CH, Reingold SC, Edan G, Filippi M, Hartung HP, Kappos L, Lublin FD, Metz LM, McFarland HF, O'Connor PW et al. Diagnostic criteria for multiple sclerosis: 2005 revisions of the "McDonald criteria." Annals of Neurology 2005;58(6):840–846.
84. Polman CH, Reingold SC, Banwell B, Clanet M, Cohen JA, Filippi M, Fujihara K, Havrdova E, Hutchinson M, Happos L et al. Diagnostic criteria for multiple sclerosis: 2010 revisions to the McDonald criteria. Annals of Neurology 2011;69:292–302.
85. McDonald WI, Compston A, Edan G, Goodkin D, Hartung HP, Lublin FD, McFarland HF, Paty DW, Polman CH, Reingold SC et al. Recommended diagnostic criteria for multiple sclerosis: Guidelines from the international panel on the diagnosis of multiple sclerosis. Annals of Neurology 2001;50(1):121–127.
86. Sahraian MA, Radue EW. MRI atlas of MS lesions. Heidelberg, Germany: Springer Publishing; 2008.
87. Miller DH, Kesselring J, McDonald WI, Paty DW, Thompson AJ. Magnetic resonance in multiple sclerosis. Cambridge, UK: Cambridge University Press; 1997.
88. Thrower B. Relapse management in MS. Neurologist 2009;15(1):1–5.
89. Burton JM, O'Conner PW, Hohol M, Beyene J. Oral versus intravenous steroids for treatment of relapses in muliple sclerosis. Cochrane Database of Systematic Reviews. 2009(3):CD006921.
90. Weinshenker B, O'Brien P, Petterson T, Noseworthy J, Lucchinetti C, Dodick D. A randomized trial of plasma exchange in acute central nervous system inflammatory demyelinating disease. Annals of Neurology 1999;46:878–886.
91. Goodman AD, Brown TR, Cohen JA, Krupp LB, Schapiro R, Schwid SR, Cohen R, Marinucci LN, Blight AD. Dose comparison trial of sustained-release fampridine in multiple sclerosis. Neurology 2008;71(15):1134–1141.
92. Goodman AD, Brown TR, Krupp LB, Schapiro R, Schwid SR, Cohen R, Marinucci LN, Blight AR. Sustained-release oral fampridine in multiple sclerosis: A randomised, double-blind, controlled trial. Lancet 2009;373(9665):732–738.

3

Stories to Inform the Art of Rehabilitation Practice

Marcia Finlayson*

CONTENTS

Rehabilitation practice is a combination of art and science.[1] This book is primarily about the science available to inform and guide multiple sclerosis (MS) rehabilitation. But the art of practice cannot be forgotten. According to Tate,[1] the art of rehabilitation reflects a mix of the creativity, critical reasoning, and empathy necessary to generate and offer potential solutions to address the complex problems that people often face. The art of rehabilitation also captures the holistic and humanistic aspects of practice and the importance of getting to know the people with whom we work. The art of practice can contribute to the development of meaningful patient–provider partnerships that support health, functioning, and well-being, regardless of a person's level of impairment.

Developing the art of rehabilitation requires observation of master clinicians, individual practice and experience, personal reflection, and the ability to listen carefully to, and learn from, patient's stories. The value of storytelling during provider–patient interactions has been recognized by several health care professions, for example, occupational therapy,[2] nursing,[3] and medicine.[4] In response, for many years, these professions (and others) have been educating their members about strategies for seeking and using client stories to inform the art of clinical practice.

Most chapters in this book describe, in some way, the everyday impact of MS, but they do not contain detailed stories about the daily experience of individuals living with MS. This chapter fills this gap by sharing the stories of two people who live with MS. Their stories are not meant to be representative of people with MS. Instead, they were selected because they illustrate—in very different ways—the breadth, depth, and temporal aspects of everyday health and functioning with MS. The overarching aims of this chapter are to present the impact of MS from an "insider" perspective; to challenge assumptions about independence, disability, and participation; to promote critical reflection about the potential role of rehabilitation for people with MS; and to encourage readers to listen to the stories of people with MS in order to develop the art of MS rehabilitation practice.

---

* In collaboration with Michael Ogg and Darlene Ptosnick.

3.1 Meet Michael: Over 4000 Miles in Three Years

Several months ago, the odometer on Michael's power wheelchair indicated that he had traveled over 4000 miles in 3 years. That translates into about 4 miles per day. Now Michael's comments about "battles for wheelchair repairs" take on a whole new meaning. His odometer is a proxy measure for the way he lives his life: on the go.

At this point, Michael has been living with progressive MS for about 14 years. When his symptoms started "I was pretty athletic then. I would go for a hard cycle and then I would notice that I had a hard time walking afterwards for about 20 min. Other people might have brushed it off, but I knew it wasn't right." Because of his level of physical activity at the time, he and his doctor initially "went the orthopedic route." After several consultations, recommendations for surgery, but no conclusive findings, "I went back to my primary care physician and said, 'I don't think this is orthopedic; I think it is neurological.'" Michael's physician referred him to a neurologist who put him through the standard diagnostic tests. "In the end, it took about two years to finally get diagnosed. I was kind of relieved—at least I knew what I was dealing with."

Now, Michael is 56 years old and divorced. His two daughters, aged 11 and 14, come to visit him several times a week as they live only about a mile away with their mother. He lives alone in a ranch-style house that he purchased after his divorce and renovated extensively to accommodate his current and future needs. He likes the small city in which he lives (~25,000 people) because it has a good infrastructure and good schools, as well as good access to public transportation. This access was a critical consideration for Michael when he was deciding where to live after his divorce. "I gave my realtor two constraints, besides cost. First was a ranch-style house so that I could make it accessible. The second one was that I wanted to be within one mile of the train station." His realtor was successful and found him a 1964 ranch-style home that was 0.7 miles from the train station. Due to this success, Michael can "get out and have a life." He is about a 50-min train ride to Manhattan, which enables him to go to the city regularly for concerts, plays, and other cultural activities. It also enables him to attend classes 2 days a week at the local university. "I'm fortunate to be able to audit classes. I do music theory in the morning and astrophysics in the afternoon—something for both sides of the brain." Michael is a physicist and professor who retired in 2004 because of his MS. He tries to stay up to date in his field by attending physics seminars when he is at the university.

Michael's MS has progressed rapidly. He explained, "Progressive MS is different. You don't suddenly have a relapse or suddenly can't use your arm. Symptoms start and they stay." He went on to share how "I went one limb at a time. It started with my left leg, followed my left arm [he is left-handed], followed by my right leg, followed by my bladder. I know I have a little bit of deficit in my right hand now. I don't know how it will go. All of my other limbs have gone from 100% to 0% in less than two years." Right now, Michael describes the movement in his right arm as "fairly reasonable," although he acknowledges that it fatigues quickly when he is holding the phone. He is also noticing changes in his right hand dexterity. "I can do things like hold pens and I type—one finger pecking at the keyboard. It works okay. I couldn't write *War and Peace* like that though." He notes that using a cell phone is very difficult because of the small keys and so "sending texts is really a painful process."

Recently, Michael purchased voice-activation software to facilitate his various computer activities. He acknowledges that he needs to be more patient and spend more time to train the program to recognize his voice and accent. He is hopeful that the effort will pay off, particularly since one of his planned projects is to write a book to summarize the extensive knowledge he developed while renovating his home. "I really want to write all of this stuff up. All of these little details that most people are not concerned about that really make a big difference—like what direction the floor joists should go to make the drain in the roll-in shower work better."

The home renovations that Michael researched, planned, and directed were extensive. "I had a go-for-broke attitude about it. I didn't want to hold back. I wanted to make sure that I had the best possible environment. I didn't want to do major work and then look back later and say, 'I wish I had done that.' I'm really glad I had that attitude." By combining his expertise in physics, the knowledge he gained through systematic research, his previous experiences doing home maintenance, and his work with an architect, Michael was able to design a space in which he hopes to stay for a long time. "I knew what I wanted and I had a blind faith that it was achievable and enough confidence in my abilities that I would eventually be able to get the answers and get this all sorted out."

To develop his plans, Michael drew on resources available through the Internet, the National Multiple Sclerosis Society, and the Abilities Expo. "Much of the DME [durable medical equipment] that I have now, I first learned about at the Expo." The architect offered "one or two ideas," but most important, he turned "my ideas into things that were code compliant." One resource that Michael found particularly helpful throughout the renovation process was the Standards Appendix from the Americans with Disabilities Act (available online).[5] "There was a wealth of information in there." For example, "One of things you need is a 60-inch turning circle for a wheelchair. That doesn't work in a 5′ × 8′ bathroom [which is what the original house had]. So we pushed back a wall to create the turning circle and a continuous floor with a roll-in shower."

Besides moving a wall to make changes in the bathroom, Michael also had a ramp installed out the back door to the deck and another one out a side door, which exits into the garage. He did not install a ramp in the front for safety reasons. "I don't think it would really be an issue, but as a general safety thing, people are advised not to advertise that there is a disabled person in the house by having a ramp at the front door." The renovations also included an overhead track system between his bedroom and bathroom to facilitate transfers during morning and evening routines. In the kitchen, Michael had lower cabinets with extra wheelchair kick space installed, as well as an induction stove top. "It is just absolutely fabulous. The reason I did it is primarily because of safety. I didn't want open flames or anything that would be a danger. For a disabled person, if they are going to do anything by themselves in the kitchen, this is the single most important nonmedical thing that they can do."

All of Michael's efforts paid off. He now has a fully accessible home in which he can live alone with the support of his personal care assistants. One comes in the morning to get him up, showered, and dressed, and another one comes in the evening to get him ready for bed. His assistants also keep Michael's house tidy, bring in and open his mail, and do his laundry. Michael also employs a yard maintenance company to take care of the grass in the summer and to clear the snow in the winter. While his personal care assistants "are just completely indispensible," their rigid scheduling is restrictive and "kills any spontaneity." He notes the same thing about paratransit (i.e., transportation for people with disabilities), which he only uses as a "last resort." Being able to get around on his own using public transportation was the reason he insisted that his realtor find him a house within a mile of the train station. "It doesn't sound like very much, but living with a disability is restricting enough. Any additional little restriction, no matter how trivial, is actually pretty tough."

Michael's inability to put a coat on himself is the reason that he leaves for the university at 7:45 a.m. on the days he has classes, even though his first class is not until 11:00 a.m. His assistant puts on his cape before she leaves, but "once it is on I can't just sit inside—it gets too warm, so I just leave." Michael makes the trek to the train station and then once at the university, he simply asks someone at the student union to take the cape off for him. "People know me there and I've learned to ask. People are very helpful." He uses the time at the student union to catch up on e-mail and do other reading. For lunch, he goes to the cafeteria but notes, "I can't use a knife and fork so I have to choose something that I can eat with one hand without making too much of a mess." It is similar when he goes out to eat at a restaurant with friends. "When I go out to a restaurant, I have three considerations. Can I get in? Can I get close to, or preferably under, the table? I don't like eating alongside the table. Last, is there anything on the menu that I can eat without making too much of a mess? This is one problem that health care professionals just don't think about."

Until a year or so ago, Michael was often staying at the university until later in the evening and would sometimes arrange to meet friends at the pub. Now, he is heading home earlier. While the rate of progression of his MS has slowed down over the past five years, Michael is noticing declines in his overall endurance. "So I just limit what I do in a day. If I try to do too much, I just collapse." He attributes some of his loss of endurance to two recent hospitalizations that left him quite deconditioned. Both were due to pressure ulcers. "A while back I was hospitalized for a stage 4* pressure ulcer and flap surgery. I just got out of recovery for bilateral stage 2s."† Although the recent stage 2 ulcers were less serious, Michael acknowledges that "they spooked me more." The stage 4 ulcer was a first occurrence, and he assumed that he would get over it and then continue on like before. "But when less than 12 months later, I was in for multiple stage 2s, it dawned on me that I was a lifer and that it was going to significantly affect how I went about my life every day."

Pressure relief is a big challenge for Michael since he is unable to shift his own weight enough when sitting in his wheelchair or lying in bed. He is currently looking to purchase a pressure relief mattress like the one he had while he was in the hospital. Michael needs his rehabilitation team to be up-to-date on the latest innovations for pressure relief and the extent to which they may or may not work in the context of his everyday life. For example, Michael's power wheelchair has a tilt-in-space feature, but achieving relief requires him to "tilt the chair back 45 degrees and then put the back down to almost horizontal. That provides relief. You can't do that everywhere every 20 min. That's the problem." Michael needs his rehabilitation team to understand this reality. He also needs them to be easily and quickly available for consultation if he starts having discomfort. Any delay in their responsiveness may mean the development of a new pressure sore.

Getting appropriate wheelchair seating and repairs and other DME is often a challenge for Michael, despite the fact that he has extensive knowledge and well-developed skills to find and evaluate resources. "Insurance is such a strange business. They will pay $50,000 for pharmaceutical therapy but will balk at spending $500 on DME. I don't get it. There seems to be this undercurrent that anyone who is trying to get DME is automatically a fraud or a scam or trying to cheat or get things that they are not entitled to. Certainly from my experience that is most definitely not the case. I have a condition, progressive MS, for which there is actually not any medical therapy. So if you are looking for bang-for-the-

---

* A stage 4 ulcer involves full-thickness skin loss that includes extensive destruction, tissue necrosis, or damage to muscle, bone, or supporting structures (e.g., tendon, joint capsule).
† A stage 2 pressure ulcer involves partial-thickness skin loss that includes the epidermis, dermis, or both. The ulcer is superficial and presents clinically as an abrasion, blister, or shallow crater.

buck, things that can make my life better, it is DME or nothing. Having the right DME makes just a huge difference for me."

Despite these challenges, "I have a very busy life." Michael enjoys hanging out at home with his daughters and "just spending time together." On the weekends, he sometimes has them with him full time. "Having a dad with a disability has made my daughters better people, more mature than is average for their age. I think it has made them much more caring and compassionate." In addition, Michael has "a good circle of friends" and is generally satisfied with his level of social engagement. "My one frustration is that almost none of my friends have houses that I can get into. That's actually very significant. If I let it, that could be a huge factor in my isolation. So I just have them over here." He continues to explain, "I don't know how long I will be able to maintain this situation. I really hope that I can keep going this way for a long time. I have spent time in nursing homes, and one of the things that did for me is make me want to stay at home as long as possible."

3.1.1 Questions for Critical Reflection

1. How would you describe Michael's priorities and challenges from the perspective of your discipline? Does your description capture the full range of his functioning from the perspective of the *International Classification of Functioning, Disability and Health* (see Chapter 1)?

2. How could you use your knowledge of Michael's story to facilitate collaborative goal setting with him?

3. If you were working with Michael, what contributions could you make to enable him to stay at home as long as possible, and how would you measure the success of your efforts?

4. How does Michael's story challenge your assumptions about the relative importance of safety, independence, and personal autonomy?

5. What did you learn from Michael's story that you could use to enhance the art of your own practice with people who are significantly disabled because of MS?

3.2 Meet Darlene: Continuation and Contribution—35 Pairs of Mittens

Ever since her grandmother taught her how to knit, Darlene has enjoyed spending her free time engaged in this pastime. She has made countless sweaters, afghans, baby blankets, and pairs of socks and mittens over the years. The process of knitting reflects a lot about Darlene and how she has coped with her MS over the years: creative, practical, and task-oriented. She explains that being diagnosed with relapsing–remitting MS at age 50 was "a shock." The doctor just said, "You have MS," and that was it. She was not given any information, explanation, or referrals elsewhere for support. So she did what she always does. Set a goal and got on with what needed to be done.

Darlene is now 72 years old and has been married for 51 years. She and her husband live in a three-bedroom bungalow in a rural community of ~5500 people. They have two grown daughters, both of whom live far away but are in touch by phone, email, and instant messaging on a regular basis. As a young woman, Darlene worked as an elementary school teacher. After her daughters were born, she stayed at home to raise them, as was the custom at the time. Once her daughters headed off to university, she began working at the local department store as a sales clerk. A few years after starting there, she became a part owner of the business together with her husband and another couple. Business was good, and she got much satisfaction out of the work and the interactions with members of the community. When they decided to retire, Darlene and her husband sold their half of the business to their partners. That was 10 years ago, right about the time that she started to share with people outside of her close family and friends about her MS.

Darlene explains why she kept her diagnosis a secret for so long: "Initially, I functioned very well. I was able to carry out a full day of activities with little difficulty." The only thing that she gave up early in the disease process was curling, a winter sport that is popular in her community. Although she had no noticeable mobility impairments at the time, walking on the ice at the curling rink was very challenging. "I really miss curling."

Over time, fatigue became a greater problem for Darlene and that was when she started resting regularly in the afternoons. Taking the rests helped her to conceal her MS from the people with whom she worked at the department store. Slowly, her mobility impairments became more noticeable, and she transitioned into a secondary progressive phase of the disease. "Really, MS didn't influence my activities until it started to affect my walking. I was tripping over my toe. But even then, I was still able to do my housework and gardening and all of those things. But it all came to a crushing end when the MS started to affect my muscle control on the right side." Over the past 10 years, Darlene has progressed from walking independently, to using a cane, to using a rollator inside the house. She now requires a scooter or wheelchair for distances more than 25 meters.

At this point, Darlene's life is fairly quiet and routine. "MS has changed my life a lot." She gets up at about 8 a.m. to catheterize herself, wash, and dress. Her morning routine takes her "at least half an hour," but she is independent. When she is ready, she joins her husband for breakfast, which he prepares. "After we have breakfast, I will sit for a while and read the newspaper." From there, Darlene spends her morning puttering around the kitchen "doing other little jobs" until about 10 or 10:30. "If there is something I can do to help get ready for lunch, like make a salad, then I will do that. Other days I might decide to make a batch of cookies. I'll mix it up and let it sit because I cannot mix it and get it on the trays all in one shot because I'm tired and I can't stand anymore. I have to cling to the counter so I won't fall over." Her husband often joins her in this task as he has recently taken an interest in learning how to bake. For the remainder of the morning, Darlene will usually read. At lunch, she and her husband will chat about "this and that" and try to catch the news on TV at noon.

Every afternoon, Darlene rests for about an hour or "sometimes up to two." The length of her rest depends on how many times she has to be up to use the toilet. Bladder symptoms are a dominant and disabling symptom for her. Some nights she is up as many as eight times to use the toilet. She has seen several urologists, all of whom have different ideas about what to do. She has tried multiple medications, none of which have been effective. "It has been so frustrating." Even self-catheterization throughout the day does not lessen the urgency she experiences when she lays down. "As soon as I lay down, I have to go to the bathroom."

After her afternoon rest, Darlene will often do stretching exercises to help manage spasticity in her legs. She also spends some time checking emails, watching TV, or talking with friends on the phone. She has a large group of friends because she has lived in the same community since she married. She has one neighbor with whom she talks almost everyday, and several other friends that she talks to several times a week. Although she does not go out of the house very much anymore, Darlene's phone calls with friends "keep me up to date with what is going on, give me some enlightenment and encouragement." About once a week, a friend will drop by. "It would be nice to have someone come more often. People are not in the habit of coming to visit because everyone is so busy." For the most part, Darlene's only outings into the community are to attend medically related appointments. A combination of mobility limitations, fatigue, and bladder symptoms "make it really difficult to get out." Despite this situation, Darlene enjoys the activities in which she engages, noting that "I enjoy my own company ... I have always enjoyed reading and knitting and doing those sorts of things ... I knit 35 pairs of mittens last winter for the Christmas Cheer Board [local charity]. That made me feel good, to have a product and make a contribution."

Darlene feels that she has a wide range of interests and notes that she has enjoyed most of them throughout her life. Learning how to use a computer is a relatively new interest and she wants to develop her skills further even though "it has been really challenging—it is hard to remember things." She notes that, for the most part, MS has not changed the things that she enjoys doing and "I pray that I will be able to keep doing them." She feels that the continuity in her core activities has helped her cope with the changes in her MS over time. She does miss being able to "just go out and visit, or just go away for the weekend, or go to the cottage." Even with fatigue and mobility problems, she was able to do these activities, but bladder problems forced her to give them up because of her concerns about being able to find and get to a bathroom on time. "Some days I just get fed up, I get really frustrated with the bladder thing, and I just want to go out. I just get fed up." Darlene explains that "MS is always in the back of your mind. You are always wondering when you will have the next blimp. It gets so upsetting and discouraging, you know, really discouraging. If there is something you want to do, you always wonder if the MS will allow you do to it."

Despite these challenges, Darlene has many good days. She describes a good day as one in which she accomplishes something—making a batch of cookies, finishing a knitting project, or reorganizing a cupboard or closet. She explains that whether or not she has a good day is dependent on "the attitude I get up with, the quality of my sleep the night before, and how I mentally accept things that day." She feels that her age is not a factor influencing her current abilities. "It is just the MS. I don't feel like I am 72. I don't feel my age. I can eat, I can enjoy my food, I enjoy my home—I just wish I could do more in it. I enjoy my family. When I am old, I won't get the same enjoyment out of these things."

3.2.1 Questions for Critical Reflection

1. How would you describe Darlene's priorities and challenges from the perspective of your discipline? Does your description capture the full range of her functioning from the perspective of the *International Classification of Functioning, Disability and Health* (see Chapter 1)?

2. How could you use your knowledge of Darlene's story to facilitate collaborative goal setting with her?

3. If you were working with Darlene, what contributions could you make to enable her to continue to engage in her valued pastimes, and how would you measure the success of your efforts?

4. How does Darlene's story challenge your assumptions about MS-related disability, aging, and health?

5. What did you learn from Darlene's story that you can use to enhance the art of your own practice with older adults who have MS?

3.3 Summary: Stories and the Art of Rehabilitation Practice

Rehabilitation is both an art and a science. The science of rehabilitation provides the necessary background to understand MS signs and symptoms, anticipate potential activity and participation restrictions, and guide the selection of appropriate assessment tools and intervention methods. The art of rehabilitation places this knowledge in the broader context of a patient's everyday life and facilitates the development of productive patient–provider relationships.

Through stories, people share and make sense of their experiences. Eliciting stories requires rehabilitation providers to develop excellent communication skills (e.g., asking open-ended questions, probing, paraphrasing, summarizing, reflecting feeling, clarifying)[6] and to use different modes of interpersonal interaction (e.g., encouraging, empathizing).[7] While taking a history, using a narrative approach[6,8] rather than a "checklist" approach can help to elicit patients' stories about their experiences with the disease.

Actively seeking and listening to the stories of people with MS can support the art of MS rehabilitation practice in several ways. First, stories provide an insider perspective on what it is like to live with MS on a day-to-day basis. This perspective provides contextual knowledge that can enhance clinical reasoning and problem solving during the course of the rehabilitation process. Second, stories offer insights into people's values, preferences, and beliefs, which can aid in the process of collaborative goal setting and tailoring rehabilitation interventions. Third, stories humanize practice and build empathy, both of which contribute to stronger patient–provider collaboration and more responsive, compassionate, and person-centered rehabilitation.[4] Finally, if we listen carefully and take the time to critically reflect, stories can challenge our assumptions about the meanings of health, disability, and independence. This, in turn, may enhance our ability to contribute fully to the lives of people with MS.

References

1. Tate DG. The state of rehabilitation research: Art or science? Archives of Physical Medicine and Rehabilitation 2006;87:160–166.
2. Frank G. Life histories in occupational therapy clinical practice. The American Journal of Occupational Therapy 1996;50(4):251–264.
3. Sakalys JA. Resorting the patient's voice: The therapeutics of illness narratives. Journal of Holistic Nursing 2003;21(3):228–241.
4. Sierpina VS, Kreitzer MJ, MacKenzie E, Sierpina M. Regaining our humanity through story. Explore 2007;3(6):626–632.
5. 2010 ADA standards for accessible design [Internet]; c2010 [cited 2011]. Available from: http:// www.ada.gov/2010ADAstandards_index.htm.
6. Haidet P, Paterniti DA. "Building" a history rather than "taking" one: A perspective on information sharing during the medical interview. Archives of Internal Medicine 2003;163(10): 1134–1140.
7. Taylor R. The intentional relationship model. Philadelphia, PA: FA Davis; 2008.
8. Kielhofner G. Model of human occupation. 4th ed. Philadelphia, PA: Wolters Kluwer/ Lippincott Williams & Wilkins; 2008.

4

An Overview of Multiple Sclerosis Rehabilitation

Jutta Hinrichs and Marcia Finlayson*

CONTENTS

Despite medical and pharmaceutical advances, there is no cure for multiple sclerosis (MS). While a growing number of medications can reduce the number of relapses and alter the course of the disease (see Chapter 2), people with MS continue to experience restrictions in their ability to participate in necessary and valued activities. As explained in Chapters 1 and 3, many factors contribute to MS-related functional restrictions, including changes in body structure and function, the physical and social environment, personal factors, what a person wants and needs to do, and the interaction of these

---

* The members of the rehabilitation team (Denisa Baxter; Neera Garga; Erin Gervais; Myrna Harden; Risha Joffe; Janice Lake; Kevin Lindland) providing services to the clients of the Out-Patient Treatment in Multiple Sclerosis (OPTIMUS) Program, Calgary.

factors with each other. Because of the complexity of MS and its consequences, efforts to prevent restrictions, restore losses already experienced, and maintain or improve existing functional abilities are paramount. As Kraft[1] wrote: "People with MS never ask for help in preventing their T cells from attacking the myelin of their central nervous system. Rather, they ask for help in improving their ability to function ... [they] desire rehabilitative care ... [it is] improvement in function that is the major plea" (p. 4). As this quote suggests, rehabilitation is a critical component of comprehensive MS care. The purpose of this chapter is to provide a broad overview of comprehensive MS rehabilitation and the disciplines and processes involved. After reading this chapter, you will be able to:

1. Outline the primary goals, potential target populations, and typical settings for MS rehabilitation,
2. Describe the general process of MS rehabilitation,
3. Describe the roles of key members of the MS rehabilitation team,
4. Identify general summary evidence and resources that support MS rehabilitation, and
5. Discuss the challenges and future directions for providing comprehensive MS rehabilitation.

By achieving these aims, this chapter sets the stage for the ones that follow. In subsequent chapters, detailed information and evidence is provided about specific assessment strategies and evidence-based intervention methods for the common impairments and activity and participation restrictions experienced by people with MS. This chapter provides a description of the general context in which these other specifics occur.

4.1 Goals, Target Populations, and Settings for MS Rehabilitation

Rehabilitation is an active, client-centered process that is goal-oriented, empowering, and usually time limited.[2,3] It involves many disciplines.[4–6] In the context of a respectful and collaborative partnership with the client and his or her family, the members of the rehabilitation team work together to enable the person with MS to

- Self-manage MS symptoms to minimize their medical, role, and emotional impact on daily life;[7]
- Maintain current abilities,[8] regain lost abilities, and maximize independence in daily activities;[9]
- Prevent deterioration and the emergence of new restrictions or secondary conditions;[10]
- Enhance participation[11] and autonomy in life roles;[12]
- Self-advocate for necessary services and supports;[13–15]
- Promote overall health, well-being, and life balance.[16]

Ultimately, MS rehabilitation is about maximizing functioning and quality of life across the full spectrum of life (e.g., physical, cognitive, emotional, social).[2–5] These overarching goals are achieved through multiple strategies, for example, task-specific training, therapeutic exercise, prescription and training on the use of adapted equipment, home modification, self-management training, health education, health promotion, consultation, counseling, and additional referral.[2-5,11] The breadth and depth of MS rehabilitation means that it is delivered across a full range of settings, including inpatient acute care; subacute, inpatient rehabilitation; outpatient rehabilitation; long-term care; home care; and community-based day programs.[2,3]

Since rehabilitation professionals have different training and areas of expertise, multidisciplinary and interdisciplinary approaches to service delivery are necessary to address the full range of issues that a person with MS faces. A *multidisciplinary* approach is one that is guided by a "gatekeeper" professional, usually a physician or MS clinic nurse, who makes decisions about which other disciplines to involve in the care of a given client based on assessment findings and the client's priorities. Involvement of these other providers is often managed through a referral process. In a multidisciplinary team, each member engages the client in individual, discipline-specific assessment, goal setting, and intervention delivery. Team members typically work independently from each other and submit findings and recommendations to other team members, directly (e.g., through team conferences) or indirectly (e.g., documentation in health care records).[17] An *interdisciplinary* team works collaboratively,[17] directly involving the client in all decision making.[18] Using an interdisciplinary approach, members of the team work together to conduct assessments, set goals, and develop intervention plans in a collaborative partnership with the client. Team members are encouraged to question, explore alternative solutions to problems, and look beyond their own disciplines toward the best and most holistic outcome for the client.[18]

Although people with the disease are the primary partners in the MS rehabilitation process, other individuals or groups may also receive services, directly or indirectly, in order to achieve therapy goals. Examples of other potential target populations for MS rehabilitation include family members, friends and other informal caregivers; personal care assistants; employers and coworkers; school teachers (for children with MS); and other health care professionals who do not specialize in MS care or rehabilitation. The services received by these other target populations often involve consultation, education, or counseling, all with the ultimate goal of supporting the everyday functioning of a person with MS.

4.2 Process of MS Rehabilitation

Over the past two decades, there has been significant development in the area of rehabilitation for people with MS. Nevertheless, the basic rehabilitation process is consistent with that followed with other patient groups. What makes MS rehabilitation unique is the population served: MS is a disease that typically strikes in young adulthood, it is chronic and progressive, and it has a variable and unpredictable course. People's needs and concerns can shift and change with little to no warning. Rehabilitation practitioners must therefore approach MS rehabilitation using a structured and iterative problem-solving process.[11]

MS rehabilitation begins with an expression of concern from a person with MS (or caregiver) to a health care provider (often a physician or MS-clinic nurse) about the impact of MS symptoms or their consequences on his or her everyday health and functioning

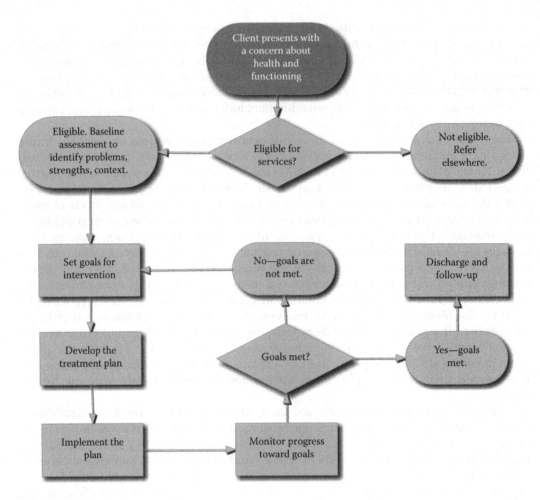

FIGURE 4.1
MS rehabilitation process.

(see Figure 4.1). Whether a person with MS is directed to rehabilitation services will depend on the concern expressed, the provider's knowledge of rehabilitation services, and the availability of services in a particular region. In some jurisdictions, people with MS are able to self-refer to rehabilitation services.

Once the person with MS comes in contact with a member of the rehabilitation team, either individually (i.e., multidisciplinary approach) or through a comprehensive rehabilitation center or program (i.e., an interdisciplinary approach), he or she will be screened for service eligibility. Eligibility criteria may include age, severity of disability, type of service needed, insurance status, geographical catchment area, or other factors. Criteria vary across providers, facilities, and health care systems. After eligibility is confirmed, a comprehensive assessment of baseline status is initiated.

4.2.1 Baseline Assessment

The purpose of the baseline assessment is to develop a current and comprehensive snapshot of the client's life at that particular point in time and how MS has influenced overall health, functioning, and the performance of everyday activities. The baseline assessment process enables the rehabilitation team to initiate a therapeutic relationship with the client, learn about his or her concerns and priorities, and gather the additional information needed to set realistic intervention goals and plans. Baseline assessment typically includes a combination of interview, observation of performance, and administration of standard assessments. Information gathered during the baseline assessment allows the members of the team to understand the client's strengths and problems as well as the contextual factors that may influence problem resolution. The following key information is sought during the baseline assessment:

- The client's MS history (e.g., type of MS, symptoms, medication regiment including history of steroid use, complementary or alternative therapies) (see Chapter 2), overall medical status, and comorbidities (see Chapter 19).
- The client's current and desired level of functioning, including what he or she wants and needs to be able to do during everyday life (see Chapters 12 through 17).
- The client's psychological well-being and coping styles (see Chapter 20).
- The client's motivators and readiness to make behavioral and lifestyle changes that may be recommended through the rehabilitation process.[19]
- The developmental stage (e.g., adolescence, early adulthood, retirement) (see Chapter 18), life roles and associated responsibilities (see Chapters 15 and 16), and general lifestyle.
- The nature and extent of the social relationships and related supports available to the client (see Chapter 22).
- The physical and cultural environments in which the client carries out his or her everyday life (see Chapters 21 and 23).

As part of the baseline assessment, information is also gathered about the client's experience with the disease and his or her perspectives on the pattern and pace of disease progression. These factors can influence a person's goals, expectations, and ability to stay active and engaged throughout the rehabilitation process.[20,21] The assessment process concludes with a careful analysis that draws on client priorities, clinical experience, theoretical and empirical knowledge, and contextual factors. The product of this analysis is a list of problems and strengths and priorities for intervention that have been developed collaboratively by the client and rehabilitation provider(s). This work provides the basis of the next step in the process: goal setting.

4.2.2 Goal Setting

Setting goals provides direction for the treatment plan and sets the criteria against which progress can be monitored and evaluated.[20] It keeps rehabilitation providers focused; guides discussions at team conferences; and provides feedback to clients, family, referral sources, and funders.[22] When goals are developed collaboratively with clients and across members of the team, there can be many benefits. Studies have found that collaborative goal setting increased the motivation and satisfaction of clients,[23] led to more targeted intervention plans,[22,23] and provided both the clients and their families with reassurance about the rehabilitation process as a whole.[22]

Achieving these benefits requires negotiation and goals that accurately reflect the client's priorities, lifestyle, interests, and values.[24] Often though, a client's priorities and those identified by the rehabilitation team through the baseline assessment do not match. In one study, researchers found that, on average, people with MS receiving inpatient rehabilitation only agreed with their service providers on 1.7 out of 5 goal priorities, and nearly 20% of clients had no concordant goals with the members of their rehabilitation team. When clients and providers have discordant goals, there is a risk of inappropriate treatment, poor communication, and low follow-through on recommendations.[25] There are several things that rehabilitation providers can do to maximize the benefits of collaborative goal setting and reduce the risk of discordant goals:

- Provide verbal and written information about the goal-setting process to clients upon admission.[26]
- Employ assessment tools that gather client perspectives on their functional performance and priorities, such as the Canadian Occupational Performance Measure[27] or the Rehabilitation Problem-Solving Form.[25]
- Educate the clients and their family members about the disease, the specific problem(s), the assessment findings, and what can realistically be achieved during the intervention process.[23]
- Increase training for rehabilitation providers on methods to support client participation in goal setting.[26,28]

When the model of service delivery is multidisciplinary rather than interdisciplinary, it is also important that individual team members are aware of all the goals set by other team members and work toward goal congruence across disciplines.[29]

To facilitate the process of writing goals, the SMART approach is commonly used.[20] SMART goals are goals that are *s*pecific, *m*easurable, *a*chievable, *r*elevant/realistic, and *t*imed.[30] Another approach is the ABCD method, which directs the goal writer to specify the *a*udience (those who will engage in the behavior), *b*ehavior (observable performance to be exhibited), *c*ondition (with what supports or cues), and *d*egree (measurable criteria for achievement).[31] Figure 4.2 depicts the process, questions, and considerations in the process of writing client-centered goals.

Depending on the setting, duration of intervention, and whether or not any follow-up will be provided on a longer-term basis, goals may be separated into those that are short term and those that are long term. What constitutes a short-term goal will depend on the setting. For example, if a client is in the hospital for a brief, acute care admission due to an exacerbation, a short-term goal may reflect a behavior that can be achieved in just one or two days. In comparison, a short-term goal in outpatient rehabilitation may extend for two weeks or more. The duration specified in a long-term goal will be equally variable. Regardless of the length of time given to achieve the goals, short-term goals should prepare the client for the achievement of a long-term goal (e.g., act as a stepping stone).

4.2.3 Intervention Planning and Implementation

Once goals for intervention have been negotiated, the next step in the rehabilitation process is to develop the intervention plan. During this step in the process, many decisions need to be made. These are some of the key decisions:

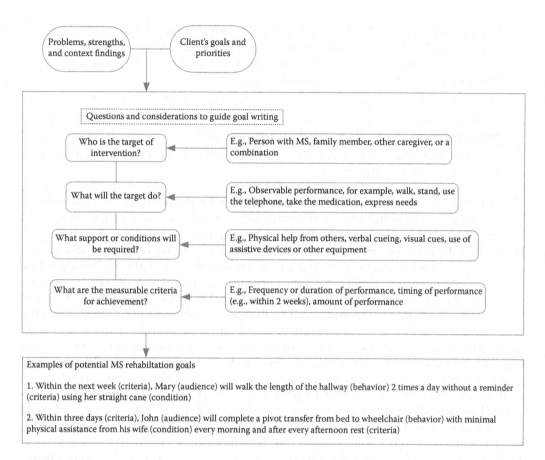

FIGURE 4.2
Writing client-centered goals.

- What is the expected duration of the intervention (total number of sessions) based on the goals to be achieved, the client's priorities and prognosis, and any other relevant factors (e.g., insurance status, geography)?
- Where will intervention be delivered (e.g., in the hospital room, rehabilitation department, in the home, or a combination of settings)?
- How often will intervention sessions occur (frequency)?
- What will the specific content focus of each intervention session be? How will the content of sessions be sequenced over time to reach short- and long-term goals?
- What specific strategies/methods can or should be employed to address the content of each session (e.g., education, exercise, device training)? How will these strategies be sequenced both within individual sessions as well as across sessions over time? How can strategies be up- or downgraded (made more or less challenging) depending on client's physical, cognitive, and emotional responses to intervention?
- How intense will the sessions be? Intensity reflects a combination of the length of the session as well as how challenging the activities within the session will be considering MS symptoms.

- What empirical evidence or clinical guidelines (e.g., National Institute of Clinical Excellence [NICE] guidelines[3] or the European Multiple Sclerosis Platform[2] recommendations) support the decisions about intervention content and methods, frequency, intensity, and expected outcomes?
- Who will be involved in each session (e.g., client only, client and family)?
- What resources will be needed for each session (e.g., space, equipment, educational materials, presence of a rehabilitation assistant)?
- Are there any contraindications for the intervention plan because of medication regiments or other comorbidities?

Making these decisions is analogous to determining the path to reach the final destination (i.e., achievement of the goal). Setting the path will be influenced by the client's and the team's hypotheses about the cause or causes of each problem to be addressed (e.g., is the problem the result of impairments of body structure and function or of restrictions imposed by the social environment?). Intervention decisions will also be influenced by the client's and the team's beliefs about the personal, environmental, or task-specific factors that have the greatest potential for successful modification to resolve the problems at hand. Finally, decisions will be influenced by the specific methods and strategies that the client and the members of the team believe will lead to the desired changes (e.g., task-specific training, exercise, education, counseling, environmental modification).[32,33] For the client, his or her beliefs about what factors can be changed will depend on previous experiences with the disease and rehabilitation, vicarious learning from others who have received rehabilitation, and personal values and preferences. For the other team members, beliefs about what factors can be changed and what methods can be applied to achieve change will be based on discipline-specific philosophy and training, guiding theories and frameworks (e.g., *International Classification of Functioning, Disability and Health* [ICF]), available empirical evidence, and clinical experience with similar clients. Since each person involved in the rehabilitation process will have different perspectives about cause and change, communication and collaboration between the client and the members of the team are critical to ensure a comprehensive plan that everyone understands.

Ultimately, a well-articulated intervention plan provides a map to goal achievement. Because of the variable nature of MS and the potential for changes in symptoms during the course of rehabilitation, this map may need to be modified as the client's symptoms, context, or priorities change.

4.2.4 Monitoring Progress and Evaluating Outcomes

Setting goals and documenting the intervention plan enables the rehabilitation team to monitor progress and outcomes and initiate discussions with the client about potential modifications as intervention proceeds. While there are many potential foci for MS rehabilitation (see Section 4.1) and goals will be individually tailored, the ultimate goal is to resolve the client's concerns about functioning.

After intervention has been initiated, periodic reassessments should be completed to determine if the client is making progress as expected and if he or she is satisfied with that progress. Is the intervention proceeding as planned? Are up- or downgrades being made with any regularity? If sessions are being upgraded regularly, the team should talk with the client and consider whether the goals and plans were ambitious enough. Alternatively, regular downgrades may suggest that the goals and plans were too ambitious. Either way, goals and plans may need to be refined and updated in collaboration with the client and then recorded in the clinical documentation.

Specific timeframes for reassessment may be determined by the team or facility administration, the referral source, or by the service funder. Informal reassessments occur almost continuously during the rehabilitation process,[20] as clients engage in the treatment process. The observations made by the provider and the feedback given by clients are used to make adjustments to intervention methods or session intensity. Formal reassessment, which involves the readministration of the same standardized instruments that were used during the baseline assessment, is the only way to definitely determine whether measurable change has occurred. Formal reassessment is indicated when goals are being reconsidered due to new symptoms, restrictions, or life roles, or when outcomes of intervention must be documented.

Comparing baseline and reassessment scores is the basis of outcome evaluations, which indicate the effectiveness of intervention.[24] Evaluation of outcomes must be based on psychometrically sound tools that are conceptually congruent with the goals being set, the client's and team's causal hypotheses for the problems, and the intervention methods employed. For example, if intervention goals and strategies are focused on activity and participation (e.g., walking in the community), then measuring changes in body structure and function (e.g., strength or lower-extremity range of motion) will not fully capture the focus of intervention. The best tools will be the ones that match the goal and the expected change and have sound psychometric properties for an MS population.[34]

4.2.5 Discharge, Follow-Up, and Periodic Re-Review

Often, when rehabilitation goals are met or the client is not making further progress (i.e., has reached a plateau), he or she is discharged from therapy services. The NICE[3] guidelines recommend that people with MS be asked if they want a regular review of their situation, and if they do, that a suitable interval be negotiated. For individuals with severe impairment, the guidelines recommend annual re-review of support needs.

Implementing these recommendations may be difficult for some rehabilitation teams and often ongoing rehabilitation follow-up is the exception rather than the rule. Despite this, it is important for the members of the rehabilitation team to discuss with the client what has been accomplished through the rehabilitation process both from the client's and the team's perspectives. If the client identifies any unmet needs that remain and that cannot be addressed through rehabilitation, referrals to other services may be required. When additional services are required, the rehabilitation team should notify the original referral source.[3]

Follow-up is the stage of linking with the person after discharge to ensure that plans are being followed, no new issues have arisen, and gains that have been achieved are being maintained. Often, follow-up can be done with a phone call a few weeks after discharge. Some services may regularly reassess clients to ensure there have been no significant functional declines that should trigger another round of goal setting and intervention delivery. For example, in comprehensive MS care centers, clients are often seen annually by their neurologists and then screened for emerging functional issues that may necessitate a course of rehabilitation. Depending on professional regulations (e.g., need for physician referral), other rehabilitation services may simply allow clients to self-refer when a new functional concern arises.

4.3 Members of the MS Rehabilitation Team and Their Primary Roles

The specific disciplines that make up the rehabilitation team vary across practice settings, facilities, and health care systems. Regardless of the team's composition or structure (multidisciplinary or interdisciplinary), the client is at the center of the process and must be considered as a collaborative partner in all efforts. Based on the client's priorities, the other members of the team work together to resolve most of the common problems faced by the client.[11] The roles and responsibilities of some of the more common members of the rehabilitation team are described below, although many others may be involved depending on client needs and priorities, symptoms, and life situation.

4.3.1 Person with MS and Family/Caregivers

The person with MS is the central member of, and partner with, the MS rehabilitation team. The activities of the other team members focus on addressing the needs and concerns identified by the client in a way that is consistent with his or her values and preferences as well as those of his or her family (or other caregivers) and their social, cultural, and environmental context. As noted previously in this chapter, rehabilitation professionals must recognize that the needs identified through their own assessment processes may not be the same as those identified by the client or the family. Ensuring that the client's priorities remain central to the rehabilitation effort requires that the client and family/caregiver(s) be active partners in the review of assessment findings; selection of treatment goals; and, to the extent possible, the selection of intervention methods and outcomes. Involving clients and families/caregivers in these processes is often referred to as shared decision making, or patient- or client-centered practice. In one study, 79% (132 out of 168) of people with MS indicated that they wanted some level of shared medical decision making.[35] Specifically, 65 of these 132 preferred shared decision making (i.e., shared responsibility for deciding best treatment), and 60 of 132 preferred informed-choice decision making (i.e., client is responsible for making the final decision about treatment after seriously considering the physician's opinion).

4.3.2 Physicians

Several different types of physicians may participate in the rehabilitation process, most common are the neurologist, physiatrist (i.e., rehabilitation specialist), and the primary care physician.

4.3.2.1 Neurologist

The neurologist is the specialist and expert in neurological diseases and is responsible for making the diagnosis of MS. The diagnostic process is described in Chapter 2. In brief, it involves the neurologist completing a thorough neurological examination, taking a patient and family history, and ordering and interpreting a series of diagnostic tests. After diagnosis, the neurologist is responsible for selecting and recommending (if appropriate) the best disease-modifying medication for the patient and determining whether additional medications may be beneficial for symptom management (e.g., pain, fatigue, urinary symptoms, etc.). The neurologist is then responsible for regularly monitoring the patient's neurological status to determine disease progression and the patient's response to treatment, and to make referrals to other members of the multidisciplinary team. Throughout the provision of care, the neurologist responds to questions about the disease process and treatment from the patient, his or her family, and other members of the rehabilitation team. Depending on the model of service delivery being used, the neurologist may be considered the leader of the professional members of the rehabilitation team.

4.3.2.2 Physiatrist

The physiatrist is the specialist and expert in physical medicine and rehabilitation. Not all people with MS who receive rehabilitation services will be referred to a physiatrist. Factors that should trigger a referral include the presence of rapid or acute worsening of functional limitations; chronic, progressive activity limitations; complex symptom-management issues; or complex functional and psychosocial issues.[36] In some settings, the physiatrist may be responsible for monitoring and managing the overall rehabilitation process, particularly when a patient's issues are complex. In these situations, the primary role of the physiatrist on the MS care team is to coordinate the medical treatments and interventions provided by the rehabilitation team that focus on the patient's activity and participation restrictions.[36]

4.3.2.3 Primary Care Physician or General Practitioner

For day-to-day health and functional concerns, a primary care physician is often the first medical practitioner that a patient will contact because of ease of communication, accessibility, and existing relationship. Depending on the health care system, primary care physicians often fulfill the role of coordinator of all medically related care, including referrals to rehabilitation providers, home support services, and other medical specialists (e.g., urologist).

4.3.3 Physical Therapist

Since many of the common impairments of MS negatively influence movement and function (e.g., limb and trunk weakness, spasticity, ataxia, balance impairments, pain, gait deficits), physical therapists play a critical role on the rehabilitation team throughout the disease course.[37,38] Physical therapy assessment in MS care evaluates limitations in strength, range of motion, balance, posture, gait, and transfers and determines their functional impact.[38] Using this information, physical therapists provide treatment aimed at developing, maintaining, and restoring maximum movement and function.[39]

To address problems in movement and function, a major component of most physical therapy interventions is client-specific exercise prescription.[40] Exercise has been shown to effectively manage many physical symptoms of MS.[41–43] For example, stretching may aid in the management of mild-to-moderate spasticity when done in conjunction with pharmaceutical treatment,[44] balance exercises can reduce the risk of falls,[45] moderate intensity resistance training can improve muscle strength,[41] and cardiovascular endurance can be improved with aerobic exercise.[41,46] Exercise programs should be a challenge to the client but not a struggle[37] because for some people, exercising can have temporary negative effects (e.g., fatigue, heat intolerance).[47] In general, moderate exertion with a focus on maintaining good quality and consistent movement may be preferable for building strength in people with MS.[38]

Exercise therapy will vary depending on its setting. Programs in an inpatient setting often require that the therapist or therapy assistant provides direct, hands-on assistance and support to the client. When the client returns home, he or she is often able to continue the exercises independently and may start to include community resources such as aquatics and yoga classes.[38,48]

When gait impairments occur, physical therapists provide gait retraining, which may include prescription and training in the use of orthotic or gait aids.[40] Examples of such aids include walking aids (e.g., straight cane, quad cane, rollator), ankle-foot orthoses, functional electrical stimulation devices or hip flexion devices.[38] Working closely with occupational therapists, physical therapists also are involved in assessing and prescribing wheelchairs[40] and contributing information needed for driving assessments (e.g., lower-extremity strength, range of motion, coordination). It is common for physical therapists to collaborate with family and caregivers on transfers, bed and wheelchair positioning, adaptive equipment, and home modifications, particularly in situations where the client's MS is advanced.

When ongoing therapy is not possible, patients should be referred to appropriate community programs to help maintain or improve their level of mobility and function. Regardless of the focus of the physical therapy intervention, a strong emphasis is placed on educating the client to self-manage his or her symptoms. For many clients, education focuses on lifestyle changes that support engagement in regular exercise and other modifications to support mobility and function. When a client follows through with physical therapy recommendations and exercise programs, he or she can gain a sense of control over MS. Physical therapy is most successful when the therapy goals and interventions are consistent with the client's priorities and ultimately influence functioning positively.[37]

4.3.4 Occupational Therapist

"Occupation" is everything that people do to occupy themselves, including looking after themselves (self-care), enjoying life (leisure), and contributing to their communities through work and other productive activities.[49] MS symptoms can restrict engagement in a wide range of occupations at any point in the disease course. Therefore, occupational therapists provide services to people with MS throughout the disease and across a full range of settings, including acute care, inpatient rehabilitation, outpatient rehabilitation, day programs, home care, and community-based services.

Occupational therapy assessment is client centered and focuses on documenting the occupations, roles, and routines the client wants or needs to do more effectively or satisfactorily and how his or her abilities and capacities (physical, cognitive, social, affective) and environment (physical, social, cultural, institutional) influence performance.[49,50] Through the assessment process, the occupational therapist identifies activity and participation restrictions and the modifiable factors contributing to those restrictions. Assessment tools include interview tools with standard procedures (e.g., Canadian Occupational Performance Measure [COPM]), self-report instruments, as well as a range of performance measures.[51]

Depending on the client's interests, needs, and goals, occupational therapy intervention may focus on improving the clients' abilities to engage in self-care, mobility (particularly upper extremity function), domestic life, leisure activities, or to maintain a productive role (e.g., worker, student, volunteer).[51,52] Achieving these goals requires that the occupational therapist identify and apply strategies and solutions that effect change at three levels:

- The person (e.g., reduce symptoms, enhance coping, improve performance capacity, increase problem-solving abilities).

- Specific occupations (e.g., modify routines and habits, modify activity demands [sequencing, timing], modify method of performance [e.g., use assistive devices]).
- The environment in which the person functions (e.g., educate caregivers, modify the home, build support networks, prescribe and train on the use of assistive technologies).

Common MS symptoms addressed by occupational therapists include, but are not limited to, fatigue;[53,54] cognition;[55–58] upper-extremity dexterity, strength, and coordination;[59,60] and balance and vision.[51] The specific intervention methods used by occupational therapists to effect change and achieve goals include, but are not limited to, therapeutic use of self, task-specific training,* consultation, education, and advocacy.[50] Through these methods, occupational therapists restore function, support adaptation, compensate for or remediate losses, and prevent secondary complications.[50] As a result, the client's occupational performance, participation, and quality of life can be maintained, restored, or enhanced.

4.3.5 Nurse

There are many roles for nurses in MS care. The domains of MS nursing have been defined as clinical practice, advocacy, education, and research.[61] In a qualitative study of the role of nurses on a multi-professional rehabilitation team, researchers identified six interconnected roles: assessment, coordination and communication, technical and physical care, therapy integration and carry-over, emotional support, and involving the family.[62] In addition, researchers noted that nurses play a particularly important role in creating an environment that supports the rehabilitation process. Fulfillment of these roles enables nurses to contribute to maximizing client choice and independent living potential.[63]

MS rehabilitation nurses offer a holistic approach to meet an individual's physical, emotional, and social needs[63] by applying knowledge and skills specific to rehabilitation as well as to MS care. Basic but essential rehabilitation nursing skills include relieving pain; helping with hygiene and mobilization; providing care to pressure areas to prevent skin breakdown and ulcers; ensuring adequate nutrition; promoting bladder and bowel care and managing incontinence; giving emotional support; and providing opportunities for adequate sleep, rest, and stimulation.[63] Rehabilitation nurses working with people with MS may practice in rehabilitation facilities, hospitals (inpatient rehabilitation units), long-term care facilities, comprehensive outpatient rehabilitation facilities, private practice, and home care agencies.[64]

Regardless of the setting, rehabilitation nurses contribute their expertise and skills to address specific impairments that are not always addressed by other team members. Bladder and bowel dysfunction is a major focus of MS rehabilitation nursing. Changes with bladder and bowel function can negatively influence many areas of life, including employment, education, socialization, sleep, intimacy, exercise, and self-esteem.[65] Utilizing nursing skills to assist with management of various symptoms not only improves function, but can also positively affect everyday life.[61] Prevention of bladder infections is important as they can make other symptoms worse and negatively affect function.[66,67]

Another important role for MS rehabilitation nurses is addressing client concerns about sexuality and intimacy. Clients' awareness and concerns about sexual issues will vary. For those clients who have concerns that they want addressed, the MS rehabilitation nurse

---

* Occupational therapists tend to use the term "occupation-based intervention," which refers to the therapeutic use of activities that are meaningful to the client during the course of intervention.

will assess the client's sexual knowledge; sexual self-view; and sexual activity, interest, and behavior.[68] Interventions will vary due to individual comfort levels, values, and beliefs. Skin integrity is monitored on an individual basis. This includes proper assessment, education for prevention of skin breakdown, and assistance with care.[69]

MS rehabilitation nurses collaborate with professionals in many other disciplines, including those beyond the rehabilitation team (e.g., home-care services). MS rehabilitation nurses are often involved in educating clients about medications and side effects, alternative therapies, nutrition and supplements, and general health issues. By emphasizing prevention through education and management, MS rehabilitation nurses promote overall wellness and enable clients to increase control and take greater responsibility for their own care.[61]

4.3.6 Social Worker

Adjusting to and coping with the symptoms and variability of MS can be very challenging for clients and their families, socially, emotionally, and financially. Adjustment challenges often arise after the initial emotional reactions of a diagnosis of MS (e.g., shock, denial, fear, anger, sadness, relief). They also can occur throughout the disease process, particularly during periods of progression and increasing disability. Adjustment challenges may revolve around changes in roles and responsibilities, family dynamics, and communication. As a member of the MS rehabilitation team, a social worker addresses these challenges by offering counseling and educational interventions and by linking clients to essential community resources that provide employment accommodations, home modifications, disability insurance, long-term care, and so on.[70] Through these interventions, the social worker seeks to promote and support client adaptation, disease management, coping, and stress management. To achieve these goals, the social worker often assists family members to identify their feelings and explore ways to engage with each other more comfortably about their concerns. This process facilitates mutual support and collaborative problem solving between the client and his or her family.

Another important role for social workers on the MS rehabilitation team is to work with clients to advocate for improved access to limited resources. This is more of an instrumental role that focuses on financial issues, such as employment accommodations and applying for and appealing short-term and long-term disability claims, government benefits, health benefits, appropriate housing, and other necessary resources. Sometimes MS cognitive impairments can limit a client's ability to make competent decisions. In these cases, the social worker often helps a client prepare for this possibility. In these situations, the social worker meets with the client and the identified support person to facilitate dialogue about his or her wishes regarding personal matters such as health care, where to live, and social activities. The social worker then supports the process of documenting these wishes in a personal directive and encourages communication about these wishes to the other members of the health care team, family, and friends, as appropriate.

In summary, the social worker works on the MS rehabilitation team to support clients in their relationships through counseling and education, advocating for instrumental needs and resources, and facilitating long-term care planning and decision making. A strength-based perspective often guides the social work process as the worker gathers and assesses individual and family needs and determines how people can build on their existing resources and strengths.

4.3.7 Speech–Language Pathologist

Speech–language pathology is a health profession dedicated to improving the quality of life of individuals with impairments and restrictions related to communication or swallowing. Speech–language pathologists work collaboratively with clients, families, and other health care providers to deliver evidence-based services that optimize clients' functional communication and swallowing abilities in meaningful contexts. Through their efforts, speech–language pathologists address the social and emotional aspects of disability, the impact of environment on the person, and a client's participation in daily life.

The speech–language pathologist provides education, assessment, treatment, and management to individuals with speech, language, fluency (stuttering), voice, cognitive, or swallowing and feeding disorders. These disorders can be congenital, developmental, or acquired in nature. Speech–language pathology services are provided in hospitals and to individuals residing at home or in continuing care facilities. Speech–language pathologists bring an in-depth understanding of the anatomy and physiology involved in speech and swallowing disorders as well as expertise in the diagnosis and management of communication disorders resulting from linguistic or cognitive impairments.

When working with people with MS, the role of the speech–language pathologist is to assess and manage communication and swallowing disabilities over the course of the disease process. Speech or communication changes are reported by 23–51% of individuals with MS.[71,72] Swallowing difficulties are reported by 33%[73] to 67% of people with MS, depending on the level of disability.[74] Consistent with the ICF, the speech–language assessment process is designed to identify underlying speech, language, and cognitive impairments and their resulting impact on activity (e.g., speech intelligibility) and life participation (e.g., parenting, vocation). Assessment also considers the impact of other impairments (e.g., vision, mobility) on an individual's communication participation.[75] Because of the potential for speech and swallowing to deteriorate over time, ongoing assessment may be necessary to monitor an individual's communicative competence.

Speech–language pathology interventions focus on improving quality of life by enhancing and maintaining communication and swallowing abilities in the context of meaningful life activities and over the course of the disease process. Treatment goals must therefore consider an individual's current and anticipated needs. To facilitate independence and active participation in client's daily routine, treatment may involve remedial techniques to improve physiology (e.g., strength, range of movement) or compensatory techniques (e.g., exaggerated articulation, modified texture foods). Spoken output may need to be augmented by using nonverbal communication strategies or devices. Modification of the environment (e.g., eliminating distractions) can also reduce barriers to successful communication.

4.3.8 Psychologist and Neuropsychologist

Psychology is the health discipline that provides assessment and treatment of cognitive and mental health concerns. A clinical psychologist focuses on mental health issues while the neuropsychologist focuses on cognition. A neuropsychologist is a recognized specialist who has "specialized training in the relationship between brain functioning and abilities such as memory, attention, language, and reasoning" (p. 32).[76]

4.3.8.1 Psychologist

The psychologist's role on the MS rehab team is to assess and then assist patients in coping with the psychological impact of their illness. Ideally, the psychologist on the MS rehabilitation team will have a background in clinical and rehabilitation psychology and expertise in psycho-diagnostic assessment and psychotherapy. The psychologist may have a variety of roles on the MS rehabilitation team, depending on the setting. These roles may include supporting the work of the other team members (e.g., providing education and psychological perspectives to ensure that the rehabilitation process acknowledges the client's psychological styles and needs; assisting in the management of difficult client behaviors); providing direct clinical services to clients, including the assessment of mental health, adjustment, and relationship concerns; and providing therapeutic interventions to address any issues that are identified.

Psychological assessment starts with an initial interview that includes a review of psychosocial history, a screening for mental health concerns (primarily adjustment, depression, and anxiety disorders), and the identification of issues requiring psychotherapeutic intervention. Specific content areas that are commonly the focus of intervention include the following:

- Adjusting to diagnosis and adapting to changing levels of disability.
- Managing the anxiety of living with the unpredictable nature of the disease and its potential downward course, and coping with a lack of control over the disease process.
- Dealing with grief over losses (e.g., physical abilities, changes in roles, and relationships).
- Overcoming interpersonal fears (e.g., of embarrassment, rejection, abandonment, etc.) that may arise from disclosing the diagnosis to others or as disability increases.
- Optimizing communication in relationships (e.g., being clear about needs/wishes and discussing sensitive topics with loved ones).
- Addressing concerns regarding sexuality and sexual functioning.
- Using psychological strategies to assist in managing pain.
- Providing cognitive-behavioral therapy for diagnosed anxiety disorders that interfere with disease self-management (e.g., needle phobia preventing use of disease-modifying therapy) or lead to avoidance of activities that could improve quality of life (e.g., fear of leaving home).
- Using cognitive-behavioral therapy approach to specifically target beliefs, thoughts, and behaviors that play a role in depressed mood.

Generally, psychotherapy is aimed at maximizing a client's psychological adjustment to living with MS so he or she can be actively involved in disease self-management; accept and find value in life; and be as involved as possible in meaningful and satisfying activities, roles, and relationships. Efforts are made to identify and overcome psychological barriers to achieving these goals, and support is provided as the client works toward enhancing his or her quality of life.

Depression is common in MS but cannot be accounted for simply on the basis of disease activity.[77] Its negative effect on quality of life and its inherent risk for suicide make it extremely important to deal with adjustment issues in order to prevent depression and to treat depression when it is present. The psychologist on the MS rehabilitation team often refers to, and works collaboratively with, a psychiatrist or family physician that can prescribe and monitor antidepressant medication.

4.3.8.2 Neuropsychologist

An MS rehabilitation team may also include a neuropsychologist, either as a core member of the team or a consultant. Some degree of cognitive dysfunction is common in individuals with MS (e.g., problems with new learning, working memory, cognitive processing speed, visual/spatial abilities, and executive functioning) and prevalence rates have been estimated to be as high as 70%.[78] While changes in cognitive functioning may be evident to the person with MS, subtle changes can be missed by health care workers and misinterpreted by family or an employer.[76] Cognitive dysfunction may have a detrimental, even dangerous, effect on the individual's capabilities in parenting, vocational, and social roles, and can diminish his or her quality of life.[79] The neuropsychologist uses interview and a carefully selected battery of standardized tests to identify areas of cognitive strengths and weaknesses. The results from this testing is used to inform the client, family, and other members of the rehabilitation team about the nature and effects of MS-related cognitive changes on the client's day-to-day functioning. This information can:

- Help the client make sense of his or her subjective experience;
- Assist the family to take cognitive dysfunction into account when faced with behavioral changes in the client;
- Help the client, family, and health care team in developing compensatory strategies for areas of difficulty;
- Document and provide support for workplace accommodations necessary to maintain employment or apply for a long-term disability leave.

4.3.9 The Importance of Team Work

Given the range of issues that people with MS may face and the different priorities that they may have for their own functioning, a team approach to MS rehabilitation has many benefits. Team members bring different knowledge and expertise, and can contribute to addressing a wide range of problems across the disease trajectory. This is advantageous, since often the same client issue can be addressed by more than one discipline, using different perspectives.[80] One example is the management of MS fatigue. If a client wants to better manage fatigue in order to complete daily tasks, an occupational therapist may provide education on energy management strategies, a physical therapist may develop a moderate aerobic exercise program for the client, a psychologist may use cognitive-behavioral therapy to help the client reframe his or her perspectives about fatigue, and the physician may prescribe medication that will help to decrease the severity of the fatigue symptom the client experiences.

A team approach is also advantageous because many of the impairments of MS are interdependent or overlap with one another.[20] For example, consider a client who has symptoms of bladder urgency and frequency. If the symptoms are mild, the client may be able to tolerate frequent and quick trips to the bathroom, but if the person also has limited mobility or transfer skills, incontinence may result. Assessment from both a nurse and physical therapist may highlight the overlap in this case, and planned interventions to address both issues may achieve the greatest functional gains.

A team approach to MS rehabilitation is also beneficial for addressing the psychosocial and emotional issues of living with MS.[79] Diagnosis can be a difficult time for many clients, accompanied by anger, despair, or anxiety. Disease variability and unpredictability may also lead to adjustment challenges. Counseling from the social worker or psychologist may be indicated before the client is ready to engage in work with other rehabilitation professionals.

In summary, a rehabilitation team working in partnership with the client can achieve more than the sum of its parts. An interdisciplinary team works together and with the client to obtain a comprehensive assessment, set shared treatment goals, and identify appropriate timing of interventions. On a multidisciplinary team, the process of sharing evaluation data and treatment goals can often highlight the "common thread" or a larger overall aim that is important to the person with MS. Regardless of whether the team operates from a multidisciplinary or interdisciplinary model of care, communication among team members (including the client) will enhance each member's understanding of the person with MS and the problems he or she is facing. It may also allow the members of the team to consider the timing of interventions and whether aspects of treatment need to occur simultaneously or in a particular sequence in order to achieve maximal benefit. This understanding can then make it easier for the members of the team to share with the client how the different components of the intervention complement each other in order to achieve the client's goals. It also allows each member of the team to reinforce work being done by other disciplines and pursue more coordinated problem-solving efforts.[29]

Members of successful teams have good communication with each other and the client, respect the scope of practice of each of their colleagues, and appreciate the advantages of different disciplinary viewpoints. When collaboration is effective, communication among team members is efficient, decision making is shared, the role of each health care provider is understood and respected, and the patient and family are active partners in the care process.[81]

4.4 Evidence and Directions for the Future

Throughout the remaining chapters in this book, authors share evidence and best practices specific to the topics on which they are focused. On a broader level, evidence also exists to support multidisciplinary or interdisciplinary rehabilitation interventions. In 2007, Khan and colleagues prepared a systematic review for the Cochrane Library that examined the effectiveness of organized, multidisciplinary rehabilitation in adults with MS.[82] The conclusions from this review were reexamined in 2011, and authors determined that the conclusions remained accurate.

To be included in the review,[82] studies had to address rehabilitation interventions delivered by two or more disciplines in conjunction with physician consultation. They could be delivered in inpatient, outpatient, home, or community settings. The authors did not distinguish between multidisciplinary or interdisciplinary approaches to intervention delivery. The interventions had to focus on limiting patient symptoms, enhancing functional independence, and maximizing participation using standardized tools, which the authors organized using the ICF framework.

Consistent with the structure of Cochrane reviews, the research team included randomized controlled trials ($N = 8$) and controlled clinical trials ($N = 2$). Comparison groups within the included studies varied from routine local services, information-only groups, wait-list groups, interventions with different levels of intensity, or interventions that were provided in different settings. The conclusions presented by the researchers provide strong evidence in some areas and identify issues that need further investigation. The positive findings from the review indicated the following:

- There is strong evidence that multidisciplinary inpatient rehabilitation leads to short-term gains in measures of activity and participation.
- There is strong evidence that low-intensity, long-term outpatient and home-based rehabilitation leads to quality-of-life gains.
- There is moderate evidence that both inpatient and outpatient rehabilitation programs are more effective that wait-list control groups for reducing activity restrictions.

The review also pointed to areas in which further investigations are required. For example, the researchers noted that there is limited evidence supporting high-intensity outpatient and home-based rehabilitation programs and their ability to improve participation and quality of life in the long term. The researchers also noted that there is limited evidence that low-intensity, long-term outpatient and home-based rehabilitation leads to benefits for caregivers. The analysis also did not provide any answers as to how much therapy or which therapy is best or whether multidisciplinary rehabilitation is cost-effective over the long term. While these findings may be discouraging for some, the researchers pointed to the challenges of implementing and managing rehabilitation trials, the need for improved outcome measurement, and larger, multisite trials. Based on the evidence that is available, the researchers indicated that all people with MS should undergo assessment for rehabilitation potential and decisions about the type and setting should be individualized based on client needs and priorities.

References

1. Kraft GH. An introduction to rehabilitation. MS in Focus 2006;7:4–5.
2. European Multiple Sclerosis Platform. Recommendations for rehabilitation services for persons with multiple sclerosis in Europe. Genoa, Italy: Associazione Italiana Sclerosi Multipla; 2004.
3. National Institute of Clinical Excellence. Multiple sclerosis: National clinical guideline for diagnosis and management in primary and secondary care. London: Royal College of Physicians of London; 2003.
4. Consortium of Multiple Sclerosis Centers. White paper: Comprehensive care in multiple sclerosis. Teaneck, NJ: Consortium of Multiple Sclerosis Centers; 2010.
5. National Clinical Advisory Board of the National Multiple Sclerosis Society. Rehabilitation: Recommendations for persons with multiple sclerosis. New York: National Multiple Sclerosis Society; 2004.
6. Freeman JA, Ford H, Mattison P, Thompson JJ, Clark F, Ridley J. Developing MS healthcare standards: Evidence-based recommendations for service providers. London: Multiple Sclerosis Society of Great Britain and Northern Ireland and the MS Professional Network; 2002.
7. Plow MA, Finlayson M, Rezac M. A scoping review of self-management interventions for adults with multiple sclerosis. PM&R: Journal of Injury, Function, and Rehabilitation 2011;3:251–262.

8. Bain LJ, Schapiro RT. Managing MS through rehabilitation. New York: National Multiple Sclerosis Society; 2009.

9. Khan F, Pallant JF, Brand C, Kilpatrick TJ. Effectiveness of rehabilitation intervention in persons with multiple sclerosis: A randomized controlled trial. Journal of Neurology, Neurosurgery, & Psychiatry 2008;79:1230–1235.

10. Cohen B. Identification, causation, alleviation, and prevention of complications (ICAP): An approach to symptom and disability management in multiple sclerosis. Neurology 2008;71:S14–S20.

11. Wade DT, de Jong BA. Recent advances in rehabilitation. British Medical Journal 2000;320:1385–1388.

12. Stevenson VL, Playford ED. Rehabilitation and MS. The International MS Journal 2007;14(3):85–92.

13. Courts NF, Buchanan EM, Werstlein PO. Focus groups: The lived experience of participants with multiple sclerosis. Journal of Neuroscience Nursing 2004;36:42–47.

14. Sweetland J, Riazi A, Cano SJ, Playford ED. Vocational rehabilitation services for people with multiple sclerosis: What patients want from clinicians and employers. Multiple Sclerosis 2007;13:1183–1189.

15. Finlayson M. Concerns among people aging with multiple sclerosis. American Journal of Occupational Therapy 2004;58:54–63.

16. Stuifbergen A, Becker H, Blozis S, Timmerman G, Kullberg V. A randomized clinical trial of a wellness intervention for women with multiple sclerosis. Archives of Physical Medical Rehabilitation 2003;84(4):467–476.

17. Dyer JA. Multidisciplinary, interdisciplinary, and transdisciplinary educational models and nursing education. Nursing Education Perspectives 2003;24(4):186–188.

18. Jessup RL. Interdisciplinary versus multidisciplinary care teams: Do we understand the difference? Australian Health Review 2007;31:330–331.

19. van den Broek MD. Why does neurorehabilitation fail? Journal of Health Trauma and Rehabilitation 2005;20(5):464–473.

20. Playford ED. Multidisciplinary rehabilitation In: A. J. Thompson, editor. Neurological rehabilitation of multiple sclerosis. London: Informa Healthcare; 2006.

21. Orsini JA, Donbovy ML. Multiple sclerosis and Parkinson's disease rehabilitation. In: R. B. Lazar, editor. Principles of neurologic rehabilitation. New York: McGraw-Hill; 1998.

22. Young CA, Manmathan GP, Ward JC. Perceptions of goal setting on a neurorehabilitation unit: A qualitative study of patients, carers and staff. Journal of Rehabilitation Medicine 2008;40:90–94.

23. Holliday RC, Cano S, Freeman JA, Playford ED. Should patients participate in clinical decision making? An optimised balance block design controlled study of goal setting in a rehabilitation unit. Journal of Neurology, Neurosurgery & Psychiatry 2007;78:576–580.

24. Whalley Hammell K. The rehabilitation process. In: M. Stokes, editor. Physical management in neurological rehabilitation. London: Elsevier; 2004.

25. Steiner WA, Ryser L, Huber E, Uebelhart D, Aeschlimann A, Stucki G. Use of the ICF model as a clinical problem-solving tool in physical therapy and rehabilitation medicine. Physical Therapy 2002;82:1098–1107.

26. Holliday RC, Antoun M, Playford ED. A survey of goal-setting methods used in rehabilitation. Neurorehabilitation & Neural Repair 2005;19:227–231.

27. Law M, Baptiste S, Carswell A, McColl MA, Polatajko H, Pollock N. The Canadian Occupational Performance Measure. 3rd ed. Canadian Association of Occupational Therapists; 1998.

28. Barnard RA, Cruice MN, Playford ED. Strategies used in the pursuit of achievability during goal setting in rehabilitation. Qualitative Health Research 2010;20:239–250.

29. Capilouto GJ. Rehabilitation settings. In: S. Kumar, editor. Multidisciplinary approach to rehabilitation. Woburn, MA: Butterworth Heinemann; 2000.

30. Bovend'Eerdt TJ, Botell RE, Wade DT. Writing SMART rehabilitation goals and achieving goal attainment scaling: A practical guide. Clinical Rehabilitation 2009;23(4):352–361.

31. Meriano C, Latella D. Occupational therapy interventions: Functions and occupations. Thorofare, NJ: Slack; 2008.
32. Bartholomew LK, Parcel GS, Kok G. Intervention mapping: A process for developing theory and evidence-based health education programs. Health Education & Behavior 1998;25:545–563.
33. Bartholomew LK, Mullen PD. Five roles for using theory and evidence in the design and testing of behavior change interventions. Journal of Public Health Dentistry 2011;71:S20–S33.
34. Joy JE, Johnston RB. Multiple sclerosis: Current status and strategies for the future. Washington, DC: National Academy Press; 2001.
35. Heesen C, Macalli M, Segal J, Kopke S, Muhlhauser I. Decisional role preferences, risk knowledge and information interests in patients with multiple sclerosis. Multiple Sclerosis Journal 2004;10:643–650.
36. McKee K, Bethoux F. Team focus: Physiatrist. International Journal of MS Care 2009;11:144–147.
37. Provance P. Physical therapy in multiple sclerosis rehabilitation. New York: National Multiple Sclerosis Society; 2008.
38. Sutliff MH. Physical therapist. International Journal of MS Care 2008;10:127–132.
39. World confederation for physical therapy [Internet]; c2007 [cited 2011. Available from: http://www.wcpt.org/.
40. Martin S, Kessler M. Neurologic interventions for physical therapy. 2nd ed. St. Louis, MO: Saunders Elsevier; 2007.
41. Dalgas U, Stenager E, Ingemann-Hansen T. Multiple sclerosis and physical exercise: Recommendations for the application of resistance-, endurance- and combined training. Multiple Sclerosis 2008;14:35–53.
42. Romberg A, Virtanen A, Ruutiainen J, Aunola S, Karppi SL, Vaara M, Surakka J, Pohjolainen T, Seppanen A. Effects of a 6-month exercise program on patients with multiple sclerosis: A randomized study. Neurology 2004;63:2034–2038.
43. Snook EM, Motl RW. Effect of exercise training on walking mobility in multiple sclerosis: A meta-analysis. Neurorehabilitation and Neural Repair 2009;23(2):108–116.
44. Brar SP, Smith MB, Nelson LM, Franklin GM, Cobble ND. Evaluation of treatment protocols on minimal to moderate spasticity in multiple sclerosis. Archives of Physical Medicine and Rehabilitation 1991;72(3):186–289.
45. Cattaneo D, Jonsdottir J, Zocchi M, Regola A. Effects of balance exercises on people with multiple sclerosis: A pilot study. Clinical Rehabilitation 2007;21:771–781.
46. Petajan JH, Gappmaier E, White AT, Spencer MK, Mino L, Hicks RW. Impact of aerobic training on fitness and quality of life in multiple sclerosis. Annals of Neurology 1996;39:432–441.
47. Schapiro RT. Managing the symptoms of multiple sclerosis. 5th ed. New York: Demos Health; 2007.
48. Brown TR, Kraft GH. Exercise and rehabilitation for individuals with multiple sclerosis. Physical Medicine and Rehabilitation Clinics of North America 2005;16(2):513–555.
49. Canadian Association of Occupational Therapists. Enabling occupation: An occupational therapy perspective. Ottawa: CAOT Publications ACE; 1997.
50. American Occupational Therapy Association. Occupational therapy practice framework: Domain and process. American Journal of Occupational Therapy 2008;62:625–683.
51. Forwell SJ, Zackowski KM. Team focus: Occupational therapist. International Journal of MS Care 2008;10:94–98.
52. Finlayson M. Clinical bulletin: Occupational therapy in multiple sclerosis rehabilitation. New York: National Multiple Sclerosis Society; 2010.
53. Finlayson ML, Preissner K, Cho CC, Plow MA. Randomized trial of a teleconference-delivered fatigue management program for people with multiple sclerosis. Multiple Sclerosis 2011;17:1130–1140.
54. Mathiowetz VG, Matuska KM, Finlayson ML, Lui P, Chen HY. One year follow up to a randomized controlled trial of an energy conservation course for persons with multiple sclerosis. International Journal of Rehabilitation Research 2007;30:305–313.

55. Shevil E, Finlayson M. Process evaluation of a self-management cognitive program for persons with multiple sclerosis. Patient Education and Counseling 2009;76:77–83.
56. Goverover Y, Chiaravalloti N, DeLuca J. Self-generation to improve learning and memory of functional activities in persons with multiple sclerosis: Meal preparation and managing finances. Archives of Physical Medicine & Rehabilitation 2008;89:1514–1521.
57. Shevil E, Finlayson M. Pilot study of a cognitive intervention program for persons with multiple sclerosis. Health Education Research 2010;25:41–53.
58. Gentry T. PDAs as cognitive aids for people with multiple sclerosis. American Journal of Occupational Therapy 2008;62:18–27.
59. Maitra K, Hall C, Kalish T, Anderson M, Dugan E, Rehak J. Five year retrospective study of inpatient occupational therapy outcomes for patients with multiple sclerosis. American Journal of Occupational Therapy 2010;64:689–694.
60. Hawes F, Billups C, Forwell S. Interventions for upper-limb intention tremor in multiple sclerosis. International Journal of MS Care 2010;12:122–132.
61. Halper J, Holland NJ. An overview of multiple sclerosis: Implications for nursing practice. In: J. Halper, N. Holland, editors. Comprehensive nursing care in multiple sclerosis. New York, NY: Demos Medical Publishing; 2002.
62. Long AF, Kneafsey R, Berry J. The role of the nurse within the multi-professional rehabilitation team. Journal of Advanced Nursing 2002;27(1):70–78.
63. Hoeman SP. Rehabilitation nursing: Prevention, intervention and outcome. 4th ed. St. Louis, MO: Mosby Elsevier; 2008.
64. Lutz BJ, Davis SM. Theory and practice models for rehabilitation nursing. In: S. P. Hoeman, editor. Rehabilitation nursing: Prevention, intervention, & outcomes. 4th ed. St. Louis, MO: Mosby Elsevier; 2008.
65. Khan F, Pallant JI, Brand C, Kilpatrick TJ. A randomised controlled trial: Outcomes of bladder rehabilitation in persons with multiple sclerosis. Journal of Neurology, Neurosurgery, & Psychiatry 2010;81(9):1033–1038.
66. Buljevac D, Flach HZ, Hop WCJ, Hijdra D, Jaman JD, Savelkoul HFJ, van der Meche FGA, van Doorn PA, Hintzen RQ. Prospective study on the relationship between infections and multiple sclerosis exacerbations. Brain 2002;125:952–960.
67. Fowler CJ, Panicker JN, Drake M, Harris C, Harrison SCW, Kirby M, Lucas M, Macleoud N, Magnall J, North A, et al. A UK consensus on the management of the bladder in multiple sclerosis. Journal of Neurology, Neurosurgery & Psychiatry 2009;80:470–477.
68. Duchene PM. Sexuality education and counseling. In: S. Hoeman, editor. Rehabilitation nursing: Prevention, intervention, & outcomes. 4th ed. St. Louis, MO: Mosby Elsevier; 2008.
69. Preston M, Tebben C, Johnson MM. Skin integrity. In: S. Hoeman, editor. Rehabilitation nursing: Prevention, intervention, & outcomes. 4th ed. St. Louis, MO: Mosby Elsevier; 2008.
70. Getting to know the MS health care team [Internet]; c2011 [cited 2011. Available from: http://www.nationalmssociety.org/living-with-multiple-sclerosis/getting-the-care-you-need/team-of-ms-professionals/getting-to-know-the-ms-health-care-team/index.aspx.
71. Hartelius L, Svensson P. Speech and swallowing symptoms associated with Parkinson's disease and multiple sclerosis: A survey. Folia Phoniatrica et Logopaedica 1994;46:9–17.
72. Beukelman DR, Kraft GH, Freal J. Expressive communication disorders in persons with multiple sclerosis: A survey Archives of Physical Medicine & Rehabilitation 1985;66:675–677.
73. Calcagno P, Ruoppolo G, Grasso MG, De Vincentis M, Paolucci S. Dysphagia in multiple sclerosis—Prevalence and prognostic factors. Acta Neurologica Scandinavica 2002;105(1):40–43.
74. De Pauw. Dysphagia in multiple sclerosis. Clinical Neurology and Neurosurgery 2002;104(4):345–351.
75. Haynes WO, Pindzola RH. Diagnosis and evaluation in speech pathology. 7th ed. Boston: Pearson/Allyn & Bacon; 2008.
76. Fuchs KL. TEAM focus: Neuropsychologist. International Journal of MS Care 2009;11:32–37.
77. Arnett PA, Barwick FH, Beeney JE. Depression in multiple sclerosis: Review and theoretical proposal. Journal of the International Neuropsychological Society 2008;14(5):691–724.

78. Rao SM, Leo GJ, Bernardin L, Unverzagt F. Cognitive dysfunction in multiple sclerosis. I. Frequency, patterns, and prediction. Neurology 1991;41:685–691.

79. Burks J, Bigley G, Hill H. Rehabilitation challenges in multiple sclerosis (Review: Management updates). Annals of Indian Academy of Neurology 2009;12(4):296–306.

80. Latella D. Teamwork in rehabilitation. In: S. Kumar, editor. Multidisciplinary approach to rehabilitation. Woburn, MA: Butterworth- Heinemann; 2000.

81. Suter E, Taylor L, Arthur N, Clinton M. Interprofessional education and practice. Final report of final report project. Calgary Alberta: Calgary Health Region; 2008.

82. Khan F, Turner-Stokes L, Nq L, Kilpatrick T. Multidisciplinary rehabilitation for adults with multiple sclerosis. Cochrane Database of Systematic Reviews 2007;(2). Art. No.: CD006036. DOI: 10.1002/14651858.CD006036.pub2.

Section II

Multiple Sclerosis Rehabilitation for Impairments in Body Structure and Function

5

Fatigue

Marcia Finlayson, Sverker Johansson, and Daphne Kos

CONTENTS

Approximately 70–90% of people with MS experience fatigue, and up to 40% of them describe it as their most disabling symptom.[1] The origin of fatigue is as complex as its definition and clinical appearance,[2] making it an intriguing MS symptom. Literature indicates that MS fatigue is not typically associated with age, gender, or the length of time with the disease, and it is only weakly related to level of disability and disease course.[3] Fatigue is a critical symptom to address during rehabilitation because it may worsen other MS symptoms (e.g., cognition) or it may increase as a consequence of others (e.g., spasticity).[3,4]

Since MS usually presents in early adulthood, the disease has a major impact on family life, work opportunities, and social relationships.[5] Rehabilitation providers play an important role in enabling people with MS to develop the routines, habits, and life skills necessary to manage fatigue and reduce its impact on their daily lives. In order to effectively fulfill this role, rehabilitation providers must understand the nature and dimensions of MS-related fatigue and how to assess its presence and impact over time. They must also be aware of the treatment strategies available, when to use them, and with whom to do so. After reading this chapter, you will be able to:

1. Describe different ways that MS-related fatigue is described in the literature,
2. Describe the impact of fatigue on the everyday lives of people with MS,
3. Discuss commonly used instruments that determine the presence of fatigue or evaluate the outcomes of fatigue-management interventions,
4. Describe evidence-based treatment strategies for managing fatigue in MS, and
5. Identify gaps in knowledge and directions for future rehabilitation research focused on managing MS fatigue.

5.1 Describing Fatigue in MS

Fatigue has been described and defined in many different ways. Definitions range from broad, subjective ones (e.g., overwhelming sense of tiredness, subjective lack of physical or mental energy)[6,7] to others that are narrow, objective, and physiologically based (e.g., exercise-induced reduction in maximal voluntary muscle force).[8] Fatigue has also been described as a common, protective response to reduce an individual's likelihood of engaging in an activity beyond his or her functional reserve,[9] either physically or mentally.[10] Three major factors have contributed to the problems of describing and defining fatigue in MS: limitations in language, types of fatigue, and dimensions of fatigue.

5.1.1 Limitations in Language

"Fatigue" is a very common word in daily language, and it captures a range of feelings and expressions, such as tiredness, sleepiness, and weakness. All individuals experience fatigue once in a while, regardless of their level of health or disability. Hence, limitations of vocabulary can be partially blamed on the challenges of defining and describing fatigue in MS. Unfortunately, this limitation often hampers the ability of rehabilitation professionals to understand exactly what their clients mean when they report fatigue,[7] which can make assessment and treatment planning challenging. Therefore, it is valuable for rehabilitation providers to learn about different types (i.e., descriptive labels) that have been used to organize thinking about MS fatigue and its presentation.

5.1.2 Types of Fatigue in MS

MS fatigue has been described or labeled in different ways that can co-occur. For example, MS fatigue can occur as a consequence of the disease process itself (primary fatigue)

TABLE 5.1

Types of Fatigue and Implications for Rehabilitation Providers

Ways of Describing Fatigue	Explanation	Potential Treatment Focus
Primary	Fatigue directly related to disease factors[3,14]	Focus on enabling the person with MS to develop skills, strategies, and lifestyle patterns that will facilitate long-term fatigue management
Secondary	Nondisease-specific factors leading to fatigue[3,14]	Collaborate with other members of the treatment team to identify and treat/manage presumed secondary cause(s) of fatigue: depression, sleep disturbances, nocturnia, physical deconditioning, stress
Acute	Short-term fatigue, induced by infection or exacerbation of disease[6]	Encourage the person with MS to temporarily take more rest and follow physician's instruction and treatment for infection or exacerbation
Chronic	Fatigue present for 50% or more days for more than 6 weeks[12]	Focus on enabling the person with MS to develop skills, strategies, and lifestyle patterns that will facilitate long-term fatigue management
Physical or motor	Inability to initiate or sustain physical activity[11]	Focus on enabling the person with MS to develop capacity, skills, and strategies to manage physically demanding activities
Cognitive	Inability to initiate or sustain mental/cognitive activity[58]	Focus on enabling the person with MS to develop capacity, skills, and strategies to manage cognitively or mentally demanding activities
Perceived/ Experienced	Feelings of lassitude (fatigue at rest) or lack of energy, interfering with initiation and completion of activities[11]	Focus on enabling the person with MS to pace and prioritize activities, enhance coping skills, and set achievable activity goals
Observed/ Physiological	Objective decline in physical or mental performance or ability to sustain activity[12]	Focus on improving physical or cognitive capacity and endurance

or as a consequence of other factors not directly related to the disease (secondary fatigue).[3] It can be acute or chronic;[6] it can be both physical and cognitive in nature.[11] It is both a subjective, experienced phenomenon that can occur in the absence of activity (e.g., fatigue at rest or asthenia) as well as a tangible, physiological experience.[11,12] Table 5.1 defines each of these MS fatigue descriptors (labels) and highlights the potential treatment focus during rehabilitation.

5.1.2.1 Primary Fatigue

Primary fatigue is hypothesized to be directly related to the disease process.[3] From the perspective of rehabilitation, it is critical to consider the extent to which a client's fatigue is primary or secondary.[13] Primary fatigue will require compensatory, educational, and consultative approaches to rehabilitation treatment, and secondary fatigue may also involve remediation (e.g., building endurance). Although research is continuing, current knowledge suggests there are three potential mechanisms for primary MS fatigue—changes in the central nervous system (including axonal loss and cortical reorganization), changes in the immune system, and endocrine influences.[14]

Individuals with MS can experience axonal loss in the brain, which may be associated with fatigue.[14] In addition, people with MS who experience fatigue have significantly higher lesion loads in the parietal lobe, internal capsule, and periventricular trigone compared to people with MS who do not experience fatigue.[2] Structural changes to the brain can alter the patterns of cerebral activation and demand higher energy in the brain,[15,16] both of which can contribute to compensatory reorganization and increased brain recruitment, as shown in functional magnetic resonance imaging (fMRI).[14] Greater lesion load has also been linked to perceived fatigue, even after controlling for physical disability and mood disorders.[2]

Although increased brain recruitment and higher energy demands are thought to contribute to greater perceived fatigue, some authors have identified only weak relationships between perceived fatigue and functional cortical reorganization.[17–21] Other studies have suggested that changes in perceived fatigue may be a warning signal preceding changes in disease activity. For example, an increase in fatigue (as measured by the Sickness Impact Profile Sleep and Rest Scale) over a 2-year period was related to progression of brain atrophy in the subsequent 6 years, suggesting that fatigue predicts the brain atrophy as opposed to being a consequence of the demyelination process.[22]

Perceived fatigue has also been linked to higher levels of some immune markers, for example, proinflammatory cytokines such as interferon-γ and TNF-α.[23,24] Fatigue at rest (asthenia) was linked to immunoactivation in one study[25] and, in another study, increased daytime sleepiness was associated with elevated proinflammatory cytokine levels.[24]

Another potential factor in primary MS fatigue is disturbances in the neuroendocrine system.[26–30] In other conditions (e.g., chronic fatigue syndrome, lupus, rheumatoid arthritis), low cortisol and low DHEA levels have been implicated in fatigue. The results in MS have been inconclusive, although work in this area continues.

5.1.2.2 Secondary Fatigue

While many disease-specific factors may influence the experience of fatigue among people with MS, secondary factors may contribute as well. Examples include sleep problems; medications; deconditioning; and depression, stress, and anxiety.[3]

5.1.2.2.1 Sleep Problems

The prevalence of sleep problems among people with MS ranges from 25% to 61%.[31–34] These rates are two to three times higher than the rates found among healthy populations.[32,33] Furthermore, sleep problems in people with MS might be related to specific MS symptoms (e.g., urinary problems, spasticity, pain); medication side effects (MS related or otherwise); or environmental factors that could affect anyone, such as light, noise, or being awakened by children during the night.[35,36]

5.1.2.2.2 Medication Side Effects

Many of the medications commonly prescribed to people with MS may provoke fatigue.[37] These medications include both the MS disease-modifying agents (e.g., interferons) as well as symptom-specific agents (e.g., some antispasmodics, analgesics for pain, some antidepressants). For this reason, rehabilitation providers need to inquire about a client's medication regime and whether any medications may have fatigue as a side effect. Learning about these side effects and their expected timing relative to medication intake may make it easier to schedule rehabilitation sessions during times that clients will be the least affected by fatiguing side effects.

5.1.2.2.3 Deconditioning

Physical deconditioning is considered a common source of secondary fatigue.[13] A person who is deconditioned will have weakness and low physical endurance and is less likely to engage in physical activities.[38,39] He or she will use greater amounts of energy to perform activities compared to individuals who are not deconditioned, which is believed to contribute to the sense of fatigue.[21] Furthermore, being physically inactive may worsen fatigue, leading to a vicious cycle.[13,40] People with MS are known to have low levels of physical activity.

Motl and colleagues[41] have presented a meta-analysis that pooled 53 effects from 13 studies of 2360 people with MS. They calculated a weighted mean effect size of −0.60 (95% CI = −0.44 to −0.77) for people with MS. They also pointed out that because the general population is not as active as guidelines for physical activity recommend, their findings emphasize the sedentary nature of people with MS. People with MS are at high risk of deconditioning, which may contribute to their perceived fatigue. Nevertheless, research on the relationship between physical activity levels and perceived fatigue is inconsistent. Motl and McAuley[42] showed that physical activity has only an indirect influence on perceived fatigue among people with MS while other studies have shown that physical exercise can reduce perceived fatigue.[43,44] A recent review of the effects of exercise therapy suggested that, overall, exercise therapy seems to have the potential to positively influence MS fatigue, despite heterogeneous findings across the studies reviewed.[45] Additional details about physical activity among people with MS are provided in Chapter 17.

5.1.2.2.4 Depression, Stress, and Anxiety

The presence of depression, stress, or anxiety may also play a role in the persistence of fatigue.[46] Depression and anxiety are frequently reported in MS and are often accompanied by fatigue.[47] In a comprehensive review of the psychology of fatigue, Bol et al.[47] noted that most cross-sectional studies investigating the relationship between fatigue and depression or anxiety have reported significant relationships between these variables.[47] Other researchers reporting on longitudinal studies have revealed that depression predicts fatigue and anxiety and vice versa, and that anxious, fatigued people with MS are more likely to become depressed.[48–50] In addition, psychological and personality factors (e.g., avoidance coping, level of negative affectivity) may interfere with this reciprocal relationship.[47] Additional details about depression and coping among people with MS are provided in Chapters 10 and 20, respectively.

5.1.2.2.5 Summary

While the distinction between primary and secondary fatigue is based on the pathogenesis, the actual expressions of fatigue cross the boundaries of this dichotomy. For example, the perceived lack of energy may result from an altered brain function, deconditioning, or both. Similarly, secondary causes of fatigue can be present in other pathologies and in healthy populations.[7] Rehabilitation providers must recognize the value of differentiating primary and secondary fatigue to inform realistic goal setting and treatment planning, but must not lose sight of the complex interplay of these different forms of fatigue and the experience of the people with whom they are working.

5.1.2.3 Acute versus Chronic Fatigue

Another important distinction to make is whether fatigue reported by a person with MS is acute or chronic. Acute fatigue is short-term in duration (<6 weeks)[6] and can be the result of a wide range of factors (e.g., MS exacerbation, infection, poor-quality sleep). Fatigue can also persist over time. When complaints are present for more than 6 weeks, fatigue is considered to be chronic. Chronic fatigue is a major challenge to the everyday functioning of people with MS. The impact of chronic fatigue is described later in this chapter. Both acute and chronic fatigue can be managed with appropriate treatment,[6] although the strategies can be quite different. The presence of chronic fatigue requires an individual to develop a wide range of habits, routines, and skills to support long-term management. Fatigue intervention strategies are described later in this chapter.

5.1.2.4 Other Types of Fatigue

While some definitions of fatigue focus on the perceived reduction of energy when performing physical or mental activity,[6] other definitions describe fatigue more objectively as a decline in physical or mental performance.[51] As in the general population, physical fatigue may be the result of exercise or high levels of physical activity among people with MS. However, in contrast to the situation of healthy persons, rest may not fully remediate the fatigue that people with MS experience.[52] Furthermore, MS-related fatigue may be easily triggered, limit physical and mental activity, worsen the subjective intensity of other symptoms, and restrict participation in everyday activities.[53] In MS, fatigue is commonly exacerbated by heat (Uhthoff's phenomenon), which is most likely the result of an impaired nerve conduction.[4] The term "cognitive (or mental) fatigue" often represents both the objective decline in cognitive performance[51,54,55] and the impact of perceived fatigue on cognitive functioning.[6,56]

As this section has illustrated, the fatigue reported by a person with MS can be described in multiple ways, and it remains unclear *if* and *how* these various types relate to each other. This lack of clarity is best illustrated by the weak correlations that have been found between the perception of fatigue and objective decline in performance, either physical, cognitive, or both.[51,57–62] Finally, it is important to acknowledge that with individual clients, the experience, causes, and descriptions of fatigue are likely to change over time.[63,64]

5.1.3 Dimensions of Fatigue

Another major consideration when describing and defining fatigue during the rehabilitation process is that, like other symptoms, it can vary on four dimensions: intensity, timing, impact, and quality.[65] These dimensions are described in the theory of unpleasant symptoms[65] and are summarized in Table 5.2 along with the potential implications for rehabilitation providers.

5.1.4 Summary

As Tables 5.1 and 5.2 suggest, fatigue is a complex symptom in MS, and how it is described and conceptualized could have a major influence on how a rehabilitation provider selects assessment tools, sets treatment goals, chooses to intervene, and determines if intervention was successful. As a result of its complexity, it is probably best to think about fatigue from a multifactorial perspective rather than to ascribe it to one mechanism. Ultimately, fatigue "may be best characterized as the absence of energy and the mind-body response to that absence (http://www.nia.nih.gov/about/events/2011/unexplained-fatigue-elderly)."[66] The complexity of fatigue means that assessment and intervention will probably be best achieved through the efforts of a multidisciplinary team, with each member contributing in his or her particular areas of expertise.

TABLE 5.2

Dimensions of Fatigue

Dimensions of Fatigue	Explanation	Implications for Rehabilitation Providers
Intensity	Severity, strength, or amount of fatigue	Reduced fatigue severity can be an important outcome of some fatigue-management interventions
Timing	Frequency of acute fatigue Duration of chronic fatigue Timing of fatigue relative to specific activities (e.g., physical activity) Variations in fatigue over the course of the day (e.g., morning versus afternoon)	Patterns of fatigue may influence scheduling of therapy appointments, suggestions for activity analysis and modification, pacing activities
Impact	Extent to which fatigue causes distress or is perceived as bothersome or restrictive for the person experiencing it	Reduced fatigue impact is a major outcome of fatigue management interventions delivered by rehabilitation providers
Quality	How the person experiencing fatigue describes the feeling of fatigue The degree to which fatigue changes with intervention	Quality of fatigue can be monitored during a treatment session as a way of gauging response to intervention

5.2 Impact of Fatigue on Everyday Life

As previously noted, fatigue is a normal human experience. Yet, reports from people with MS make it abundantly clear that MS fatigue is different from anything else they have previously experienced. Statements from people without MS such as "I'm tired, too" dismiss the unique experience of MS fatigue and its impact on the everyday lives of the people who experience it. Respecting and responding to the unique experience of MS fatigue is critical for rehabilitation providers if they are going to be effective in assessing MS fatigue and selecting, implementing, and evaluating comprehensive interventions to reduce the impact of this symptom on the lives of their clients. This section will draw on qualitative research with people who have MS as well as descriptions of MS fatigue and its impact, which have been gathered through a fatigue-management trial for people with MS conducted by the first author and her colleagues.*

5.2.1 Experience of MS Fatigue

It is constant tiredness. I wake up tired; I go to bed tired. I'm tired all of the time.[*]

Qualitative studies focusing explicitly on understanding MS fatigue illustrate that this symptom can come on suddenly, is very unpredictable, and often has little or no relationship to actual energy output. Descriptions provided by the participants in these studies indicate that MS fatigue is all encompassing, ever present, and multidimensional.[67–70] MS fatigue influences every decision that a person with MS makes[69] and contributes to worry

* Finlayson, M. Principal Investigator—Effectiveness of a Teleconference Delivered Fatigue Management Program for people with MS. Funded by the National Institute of Disability and Rehabilitation Research, 2007–2010. Grant# H133G070006.

TABLE 5.3

Descriptors of MS Fatigue Provided by People with MS Participating in a
Fatigue-Management Trial[a]

Categories	Descriptions of MS Fatigue by Participants
Physical	"It's a constant heaviness; like walking in cement."
	"My body just shuts down."
	"It's like having a 300-pound weight on each arm."
Cognitive	"I can't wrap my head around complex topics."
	"I feel brain dead."
	"It's cloudy thinking."
Affective	"It's debilitating, frustrating, and scary."
	"It's extremely devastating."
Temporal	"It's a time waster."
	"It never goes away."

[a] Finlayson, M. Principal Investigator—Effectiveness of a Teleconference Delivered
Fatigue Management Program for people with MS. Funded by the National Institute
of Disability and Rehabilitation Research, 2007–2010. Grant# H133G070006.

and concern about when it will happen again.[67] Participants in the study by Flensner et al.[67] described the time-consuming nature of MS fatigue—its presence required planning and scheduling to accommodate it; the need to slow down; and finding ways to do activities in different, more energy-efficient ways. These descriptions go beyond the classifications and dimensions previously presented in this chapter and frame them from the perspective of personal experience. Data from the first author's research* support these additional descriptors and highlight four ways that MS fatigue is often described by people with MS: as physical, cognitive, affective, or temporal. Examples are provided in Table 5.3.

5.2.2 Impact of Fatigue on Other Body Functions

According to people with MS, fatigue contributes to deterioration in other body functions and increases the impact of other MS symptoms.[67–69,71] For example, during the course of qualitative interviews, people with MS have linked MS fatigue to increased pain,[67] feelings of weakness,[68] cognitive impairment (particularly memory, concentration, and problem solving)[68,69,71] and worsening of communication symptoms such as dysarthria and slurring.[71] Fatigue has also been linked to deterioration of gait and balance.[69] The qualitative findings from the first author's research are consistent with both basic science and epidemiological work related to the fatigue experience in MS (see Section 5.1).

5.2.3 Impact of Fatigue on Activity and Participation

I feel like my world is shrinking. I have only so much energy during the day. The enjoyable things go to the bottom of the list. After the chores, I don't have any more energy. It's discouraging.

---

* Finlayson, M. Principal Investigator—Effectiveness of a Teleconference Delivered Fatigue Management Program for people with MS. Funded by the National Institute of Disability and Rehabilitation Research, 2007–2010. Grant# H133G070006.

MS fatigue can negatively influence any aspect of activity and participation, particularly employment,[67,69,72] social and leisure participation,[67,69] and family and household responsibilities.[67,68] Participants in Finlayson's trial* described giving up or significantly reducing their participation in a wide range of valued activities because of fatigue, including driving, shopping for fun, gardening, traveling, sporting activities (e.g., golfing, bowling, camping) and attending various cultural events (e.g., going to movies, theater, or symphony). They also described changing from being an active participant in some activities to being a passive observer (e.g., involvement in a child's after-school activities).

5.2.4 Impact of Fatigue on Coping and Sense of Self

MS fatigue is relentless, exacerbates other MS symptoms, and typically leads to significant restrictions in valued activities. Changes in valued activities can contribute to unwelcome shifts in people's sense of self, perceptions of worthiness, and their beliefs in their ability to contribute in a meaningful way within their homes and communities. Participants in the study by Flensner et al.[67] described how MS fatigue made them feel useless, isolated, full of sorrow and shame, and anger. They also described feeling helpless and dependent on others to do basic tasks. Similarly, Olsson et al.[68] described how their participants expressed disappointment in themselves and felt insecure, anxious, and frustrated because of their inability to do the things that they wanted or were expected to do because of their MS fatigue. Together, all of these feelings contribute to stress, which then further worsens the fatigue.[69] As one person with MS explained, "MS fatigue is very stressful, both physically and mentally. It's hard to cope. It's hard to keep going."* Rehabilitation providers need to consider coping skills and resources as they develop and monitor fatigue-management interventions because emotion-focused coping and low self-efficacy may make treatment progress difficult. Chapter 20 focuses on coping and the development of coping skills among people with MS.

5.2.5 Impact of Fatigue on Caregivers and Family Members

In order to manage the everyday consequences of MS, up to 80% of the people with the disease seek some type of physical, emotional, and instrumental support from informal caregivers.[73–75] While there is a growing body of literature on MS caregiving, its nature, and its consequences (see Chapter 22), there is relatively little documented about the specific impact that MS fatigue has on the caregivers and family members of people with MS. In two different studies, researchers did not find a relationship between fatigue measures of people with MS and negative health states or distress on the part of their caregivers.[76,77] Yet, in one of these studies, researchers did find that higher ratings of fatigue severity had a moderate ($r = 0.33$, $p < 0.05$) and significant relationship with caregiver's life upset as measured by one of the subscales of the Relative Stress Scale.[76] Consequently, while rehabilitation providers should be sensitive to the additional physical and emotional demands that MS fatigue may place on caregivers and family members, the specific effects are not well understood.

---

* Finlayson, M. Principal Investigator—Effectiveness of a Teleconference Delivered Fatigue Management Program for people with MS. Funded by the National Institute of Disability and Rehabilitation Research, 2007–2010. Grant# H133G070006.

5.3 Fatigue Assessment Strategies

Despite the multiple ways of describing fatigue, its measurement among people with MS has largely been based upon self-report.[13] Several methods to measure objective, physiological fatigue have been presented in the literature, for example, walking more slowly, poor-quality gait,[78] and cognitive tests.[57] Nevertheless, these measures have had only weak associations with self-reported fatigue. Perhaps, as a consequence, there is not yet a standardized and broadly accepted approach that can be easily and consistently applied in everyday clinical practice to measure objective, physiological fatigue due to MS.

Self-report questionnaires are used extensively to understand and monitor the full range of fatigue experiences. In recent years, several papers have been published that identify questionnaires suitable for use with people who have chronic illness[79] or that critically examine specific self-report fatigue assessment scales.[80–82] Although 252 different ways of measuring perceived fatigue have been identified in the literature, there is as yet no gold standard.[79] Table 5.4 presents seven of the most common fatigue assessment instruments used in MS practice and research.

Because there is a wide range of fatigue assessment instruments, clinicians and researchers must select carefully before using one in their settings. First, consideration must be given to the definition of fatigue being used and the extent to which an instrument matches conceptually. Consideration must also be given to the psychometric properties (reliability, validity), applicability to the specific patient population, and clinical utility of the instrument.[80] Clinical utility incorporates issues such as cost, availability, and time to learn, administer, and score the instrument. If an instrument is going to be used to measure outcomes of therapy, then responsiveness is also a critical factor.[83] Table 5.5 summarizes the major psychometric features of the fatigue assessment instruments that are described in Table 5.4. A comprehensive history will also be needed to identify potential sources leading to secondary fatigue so that they can be addressed through appropriate referrals and intervention.[84] This history can also provide information to determine if fatigue is acute or chronic in nature, and whether the fatigue is primarily experienced as a physical phenomenon, a cognitive one, or both.

Through the use of these various instruments, people with MS can be screened to determine the nature and extent of fatigue and the need for fatigue management intervention. The information generated from fatigue assessment instruments can also be used to set targeted goals for intervention and to track progress towards intervention outcomes. Table 5.6 gives examples of potential fatigue assessment findings and how treatment can be focused to address them. Details on treatment options are provided in the next section.

When tracking progress in therapy, knowing an instrument's minimal detectable change (MDC) and minimal clinically important difference (MCID) is useful. The MDC is an estimate of how much a client's score needs to change to ensure that it is not simply measurement error.[85] Unlike an MCID, the MDC is based on statistical analysis only and does not take into account the amount of change that is considered important by client or clinician.[86] A recent study from a Dutch research team[87] calculated the MDC for the Modified Fatigue Impact Scale (MFIS) and the Fatigue Severity Scale (FSS) based on data from 43 ambulatory people with MS (Expanded Disability Status Scale [EDSS] ranging from 1.0 to 6.5, median = 3.5). Their findings indicated that the MFIS had to change 19.2% and the FSS had to change 20.7% between two measurement points in order for a clinician to be 95% confident that the change was not simply due to measurement error. In other words, when

TABLE 5.4

Purpose and Key Features of Seven Self-Report Fatigue Measures Useful in MS Rehabilitation

Tool	Purpose	Populations Studied	Aspects of Fatigue Addressed
Fatigue Impact Scale (FIS)	To assess the problems in patients' QOL that they attribute to their symptoms of fatigue in order to improve the understanding of the effects of fatigue on the lives of patients with chronic disease	MS (chronic-progressive, relapsing–remitting, exacerbating–remitting, and benign)[114] Chronic fatigue, mild hypertension, and clinically definite or probable MS[126]	Physical impact of fatigue Psychosocial impact of fatigue Cognitive impact of fatigue
Modified Fatigue Impact Scale (MFIS)	To assess the effects of fatigue in terms of physical, cognitive, and psychosocial functioning	MS (relapsing–remitting, secondary or primary chronic progressive)[165,166] MS with an EDSS of 3.0–8.5[167] and of 0–8.5[168]	Physical impact of fatigue, cognitive impact of fatigue, and psychosocial impact of fatigue
Fatigue Severity Scale (FSS)	To assess disabling fatigue across people with MS and systemic lupus erythematosus in order to facilitate research and patient treatment	MS (chronic-progressive) and systemic lupus erythmatosis[169]	Severity[169] Modality, severity, frequency, impact on daily life[80]
Multidimensional Fatigue Inventory (MFI)	Offers a shorter assessment of fatigue that is multidimensional and does not contain any somatic items that induce the risk of contamination of fatigue with somatic illness	Cancer[170–173] Chronic fatigue[173] Chronic fatigue-like[174] Students[173] Army recruits[173] The "chronically unwell"[174] The "tired" and "moderately tired"[175]	General fatigue, physical fatigue, reduced activity, reduced motivation, and mental fatigue
Checklist of Individual Strength (CIS)	To measure several aspects of fatigue and provide data on behavior	MS and chronic subjective fatigue[176] Chronic fatigue syndrome[176] Fatigued employees, pregnant women, lumbar disc surgery patients, chronic fatigue syndrome, and veterans[177] Employees[178,179]	Subjective experience of fatigue, concentration, motivation, and physical activity
Visual Analogue Scale (VAS)	To assess the therapeutic effect of clinical interventions for individuals with fatigue	Sleep disorder[180] Cancer[181]	Fatigue severity[182,183]
Fatigue Descriptive Scale (FDS)	To measure severity and to define characteristics of MS fatigue. The FDS was proposed to evaluate specifically the symptom of fatigue, that is, the subjective complaint of tiredness	MS (relapsing–remitting)[184] MS (relapsing–remitting and secondary progressive)[185]	Fatigue initiative, modality, frequency, severity, and Uhthoff's phenomenon (influence of heat on fatigue)

TABLE 5.5

Summary of the Psychometric Properties of Fatigue Measures Useful in MS Rehabilitation

Tool	Reliability		Validity			Sensitivity to Change
	Test–Retest	Internal Consistency	Content	Discriminant	Concurrent	
FIS	Excellent[186-188]	Excellent[126,186,187]	Excellent[126]	Excellent[126,186,189]	Adequate[126,188,189]	Adequate[114,188]
MFIS	Adequate[87,167,190]	Adequate[167,190,191]	Adequate[87]	Excellent[166,190,192]	Adequate[87,168,190]	Adequate[87,167]
FSS	Excellent[169,193,194]	Excellent[169,193,194]	Adequate[169]	Excellent[80,169,193]	Poor[11,165,169,195] to Adequate[87,194]	Poor[80,87,194] to Adequate[169,193]
MFI	Adequate[170,196-198]	Adequate[171-175,197,199,200]	Adequate[172,173]	Excellent[173-175]	Adequate[170,173,196,200]	Varies by subscale[173,197,198]
CIS	Adequate[87,201,202]	Excellent[179,202,203]	Adequate[178,202,203]	Excellent[202-204]	Adequate[202-204]	Poor[87]
VAS	Unknown	Adequate[180,181]	Excellent[180]	Poor[205] to Adequate[206]	Adequate[180,181,205,206]	Poor[180]
FDS	Adequate[184]	Unknown	Adequate[184]	Unknown	Adequate[185]	Unknown

Note: Ratings based on recommendations from Law and MacDermid.[207]

TABLE 5.6

Examples of the Potential Relationship among Fatigue Assessment Findings and Potential Rehabilitation Focus

Examples of Potential Assessment Findings	Potential Treatment Focus
History indicates new, recent work stress; high anxiety; and deteriorating sleep quality. FIS indicates fatigue impact is both physical and psychosocial.	Relaxation and stress reduction Education on sleep hygiene Energy-management education focused on planning and pacing
History indicates chronic fatigue, low levels of physical activity, and poor nutritional habits. Patient is overweight. Patient reports high levels of perceived exertion after walking one city block.	Physical-activity tolerance Nutrition education
History indicates chronic fatigue without obvious secondary causes. Patient reports high levels of anxiety about fatigue and low confidence about being able to manage it. MFIS indicates fatigue impact is primarily cognitive in nature.	Application of efficacy-enhancing methods during therapy to build confidence needed to manage fatigue Identification, analysis, and modification of cognitively demanding tasks Reduction of environmental distracters during cognitive tasks
History indicates chronic fatigue that is exacerbated by warm weather. MFIS indicates fatigue impact is primarily physical in nature, although patient reports being physically active.	Education on heat sensitivity and its management Education to build self-monitoring skills specific to physical activity across environments

tracking progress in therapy, a clinician would have to have seen changes at these levels or higher to be sure improvements were real. Further research will be required to determine if the MDC also reflects meaningful change from the perspective of clients and clinicians (i.e., MCID).

In the future, self-reported assessment of fatigue may become more consistent because of the Patient-Reported Outcomes Measurement System (PROMIS) initiative, established in 2004 by the US National Institutes of Health. The focus of PROMIS is on developing "efficient, consistent, well-validated" ways to measure patient-reported symptoms that are known to be difficult to measure (e.g., fatigue). The preliminary item bank on fatigue contains 95 items with 55 addressing the fatigue experience and 40 addressing fatigue impact. Testing and calibration continue, and eventually it will be possible to publicly access these instruments through computerized adaptive tests.[88]

5.4 Treatment Planning: Interventions to Address Fatigue

There are several recognized approaches to addressing fatigue, including medication, patient education and self-management, exercise, temperature regulation, and strategies to improve sleep quality. Each of these strategies and its supporting evidence is summarized below.

5.4.1 Pharmaceutical Approach

The underlying assumption of the pharmaceutical approach to fatigue management in MS is that fatigue represents a biochemical problem that medication can remediate either partially or fully. Typically only physicians, physician assistants, nurse practitioners, and pharmacists will be directly involved in the selection and prescription of pharmaceutical agents to manage fatigue. Nevertheless, all rehabilitation providers should have a basic working knowledge of the medications that can be used to address this important symptom and, in particular, the common side effects of these agents since they may influence the process and progress of therapy.

The primary pharmacological agents used to fight fatigue are amantadine, pemoline, potassium channel blockers (4-aminopyridine), antidepressants, and modafinil.[7,89,90] The efficacy of these agents, however, is not fully established, partly due to lack of high-quality research.[89–91]

Amantadine is an antiviral agent and is widely used to treat MS-related fatigue. Common side effects include insomnia, nausea, dizziness, ankle edema, concentration problems, nervousness, and headaches. Rates of these side effects range from 10% to 57% of people using amantadine and can negatively influence its tolerability. Overall, the studies of amantadine have shown small and inconsistent decreases in fatigue with uncertain clinical relevance.[89,91–96]

Pemoline is a central nervous system stimulant and studies of the drug have shown a small effect on fatigue in MS.[89,91,97] The most common side effects were anorexia, irritability, nausea, and insomnia; 25% of participants did not tolerate the drug well. Potassium channel blockers like 4-aminopyridine are used for improving physical functioning in MS (e.g., improving gait) and have also been investigated for potential use to decrease fatigue. However, none of these measures proved superior when compared with a placebo.[98,99] Reported side effects were acute encephalopathy, confusion, and seizures. Antidepressants have also been studied in people with MS-related fatigue and have shown some benefit.[89] Mohr et al.[100] found a positive effect of sertraline on fatigue severity in subjects with relapsing–remitting MS. It is very likely that this drug treated fatigue secondary to depression. Users of antidepressants reported adverse effects such as anorexia, anxiety, insomnia, gastrointestinal complaints, and increased spasticity. Modafinil is a wake-promoting agent used in narcolepsy and may be useful in the reduction of MS-related fatigue, particularly when fatigue is associated with excessive daytime sleepiness.[28,101,102] Study results have been inconsistent, however. Common side effects include insomnia, headache, nausea, nervousness, and hypertension.

5.4.2 Patient Education and Self-Management

The underlying assumption of patient education and self-management approaches for addressing MS fatigue is that an informed patient will make choices and engage in behaviors that lead to lower levels of perceived fatigue and reduced fatigue impact. Patient education may address a range of behaviors recognized as influencing fatigue: physical activity and exercise, temperature regulation, stress management, eating habits, and application of energy-management principles (formerly known as energy conservation). There are several examples of comprehensive patient education interventions that have addressed several of these behaviors in a single intervention and reported improvements in fatigue severity or fatigue impact among people with MS.[103–107] Because of the potential diversity of topics that can be addressed through patient education, a full range of rehabilitation providers can be involved in delivery of these programs, including occupational and physical therapists, nurses, psychologists, physician assistants, and so forth. Due to different training and areas of expertise, each of these disciplines has a unique perspective to contribute to managing fatigue in MS.

Regardless of the content focus, traditional clinical patient education involves a planned, systematic, sequential, and logical process of teaching and learning[108] that focuses on motivating the individual to engage in a process of behavioral change by influencing his or her knowledge, beliefs, and attitudes.[109] Theoretical models that often guide the development and delivery of patient education include the health belief model,[110] the theory of reasoned action,[111] the transtheoretical model of behavior change,[112] and social cognitive theory.[113] While each of these theories has unique aspects, each one acknowledges that the likelihood of a client's making lasting behavioral changes is influenced by a range of internal (e.g., perceptions of susceptibility and benefits of change, confidence in abilities, perceived control) and external factors (e.g., social and cultural barriers to change, environmental context). Failure to acknowledge and address these factors during the intervention process will compromise the likelihood of success.

Both social cognitive theory and the transtheoretical model of behavior change propose specific methods that can be used by rehabilitation providers to support and encourage behavioral change. A few of these processes are summarized in Table 5.7.

Across the literature on interventions that explicitly use an educational approach to improve fatigue management, the largest body of work focuses on educating people with MS to select and apply energy management principles.[114–122] Most of this work has its origin in the program *Managing Fatigue: A 6-Week Program for Energy Conservation.*[123] Table 5.8 provides examples of strategies taught in this program and, more generally, under the rubric of energy management.

Some of the unique features of the *Managing Fatigue* program and its adaptations (e.g., teleconference,[118,119] self-study modules,[122] Internet version[120]) are its focus on translation to

TABLE 5.7

Examples of Methods That Support Behavioral Change

Method and Definition	Potential Application to Educational Efforts for Fatigue Management
Consciousness raising—Learning new information to support behavioral change	Providing basic information about fatigue in MS, including knowledge about primary versus secondary fatigue
Social modeling and vicarious experience—watching or hearing about other people's successes with behavioral change	Offering education in a group format in order to provide opportunities for clients to share their successes with each other; using videos of clients trying specific fatigue management strategies
Goal setting—Making a firm commitment to change	Involving clients in the goal-setting process; having clients record the goal and report back on progress at the next session
Helping relationships—Finding and using social supports to facilitate behavioral change	Involving family and friends in the therapy process so that they can support the person with MS outside of therapy sessions
Mastery experiences—Experiencing success to motivate continued efforts for behavioral change	Grading activities and experiences during therapy to support small successes that patients can build on over time. Ensuring that patients do not fail, particularly early in the therapy process.
Environmental reevaluation—Realizing the positive impact of the healthy behavior on one's proximal environment (physical and social)	Asking clients about the influence of their fatigue management strategies on their relationships with family, friends, and so on
	Encouraging clients to journal about their successes and the factors that supported success

Note: For further reading see Bandura,[113] Prochaska and DiClemente,[112] and Redding et al.[208]

TABLE 5.8

Common Habits, Routines, and Lifestyle Changes for Energy Management

- Rest before becoming fatigued.
- Alternate periods of work with periods of rest.
- Use tools and technologies to reduce energy demands for specific tasks.
- Adjust task priorities by making active choices about how to spend available energy.
- Change height of work stations to minimize energy expenditures during everyday activities.
- Organize work spaces to minimize energy expenditures during everyday activities.
- Use proper body mechanics to minimize energy expenditures during everyday activities.
- Modify or eliminate steps in an activity to minimize energy demands.
- Ask for assistance.
- Delegate some or all of an activity to another person.
- Reduce the frequency of performing specific activities.
- Plan and organize activities in order to conduct energy-demanding activities during high-energy times of the day.

everyday life through home-based practice activities and its application of principles from social cognitive theory, particularly efficacy-enhancing strategies such as vicarious learning, mastery, and social modeling. In addition, the program and its adaptations are consistent with recent health care trends to build self-management skills among people with chronic illnesses.[124] Unlike a traditional patient education approach, the self-management approach recognizes the patient as the expert on his or her own condition and focuses on empowerment rather than compliance and adherence.[125] Self-management skills include problem solving, decision making, finding and using resources, communicating effectively with providers, goal setting, and self-tailoring (i.e., customizing strategies and behaviors to fit personal lifestyles).[124] In the context of fatigue intervention, using a self-management approach allows people with MS to increase their confidence in their ability to manage fatigue, which, in turn, enables them to make behavior changes that support a reduction in fatigue impact and severity and an increase in quality of life.

Evidence in support of energy management education has been growing in volume and quality over the past 10 years. The strongest evidence to date comes from a randomized control trial of 169 people by Mathiowetz et al.[115,116] Both efficacy and effectiveness were evaluated. Efficacy refers to whether an intervention works when delivered as intended, which in this study was defined as attending at least five of the six sessions. Effectiveness refers to whether an intervention works in the "real world," that is, will people obtain all of the benefits of the intervention in spite of absences or other limiting factors? Using an intent-to-treat analysis, the researchers found that the program was both efficacious and effective. Participants experienced a significant reduction in fatigue impact (effect sizes [ES] ranged from 0.57 to 0.83 on the Fatigue Impact Scale[126]), improvements in some aspects of health-related quality of life as measured by the SF-36[127] (Vitality–ES = 1.14; Role Physical–ES = 0.63; Mental Health–ES = 0.60; Social Function–ES = 0.42; Role Emotional–ES = 0.40), and increased self-efficacy for managing fatigue (ES = 1.72) (Self-Efficacy for Managing Fatigue Scale[128]).

The study by Mathiowetz et al.[116] also found that participants changed behaviors in relation to their fatigue management, which likely contributed to the fact that the effects of the course lasted for up to a year after its completion. Yet, some strategies were easier for participants to incorporate into their daily lives than others. Additional analyses of study data found that rest strategies and delegation of tasks to others were used most often and rated most effective, followed by strategies that required participants to modify their activity priorities and their standards for activity performance.[129] Later work by Holberg and Finlayson[70] involving a follow-up of participants from the teleconference modification of the program indicated that factors such as level of disability, prior fatigue experiences, environmental supports and barriers, and sense of self influenced which energy management strategies participants were willing to try, were considering using, or were continuing to use over time. Together, the studies of Matuska et al.[129] and Holberg and Finlayson[70] suggest that rehabilitation providers must offer a range of energy management strategies to their clients as not all strategies will be acceptable or effective for everyone. The findings of Holberg and Finlayson[70] further suggested that strategy use needs to evolve over time as MS and fatigue experiences change.

5.4.3 Exercise

Among people with MS, health problems related to physical inactivity are common, for example, reductions in muscle strength[130] and maximal oxygen consumption.[131] The benefits of physical exercise for people with MS include improvements in muscle power and exercise-tolerance functions, increased mobility-related activities,[132] and better quality of life.[133,134] The underlying assumption for using exercise as part of rehabilitation for MS fatigue is that individuals who engage in physical exercise will be better able to maintain or improve body functions. Consequently, they may also reduce the risk of perceived fatigue or of perceiving any increase in fatigue that is already present.[135,136]

Despite the known benefits of exercise, most people with MS need guidance in order to succeed in, and adhere to, a physical exercise program,[135] particularly because impairments vary within individuals over time.[137–140] When prescribing exercise programs, many factors must be considered,[40,141] and behavior change methods should be employed to support success (see Table 5.7). Since several other chapters in this text address physical activity and exercise (see Chapters 7, 13, and 17), the focus in this chapter is on exercise knowledge specific to fatigue.

Of the major forms of physical exercise therapy recommended for people with MS, endurance and strengthening exercise interventions have been studied for their capacity to influence fatigue; however, fatigue effects are often studied as a secondary outcome and in nonfatigued MS populations.[45] Furthermore, the reported findings are heterogeneous, and since there are almost no studies comparing different types of exercise interventions, it is not clear which type is most effective.[45]

5.4.3.1 Endurance Exercise

Endurance exercise interventions have been performed in people with mild and moderate MS with bicycle ergometry,[43,142–144] arm/leg ergometry,[145] and treadmill walking.[146] While Rietberg et al.[132] concluded that the evidence for positive changes in *endurance* after exercise is strong, findings regarding changes in fatigue have been somewhat inconsistent.

Reduced fatigue impact has been reported using the Multidimensional Fatigue Inventory (MFI)[43] or Modified Fatigue Impact Scale (MFIS).[143] Reduced fatigue severity has also been found using the Fatigue Severity Scale (FSS) in one study,[144] but other studies report no change in this measure.[142,145,146] Differences in fatigue-related findings across the different endurance exercise studies is likely the result of differences in the interventions (e.g., type, frequency, intensity, length), extent of baseline fatigue of participants, and differences in the fatigue measurement.[45,132] The lack of change and the fact that fatigue severity did not increase has been suggested to imply that the intervention was well tolerated.[146] Overall, studies reporting positive effects of exercise on perceived fatigue tended to use multidimensional rather than unidimensional fatigue measures (e.g., MFIS versus FSS). This observation has implications for the selection of fatigue assessment tools that are intended to capture outcomes in response to therapy.

On the basis of existing evidence, the recommendations for endurance training in people with mild or moderate MS are as follows: An initial frequency of 2–3 sessions per week and low-to-moderate intensity of 50–70% of maximal oxygen consumption or 60–70% of maximal heart rate during 10–40 min is optimal. Progression over months is achieved either by a longer duration of sessions or by adding an extra session per week. After a period of 2–6 months with exercise on low-to-moderate intensity, a higher intensity can be tested if tolerated.[147] These considerations regarding intensity, frequency, and duration are necessary to prevent any increase in intensity or duration of perceived or observed fatigue.

5.4.3.2 Resistance Exercise

Research indicates that among people with mild-to-moderate MS, progressive resistance exercises are well tolerated and muscle strength can be improved.[148–152] In addition, these studies have reported that progressive resistance exercises reduce fatigue severity as measured with the FSS[153] or the MFIS[152] and as described qualitatively by people with MS.[154] Yet, a study on combined resistance and endurance exercise reported positive effects on gait speed and aerobic capacity but not on fatigue as measured with the FSS.[60,155] This finding indicates that more research is needed to understand the best exercise interventions for fatigue among people with MS.

On the basis of existing evidence, the recommendations for progressive resistance training in people with mild or moderate MS are that a training frequency of 2–3 days per week will result in relevant improvements. A program of 4–8 exercises in 1–3 sets with intensities of 15 repetition maximum (RM) during the initial sessions is recommended. The intensity can be progressively increased over weeks and months to 3–4 sets of 8–10 RM. Rest periods in the range of 2–4 min between sets and exercises are recommended. The program should contain exercises for the whole body. Larger muscle group exercises should precede smaller muscle group exercises.[147] Additional information about developing and implementing strengthening programs for people with MS is provided in Chapter 7.

5.4.3.3 Other Considerations

To date, the research and evidence about physical exercise for people with MS have focused primarily on individuals with mild-to-moderate disease. This has left a gap in knowledge about the effects of endurance and resistance exercises for people with severe MS. For this latter group, the literature emphasizes the importance of promoting physical activity through everyday activities such as mobility and self-care.[40,141]

For all people with MS, exercise prescriptions must take into consideration the risk of heat intolerance and temporary increases in fatigue in order to limit risks for negative exercise experiences.[40,141] Overall, the potential for physical exercise to reduce fatigue is only partly understood. A range of studies have reported decreased fatigue after endurance and resistance exercise, but conclusive evidence to inform protocols with specific detail about frequency, intensity, and length of the interventions is needed.[45] For this reason, it is important that a rehabilitation professional with expertise in exercise and exercise prescription (e.g., physical therapist) should be consulted when planning for and starting an exercise program for a person with MS. These exercises in these programs need to progress gradually to minimize risks (e.g., provoking heat intolerance). Goals need to be clear, realistic, and achievable. Strategies to support behavior change and the integration of exercise into everyday patterns of behavior can be facilitated using the strategies described in Table 5.7. A study of people with MS in which physical exercise was combined with efficacy-enhancing methods showed promising results by reducing fatigue and increasing the exercise adherence.[156] Finally, exercise programs focused on reducing fatigue must be monitored and adjusted over time, and consideration should be given to group training to provide social support and motivate ongoing behavior maintenance.

5.4.4 Temperature Regulation

As many as 60–80% of people with MS[157] find that an increase in their core body temperature triggers symptoms such as fatigue, muscle weakness, or reduced vision.[158,159] A generally accepted view[159] is that the rise in body temperature influences sodium channels and the electrical current necessary for depolarization of the axon,[158,160] and thus frequency of nerve conduction decreases or slows down.[158] This slowing may worsen symptoms.[160–162] Another hypothesis explaining the aggravation of fatigue by heat is that delayed voluntary muscle innervations require a compensatory increase in central excitatory mechanisms, which, in turn, may require greater energy expenditure to achieve a given level of physical activity. This sequence of events may exacerbate fatigue.[4]

Factors raising core body temperature can be extrinsic (e.g., environmental heat and humidity) or intrinsic (e.g., bodily response to physical exercise).[158] Therefore, heat intolerance must be considered and incorporated into fatigue-management education and into the planning and prescription of physical activity programs.[40,141] Rehabilitation providers must carefully analyze the environment(s) in which the client will exercise and provide suggestions on ways to minimize the risk of Uhthoff's phenomenon—the worsening of fatigue or other impairments due to a rise in core body temperature—from occurring. Although there are several temperature regulation strategies that have been identified in the literature (e.g., cooling vests, air conditioning, cool showers), there are few empirical studies evaluating their influence on fatigue among people with MS. In a recent review article, Davis et al.[163] reported some evidence to support use of cooling strategies for people with MS and specifically mentioned using air conditioning, working out of doors in cooler times of day, wearing cooling vests, engaging in pre-cooling before exercise, drinking cold beverages, and taking cool pools and baths (<85°F or 29°C). Since there has yet to be comparative research supporting one strategy over another, it is best for rehabilitation providers to educate clients about the range of possibilities and work with them in a systematic way to identify those that work best for the client.

5.4.5 Sleep

The underlying assumption of sleep-related interventions is that a better quality sleep will reduce MS-related fatigue and provide more energy during waking hours. A broad range of sleep interventions have been discussed in the medical literature, including pharmaceutical treatments, education on sleep hygiene, cognitive behavioral therapies, and relaxation methods.[164] Despite this range, there are very few examples of sleep-related interventions that have been tested for effectiveness in managing MS-related fatigue and that fit within the scope of rehabilitation practice. Physicians, nurse practitioners, and physician assistants may be able to address factors contributing to sleep disturbances (e.g., nocturia, spasticity, pain, anxiety) through pharmaceutical interventions. Other rehabilitation providers

TABLE 5.9

Basic Sleep Hygiene Strategies

- Maintain consistent time for going to bed and rising each morning.
- Keep the bedroom quiet, dark, relaxing, and at a comfortable temperature.
- Have a comfortable bed and associate it with sleep rather than activities such as reading, watching TV, or listening to music.
- Keep the bedroom free of TVs, computers, and other electronic gadgets.
- Avoid exercising within a few hours of bedtime.
- Avoid large meals before bedtime.
- Establish a consistent and relaxing bedtime routine.
- Avoid stimulants close to bedtime (e.g., caffeine, nicotine, alcohol).
- Ensure adequate exposure to natural light during the day to promote a healthy sleep–wake cycle.

Note: On the basis of information from Centers for Disease Control[209] and the National Sleep Foundation website.[210]

can offer education on sleep hygiene in hopes of supporting habits, routines, and behaviors consistent with good-quality sleep. While Table 5.9 summarizes key sleep hygiene strategies, it is important to note that there has yet to be a high-quality study to evaluate the extent to which education on sleep hygiene can influence the fatigue that people with MS experience.

5.5 Managing MS Fatigue across Rehabilitation Settings

MS fatigue management interventions will vary in focus and intensity depending on the client's medical status and the setting where rehabilitation is provided. People with MS are most likely to require inpatient rehabilitation after an exacerbation of symptoms. During an exacerbation, a sudden increase in fatigue can be expected (i.e., acute fatigue). The focus in the acute phase is, therefore, to treat the exacerbation in hopes of relieving the severity of fatigue symptoms. Treatment will be pharmaceutical and directed by the physician or neurologist. Once the acute phase of the exacerbation is under control, rehabilitation providers must focus on enabling the patient to regain lost functions and manage any new symptoms that have appeared. In other words, treatment plans will likely shift in focus from reducing fatigue severity to managing fatigue impact, increasing endurance, and reducing the risk of secondary contributors to fatigue.

Assessment during an inpatient stay must consider the setting to which the patient will return, daily activity patterns and demands in these settings, and the skills and resources the client will be required to manage after discharge. Understanding a client's context is critical for selecting specific energy management strategies for the individual client, prescribing exercises that can be maintained after discharge, and providing guidance or referral to reduce secondary sources of fatigue. While inpatient rehabilitation intervention is most often provided on a one-to-one basis, the inclusion of family members is encouraged to maximize ongoing management after discharge.

In outpatient and community-based rehabilitation settings (e.g., home care), rehabilitation providers have a greater opportunity to observe patients in the course of their everyday lives. If possible, home visits can provide a very useful source of assessment information to guide interventions focused on reducing the impact of fatigue on daily life. For clients who experience chronic fatigue, interventions to maximize self-care and domestic life involvement (Chapters 14 and 15), maintain employment and productivity (Chapter 16), and promote ongoing physical activity and leisure participation (Chapter 17) will be important. Offering fatigue-specific interventions in group settings in the community (e.g., energy management education groups) may enable clients to develop the self-management skills necessary for ongoing, successful management of their fatigue symptoms.

5.6 Gaps in Knowledge and Directions for Future Research

As the contents of this chapter have illustrated, MS fatigue is a complex phenomenon. The ability to successfully manage this symptom during the rehabilitation process will require interdisciplinary research at both the basic and applied levels. From a basic science perspective, further knowledge about the pathophysiology of MS fatigue and its underlying causes may lead to more effective pharmaceutical interventions and offer rehabilitation providers with clear directions to target intervention.

From an applied perspective, assessments that capture observed physiological fatigue and are relevant and practical in a clinical setting would be valuable. While self-reports will always have a role in fatigue assessment and rehabilitation practice, more objective evaluations would be useful when there are questions about workplace accommodations because of fatigue. For self-report assessments, research to establish MCID would also be extremely valuable.

From an intervention perspective, much knowledge has been developed over the last 10–15 years, but many questions still remain. Research on the efficacy and effectiveness of sleep hygiene and cooling interventions for reducing fatigue among people with MS are needed. More research is also needed on the use of exercise to manage MS fatigue, particularly among people with severe disability. Future studies need to provide clear guidelines on types of exercises as well as on intensity and frequency of exercise over time. Educational and self-management interventions targeting MS fatigue also require further work, particularly in terms of the efficacy and effectiveness of using these approaches on a one-to-one basis or during inpatient rehabilitation stays. Further research is also needed in terms of specific energy management strategies, how people incorporate them into their lives, what barriers exist and what supports are needed to establish exercise routines over time.

In conclusion, fatigue is a common and disabling symptom for people with MS, and rehabilitation providers have much to offer in terms of reducing the impact of fatigue on daily life. This chapter has described different classifications of the fatigue observed among people with MS, described the impact of this symptom on everyday life, reviewed commonly used assessments of fatigue, and described the major types of evidence-based fatigue management interventions that can be applied by rehabilitation providers.

Acknowledgment

Special acknowledgment to Meghan Ginter.

References

1. Bergamaschi R, Romani A, Versino M, Poli R, Cosi V. Clinical aspects of fatigue in multiple sclerosis. Functional Neurology 1997;12(5):247–251.

2. Colombo B, Martinelli Boneschi F, Rossi P, Rovaris M, Maderna L, Filippi M, Comi G. MRI and motor evoked potential findings in nondisabled multiple sclerosis patients with and without symptoms of fatigue. Journal of Neurology 2000;247:506–509.

3. Kos D, Kerckhofs E, Nagels G, D'hooghe MB, Ilsbroukx S. Origin of fatigue in multiple sclerosis: Review of the literature. Neurorehabilitation and Neural Repair 2008;22:91–100.

4. Bakshi R. Fatigue associated with multiple sclerosis: Diagnosis, impact and management. Multiple Sclerosis 2003;9:219–227.

5. Clanet M, Brassat D. The management of multiple sclerosis patients Current Opinion in Neurology 2000;13(3):263–270.

6. Multiple Sclerosis Council for Clinical Practice Guidelines. Fatigue and multiple sclerosis: Evidence-based management strategies for fatigue in multiple sclerosis. Washington, DC: Paralyzed Veterans of America; 1998.

7. Krupp LB. Fatigue. Philadelphia: Elsevier Science; 2003.

8. Kalkman JS, Zwartz MJ, Schillings ML, van Engelen BG, Bleijenberg G. Different types of fatigue in patients with facioscapulohumeral dystrophy, myotonic dystrophy and HMSN-I. Experienced fatigue and physiological fatigue. Neurological Sciences 2008;29(S2):S238–S240.

9. Tralongo P, Respini D, Ferra F. Fatigue and aging. Critical Reviews in Oncology/Hematology 2003;48:S57–S64.

10. Schanke AK, Stanghelle JK. Fatigue in polio survivors. Spinal Cord 2001;39:243–251.

11. Schwid SR, Covington M, Segal BM, Goodman AD. Fatigue and multiple sclerosis: Current understanding and future directions. Journal of Rehabilitation Research and Development 2002;39(2):211–224.

12. Krupp L, Serafin D, Christodoulou C. Multiple sclerosis-associated fatigue. Expert Review of Neurotherapeutics 2010;10(9):1437–1447.

13. MacAllister WS, Krupp LB. Multiple sclerosis-related fatigue. Physical Medicine and Rehabilitation Clinics of North America 2005;16:483–502.

14. Braley T, Chervin R. Fatigue in multiple sclerosis: Mechanisms, evaluation, and treatment. Sleep 2010;33(8):1061–1067.

15. Roelcke U, Kappos L, Lechner-Scott J, Brunnschweiler H, Huber S, Ammann W, Plohmann A, Dellas S, Maguire RP, Missimer J et al. Reduced glucose metabolism in the frontal cortex and basal ganglia of multiple sclerosis patients with fatigue: A F-18-fluorodeoxyglucose positron emission tomography study. Neurology 1997;48:1566–1571.

16. Tartaglia MC, Narayanan S, Arnold DL. Mental fatigue alters the pattern and increases the volume of cerebral activation required for a motor task in multiple sclerosis patients with fatigue. European Journal of Neurology 2008;15(4):413–419.

17. Filippi M, Rocca MA, Colombo B, Falini A, Codella M, Scotti G, Comi G. Functional magnetic resonance imaging correlates of fatigue in multiple sclerosis. NeuroImage 2002;15:559–567.

18. Leocani L, Colombo B, Magnani G, Martinelli-Boneschi F, Cursi M, Rossi P, Martinelli V, Comi G. Fatigue in multiple sclerosis is associated with abnormal cortical activation to voluntary movement-EEG evidence. Neuroimage 2001;13:1186–1192.

19. Liepert J, Mingers D, Heesen C, Baumer T, Weiller C. Motor cortex excitability and fatigue in multiple sclerosis: A transcranial magnetic stimulation study. Multiple Sclerosis 2005;11:316–321.

20. Schubert M, Wohlfarth K, Rollnik JD, Dengler R. Walking and fatigue in multiple sclerosis: The role of the corticospinal system. Muscle & Nerve 1998;21:1068–1070.

21. White AT, Lee JN, Light AR, Light KC. Brain activation in multiple sclerosis: A BOLD fMRI study on the effects of fatiguing hand exercise. Multiple Sclerosis 2009;15(5):580–586.

22. Marrie RA, Fisher E, Miller DM, Lee JC, Rudick RA. Association of fatigue and brain atrophy in multiple sclerosis. Journal of Neurological Sciences 2005;228(2):161–166.

23. Flachenecker P, Bihler I, Weber F, Gottschalk M, Toyka KV, Rieckmann P. Cytokine mRNA expression in patients with multiple sclerosis and fatigue. Multiple Sclerosis 2004;10:165–169.

24. Heesen C, Nawrath L, Reich C, Bauer N, Schulz KH, Gold SM. Fatigue in multiple sclerosis: An example of cytokine mediated sickness behaviour? Journal of Neurology, Neurosurgery & Psychiatry 2006;77:34–39.

25. Iriarte J, Subira ML, Castro P. Modalities of fatigue in multiple sclerosis: Correlation with clinical and biological factors. Multiple Sclerosis 2000;6(2):124–130.

26. Gottschalk M, Kümpfel T, Flachenecker P, Uhr M, Trenkwalder C, Holsboer F, Weber F. Fatigue and regulation of the hypothalamo–pituitary–adrenal axis in multiple sclerosis. Archives of Neurology 2005;62:277–280.

27. Téllez N, Comabella M, Julià E, Río J, Tintoré M, Brieva L, Nos C, Montalban X. Fatigue in progressive multiple sclerosis is associated with low levels of dehydroepiandrosterone. Multiple Sclerosis 2006;12:487–494.

28. Rammohan KW, Rosenberg JH, Lynn DJ, Blumenfeld AM, Pollak CP, Nagaraja HN. Efficacy and safety of modafinil (provigil) for treatment of fatigue in multiple sclerosis: A two centre phase 2 study. Journal of Neurology 2002;72:179–183.

29. Zifko UA, Rupp M, Schwarz S, Zipko HT, Maida EM. Modafinil in treatment of fatigue in multiple sclerosis: Results of an open-label study. Journal of Neurology 2002;249: 983–987.

30. Zellini F, Niepel G, Tench CR, Constantinescu CS. Hypothalamic involvement assessed by T1 relaxation time in patients with relapsing–remitting multiple sclerosis. Multiple Sclerosis 2009; 15(12):1442–1449.

31. Bamer AM, Johnson KL, Amtmann D, Kraft GH. Prevalence of sleep problems in individuals with multiple sclerosis. Multiple Sclerosis 2008;14:1127–1130.

32. Clark CM, Fleming JA, Li D, Oger J, Klonoff H, Paty D. Sleep disturbance, depression, and lesion site in patients with multiple sclerosis. Archives of Neurology 1992;49:641–643.

33. Lobentanz IS, Asenbaum S, Vass K, Sauter C, Klösch G, Kollegger H, Kristoferitsch W, Zeitlhofer J. Factors influencing quality of life in multiple sclerosis patients: Disability, depressive mood, fatigue and sleep quality. Acta Neurologica Scandinavica 2004;110:6–13.

34. Merlino G, Fratticci L, Lenchig C, Valente M, Cargnelutti D, Picello M, Serafini A, Dolso P, Gigli GL. Prevalence of "poor sleep" among patients with multiple sclerosis: An independent predictor of mental and physical status. Sleep Medicine 2009;10:26–34.

35. Stanton BR, Barnes F, Silber E. Sleep and fatigue in multiple sclerosis. Multiple Sclerosis 2006;12: 481–486.

36. Tachibana N, Howard RS, Hirsch NP, Miller DH, Moseley IF, Fish D. Sleep problems in multiple sclerosis. European Neurology 1994;34:320–323.

37. Gottberg K, Gardulf A, Fredrikson S. Interferon-beta treatment for patients with multiple sclerosis: The patients' perceptions of the side-effects. Multiple Sclerosis 2000;6:349–354.

38. Motl RW, Goldman MD, Benedict RH. Walking impairment in patients with multiple sclerosis: Exercise training as a treatment option. Journal of Neuropsychiatric Disease and Treatment 2010;16(6):767–774.

39. Motl RW, Goldman M. Physical inactivity, neurological disability, and cardiorespiratory fitness in multiple sclerosis. Acta Neurologica Scandinavica 2011;123:98–104.

40. Heesen C, Romberg A, Gold S, Schulz KH. Physical exercise in multiple sclerosis: Supportive care or a putative disease-modifying treatment. Expert Review of Neurotherapeutics 2006;6: 347–355.

41. Motl RW, Gosney JL. Effect of exercise training on quality of life in multiple sclerosis: A meta-analysis. Multiple Sclerosis 2008;14:129–135.

42. Motl RW, McAuley E. Pathways between physical activity and quality of life in adults with multiple sclerosis. Health Psychology 2009;28(6):682–689.

43. Oken BS, Kishiyama S, Zajdel D, Bourdette D, Carlsen J, Haas M, Hugos C, Kraemer DF, Lawrence J, Mass M. Randomized controlled trial of yoga and exercise in multiple sclerosis. Neurology 2004;62:2058–2064.

44. Dalgus U, Stenager E, Jakobsen J, Peterson T, Hansen H, Knudsen C, Overgard K, Ingemann-Hansen T. Fatigue, mood and quality of life improve in MS patients after progressive resistance training. Multiple Sclerosis 2010;16(4):480–490.

45. Andreasen AK, Stenager E, Dalgas U. The effect of exercise therapy on fatigue in multiple sclerosis. Multiple Sclerosis Journal 2011;17:1041–1054.

46. Bakshi R, Shaikh ZA, Miletich RS, Czarnecki D, Dmochowski J, Henschel K, Janardhan V, Dubey N, Kinkel PR. Fatigue in multiple sclerosis and its relationship to depression and neurologic disability. Multiple Sclerosis 2000;6:181–185.

47. Bol Y, Duits AA, Hupperts RM, Vlaeyen JW, Verhey FR. The psychology of fatigue in patients with multiple sclerosis: A review. Journal of Psychosomatic Research 2009;66(1):3–11.

48. Brown RF, Valpiani EM, Tennant CC, Dunn SM, Sharrock M, Hodgikinson S, Pollard JD. Longitudinal assessment of anxiety, depression, and fatigue in people with multiple sclerosis. Psychology and Psychotherapy 2009;82(Pt 1):41–56.

49. Patrick E, Christodoulou C, Krupp LB, New York State MS Consortium. Longitudinal correlates of fatigue in multiple sclerosis. Multiple Sclerosis 2009;15(2):258–261.

50. Johansson S, Ytterberg C, Hillert J, Widén Holmqvist L, von Koch L. A longitudinal study of variations in and predictors of fatigue in multiple sclerosis. Journal of Neurology, Neurosurgery & Psychiatry 2008;79:454–457.

51. Schwid SR, Tyler CM, Scheid EA, Weinstein A, Goodman AD, McDermott MP. Cognitive fatigue during a test requiring sustained attention: A pilot study. Multiple Sclerosis 2003;9: 503–508.

52. Christodoulou C. The assessment and measurement of fatigue. In: J. Deluca, editor. Fatigue as a window to the brain. Cambridge, MA: MIT Press; 2005.

53. Deluca J. Fatigue as a window to the brain. Massachusettes: The MIT Press; 2005.

54. Bryant D, Chiaravalloti ND, Deluca J. Objective measurement of cognitive fatigue in multiple sclerosis. Rehabilitation Psychology 2004;49:114–122.

55. Paul RH, Beatty WW, Schneider R, Blanco CR, Hames KA. Cognitive and physical fatigue in multiple sclerosis: Relations between self-report and objective performance. Applied Neuropsychology 1998;5:143–148.

56. Barak Y, Achiron A. Cognitive fatigue in multiple sclerosis: Findings from a two-wave screening project. Journal of Neurological Sciences 2006;245:73–76.

57. Beatty WW, Goretti B, Siracusa G, Zipoli V, Portaccio E, Amato MP. Changes in neuropsychological test performance over the workday in multiple sclerosis. Clinical Neuropsychologist 2003;17:551–560.

58. Krupp LB, Elkins LE. Fatigue and declines in cognitive functioning in multiple sclerosis. Neurology 2000;55:934–939.

59. Kos D, Kerckhofs E, Nagels G, Geentjens L. Cognitive fatigue in multiple sclerosis: Comment on Schwid SR, Tyler CM, Scheid EA, Weinstein A, Goodman AD and McDermott MR. Multiple Sclerosis 2004;10(3):337.

60. Surakka J, Romberg A, Ruutiainen J, Aunola S, Virtanen A, Karppi SL, Mäentaka K. Effects of aerobic and strength exercise on motor fatigue in men and women with multiple sclerosis: A randomized controlled trial. Clinical Rehabilitation 2004;18:737–746.

61. Sheean GL, Murray NM, Rothwell JC, Miller DH, Thompson AJ. An electrophysiological study of the mechanism of fatigue in multiple sclerosis. Brain 1997;120(Pt 2):299–315.

62. Sharma KR, Kent-Braun J, Mynhier MA, Weiner MW, Miller RG. Evidence of an abnormal intramuscular component of fatigue in multiple sclerosis. Muscle & Nerve 1995;18:1403–1411.

63. Téllez N, Río J, Tintoré M, Nos C, Galán I, Montalban X. Fatigue in multiple sclerosis persists over time: A longitudinal study. Journal of Neurology 2006a;253:1466–1470.

64. Schreurs KMG, de Ridder DTD, Bensing JM. Fatigue in multipe sclerosis—Reciprocal relationships with physical disabilities and depression. Journal of Psychosomatic Research 2002;53:775–781.

65. Lenz ER, Pugh LC, Milligan RA, Gift A. The middle-range theory of unpleasant symptoms: An update. Advances in Nursing Science 1997;19(3):14–27.

66. National Institute on Aging. Unexplained fatigue in the elderly: An exploratory workshop sponsored by the National Institughe on Aging—Workshop summary, June 25–26, 2007. National Institute on Aging [Serial Online] [Internet]. [revised 2007:July 19, 2009].

67. Flensner G, Ek A, Söderhamn O. Lived experience of MS-related fatigue: A phenomenological interview study. International Journal of Nursing Studies 2003;40:707–717.

68. Olsson M, Lexell J, Söderberg S. The meaning of fatigue for women with multiple sclerosis. Journal of Advanced Nursing 2005;49(1):7–15.

69. Stuifbergen A, Rogers S. The experience of fatigue and strategies of self-care among persons with multiple sclerosis. Applied Nursing Research 1997;10(1):2–10.

70. Holberg C, Finlayson M. Factors influencing the use of energy conservation strategies by persons with multiple sclerosis. American Journal of Occupational Therapy 2007; 61(1):96–107.

71. Blaney BE, Lowe-Strong A. The impact of fatigue on communication in multiple sclerosis: The insider's perspective. Disability and Rehabilitation 2009;31(3):170–180.

72. Yorkston KM, Johnson K, Klasner ER, Amtmann D, Kuehn CM, Dudgeon B. Getting the work done: A qualitative study of individuals with multiple sclerosis. Disability and Rehabilitation 2003;25(8):369–379.

73. Finlayson M, Cho C. A descriptive profile of caregivers of older adults with MS and the assistance they provide. Disability and Rehabilitation 2008;30(24):1848–1857.

74. Carton H, Loos R, Pacolet J, Versieck K, Vlietinck R. A quantitative study of unpaid caregiving in multiple sclerosis. Multiple Sclerosis 2000;6:274–279.

75. Buchanan RJ, Radin D, Chakravorty BJ, Tyry T. Informal caregiving to more disabled people with multiple sclerosis. Disability & Rehabilitation 2009;31:1244–1256.

76. Figved N, Myhr KM, Larsen JP, Aarsland D. Caregiver burden in multiple sclerosis: The impact of neuropsychiatric symptoms. Journal of Neurology, Neurosurgery & Psychiatry 2007;78(10): 1097–1102.

77. Pozzilli C, Palmisano L, Mainero C, Tomassini V, Marinelli F, Ristori G, Gasperini C, Fabiani M, Battaglia MA. Relationship between emotional distress in caregivers and health status in persons with multiple sclerosis. Multiple Sclerosis Journal 2004;10(4):442–446.

78. Morris ME, Cantwell C, Vowels L, Dodd K. Changes in gait and fatigue from morning to afternoon in people with multiple sclerosis. Journal of Neurology, Neurosurgery & Psychiatry 2002;72(3):361–365.

79. Hjollund NH, Andersen JH, Bech P. Assessment of fatigue in chronic disease: A bibliographic study of fatigue measurement scales. Health and Quality of Life Outcomes 2007;5(12), doi:10.1186/1477-7525-5-12.

80. Kos D, Kerckhofs E, Ketelaer P, Duportail M, Nagels G, D'Hooghe M, Nuyens G. Self-report assessment of fatigue in multiple sclerosis: A critical evaluation. Occupational Therapy in Health Care 2003;17(3–4):45–62.

81. Taylor R, Jason L, Torres A. Fatigue rating scales: An empirical comparison. Psychological Medicine 2000;30:849–856.

82. Mota DC, Pimenta CA. Self-report instruments for fatigue assessment: A systematic review. Research and Theory for Nursing Practice: An International Journal 2006; 20(1):49–78.

83. DePoy E, Gitlin LN. Introduction to research: Understanding and applying multiple strategies. 4th ed. St. Louis, MO: Mosby; 2011.

84. Rosenthal TC, Majeroni BA, Pretorius R, Malik K. Fatigue: An overview. American Family Physician 2008;78(10):1173–1179.

85. Finch E, Brooks D, Stratford PW, Mayo NE. Physical rehabilitation outcome measures: A guide to enhanced clinical decision making. 2nd ed. Philadelphia, PA: Lippincott Williams & Wilkins; 2002.

86. McDowell I. Measuring health: A guide to rating scales and questionnaires. 3rd ed. New York: Oxford University Press; 2006.

87. Rietberg MB, Van Wegen EE, Kwakkel G. Measuring fatigue in patients with multiple sclerosis: Reproducibility, responsiveness and concurrent validity of three Dutch self-report questionnaires. Disability and Rehabilitation 2010;32(22):1870–1876.

88. PROMIS: Dynamic tools to measure health outcomes from the patient perspective [Internet] [cited 2011]. Available from: www.nihpromis.org.

89. Branas P, Jordan R, Fry-Smith A, Burls A, Hyde C. Treatments for fatigue in multiple sclerosis: A rapid and systematic review. Health Technology Assessment 2000;4:1–61.

90. Lange G, Cook DB, Natelson BH. Rehabilitation and treatment of fatigue. In: J. Deluca, editor. Fatigue as a window to the brain. Massachusetts: The MIT Press; 2005.

91. Lee D, Newell R, Ziegler L, Topping A. Treatment of fatigue in multiple sclerosis: A systematic review of the literature. International Journal of Nursing Practice 2008;14(2):81–93.

92. Krupp LB, Coyle PK, Doscher C, Miller A, Cross AH, Jandorf L, Halper J, Johnson B, Morgante L, Grimson R. Fatigue therapy in multiple sclerosis: Results of a double-blind, randomized, parallel trial of amantadine, pemoline, and placebo. Neurology 1995;45:1956–1961.

93. Rosenberg GA, Appenzeller O. Amantadine, fatigue, and multiple sclerosis. Archives of Neurology 1988;45:1104–1116.

94. Cohen RA, Fisher M. Amantadine treatment of fatigue associated with multiple sclerosis. Archives of Neurology 1989;46:676–680.

95. Pucci E, Branas F, D'Amico R, Giuliani G, Solari A, Taus C. Amantadine for fatigue in multiple sclerosis. Cochrane Database of Systematic Reviews. 2007;(1). Art. No.: CD002818. DOI: 10.1002/14651858.CD002818.pub2.

96. The Canadian MS Research Group. A randomized controlled trial of amantadine in fatigue associated with multiple sclerosis. Canadian Journal of Neurological Sciences 1987;14:273–278.

97. Weinshenker BG, Penman M, Bass B, Ebers GC, Rice GP. A double-blind, randomized, crossover trial of pemoline in fatigue associated with multiple sclerosis. Neurology 1992;42:1468–1471.

98. Goodman AD, Cohen JA, Cross A, Vollmer T, Rizzo M, Cohen R, Marinucci L, Blight AR. Fampridine-SR in multiple sclerosis: A randomized, double-blind, placebo-controlled, dose-ranging study. Multiple Sclerosis 2007;13(3):357–368.

99. Rossini PM, Pasqualetti P, Pozzilli C, Grasso MG, Millefiorini E, Graceffa A, Carlesimo GA, Zibellini G, Caltagirone C. Fatigue in progressive multiple sclerosis: Results of a randomized, double-blind, placebo-controlled, crossover trial of oral 4-aminopyridine. Multiple Sclerosis 2001;7(6):354–358.

100. Mohr DC, Hart SL, Goldberg A. Effects of treatment for depression on fatigue in multiple sclerosis. Psychosomatic Medicine 2003;65(4):542–547.

101. Stankoff B, Waubant E, Confavreux C, Edan G, Debouverie M, Rumbach L, Moreau T, Pelletier J, Lubetzki C, Clanet M et al. Modafinil for fatigue in MS: A randomized placebo-controlled double-blind study. Neurology 2005;64:1139–1143.

102. Littleton ET, Hobart JC, Palace J. Modafinil for multiple sclerosis fatigue: Does it work? Clinical Neurology and Neurosurgery 2010;112(1):29–31.

103. Kos D, Duportial M, D'hooghe MB, Nagels G, Kerckhofs E. Multidisciplinary fatigue management program in multiple sclerosis: A randomized clinical trial. Multiple Sclerosis 2007;13: 996–1003.

104. Plow MA, Mathiowetz V, Lowe DA. Comparing individualized rehabilitation to a group wellness intervention for persons with multiple sclerosis. American Journal of Health Promotion. 2009;24(1):23–26.

105. Wassem R, Dudley W. Symptom management and adjustment of patients with multiple sclerosis: A 4-year longitudinal intervention study. Clinical Nursing Research 2003;12(1):102–117.

106. Navipour H, Madani H, Mohebbi MR, Navipour R, Roozbayani P, Paydar A. Improved fatigue in individuals with multiple sclerosis after participating in a short-term self-care programme. Neurorehabilitation 2006;21(1):37–41.

107. Bombardier CH, Cunniffe M, Wadhwani R, Gibbons LE, Blake KD, Kraft GH. The efficacy of telephone counseling for health promotion in people with multiple sclerosis: A randomized controlled trial. Archives of Physical Medicine & Rehabilitation 2008;89(10):1849–1856.

108. Whitehead D. Health promotion and health education: Advancing the concepts. Journal of Advanced Nursing 2004;47(3):311–320.

109. Dreeben O. Patient education in rehabilitation. Boston, MA: Jones & Bartlett; 2010.

110. Becker MH. The health belief model and sick role behavior. Health Education Monographs 1974;2:409–419.

111. Ajzen I, Fishbein M. Understanding attitudes and predicting social behavior. Englewood Cliffs, NJ: Prentice-Hall; 1980.

112. Prochaska J, DiClemente C. Stages and processes of self-change of smoking: Toward an integrative model of change. Journal of Consulting and Clinical Psychology 1983;51(3):390–395.

113. Bandura A. Social foundations of thought and action: A social cognitive theory. Englewood Cliffs, NJ: Prentice-Hall; 1986.

114. Mathiowetz V, Matuska KM, Murphy ME. Efficacy of an energy conservation course for persons with multiple sclerosis. Archives of Physical Medicine & Rehabilitation 2001;82(4):449–456.

115. Mathiowetz VG, Finlayson ML, Matuska KM, Chen HY, Luo P. Randomized controlled trial of an energy conservation course for persons with multiple sclerosis. Multiple Sclerosis 2005; 11(5):592–601.

116. Mathiowetz VG, Matuska KM, Finlayson ML, Luo P, Chen HY. One-year follow-up to a randomized controlled trial of an energy conservation course for persons with multiple sclerosis. International Journal of Rehabilitation Research 2007;30(4):305–313.

117. Vanage SM, Gilbertson KK, Mathiowetz V. Effects of an energy conservation course on fatigue impact for persons with progressive multiple sclerosis. American Journal of Occupational Therapy 2003;57(3):315–323.

118. Finlayson M. Pilot study of an energy conservation education program delivered by telephone conference call to people with multiple sclerosis. Neurorehabilitation 2005;20(4):267–277.

119. Finlayson ML, Preissner K, Cho CC, Plow MA. Randomized trial of a teleconference-delivered fatigue management program for people with multiple sclerosis. Multiple Sclerosis 2011;17(9): 1130–1140.

120. Ghahari S, Packer TL, Passmore AE. Effectiveness of an online fatigue self-management programme for people with chronic neurological conditions: A randomized controlled trial. Clinical Rehabilitation 2010;24:727–744.

121. Sauter C, Zebenholzer K, Hisakawa J, Zeitlhofer J, Vass K. A longitudinal study on effects of a six-week course for energy conservation for multiple sclerosis patients. Multiple Sclerosis 2008;14(4):500–505.

122. Lamb AL, Finlayson M, Mathiowetz V, Chen HY. The outcomes of using self-study modules in energy conservation education for people with multiple sclerosis. Clinical Rehabilitation 2005;19(5):475–481.

123. Packer TL, Brink N, Sauriol A. Managing fatigue: A 6-week course for energy conservation. Tucson, AZ: Therapy Skill Builders; 1995.

124. Lorig KR, Holman H. Self-management education: History, definition, outcomes, and mechanisms. Annals of Behavioral Medicine 2003;26:1–7.

125. Bodenheimer T, Lorig K, Holman H, Grumbach K. Patient self-management of chronic disease in primary care. JAMA: Journal of the American Medical Association 2002;288:2469–2475.

126. Fisk JD, Ritvo PG, Ross L, Haase DA, Marrie TJ, Schlech WF. Measuring the functional impact of fatigue: Initial validation of the fatigue impact scale. Clinical Infectious Diseases 1994; 18:S79–83.

127. Ware J, Sherbourne C. The MOS 36-item short-form health survey (SF-36). Medical Care 1992;30:473–483.

128. Liepold A, Mathiowetz V. Reliability and validity of the self-efficacy for performing energy conservation strategies assessment for persons with multiple sclerosis. Occupational Therapy International 2005;12(4):234–249.

129. Matuska K, Mathiowetz V, Finlayson M. Use and perceived effectiveness of energy conservation strategies for managing multiple sclerosis fatigue. American Journal of Occupational Therapy 2007;61:62–69.

130. Ng AV, Miller RG, Gelinas D, Kent-Braun JA. Functional relationships of central and peripheral muscle alterations in multiple sclerosis. Muscle & Nerve 2004;29:843–852.
131. Tantucci C, Massucci M, Piperno R, Grassi V, Sorbini CA. Energy cost of exercise in multiple sclerosis patients with low degree of disability. Multiple Sclerosis 1996;2:161–167.
132. Rietberg MB, Brooks D, Uitdehaag BMJ, Kwakkel G. Exercise therapy for multiple sclerosis. Cochrane Database of Systematic Reviews. 2004;(3). Art. No.: CD003980. DOI: 10.1002/14651858. CD003980.pub2.
133. Motl RW, Snook EM. Physical activity, self-efficacy, and quality of life in multiple sclerosis. Annals of Behavioral Medicine 2008;35:111–115.
134. Stuifbergen AK, Blozis SA, Harrison TC, Becker HA. Exercise, functional limitations, and quality of life: A longitudinal study of persons with multiple sclerosis. Archives of Physical Medicine and Rehabilitation 2006;87:935–943.
135. White LJ, Dressendorfer RH. Exercise and multiple sclerosis. Sports Medicine 2004;34: 1077–1100.
136. Sutherland G, Anderson MB. Exercise and multiple sclerosis: Physiological, psychological and quality of life stress. Journal of Sports Medicine & Physical Fitness 2001;41:421–432.
137. Amato MP, Ponziani G, Siracusa G, Sorbi S. Cognitive dysfunction in early-onset multiple sclerosis: A reappraisal after 10 years. Archives of Neurology 2001;58:1602–1606.
138. Arnett PA, Randolph JJ. Longitudinal course of depression symptoms in multiple sclerosis. Journal of Neurology, Neurosurgery & Psychiatry 2006;77:606–610.
139. Gulick EE. Symptom and activities of daily living trajectory in multiple sclerosis: A 10-year study. Nursing Research 1998;47:137–146.
140. Ytterberg C, Johansson S, Andersson M, Widén Holmqvist L, von Koch L. Variations in functioning and disability in multiple sclerosis. A two-year prospective study. Journal of Neurology 2008;255:967–973.
141. Petajan JH, White AT. Recommendations for physical activity in patients with multiple sclerosis. Sports Medicine 1999;27:179–191.
142. Mostert S, Kesselring J. Effects of a short-term exercise training program on aerobic fitness, fatigue, health perception and activity level of subjects with multiple sclerosis. Multiple Sclerosis 2002;8:161–168.
143. Rasova K, Havrdova E, Brandejsky P, Zálišová M, Foubikova B, Martinkova P. Comparison of the influence of different rehabilitation programmes on clinical, spirometric and spiroergometric parameters in patients with multiple sclerosis. Multiple Sclerosis 2006;12:227–234.
144. Cakt BD, Nacir B, Gene H, Saracoglu M, Karagoz A, Erdem HR, Ergun U. Cycling progressive resistance training for people with multiple sclerosis: A randomized controlled study. American Journal of Physical Medicine & Rehabilitation 2010;89:446–457.
145. Petajan JH, Gappmaier E, White AT, Spencer MK, Mino L, Hicks RW. Impact of aerobic training on fitness and quality of life in multiple sclerosis. Annals of Neurology 1996;39:432–441.
146. Newman MA, Dawes H, van den Berg M, Wade DT, Burridge J, Izadi H. Can aerobic treadmill training reduce the effort of walking and fatigue in people with multiple sclerosis: A pilot study. Multiple Sclerosis 2007;13:113–119.
147. Dalgas U, Stenager E, Ingemann-Hansen T. Multiple sclerosis and physical exercise: Recommendations for the application of resistance-, endurance- and combined training. Multiple Sclerosis 2008;14:35–53.
148. Dalgas U, Stenager E, Jakobsen J, Petersen T, Hansen H, Knudsen C, Overgard K, Ingemann-Hansen T. Resistance training improves muscle strength and functional capacity in multiple sclerosis. Neurology 2009;73:1478–1484.
149. DeBolt LS, McCubbin JA. The effects of home-based resistance exercise on balance, power, and mobility in adults with multiple sclerosis. Archives of Physical Medicine & Rehabilitation 2004;85:290–297.
150. Gutierrez GM, Chow JW, Tillman MD, McCoy SC, Castellano V, White LJ. Resistance training improves gait kinematics in persons with multiple sclerosis. Archives of Physical Medicine & Rehabilitation 2005 Sep;86(9):1824–1829.

151. Taylor NF, Dodd KJ, Prasad D, Denisenko S. Progressive resistance exercise for people with multiple sclerosis. Disability Rehabilitation 2006;28:1119–1126.

152. White LJ, McCoy SC, Castellano V, Gutierrez G, Stevens JE, Walter GA, Vandenborne K. Resistance training improves strength and functional capacity in persons with multiple sclerosis. Multiple Sclerosis 2004;10:668–674.

153. Dalgas U, Stenager E, Jakobsen J, Petersen T, Hansen HJ, Knudsen C, Overgaard K, Ingemann-Hansen T. Fatigue, mood and quality of life improve in MS patients after progressive resistance training. Multiple Sclerosis 2010;16(4):480–490.

154. Dodd KJ, Taylor NF, Denisenko S, Prasad D. A qualitative analysis of a progressive resistance exercise programme for people with multiple sclerosis. Disability & Rehabilitation 2006; 28:1127–1134.

155. Romberg A, Virtanen A, Ruutiainen J, Aunola S, Karppi SL, Vaara M, Surakka J, Pohjolainen T, Seppänen A. Effects of a 6-month exercise progam on patients with multiple sclerosis: A randomized study. Neurology 2004;63:2034–2038.

156. McAuley E, Motl RW, Morris KS, Hu L, Doerksen SE, Elavsky S, Konopack JF. Enhancing physical activity adherence and well-being in multiple sclerosis: A randomised controlled trial. Multiple Sclerosis 2007;13:652–659.

157. Guthrie TC, Nelson DA. Influence of temperature changes on multiple sclerosis: Critical review of mechanisms and research potential. Journal of Neurological Sciences 1995;129:1–8.

158. Baker DG. Multiple sclerosis and thermoregulatory dysfunction. Journal of Applied Physiology 2002;92:1779–1780.

159. Compston A, McDonald IR, Noseworthy J, Lassmann H, Miller DH, Smith KJ, Wekerle H, Confavreux C. McAlpine's multiple sclerosis. 4th ed. London: Churchill Livingstone Elsevier; 2006.

160. Smith KJ, McDonald WI. The pathophysiology of multiple sclerosis: The mechanisms underlying the production of symptoms and the natural history of the disease. Philosophical Transactions of the Royal Society B: Biological Sciences 1999;354:1649–1673.

161. Rasminsky M, Sears TA. Internodal conduction in undissected demyelinated nerve fibres. Journal of Physiology 1972;227:323–350.

162. Schauf CL, Davis FA. Impulse conduction in multiple sclerosis: A theoretical basis for modification by temperature and pharmacological agents. Journal of Neurology, Neurosurgery & Psychiatry 1974;37:152–161.

163. Davis SL, Wilson TE, White AT, Frohman EM. Thermoregulation in mulitple sclerosis. Journal of Applied Physiology 2010;109(5):1531–1537.

164. Saddichha S. Diagnosis and treatment of chronic insomnia. Annals of Indian Academy of Neurology 2010;13(2):94–102.

165. Flachenecker P, Kümpfel T, Kallmann B, Gottschalk M, Grauer O, Rieckmann P, Trenkwalder C, Toyka KV. Fatigue in multiple sclerosis: A comparison of different rating scales and correlation to clinical parameters. Multiple Sclerosis 2002;8:523–526.

166. Téllez N, Río J, Tintoré M, Nos C, Galán I, Montalban X. Does the Modified Fatigue Impact Scale offer a more comprehensive assessment of fatigue in MS? Multiple Sclerosis 2005; 11:198–202.

167. Kos D, Kerckhofs E, Carrea I, Verza R, Ramos M, Jansa J. Evaluation of the Modified Fatigue Impact Scale in four different European countries. Multiple Sclerosis 2005;11:76–80.

168. Ritvo PG, Fischer JS, Miller DM, Andrews H, Paty DW, LaRocca NG. Multiple sclerosis quality of life inventory: Technical supplement. New York, NY: National Multiple Sclerosis Society; 1997.

169. Krupp LB, LaRocca NG, Muir-Nash J, Steinberg AD. The Fatigue Severity Scale: Application to patients with multiple sclerosis and systemic lupus erythematosus. Archives of Neurology 1989;46:1121–1123.

170. Fillion L, Gélinas C, Simard S, Savard J, Gagnon P. Validation evidence for the French Canadian adaptation of the Multidimensional Fatigue Inventory as a measure of cancer-related fatigue. Cancer Nursing 2003;26(2):143–154.

171. Hagelin CL, Wengström Y, Runesdotter S, Fürst CJ. The psychometric properties of the Swedish Multidimensional Fatigue Inventory MFI-20 in four different populations. Acta Oncologica 2007;46:97–104.

172. Schneider RA. Reliability and validity of the Multidimensional Fatigue Inventory (MFI-20) and the Rhoten Fatigue Scale among rural cancer outpatients. Cancer Nursing 1998;21(5):370–373.

173. Smets EM, Garssen B, Bonke B, De Haes JC. The multidimensional fatigue inventory (MFI) psychometric qualities of an instrument to assess fatigue. Journal of Psychosomatic Research 1994;39(5):315–325.

174. Lin JMS, Brimmer DJ, Maloney EM, Nyarko E, BeLue R, Reeves WC. Further validation of the multidimensional fatigue inventory in a US adult population sample. Population Health Metrics 2009;7:18.

175. Gentile S, Delarozière JC, Favre F, Sambuc R, San Marco JL. Validation of the French "multidimensional fatigue inventory" (MFI 20). European Journal of Cancer Care 2003;12:58–64.

176. Vercoulen JH, Hommes OR, Swanink CM, Jongen PJ, Fennis JF, Galama JM, van der Meer JW, Bleijenberg G. The measurement of fatigue in patients with multiple sclerosis. Archives of Neurology 1996;53:642–649.

177. Bültmann U, de Vries M, Beurskens AJ, Bleijenberg G, Vercoulen JH, Kant I. Measurement of prolonged fatigue in the working population: Determination of a cutoff point for the checklist individual strength. Journal of Occupational Health Psychology 2000;5(4):411–416.

178. Andrea H, Beurskens AJ, Kant I, Davey GC, Field AP, van Schayck CP. The relation between pathological worrying and fatigue in a working population. Journal of Psychosomatic Research 2004;57:399–407.

179. Janssen N, Kant I, Swaen G, Janssen P, Schroer C. Fatigue as a predictor of sickness absence: Results from the Maastricht cohort study on fatigue at work. Occupational and Environmental Medicine 2003;60:i71–i76.

180. Lee KA, Hicks G, Nino-Murcia G. Validity and reliability of a scale to assess fatigue. Psychiatry Research 1991;36:291–298.

181. Winstead-Fry P. Psychometric assessment of four fatigue scales with a sample of rural cancer patients. Journal of Nursing Measurement 1998;6(2):111–122.

182. Dittner AJ, Wessely SC, Brown RG. The assessment of fatigue: A practical guide for clinicians and researchers. Journal of Psychosomatic Research 2004;56:157–170.

183. Mota DC, Pimenta CA. Self-report instruments for fatigue assessment: A systematic review. Research and Theory for Nursing Practice: An International Journal 2006;20:49–78.

184. Iriarte J, de Castro P. Proposal of a new scale for assessing fatigue in multiple sclerosis. Neurologia 1994;9(3):96–100.

185. Iriarte J, Katsamakis G, De Castro P. The fatigue descriptive scale (FDS): A useful tool to evaluate fatigue in multiple sclerosis. Multiple Sclerosis 1999;5:10–16.

186. Armutlu K, Keser I, Korkmaz N, Akbiyik DI, Sümbüloğlu V, Güney Z, Karabudak R. Psychometric study of Turkish version of the fatigue impact scale in multiple sclerosis patients. Journal of the Neurological Sciences 2007;255:64–68.

187. Debouverie M, Pittion-Vouyovitch S, Louis S, Guillemin F. Validity of a French version of the fatigue impact scale in multiple sclerosis. Multiple Sclerosis 2007;13:1026–1032.

188. Mathiowetz V. Test–retest reliability and convergent validity of the fatigue impact scale for persons with multiple sclerosis. The American Journal of Occupational Therapy 2003; 57(4):389–395.

189. Flensner G, Ek AC, Söderhamn O. Reliability and validity of the Swedish version of the fatigue impact scale (FIS). Scandinavian Journal of Occupational Therapy 2005;12:170–180.

190. Kos D, Kerckhofs E, Nagels G, D'Hooghe BD, Duquet W, Duportail M, Ketelaer P. Assessing fatigue in multiple sclerosis: Dutch modified fatigue impact scale. Acta Neurologica Belgica 2003;103:185–191.

191. Ritvo PG, Fischer JS, Miller DM, Andrews H, Paty D, LaRocca N. Multiple sclerosis quality of life inventory: A user's manual. New York, NY: National Multiple Sclerosis Society; 1997.

192. Fachenecker P, Kümpfel T, Kallmann B, Gottschalk M, Grauer O, Rieckmann P, Trenkwalder C, Toyka KV. Fatigue in multiple sclerosis: A comparison of different rating scales and correlation to clinical parameters. Multiple Sclerosis 2002;8:523–526.

193. Armutlu K, Korkmaz NC, Keser I, Sumbuloglu V, Akbiyik DI, Guney Z, Karabudak R. The validity and reliability of the Fatigue Severity Scale in Turkish multiple sclerosis patients. International Journal of Rehabilitation Research 2007;30:81–85.

194. Valko PO, Bassetti CL, Bloch KE, Held U, Baumann CR. Validation of the Fatigue Severity Scale in a Swiss cohort. Sleep 2008;31(11):1601–1607.

195. Mills RJ, Young CA, Nicholas RS, Pallant JF, Tennant A. Rasch analysis of the Fatigue Severity Scale in multiple sclerosis. Multiple Sclerosis 2009;15:81–87.

196. Ericsson A, Mannerkorpi K. Assessment and fatigue in patients with fibromyalgia and chronic widespread pain. Reliability and validity of the Swedish version of the MFI-20. Disability and Rehabilitation 2007;29(22):1665–1670.

197. Meek PM, Nail LM, Barsevick A, Schwartz AL, Stephen S, Whitmer K, Beck SL, Jones LS, Walker BL. Psychometric testing of fatigue instruments for use with cancer patients. Nursing Research 2000;49(4):181–190.

198. van Tubergen A, Coenen J, Landewé R, Spoorenberg A, Chorus A, Boonen A, van der Linden S, van der Heijde D. Assessment of fatigue in patients with ankylosing spondylitis: A psychometric analysis. Arthritis & Rheumatism 2002;47:8–16.

199. Fürst CJ, Ahsberg E. Dimensions of fatigue during radiotherapy: An application of the Multidimensional Fatigue Inventory. Supportive Care in Cancer 2001;9:355–360.

200. Smets EM, Garssen B, Cull A, de Haes JC. Application of the multidimensional fatigue inventory (MFI-20) in cancer patients receiving radiotherapy. British Journal of Cancer 1996;73: 241–245.

201. Swaen GM, Van Amelsvoort LG, Bültmann U, Kant IJ. Fatigue as a risk factor for being injured in an occupational accident: Results from the Maastricht cohort study. Occupational and Environmental Medicine 2003;60:i88–i92.

202. Aratake Y, Tanaka K, Wada K, Watanabe M, Katoh N, Sakata Y, Aizawa Y. Development of Japanese version of the checklist of individual strength questionnaire in a working population. Journal of Occupational Health 2007;49:453–460.

203. Vercoulen JH, Swanink CM, Fennis JF, Galama JM, van der Meer JW, Bleijenberg G. Dimensional assessment of chronic fatigue syndrome. Journal of Psychosomatic Research 1994;38(5): 383–392.

204. Beurskens AJHM, Bültmann U, Kant IJ, Vercoulen J, Bleijenberg G, Swaen GMH. Fatigue among working people: Validity of a questionnaire measure. Occupational and Environmental Medicine 2000;57:353–357.

205. LaChapelle DL, Finlayson MA. An evaluation of subjective and objective measures of fatigue in patients with brain injury and healthy controls. Brain Injury 1998;12(8):649–659.

206. Drakulic SD, Jevdjic JD, Drulovic JS. Fatigue in multiple sclerosis: Correlation to clinical parameters. Medicus 2006;7(2):57–60.

207. Law M, MacDermid J, editors. Evidence-based rehabilitation: A guide to practice. 2nd ed. Thorofare, NJ: SLACK Incorporated; 2008.

208. Redding CA, Rossi JS, Rossi SR, Velicer WF, Prochaska JO. Health behavior models. International Electronic Journal of Health Education. 2000;3:180–193.

209. Sleep and sleep disorders [Internet] [cited 2011]. Available from: http://www.cdc.gov/sleep/.

210. Sleep Hygiene [Internet]; c2011 [cited 2011]. Available from: http://www.sleepfoundation.org/article/ask-the-expert/sleep-hygiene.

6

Balance Disorders

Davide Cattaneo and Johanna Jonsdottir

CONTENTS

Multiple sclerosis (MS) is an inflammatory disease of the central nervous system (CNS) that leads to destruction of myelin, oligodendrocytes, and axons. Damage within the CNS can contribute to disorders of coordination, strength, sensation, and balance among people with MS. Often, balance disorders are one of the initial symptoms of the disease.[1] Epidemiological studies have found that 23% of people with MS show evidence of cerebellar and brainstem involvement at disease onset and that this figure increases to 82% after longstanding illness.[1] Furthermore, up to 70% of people with MS experience ataxia and sensory system involvement.[2] Since balance disorders can negatively affect activities of daily living, participation, and quality of life, understanding the underlying impairments and intervening appropriately is imperative. The present chapter focuses on balance disorders among people with MS who are mobile, with or without an aid, or who, at minimum, are able to stand. After reading this chapter, you will be able to:

1. Review basic principles of balance control, highlighting differences between people with MS and healthy controls,

2. Describe the impact of balance problems on everyday life by summarizing available data about fall frequency, risk factors for falls, and fear of falling in people with MS,

3. Summarize current knowledge about balance assessments and the tools and procedures that can provide appropriate, meaningful data to guide treatment planning during the rehabilitation process, and

4. Offer implementation guidelines for balance rehabilitation interventions that aim to improve balance control in daily life and reduce frequency of falls.

6.1 Balance Control in People with and without MS

Adequate balance relies on the interaction of many body systems in relation to the demands of the environment and a specific task.[3,4] The balance system controls body posture in relation to the base of support (BoS), which is defined by the boundaries of the feet. The goal of the balance system is to keep a person's center of mass (CoM) within his or her BoS. The limits of stability beyond which the person is at risk for falling is considered to be the outer limits of the BoS. The CoM of each body segment contributes to the CoM of the whole body. Postural orientation aims to control posture and inertia of these body segments to ensure correct position of the CoM of the whole body in space.

Input from visual, somatosensory, and vestibular systems (see Figure 6.1) provide information about the body in space. Input is processed and integrated by sensory strategies that give more weight or credence to the most reliable information for a given task and environmental condition. Next, a body schema is generated that is defined as "a combined standard against which all subsequent changes of posture are measured."[5] Changes in balance control in MS may be due to reduced ability of the visual, somatosensory, or vestibular systems to collect reliable information; to impaired central integration[6]; or both, which can lead to deficits in postural reaction.

Once information has been gathered through the visual, somatosensory, and vestibular systems and integrated centrally, adequate motor responses are needed to maintain balance. These responses are organized into motor strategies. Because of muscle weakness and

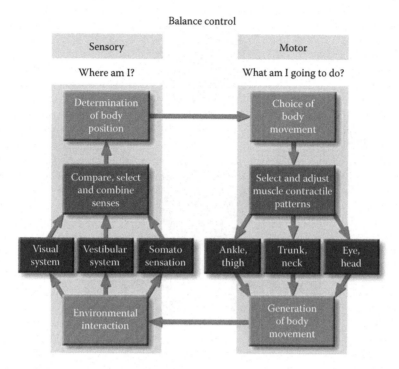

FIGURE 6.1
Sensory-motor control of balance. (Courtesy of Neurocom.)

spasticity, people with MS may have deficits in their ability to execute motor strategies, which may further compromise their ability to maintain balance.

Since people with MS can present with a number of impairments and disabilities affecting the balance control system, a theoretical framework can facilitate clinical reasoning and enable a clinician to focus on the most important features of balance control during rehabilitation. Experimental evidence has identified five key aspects of balance control (see Figure 6.2), each of which is summarized in the following paragraphs.

6.1.1 Biomechanical Constraints and Musculoskeletal Components

Understanding difficulty in balance control requires basic knowledge of the body's mechanical characteristics since the dimensions of the BoS, the position of the CoM, and the mobility and strength of ankle plantarflexors greatly influence body stability. Biomechanical constraints on balance performance are represented by physical characteristics of body segments, muscles, and so on. Biomechanical variables, such as a person's height and weight, influence the motion of the CoM.[7] The CoM is located at about 55% of a person's height when he or she is in an upright position. The average position of the CoM on the ground falls about 25 mm in front of the ankle. Since the human body is inherently unstable, the CoM sways randomly around this average position, although the small amplitude of these sways is not visually detectable. They are approximately 5 mm in the sagittal plane and 2.5 mm in the frontal plane when a person's eyes are open and he or she is standing on a firm surface.

The dimension, shape, and nature of BoS influence balance performance greatly. Feet width and the area of the BoS influence the direction and amount of CoM sway. In

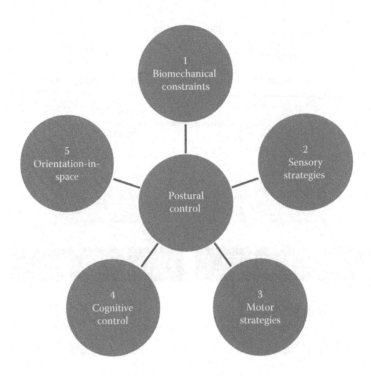

FIGURE 6.2
Postural control system. (Adapted from Shumway-Cook A, Woollacott M. Motor control: Theory and practical applications. 2nd ed. Philadelphia, PA: Lippincott Williams and Wilkins; 2001.)

normal stance, the direction and amount of sway are small and mainly in the antero-posterior direction. In unipedal stance, sway is greater and mainly in the medio-lateral direction. People with balance impairments and people with MS tend to widen their BoS in static and dynamic tasks. They tend to show poor balance when standing with feet together[8] or in tandem or stride position.[9] Moreover, people with significant balance disorders tend to increase the width of their BoS during gait and to reduce stride length[10] as well as hip, knee, and ankle extension to reduce the distance between the CoM and the BoS.

Maintaining an upright position is the result of active muscle effort. In quiet standing position, the line of gravity passes in front of the ankle and knee joints, slightly behind the hip joints, and in front of the thoracic and neck segments (see Figure 6.3). This line illustrates the necessity of muscle activation to control the torque on the body that results from the natural pull of gravity. Calf muscles and hip and trunk extensors actively maintain the body in upright position. Weakness of other muscular groups (e.g., quadriceps, hip abductors, ankle dorsiflexors) may cause imbalance during activities such as gait and stair climbing. Muscle weakness is a common impairment in people with MS.[11] Nevertheless, the relationship between weakness and balance has been assessed in only one study. Chung, Remelius, Van Emmerik, and Kent-Braun[12] studied knee extensors and ankle dorsiflexors in people with MS who had mild to moderate impairment (EDSS range: 2–6). The researchers found no difference between healthy controls and people with MS except for peak power of knee extensors, which reflects the difficulty that people with MS have contracting muscles at high velocity. More important is that the researchers did not find correlations

FIGURE 6.3
Relationship between gravity vector and body joints.

between lack of muscle power and asymmetry of posture. Apparently, individuals with reduced power in one lower limb did not shift their center of mass toward the sound side.

Spasticity is a common symptom in MS that results from neuronal damage in the long fiber tracts of the CNS (see Chapter 7). Although there is no evidence of a causal effect of spasticity on balance control, spasticity and balance are correlated when measured by clinical scales.[13,14] Two papers have addressed the correlation between spasticity and balance in quiet standing.[15,16] In particular, Sosnoff et al.[15] categorized spasticity using the ratio between H reflex and M wave, both of which are spinal reflexes indicative of alpha motor neuron excitability that occur after electrical stimulation of peripheral nerves. Sosnoff and colleagues compared balance performance of people with MS with moderate versus low levels of spasticity in quiet standing using a stabilometric platform and found that those with moderate spasticity had increased velocity, amplitude, and area of sway in medio-lateral plane.

Fatigue is another common symptom in MS that may influence postural control. Fatigue may be defined as a physical or mental weariness resulting from exertion and the inability to continue activities or exercise with the same level of intensity, which results in a deterioration of performance.[17] In 28–40% of people with MS, fatigue is the most disabling symptom and also one of the most common. According to Lapierre and colleagues,[18] 78–90% of people with MS experience fatigue, and 40% were fatigued every day across a 30-day period. Although increased fatigability is believed to lead to balance disorders, no correlations have been found between fatigue and fall risk, and people with MS do not report increased fatigue prior to falls.[14] Furthermore, among people with MS who have moderate impairments, no differences were found in standing balance between morning and afternoon, even though subjects reported a marked increase in perceived level of fatigue in the afternoon.[9]

6.1.2 Sensory Strategies

To maintain balance, the CNS must perceive body position and movement (see the sensory section of Figure 6.1), which is achieved by (1) body receptors that collect somatosensory, vestibular, and visual information, and (2) the ability of the CNS to select and integrate reliable information about the body's position in space and its position in relation to the external environment. Consequently, balance disorders in people with MS may be due to reduced ability of the body receptors to correctly collect information or the ability of the CNS to interpret that information.

The somatosensory system is made up of several receptors (e.g., muscle spindles, cutaneous and joint receptors) and provides information that allows the CNS to monitor the position of body segments, movement, and internal and external forces acting on the body.[19] The vestibular system provides information regarding inertial forces and the position and movement of the head with respect to gravity. The input from this system provides information about head–trunk position via neck proprioceptors. Finally, visual inputs provide information about the position and motion of the head and eyes with respect to the surrounding environment. They also provide information about the altitude of the objects in the external world relative to the ground, thus increasing the individual's perception of verticality.

The specific contributions of these three sensory systems to balance control across different tasks is unknown. There is also no agreement on the importance of individual sensory inputs for maintaining upright balance. It has been suggested that a person standing in a stationary, visual scene on a fixed-support surface relies on all three sensory systems equally, while others argue that vision may be more critical.[3,20] Conversely, the sensory systems involved in postural stance become more or less important as changes occur in the environmental conditions or task demands. For example, in case of perturbation of the BoS (see Figure 6.7), people appear to rely primarily on the proprioceptive system for balance.

Even if sensory systems are not excessively damaged, people with MS may experience balance disorders because of difficulty selecting and integrating sensory information about the task and the environment. These processes allow the balance system to adapt its output to a variety of demands. To assess the ability of the CNS to select and integrate sensory information, the person is usually required to maintain upright balance in different sensory contexts, for example, while standing in an upright position on a foam mat. In this condition, somatosensory inputs are less reliable and visual-vestibular information tends to be preferred.

Efficiency of sensory systems and sensory strategies in MS have been investigated in several studies using a stabilometric platform in conditions in which the sensory information available is manipulated.[21–23] These devices quantify postural sway during quiet standing by displacements of the center of foot pressure (CoP). In quiet stance, the CoP sway broadly reflects body sway. Cattaneo and colleagues[23] demonstrated that more than 50% of people with MS had abnormal stabilometric assessment scores in quiet standing, as reflected by body sway or velocity of sway values that were two standard deviations higher than the mean values of healthy subjects, even in the easiest sensory condition (eyes open, firm surface).[23] Balance performance was even more likely to be abnormal when sensory inputs were altered (e.g., eyes closed, foam pads under feet), which also revealed impairments in sensory strategies. Figure 6.4 illustrates the results of stabilometric assessment in the upright position of 53 people with MS. Balance performance was assessed by measuring the length of CoP displacement over a 30-s period. Greater instability produced higher values of CoP displacement, which reflected worse balance control.

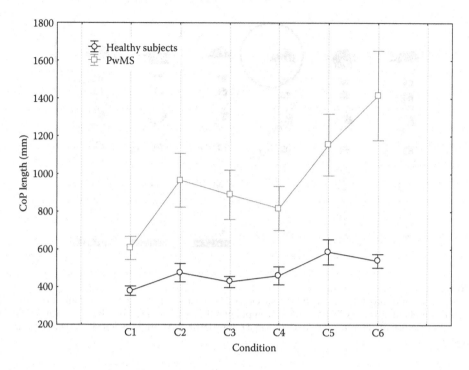

FIGURE 6.4

CoP displacements in upright balance performance in six sensory conditions. C1: eyes open, firm surface; C2: eyes closed, firm surface; C3: dome, firm surface; C4: eyes open, compliant surface; C5: eyes closed, compliant surface; C6: dome, compliant surface. Vertical bars refer to 95% coefficient interval. (From Cattaneo D, Jonsdottir J. Multiple Sclerosis 2009;15(1):59–67. With permission.)

In a study by Cattaneo and colleagues,[23] balance was assessed under six sensory conditions (see Figure 6.5): eyes open (C1), eyes closed (C2), and eyes open with a visual conflict dome on a firm surface (C3). This latter condition was achieved by a surrounding visual device (depicted in Figure 6.5 as black circles) that swayed consensually with subject's body sway to provide a mismatch between visual and somatosensory–vestibular information. The other three sensory conditions (C4–C6) were the same but with the addition of foam pads placed under the subject's feet to reduce the reliability of somatosensory information.

The impact of impairment of the sensory strategies on balance was evident with the alteration of a single sensory input (e.g., eyes closed, condition C2), which led to an increase in sway or velocity of sway and abnormal balance performance scores for 80% of the subjects. These findings reflect the difficulty the participants had in switching from visual cues to somatosensory and vestibular cues. The alteration of two sensory inputs (e.g., eyes closed, compliant surface, condition C5) led to a sharp increase in abnormal scores for almost all participants, and many of them fell during the test. The high number of subjects with abnormal scores in this condition reflects the difficulty that people with MS have in using vestibular information and in maintaining balance with only one reliable sensory cue.

Several studies have been carried out to assess motor control after postural perturbations (for a review, see reference 6). People with MS have problems maintaining their balance when the platform support surface is moved suddenly, which indicates that they have

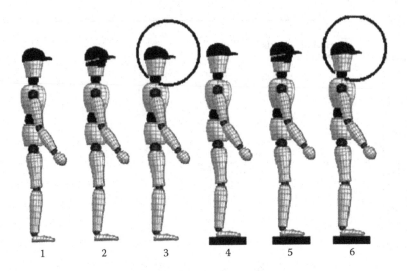

FIGURE 6.5
The six sensory conditions. C1: eyes open, firm surface; C2: eyes closed, firm surface; C3: dome, firm surface; C4: eyes open, compliant surface; C5: eyes closed, compliant surface; C6: dome, compliant surface. Vertical bars refer to 95% coefficient interval. (From Cattaneo D, Jonsdottir J. Multiple Sclerosis 2009;15(1):59–67. With permission.)

delayed automatic postural responses. These studies also found a relationship between the postural response delays and delays in spinal somatosensory conduction, measured by somatosensory-evoked potential latencies. This relationship suggests a possible involvement of the somatosensory network, which may contribute to slowed or incorrect perception of body segments and inadequate or delayed motor responses.

During balance assessment, instability after postural perturbation or instability in altered sensory conditions should lead therapists to consider treatments for improvement of sensory strategies (see Section 6.5).

6.1.3 Motor Strategies and Synergies

The human body is a chain of unstable segments connected with joints. To maintain upright balance, the CNS must implement motor strategies to control these segments (see the motor section of Figure 6.1). Clinically, it is important to understand how the CNS plans these strategies and how the performance of people with MS compares to healthy controls. Indeed problems with movement execution should lead therapists to consider rehabilitation focusing on the impaired motor strategy and treatment tailored to that specific balance disorder.

Motor strategies involved in the control of upright balance and the avoidance of falls have been extensively studied.[24–26] With respect to upright balance, three different strategies have been identified—ankle strategy, hip strategy, and stepping strategy.

6.1.3.1 Ankle Strategy

The ankle strategy is characterized by pendulum-like movements pivoting about the ankle joints; the other joints are thought to be motionless (see Figure 6.6). This strategy is normally adopted during daily-life activities, and it is mainly used when an individual is on a firm

FIGURE 6.6
The ankle strategy.

surface with the CoM close to the center of the BoS. This strategy uses forces developed at the ankle level to keep the CoM inside the BoS. The movement of the pendulum is quite small and not visually detectable. Since the ankle strategy requires a firm base of support and the ability to generate adequate force at ankle joints, people with MS who have bilateral weakness of ankle plantarflexors and impaired proprioception may experience difficulties carrying out this strategy.

6.1.3.2 Hip Strategy

The hip strategy uses shear forces to keep the CoM inside of the BoS (see Figure 6.7). It is characterized by a large bending at the hips, which results in a repositioning of the CoM. This strategy is used mainly when the CoM is close to the limits of stability or when the BoS is small. The hip strategy requires fast movements of axial segments and the collection of reliable visuo-vestibular information. Therefore, this strategy can be difficult for people with MS who have vestibular disorders. The hip strategy is sometimes seen in individuals with distal muscle weakness and in people who have cerebellar disorders. In the latter case, a high frequency of postural correction at hip level can be observed along with tremor of trunk and head.

6.1.3.3 Stepping Strategy

The stepping strategy is characterized by fast movement of the lower limb in the direction of the falling CoM (see Figure 6.8). In contrast to the ankle and hip strategies, which tend

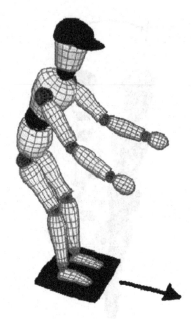

FIGURE 6.7
The hip strategy is used to control sudden forward movement of BoS.

to maintain the CoM inside the BoS, the stepping strategy moves the BoS to the location where the CoM is going to fall in order to regain balance. It is used when the CoM is close to the limits of stability or is already outside the BoS.

The stepping strategy demands fast and prompt movements of the lower limb and the correct altitude of the swinging foot with respect to the ground. Consequently, weakness

FIGURE 6.8
The stepping strategy.

of hip flexors and spasticity of calf muscles may reduce the reliability of this strategy. People with MS tend to rely more heavily on the stepping strategy when the CoM is closer to the limits of BoS, which avoids the need to use the hip strategy and a fast movement of the trunk and pelvis. However, during the use of stepping strategy, people with MS tend to take multiple, shorter steps with an increased amplitude of the distance between feet. This suggests difficulty in the control of medio-lateral sway. To our knowledge, no studies have been done to assess and specifically treat motor strategies in MS and, thus, only clinical information is available.

6.1.4 Cognitive Control

To maintain an upright stance, the CNS must take into account environmental hazards and prior experiences to select the appropriate motor response for a particular environmental condition and task. In healthy subjects, sensory collection, interpretation, and action work together so that balance can be maintained with a minimum of conscious attention. This automatic process requires cognitive resources and the involvement of cortical areas. Current hypotheses suggest that the CNS uses the information learned from past experiences to define a modified neuromotor state to prime postural response strategies, thereby anticipating incoming balance hazards and optimizing postural responses for a given environmental context.[27]

The need for a more cognitive and voluntary control of balance can come into play after damage to the CNS; this increase of cognitive control increases the demand for adequate levels of attention, memory, and problem-solving skills. Estimates suggest that 43–65% of people with MS have cognitive deficits in the areas of attention, information-processing speed, working memory, visuospatial memory, and executive functions,[28] all of which have been identified as fall risk factors in other populations.[14]

The involvement of higher-order cognitive functions was addressed recently in a paper focusing on balance control.[29] Authors assessed subjects' walking speed and other gait variables related to falls under single- and dual-task conditions. In the dual-task condition, individuals had to walk while listening to sequences of digits and were required to repeat each sequence in order.[29] Compared to healthy controls, people with MS showed greater decrements in walking speed in sessions in which the dual task was added. Three hypotheses have been put forward to explain this reduction of speed:

1. Disorders in sensory-motor functioning among people with MS lead to conscious control of tasks requiring attention, but since the working memory system has limited capacity, the dual-task reduces attention devoted to movement control resulting in a worsening of performance.

2. Working memory may be reduced in people with MS, and this may result in an overload of working memory even for previously normal levels of motor and cognitive content.

3. Attentional deficit may lead to difficulty in controlling concurrent demands.

6.1.5 Orientation-in-Space

The term "orientation-in-space" refers to an individual's ability to recognize the position of body segments in space and to represent the spatial properties of the environment in order to locate objects and navigate effectively.[30,31] To maintain balance, multimodal sensory information must be integrated and interpreted with respect to a stable frame of reference that is relevant to the postural task (see Figures 6.1 and 6.2). Motor and sensory strategies and cognitive processes are involved in this integration and provide the ability to orient. While there is limited knowledge about this aspect of balance among people with MS, problems with orientation-in-space may explain why people who appear to have normally functioning sensory systems experience dizziness. The Dizziness Handicap Index (DHI) may provide information in this regard (see Section 6.3).

6.2 Impact of Balance Problems on Everyday Life

Reductions in postural control (see Figure 6.2) can compromise the sensory-motor flow depicted in Figure 6.1 and lead to an increased level of unbalance during the execution of static and dynamic tasks. At the physical level, impaired postural control may lead to increased energy costs, musculoskeletal pain, decreased flexibility, and deconditioning, and create increased difficulties with activities of daily living. These problems may contribute to falls and fear of falling,[32] decreased overall activity, and participation restriction.[33] People with balance disorders are often perceived by others as being drunk, which can contribute to social isolation. Consequently, the assessment of fall frequency and fall risk factors may allow clinicians to develop rehabilitation programs that improve balance and subsequently quality of life.

6.2.1 Mobility Disorders, Fall Frequency, Fall Risk Factors, and Injurious Falls

6.2.1.1 Mobility Disorders

In two published studies, Confavreux et al.[34,35] addressed the impact of MS on mobility in more than 1800 people. The median times from the onset of MS to the onset of mild walking disability (EDSS score = 4) was 11.4 and 0.0 years for the relapsing–remitting and the progressive form of MS, respectively. Moderate walking disability (EDSS score = 6) arose after 23.1 and 7.1 years, respectively. After 33.1 and 13.4 years, on average, people with these types of MS must use a wheelchair.

Iezzoni et al.[36] conducted a 30-min telephone survey that focused on mobility distress with 703 community-dwelling people with MS. The most frequent areas of mobility distress were difficulties with walking (79.3%) and standing (69.9%), the increased effort needed to walk (59.0%), and needing to hold onto furniture when walking indoors (40.0%). These functional disorders likely contributed to the adoption of mobility aids since 60.5% of interviewed subjects used at least one mobility aid. Among this group, 56.7% used a cane, 63.4% used a manual wheelchair, 36.7% used a power wheelchair, and 32.2% used a scooter (note: multiple responses allowed).

6.2.1.2 Fall Frequency

In the past 10 years, three studies (two retrospective, one prospective) have examined frequency of falls and fall risk factors among people with MS. The largest study was carried out by telephone interview on a sample of 1089 people with MS, aged 45–90 years. In this group, 52.2% of participants reported a fall in the past 6 months.[37] The second retrospective study[13] assessed a random sample of 50 people (10% of a regional population) through interviews with people who had MS and caregivers in their own homes. Researchers reported that 52% of subjects had fallen in the 2 months prior to the assessment. In the prospective study,[14] researchers found that over a 3-month period, 63% of 76 subjects fell. Results from these three studies demonstrate that people with MS experience higher fall rates compared to older adult populations, of whom one-fourth to one-third experience a fall over 12 months.[38] Compared to people with other neurological pathologies, people with MS have higher fall rates than people with stroke, but lower than people with Parkinson's disease.[39] In all of these studies, a fall was defined as an unintentional event that resulted in a person coming to rest on the ground or on another lower level.

6.2.1.3 Risk Factors

Clinicians are often interested in knowing fall risk factors in order to identify individuals who may be prone to fall. Balance disorders or the presence of other risk factors should trigger a clinician to design a treatment plan that includes fall prevention strategies (see Section 6.5). A risk factor is any characteristic of a person with MS, a situation, or the environment that increases the likelihood that a person will eventually fall.

A study by Cattaneo et al.[13] collected information regarding the use of a walking aid (cane or walker), walking skills (Ambulation Index), and static balance (Equiscale). Balance disorders were associated with fall risk: an increase of 1 point on the Equiscale test increased the fall risk by 35%. Similarly, work by Finlayson and colleagues[37] found that subjects who were concerned about their balance and mobility experienced greater likelihood of a fall. This study also illustrated the heterogeneous nature of falls because fall risk was also increased by cognitive symptoms, bladder dysfunction, worsening of symptoms over the past year, and being male. In contrast, Nilsagård[14] found that falls were not correlated with balance disorder since odds ratios, calculated from tests assessing balance skills [Berg Balance Scale (BBS), Timed Up and Go (TUG) and 12-item Multiple Sclerosis Walking Scale), were close to 1. This finding may be due to low levels of disability among the sample, which had a median EDSS score of 4 (nonfallers) and 5 (fallers). This low level of disability resulted in a ceiling effect on the BBS. Other risk factors found to be associated with falls by the Nilsagård team[14] were history of falling, use of walking aids, EDSS score, and proprioceptive disorders.

6.2.1.4 Fall Circumstances

The circumstances of a fall may provide information about the nature of falls and on preventive actions that can be adopted to avoid similar situations in the future. In a study by Nilsagård and colleagues,[14] researchers found that 25% of falls occurred during kitchen or cleaning activities and 43% occurred in the afternoon.

6.2.1.5 Injurious Falls

Some MS symptoms, such as dysphagia and loss of balance, have the potential to be life threatening. High susceptibility to trauma is a combination of high incidence of falls, failure of protective mechanisms (e.g., arm protection), and other factors such as osteoporosis. Peterson and colleagues[40] investigated the risk for fall-related injuries among 354 people with MS. In the study, injurious falls were defined as "falls that lead a person to seek medical attention from a physician, medical office, hospital emergency department, or paramedic."[40] Study results indicated that half of the participants received medical care for a fall-related injury. Hip and lower-limb fractures were the most common injuries (52% of the total number of fractures), followed by soft tissue injuries (e.g., bruises, lacerations not requiring stitches, strains, sprains, and dislocations). The risk factors associated with increased likelihood of experiencing an injurious fall were fear of falling and osteoporosis.

6.2.1.6 Fear of Falling and Balance Confidence

Even in the absence of falls or injuries, balance disorders can lead to fear of falling and reduced balance confidence. Fear of falling has been defined as low perceived self-efficacy at avoiding falls during essential activities of daily living.[41] The question "Are you concerned about falling?" is usually posed in studies addressing fear of falling.[32] Since people with MS tend to be sedentary and to function below their physical capacity,[42] fear of falling may be a factor that further reduces an individual's level of physical activity, which may then contribute to deconditioning and further fall risk. Among older adults, activity curtailment due to fear of falling actually increases fall risk.[43] Similarly, Finlayson et al.[37] demonstrated that fear of falling is associated with falling among people with MS. Peterson and colleagues studied a cohort of 1064 people with MS in which 63.5% of the subjects were concerned about falling.[32] Within the group who reported fear of falling, 82.6% reported curtailing activity because of their fear. Women, individuals who reported a fall in the previous 6 months, those who used walking aids, and individuals with high symptom interference were more likely to report fear of falling.

In comparison with fear of falling, balance confidence reflects an individual's evaluation of his or her own balance skills. The Activities-specific Balance Confidence (ABC) scale has been used in several studies to measure balance confidence.[44] The ABC scale asks subjects to respond to 16 items by pointing to a numerical value between 0% and 100%. The stem for each item is: "How confident are you that you will not lose your balance or become unsteady when you ..." Examples of items include walking around the house, walking up and down stairs, bending down and picking up a slipper from the floor, standing on a chair and reaching for something, and stepping onto or off of an escalator while holding parcels. In a study of 63 ambulatory people with MS, participants reported limited balance confidence: the ABC score was 51.4 points (±25.8 SD) out of a total score of 100 points. In addition, ABC scores were significantly different between fallers and nonfallers; the two groups scored 61.1 (±25.3) and 36.9 (±19.2) points, respectively.[45]

6.3 Balance Assessment Strategies

As already described, the balance system is a complex interaction of subsystems that modulate the sensory-motor state according to task and environmental demands. Consequently, the assessment of balance systems requires an interdisciplinary approach that takes into account functional, environmental, cognitive, psychological, and social variables. The assessment procedure can be divided into four components: a general history, performance-based assessment of skills, assessment of use of sensory-motor skills during the execution of balance tasks, and assessment of impairments that may influence balance performance.

6.3.1 General History

The first component of a balance assessment involves gathering information about the person's abilities to perform activities of daily living. It is important to collect both the person's and caregiver's point of view with respect to the social, psychological, and environmental variables that may be related to or influence balance disorders. Particular attention should be paid to collecting information about the circumstances, place, and time of any reported falls; what the person was doing at the time of the fall; and whether he or she was experiencing any symptoms such as dizziness or oscillopsia. Table 6.1 shows factors that may be included in the general assessment.

Several self-report scales are available to gather general information during this component of the assessment process. Self-report questionnaires allow a structured collection of information regarding different aspects of balance disorders. Reliability of collected data depends on memory, cognitive and psychological factors, and speech disorders; therefore, the presence of a caregiver is sometimes useful for completing the questionnaire. The following scales were developed for older adults but have also been used in studies involving people with MS: the Falls Efficacy Scale (FES), the ABC scale, the Falls Control Scale (FCS), and the Falls Management Scale (FMS).

The FES rates subjects' degree of confidence in carrying out several everyday indoor activities without falling. It comprises 16 items assessing walking on slippery, uneven, or sloping surfaces; visiting friends or relatives; going to a social event; or going to a place with crowds. The items are scored on a 4-point Likert scale ranging from not at all sure (1 point) to very sure (4 points). Higher values reflect greater confidence in being able to perform activities without falling. Among older adults, a score of lower than 70 on the FES is thought to indicate a significant fear of falling.[41]

The ABC scale requires the subject to rate his or her perceived level of balance confidence while performing 16 daily-living activities.[44] The ABC scale inquires about outdoor as well as indoor activities. The potential total score is 100 points, which reflects a high level of confidence. The test–retest reliability of the ABC scale among people with MS is reported at 0.92,[45] and associations with several static and dynamic balance tests have been reported with correlations ranging from 0.38 to 0.54.[46] A total score of fewer than 75 points is considered abnormal in subjects aged 65.6 ± 7.6 years.[47]

Clinicians interested in understanding the subject's level of disability and handicap could consider using the DHI. It is a multidimensional self-administered scale that quantifies the level of disability in physical, emotional, and functional subscales. Scores range

TABLE 6.1

Factors to Consider When Gathering General History for Balance Assessments

Environmental and social factors	*Cognitive and psychosocial factors*
• Activities of daily living and activity curtailment	• Attention and concentration
• Residence	• Disorientation
• Environmental hazards	• Confusion
• Presence of caregiver	• Inability to follow directions
• Equipment in use	• Fear of falling
• Work and lifestyle	
Functional and impairment factors	*Medical considerations*
• Balance deficit	• Generalized pain
• Gait deficit	• Orthostatic blood pressure
• Mobility impairment	• Anxiolytic drugs and antidepressants
• Muscle weakness, spasticity, tremor	• High alcohol consumption
• Visual screening test	• Low/high body mass
• Sensory impairment	
• Incontinence	

from 0 to 100, where 100 means a high level of disability and handicap. The test–retest reliability of the scale among people with MS was 0.90,[45] and weak associations with static and dynamic balance tests were reported with correlations ranging from 0.32 to 0.39.[46]

The final two scales are useful for understanding the subject's perceived ability to avoid falls and cope with them if they occur. The FCS requires the person to rate his or her sense of control over preventing falls. It includes four items on a 5-point scale, where 1 indicates "strongly disagree" and 5 indicates "strongly agree." Higher scores reflect a greater sense of control over fall prevention.[48] The FMS requires the person to rate his or her ability to manage risk of falls or actual falls. FSM is a five-item scale, each item ranging from 1, indicating "not sure at all," to 4, indicating "very sure." Higher scores reflect greater confidence in managing risk of falls.[48]

6.3.2 Performance-Based Assessment of Skills

The second assessment component focuses on the person's skills. Balance disorders can be thought of as difficulties in task execution during activities of daily living. For example, difficulties going to the mall can be due to difficulties moving from the wheelchair to car seat or walking and turning the head to look at displays. Performance-based tests can help clinicians to assess the person's skills related to balance control.

Clinical observation can provide a great deal of information. In general, subjects with balance disorders in an upright position show a wider BoS, flexion at ankle-knee and hip level, and the trunk is often bent forward. The gait pattern is characterized by three factors:

1. Increased hip and knee flexion and ankle plantarflexion at heel contact to reduce the distance between the CoM and the ground[10]
2. Reduced speed of progression, shorter strides, prolonged double support to increase stability at the expense of speed
3. Contraction of agonist and antagonist muscles to increase joint stiffness

In addition, reduced head and trunk movements are generally observed, although some subjects show ataxia of axial and distal segments.

In addition to observations, clinical tests can provide further information. A wide variety of tasks can be addressed depending on the subject's difficulties in carrying out activities of daily living (based on information gathered in the general history). There are several scales that can be used that allow for standardization of assessment and the possibility to score and compare subjects' performance over time. For example, the Equiscale[49] and the Six-Step Spot Test[50] were developed to address balance disorders among people with MS. Several other commonly used tools are described in the following paragraphs.

The BBS and Dynamic Gait Index (DGI) can be used to rate a subject's performance in tasks related to static and dynamic balance. Both scales provide a final total score; however, the analysis of each item can also be useful for implementing a more tailored treatment plan to address balance disorders. The BBS rates performance from 0 (cannot perform) to 4 (normal performance) on 14 items (see Table 6.2).[51] The scale is a valid and reliable instrument for people with MS who are able to stand unsupported for at least 30 s.[45,46] The total score of the BBS correlated with tests assessing dynamic balance and balance perception. The reliability of the BBS was high with intraclass correlation coefficients of 0.96 for intrarater and interrater reliability. Moreover, the BBS discriminated between fallers and nonfallers. A cut-off score of 44 is an established criterion to identify people with MS at high risk of falling.[45]

TABLE 6.2

Items of the Berg Balance Scale

1. Sitting to standing
2. Standing unsupported
3. Sitting unsupported
4. Standing to sitting
5. Transfers from bed to chair
6. Standing with eyes closed
7. Standing with feet together
8. Reaching forward
9. Retrieving object from floor
10. Turning to look behind
11. Turning 360 degrees
12. Placing alternate foot on stool
13. Standing with one foot in front
14. Standing on one foot

Source: Adapted from Berg KO. Physiotherapy Canada 1989;41(6):304–311.

Figure 6.9 reports mean performance of a sample of 102 subjects with mild-to-moderate MS (mean BBS total score: 47.6 ± 8.4). People with MS showed poorer performance in items requiring upright balance on a small base of support (items 13 and 14) and during weight shifting in the medio-lateral plane (item 12). Sensory and vestibular impairments were highlighted by poor performance in tasks involving head (item 7) and whole-body rotation (item 11). Individual performances are depicted in Figure 6.10. The high variability

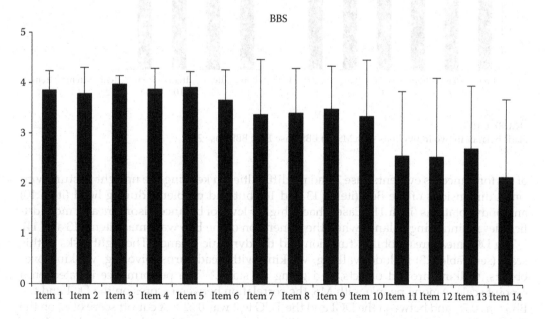

FIGURE 6.9
Berg Balance Scale among people with MS.

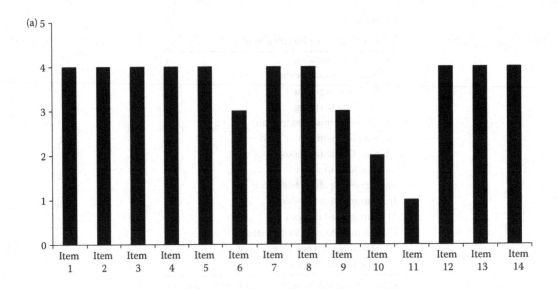

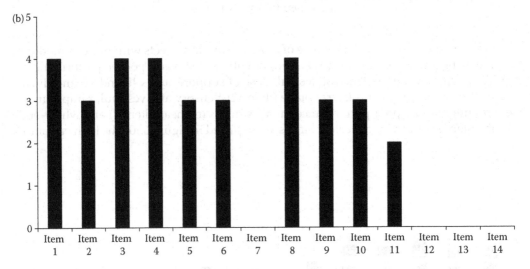

FIGURE 6.10
Static balance in two individuals with MS. (a) BBS Case 1. (b) BBS Case 2.

of performances is evident: Case 1 had no difficulties in keeping the upright posture with small dimensions of the BoS (items 13 and 14), but had problems during head (item 10) and body rotations (item 11). Case 2 had a higher level of balance disorders and more difficulties maintaining balance when the dimension of the BoS was small (items 13 and 14).

The DGI measures mobility function and the dynamic balance. The eight tasks of this scale (see Table 6.3) include walking, walking with head turns, pivoting, walking over objects, walking around objects, and going up stairs.[52] The performance is rated on a 4-point scale. Among people with MS, the correlation between scores on the DGI and the BBS was 0.78, and between the DGI and the TUG test was 0.62.[45] A cut-off score of 12 on the DGI has been established as a criterion to identify people with MS with a high risk of falling. Reliability was good: the intraclass correlation coefficients were 0.85 and 0.94, respectively,

TABLE 6.3

Items of the Dynamic Gait Index

1. Gait, level surface
2. Change in gait speed
3. Gait with horizontal head turns
4. Gait with vertical head turns
5. Gait and pivot turn
6. Step over obstacle
7. Step around obstacles
8. Stairs

Source: Adapted from Shumway-Cook A, Woollacot M. Motor control: Theory and practical applications. Baltimore, MD: Williams & Williams; 1995.

for intrarater and interrater reliability.[46,53] Figure 6.11 depicts mean performance of a sample of 102 subjects with mild to moderate MS (mean total DGI score: 15.5 ± 6.5). Dynamic balance was impaired, especially during head rotation in the sagittal and horizontal planes (items 3 and 4). Those findings, combined with the findings obtained with measurements of head motion during walking,[45] suggest impairment in trunk–head control, difficulty in maintaining balance when visuo-vestibular information is challenged by head movements, or both.

Two additional tests—the TUG and the 6-min walk test (6MWT)—are also commonly used during balance assessment. Both provide fast information about mobility and dynamic balance skills although no information regarding the quality of task execution is recorded. The TUG evaluates dynamic balance. It requires the subjects to stand up from a chair, walk 3 m, turn around, walk back to the chair, and be seated. The subject is timed from the moment he or she lifts the pelvis from the chair until he or she returns with the pelvis in the chair.[54] The validity and reliability of the instrument were tested on a cohort

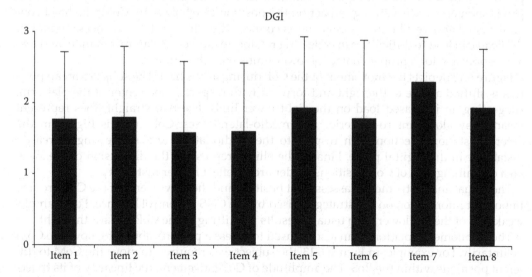

FIGURE 6.11
Dynamic Gait Index in 102 people with MS.

of people with MS.[45] The results showed good correlation between TUG and BBS (–0.61). However, TUG was unable to distinguish between fallers and nonfallers. Good interrater (0.96) and intrarater (0.93) reliability have been found in older adults,[54] but no data on people with MS are currently available. For the 6MWT, the subject is asked to walk for a period of 6 min, and the distance walked is recorded. It has been validated as an outcome measure for people with MS,[55] and it has been found to be highly reliable for people with mild to moderate MS (EDSS 2–6.5).

For persons with balance disorders in sitting, the Trunk Impairment Scale (TIS) has been validated among people with MS. The TIS assesses static and dynamic sitting balance and trunk coordination in a sitting position. Each item is rated on a 2-, 3-, or 4-point ordinal scale. The total score ranges between 0 (worst score) and 23 (best score), indicating a good performance. Verheyden and colleagues[56] reported a correlation of –0.85 between the scores of the TIS and Expanded Disability Status Score. Reliability is also good—the intraclass correlation coefficients were 0.97 and 0.95 for interrater and test–retest agreement, respectively.

6.3.3 Assessment of Sensory-Motor Strategies Used to Accomplish Tasks

The third assessment component focuses on evaluating the person's use of sensory-motor strategies during the execution of balance tasks. Motion analysis systems and stabilometric platforms can be used to assess balance skills during static and walking tasks. Motion analysis has the advantage of quantifying the quality of movement and the relevant motion of associated body segments. Stabilometric platforms can be used to understand balance performance, the sensory-motor strategies used, and to identify the underlying sensory-motor impairments. These platforms can be used to assess upright balance in static and dynamic conditions. In static conditions, the platform surface is held motionless, whereas in dynamic conditions the platform surface is tilted or translated under the subject's feet.

Static-platform assessment usually requires subjects to stand motionless for 20–60 s. The device measures the position of center of pressure (CoP) that in quiet standing is similar to position of the CoM, indicating the position of subject's weight on the BoS. The trunk and pelvis segments have a strong impact on the position of whole-body CoM; the head contains visual and vestibular systems and is connected to the trunk by the neck; tremor of the head can lead to difficulties in collecting reliable visuo-vestibular information; and feet are responsible for a proper contact of lower limbs with the ground.

Figure 6.12 depicts the movement of the CoP during a 30-s trial. The subject's mean position is shifted a little to the right and forward with respect to the center of the platform, suggesting an increased load on the right lower limb. The two straight lines represent mean sway along antero-posterior and medio-lateral axes: CoP sway is higher in the antero-posterior direction with respect to the medio-lateral one, suggesting increased instability in the sagittal plane. Finally, the ellipse represents the dimension of the sway area containing 95% of CoP positions; wider areas reflect higher instability.

The visual and instrumental assessment position and the movements of the CoP provide important information on the strategies used by the CNS to control balance. For example, weakness of the left lower limb usually results in shifting of the CoP toward the right.

The stabilometric platform can also be used to assess a person's ability to move the CoM voluntarily. For example, in Figure 6.13, the subject was required to move the CoM to different positions within the BoS. The amplitude of CoP movement, the linearity of its trajectory, and the amount of sway in particular zones of BoS are usually assessed to understand the ability to explore the whole BoS, promote correct muscular synergies to control the

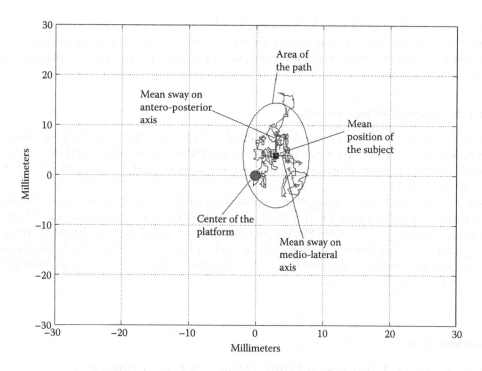

FIGURE 6.12
Stabilometric assessment of body sway in a subjects with MS. CoP path for a 30-s trial.

trajectory of the CoM, and recruit muscles to keep an upright posture when the CoM is close to the boundaries of the BoS. Stabilometric platforms have been used to address strategies for controlling balance during sudden movement of the BoS (e.g., see Nashner[57]) or in specific sensory contexts.

Sensory strategies are usually assessed by the Sensory Organization Test (SOT).[25] This test, usually performed on a stabilometric platform, has been adapted by Shumway-Cook and colleagues to be used in clinical practice.[58] The SOT assesses the subject's upright performance in six sensory contexts (see Figure 6.5). Although the SOT is widely used, it has some limitations, as standing upright requires more than intact sensation. The test assumes that the observed performance in different conditions depicts the effect of the different sensory modalities used without taking into account strength or endurance disorders or

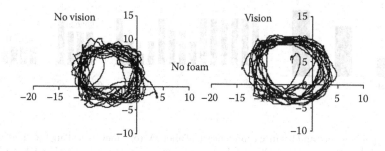

FIGURE 6.13
Moving the CoM voluntarily.

their effects on the performance. Moreover, unlike somatosensory and visual cues, vestibular information cannot be altered, thus making the procedure incomplete. Results of this test do, however, give useful information and can be used to help in planning the rehabilitation program.

6.3.4 Assessment of Impairments

The last component of the balance assessment involves evaluation of the person's impairments. Range of motion, spasticity, and tremor are addressed to understand the influence they may have on the person's ability to carry out sensory-motor strategies and balance capacity in daily life.

Muscle force can be rated using manual muscle testing[59] or isokinetic instruments, while spasticity is usually rated by the Modified Ashworth Scale (MAS) or the Multiple Sclerosis Spasticity Scale (MSSS-88). Further information about the assessment of strength and spasticity is available in Chapter 7. It is important to note that when spasticity of quadriceps and calf muscles was tested in people with MS, the MAS was unable to distinguish between fallers and nonfallers.[13]

6.4 Assessment Findings and Implications: A Case Example

Maria is a 46-year-old woman with MS who is experiencing problems with balance. Figure 6.14, panel 1(a), presents Maria's total scores for the BBS, DGI, DHI, and ABC. The maximum score for each scale indicates that good performance is 56, 24, 100, and 100 points, respectively (see assessment section, above). Maria's scores reveal reduced balance skills, a high level of handicap, and low-balance confidence during activities of daily living. She did not report a recent fall, but reported fear of falling.

The singular items of Maria's BBS are reported in panel 1(b). She was able to transfer from chair to bed (item 5) and to stand upright during eyes open and closed conditions (items 2 and 6). Nevertheless, she had difficulties carrying out tasks in which her feet were close together or in tandem position (items 7, 13) or where she had to move her CoM in either the antero-posterior or the medio-lateral plane (items 8 and 12). Her axial strategy,

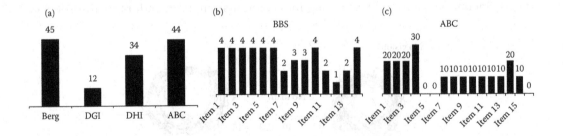

FIGURE 6.14
Maria's balance performance and balance confidence. (a) Panel 1A. (b) Panel 1B. (c) Panel 1C. Berg: total score of Berg Balance Scale [arbitrary unit (au)]; DGI: total score of Dynamic Gait Index [au]; DHI: total score of Dizziness Handicap Inventory [au]; ABC: total score of Activity Balance Confidence [au]. Histograms of panels B and C represent ordered scale items.

which requires coupling of movements of trunk and lower limbs to maintain the whole-body CoM within the center of her BoS, was impaired and likely contributed to the difficulties she had picking up an object from the floor (item 9). Her difficulty with whole-body rotation was highlighted by a low score on item 11.

With respect to balance confidence, Figure 6.14, panel 1(c), shows Maria's scores from the ABC scale. She reported poor balance confidence in almost all activities except for walking at home (items 1 and 2). Outdoor activities were perceived as the most hazardous. The results from her SOT indicated that she had difficulties when somatosensory information was removed, and increasing difficulty with her eyes closed, indicating poor ability in using vestibular information.

Given Maria's assessment findings, specific interventions could be aimed at improving her axial strategy to increase confidence in picking up objects and increasing her balance in the sit-to-stand task, where her balance was challenged. Maria's perception of her position of CoM and her ability to shift her CoM within her BoS should be addressed in different BoS configurations for both static and dynamic tasks. Her sensory system can be challenged (see Section 6.5) by asking her to keep her CoM in a particular zone of the BoS during movements of the head, movements of head with eyes staring at a stationary target, movements of eyes with head still, and combined movements of head and eyes. With respect to activities of daily living, Maria's confidence in her balance should be addressed in order to decrease fear of falling and increase participation. A specific intervention in the community setting along with the prescription of walking aids could be implemented. Several home modifications can also be considered, including removing hazards to reduce the risk of tripping or slipping, adding grab bars in the bathroom, and placing chairs in strategic locations around the house to provide rest. In addition, education and self-management strategies can be used to enable Maria to recognize and avoid hazardous activities that may increase her risk of falling. Further details about specific treatment options for people like Maria who have balance disorders are presented in the next section.

6.5 Treatment Planning: Interventions for Balance Impairments

Balance disorders manifest themselves in both quiet standing and dynamic activities and lead to problems engaging in activities of daily living and participating in society. Balance disorders in MS are a functional disability arising from a number of neuromotor abnormalities, such as muscular weakness, sensory-motor disorders, and general deconditioning. To decrease disability and improve participation, those impairments and other factors contributing to the decreased participation need to be addressed. Consequently, the treatment of balance disorders should be multifactorial and based on a multimodal assessment of balance functioning in real life. Common overarching goals for the treatment of balance disorders include decreasing disability by increasing balance, strength, and mobility, and improving participation by decreasing falls, fall risks, and fear of falling. Many different approaches have been used to achieve these goals, including balance-specific exercises, training of postural strategies, sensory-organization training, environmental modifications paired with education, and general aerobic and strengthening exercise programs.[33,60,61] Various health professionals participate in implementing these approaches.

6.5.1 Exercise

The general physical condition of people with MS is often poor, which contributes to balance disorders and reduced participation. With emerging evidence of the modulatory role of exercise on neuronal growth factors in reducing damage due to neurodegenerative diseases, exercise activity has become even more important for people with MS.[62]

There is significant evidence suggesting that exercise programs, including aerobics, flexibility, strengthening exercises, and yoga, have a beneficial effect on both disease symptoms and general fitness of people with MS with mild to moderate disability, with positive effects on quality of life. Reviews also indicate that the content and benefits of such exercise programs differ according to level of disability and mobility level of people with MS, with subjects with mild impairments benefiting more than those with severe impairments.[60,61,63-66] Chapter 17 addresses this issue in detail.

6.5.2 Strengthening Exercises

Strength of antigravity muscles is important in postural control and mobility. Studies examining the effectiveness of strengthening programs for people with MS indicate that functional improvement may be achieved and that neuromuscular capacity in MS can be improved even when there is underlying neurological damage.[63,67,68] Muscle weakness contributes to impaired mobility and balance disorders; strengthening exercises in functional contexts may assist in improving balance and gait. Evidence about improving strength among people with MS is summarized in Chapter 7.

6.5.3 Endurance Exercise

General deconditioning in people with MS may increase the sensation of fatigue and lead to less efficient sensory-motor control. Exercises focusing on endurance for balance-relevant tasks may improve balance, reduce exertion, and increase an individual's perception of his or her ability to carry out activities of daily life. To date, only few studies have incorporated specific balance training of people with MS in an exercise program.[68-71]

In one study, researchers incorporated balance training in a home-based resistance exercise program promoting lower limb strength training and mobility for people with MS with EDSS scores ranging from 1.0 to 6.5.[68] While there was an increase in leg-power measures, there were no corresponding changes in stabilometric balance measures and mobility. In comparison, Cantalloube[69] demonstrated improvement in standing balance (stabilometry) with opened and closed eyes, velocity of gait, strength of lower extremities, and in Functional Independence Measure™ values following an exercise program that included a balance component. De Souza and Worthington[70] similarly included a balance component in their longitudinal exercise program of 40 people with MS with moderate disability (mean EDSS = 6.0). The program involved outpatient exercise classes (frequency unspecified) and home-based physical therapy. After 1.5 years, 50% of participants were believed to have improved activities of daily living (ADL) and balance. The investigators found a dose–response relationship between therapy and ADL improvement and determined that functionally oriented movement exercises were more beneficial than balance-specific exercises. However, the positive effect wore off after the first 6 months, and active range of motion (a clinical measure of muscle efficiency) declined significantly in 75% of the subjects.[33] In another study evaluating the impact of an MS physical telerehabilitation system on functional status and patient acceptance, significant improvements in gait mobility and the BBS were found while self-reported outcome measures did not show improvement.[71] Stabilometric measures taken during quiet standing tended not to improve. This particular intervention consisted of a 12-week program of exercises customized for each participant following his or her initial evaluation by a physical therapist, and included functional strengthening, stretching, and balance activities.

Although it is apparent that balance can be positively influenced by exercise and rehabilitation programs, balance dysfunction and falls remain a major problem for people with MS. There is an urgent need for improved assessment and treatment programs and complementary research programs to address these ongoing problems more adequately. Accurate assessment of the individual with MS is imperative to determine which functions to target with training. The training must target the functions considered important to the individual, and be effective in restoring the function and increasing participation.

6.5.4 Rehabilitation of Sensory-Motor Strategies and Balance

Addressing sensory-motor strategies was an integral part of balance rehabilitation in two recent trials.[46,72,73] Balance exercises typically include training balance under challenging sensory and dynamic conditions with the goal of improving sensory strategies so that the individual can maintain balance in different environmental contexts (e.g., different lighting, type of ground). Exercises can include balancing under conditions of altered somatosensory input (e.g., foam cushions under feet), reduced visual input (e.g., moving eyes with head still, closing eyes), or with the stimulation of vestibular system (e.g., exercises done with head turning). The tasks can be made more challenging by reducing the BoS, increasing the number of segments to control, exercising in quiet or busy environmental conditions, and using static or dynamic balance exercise. Dynamic balance training includes walking with head turns looking at a stationary target, walking with horizontal or vertical eye movements, or performing a secondary motor task while walking. Adding secondary cognitive tasks can further challenge dynamic balance.

When designing a treatment plan to address balance disorder, consideration must be given to whether or not the impaired sensory system (i.e., somatosensory, visual, vestibular) has the potential to recover or not. If it does, treatment can be directed toward improving the quality of information that the system receives and promoting better integration of that information. In this approach, training aims to facilitate the use of the impaired system in balance control, often by inhibiting the use of other systems during balance exercises. If the impaired sensory system is considered beyond recovery, treatment aims to facilitate the remaining, more intact systems to compensate for the information that is missing in order to minimize impact on balance control. Even if no algorithms are available to predict the recovery of a sensory system, an accurate assessment of sensory disorders is of utmost importance in deciding which approach to use.

Stabilometric and clinical tests can be used to assess balance control in different sensory contests (SOT); after the assessment, it is possible to assess a person's dependence on a particular input; for example, if a person increases his or her sway in an eyes-closed condition, it means he or she is dependent upon visual input. Rehabilitation procedures aimed at reducing the dependence on a particular input and increasing the recovery of the damaged ones can be pursued in several ways, as described in Table 6.4.

Specific intervention for the improvement of sensory-motor strategies to control static and dynamic balance in balance disorders of people with MS has been implemented with some success.[46,72–75] It appears that persons with highly compromised balance can improve their ability to use the damaged sensory system[46] (unpublished data). A pilot study by

TABLE 6.4

Training Sensory Strategies

Goal	Methods
Reducing dependence on visual information for balance control	The manipulation of visual information can be achieved by • Varying visual conditions: eyes open, closed, dim lighting, glasses that reduce sight, visual motion • Creating conflict of information between the perception of movement of the retina and somatosensory and vestibular information • Varying head orientation and movement • Asking the person with MS to follow moving objects with the eyes
Reducing dependence on information from the somatosensory system for balance control	The manipulation of somatosensory information can be achieved at the sole and ankle level by • Varying surface conditions: carpet, foam, incline, and tilting surfaces. These alternate surfaces reduce the reliability of the information from the ankles and soles about the CoM and create a conflict with other incoming sensory information • Using vibrating stimulators that can alter proprioceptive information
Reducing dependence on information from the vestibular system for balance control	The manipulation of vestibular information is more complicated than manipulating that from the visual or proprioceptive systems. • Varying the head orientation and movement can challenge these receptors although the vestibular system is relatively functional also at high rotational frequencies

Cattaneo and colleagues[76] evaluated the effects of balance rehabilitation in people with MS using two approaches to improving balance: one focused purely on motor retraining and the other on integrated sensory-motor retraining. A third control group had the usual rehabilitation care. The emphasis for the two protocols was on recovery and reintegration of the impaired sensory systems. Exercises gradually progressed from body stability exercises to gait exercises in variable environments. Both approaches resulted in a reduction of the number of falls during stabilometric evaluation with SOT. Similarly, both the experimental groups had a clinically significant improvement on the BBS (more than 4 points improvement on the total score) while the control group showed no relevant changes. No differences in static balance and stabilometric measures were found between the two experimental groups, but the group treated with sensory strategies had a significant improvement in dynamic balance as measured by the DGI when compared to the non-sensory-strategies group and the control group (more than 2 points). This finding suggests that rehabilitation for dynamic balance may need to put more emphasis on sensory strategies.

Since balance is more challenged during dynamic activities than during stance, sensory conflict may occur more frequently during dynamic activities; therefore, efficient integration of information may be even more important. Prosperini and colleagues[73] used a novel visuo-proprioceptive paradigm to improve balance and reduce risk of falls in people with MS. They provided visual feedback on a computer screen during exercises that included static and dynamic exercises in both double- and single-leg stance. Twenty-eight subjects finished the six-week training protocol, which was found to improve balance and reduce falls in those with MS. Researchers observed an improvement in the stabilometric test in the eyes-closed condition/feet apart, eyes-open and eyes-closed, feet together conditions, and in monopodalic stance.

Ultimately, when there are sensory impairments or conflicts in the use of sensory information, the retraining of sensory strategies may be an essential component in improving balance and in particular, dynamic balance. Specificity of assessment and understanding of the underlying problem is very important in planning the rehabilitation approach in order to achieve relevant results.

6.5.5 Other Therapeutic Approaches to Exercise and Balance

Community-based programs such as tai chi, yoga, aquatics, and Feldenkrais may be helpful in maintaining gait and balance function in people with MS.[77] Hippotherapy may also have some promise for the treatment of balance disorders.[78] In a pilot study carried out on a group of people with MS, a 14-week hippotherapy program resulted in a gain of 9 points on the BBS and of 5 points on the Tinetti Performance Oriented Mobility Assessment. In that study, persons with primary progressive MS demonstrated the greatest amount of change compared to other subtypes of MS.[79] However, thus far, data are limited and further research is needed to understand the potential of hippotherapy in improving balance and for whom it may be most indicated.

6.5.6 Mobility Device Prescription and Self-Management Training

Balance is affected by intrinsic and extrinsic factors. Extrinsic factors that can influence balance control and reduce the risk of falling are modification of the home environment (see Chapter 21) and the prescription of assistive devices, such as an ankle-foot orthoses (AFOs), cane, or a walker (see Chapter 13). However, some people with MS may be reluctant to use assistive equipment and often need training and encouragement from the health professionals. In addition, assistive equipment may improve some functions but lead to decrements of others. For example, AFOs can increase static balance in people with MS who have mild strength problems and balance disorders while at the same time limiting dynamic balance as measured by the time taken to walk 10 m.[80] In these cases, rehabilitation professionals must find the right compromise between increased mobility and safety.

The prevention of falls is an important objective in the rehabilitation of people with MS, especially since falls and fear of falling can lead to alterations in mobility and decreased participation. An improvement in static and dynamic equilibrium is an integral factor in reducing falls. In addition, evaluation of home hazards and counseling should be implemented in order to reduce risk of falls. Finlayson and colleagues[81] conducted a pilot feasibility study of a self-management program to increase knowledge of risk factors for falls, increase knowledge and skills to manage falls, and change behaviors to reduce personal fall risk. Participants engaged in a 12-h group program facilitated by an occupational therapist. Group discussions, activities, and take-home exercises enabled participants to examine how behavior, attitudes, activity, MS symptoms, and the environment might influence falls and how these risk factors could be modified to reduce fall risk. The results of the study were encouraging since the program was able to provide changes in knowledge, skills, and behaviors related to fall risk.

6.6 Conclusions and Guidelines for an Approach to Balance Rehabilitation and Training

Generally, the goal of rehabilitation for balance disorders among people with MS is to decrease disability due to balance disorders by improving balance and mobility, and to improve participation by improving function and decreasing falls and fear of fall. Problems with balance, mobility, and function are interrelated, complex, and long term. They constantly evolve throughout the course of the disease and must therefore be considered from a long-term perspective. This perspective may require multidisciplinary rehabilitation in different settings and at different times, depending on the principal problems. Table 6.5 summarizes general principles to guide rehabilitation.

Several studies have demonstrated that in the early stages of MS, maintenance of mobility and function, including balance functions, can be achieved by relatively straightforward interventions. At these earlier levels of the disease (EDSS 0–6.0) there is enough evidence to recommend participation in endurance training at low to moderate intensity and resistance training at moderate intensity.[60,61,82,83] The intervention at the earlier stages

TABLE 6.5

Main Principles in Mobility and Balance Training for People with MS

Assess comprehensively and accurately the individual's balance skills and underlying impairments with appropriate and sensitive measurement instruments:

- Assess the environmental context in which the individual lives
- Assess the individual's attitudes toward risk of falls
- Assess the devices used in daily living

Set realistic short-term and long-term goals in accordance with the needs of the person with MS, with the goal of affecting quality of life:

- Select a variety of rehabilitation options that target the individual's areas of dysfunction and take into account his or her areas of strength and interests.
- Select the appropriate level at which to train balance in order to challenge the ability of the individual (in functional tasks). Demands of the task should drive the impairment training (strengthening and endurance exercises, flexibility, postural alignment).

Progressively increase the complexity of demands on the balance system in order to continuously challenge the individual:

- Increase the complexity of the demands, change the base of support, progress to more dynamic tasks, sensory manipulation, dual tasks
- Increase the variability of practice in functional tasks
- Practice in increasingly varied contexts

Reassess balance skills regularly to provide feedback to the person with MS and health professionals, who can then accordingly reset goals of training:

- Reassess as short-term goals are met
- Reassess periodically for long-term goals to take into account the progressive nature of MS and meet the changing needs of the person with MS
- Give strategies to maximize both supervised and unsupervised training opportunities
- Give counseling and strategies to cope with balance disorders and falls
- Give counseling for home and environmental modifications to reduce fall hazards

is primarily focused on prevention to maintain balance and reduce risk for falls, and maintain overall conditioning. A regular community exercise program that is appropriate for the individual needs of the person with MS may be enough at this stage, although regular reassessment of needs is vital. However, as function and participation become compromised, a more comprehensive and intensive multidisciplinary approach is generally necessary. At this stage, interventions often need to address the impairment (e.g., sensory-motor training[73,74,77]) the functional activities (e.g., assistive devices, promotion of physical activity), and the environment (e.g., home hazards, accessibility of buildings and public transportation).

Adherence to exercise programs can be a problem, especially if the program is unsupervised (home based or community based) and also positive effects of the treatment tend to fade in a few weeks.[70,84,85] Similarly, there may be a fade-out of the treatment effects when training periods are longer.[33,83] For people with MS, symptoms of fatigue and disability can reduce motivation to exercise. Self-motivation of the individual is imperative, as he or she needs to remain motivated and see changes in functional level and quality of life. It is important to reevaluate regularly the individual's functional level and reset therapeutic goals as appropriate to keep the exercise program challenging and fruitful. The exercise environment needs to be comfortable for persons with disabilities and provide an opportunity for social interaction. Physical activity for the purpose of pleasure should be emphasized. There is an urgent need for studies with longer follow-up periods to better understand how this problem of lack of motivation and longitudinal-treatment effectiveness can be circumvented.

In general, there appear to be few contraindications to exercise and rehabilitation for people with MS. The positive outcomes have tended, however, to be specific to the intervention used: balance rehabilitation resulted in better balance, aerobic exercises resulted in better fitness, and strengthening exercises increased strength, although benefit in quality of life and gait speed was apparent in many studies with different approaches. More studies combining approaches are recommended in order to achieve a more general exercise effect. Many of the studies cited have a problem with blinding and not having a control group or else a lack of follow up that makes the results questionable and the need for more rigorous studies imperative.

There is a need for intervention strategies that are appropriate and interesting for people with MS to keep them motivated, and those should be regularly updated according to positive or negative changes in function. Assessment needs to be accurate in order to identify balance disorders, and exercise activity needs to be challenging enough to bring about changes in rehabilitation or exercise programs aimed at improving functional ability and preventing falls, thereby increasing participation.

References

1. Kurtzke JF. Demyelinating diseases. In: J. C. Koetsier, editor. Epidemiology of multiple sclerosis in handbook of clinical neurology. Amsterdam: Elsevier Science Publishers; 1985.
2. Martyn C. The epidemiology of MS. In: W. B. Matthews, editor. McAlpine's multiple sclerosis. 2nd ed. Edinburgh, UK: Churchill Livingstone; 1991.
3. Shumway-Cook A, Woollacot M. Motor control: Theory and practical applications. Baltimore, MD: Williams & Williams; 1995.
4. Horak FB. Postural orientation and equilibrium: What do we need to know about neural control of balance to prevent falls? Age and Ageing 2006;35:ii7–ii11.

5. Di Fabio RP, Emasithi A. Aging and the mechanisms underlying head and postural control during voluntary motion. Physical Therapy 1997;77(5):458–475.
6. Cameron MH, Lord S. Postural control in multiple sclerosis: Implications for fall prevention. Current 2010;10(5):407–412.
7. Chiari L, Rocchi L, Cappello A. Stabilometric parameters are affected by anthropometry and foot placement. Clinical Biomechanics 2002;17(9–10):666–677.
8. Ramdharry GM, Marsden JF, Day BL, Thompson AJ. De-stabilizing and training effects of foot orthoses in multiple sclerosis. Multiple Sclerosis 2006;12(2):219–226.
9. Frzovic D, Morris ME, Vowels L. Clinical test of standing balance: Performance of persons with multiple sclerosis. Archives of Physical Medicine and Rehabilitation 2000;81(2):383–390.
10. Benedetti MG, Piperno R, Simoncini L, Bonato P, Tonini A, Giannini S. Gait abnormalities in minimally impaired multiple sclerosis patients. Multiple Sclerosis 1999 5;(5)363–368.
11. Lambert CP, Archer RL, Evans WJ. Muscle strength and fatigue during isokinetic exercise in individuals with multiple sclerosis. 2001;33(10):1613–1619.
12. Chung LH, Remelius JG, Van Emmerik RE, Kent-Braun JA. Leg power asymmetry and postural control in women with multiple sclerosis. Medicine & Science in Sports & Exercise 2008;40(10):1717–1724.
13. Cattaneo D, De Nuzzo C, Fascia T, Macalli M, Pisoni I, Cardini R. Risks of falls in subjects with multiple sclerosis. Archives of Physical Medicine and Rehabilitation 2002;83(6):864–867.
14. Nilsagard Y, Lundholm C, Denison E, Gunnarsson LG. Predicting accidental falls in people with multiple sclerosis: A longitudinal study. Clinical Rehabilitation 2009;23(3):259–269.
15. Sosnoff JJ, Shin S, Motl RW. Multiple sclerosis and postural control: The role of spasticity. Archives of Physical Medicine and Rehabilitation 2010;91(1):93–99.
16. Rougier P, Faucher M, Cantalloube S, Lamotte D, Vinti M, Thoumie P. How proprioceptive impairments affect quiet standing in patients with multiple sclerosis. Somatosensory and Motor Research 2007;24(1):41–51.
17. Evans WJ, Lambert CP. Physiological basis of fatigue. American Journal of Physical Medicine and Rehabilitation 2007;86:S29–S46.
18. Lapierre Y, Hum S. Treating fatigue. International MS Journal 2007;14(2):64–71.
19. Fitzpatrick R, McCloskey DI. Proprioceptive, visual and vestibular thresholds for the perception of sway during standing in humans. Journal of Physiology 1994;478(Pt.1):173–186.
20. Peterka RJ. Sensorimotor integration in human postural control. Journal of Neurophysiology 2002;88(3):1097–1118.
21. Nelson SR, Di Fabio RP, Anderson JH. Vestibular and sensory interaction deficits assessed by dynamic platform posturography in patients with multiple sclerosis. Annals of Otology, Rhinology, Laryngology 1995;104:62–68.
22. Frohman EM, Zhang H, Dewey RB, Hawker KS, Racke MK, Frohman TC. Vertigo in MS: Utility of positional and particle repositioning maneuvers. Neurology 2000;55:1566–1569.
23. Cattaneo D, Jonsdottir J. Sensory impairments in quiet standing in subjects with multiple sclerosis. Multiple Sclerosis 2009;15(1):59–67.
24. Winter DA. Human balance and posture control during standing and walking. Gait & Posture 1995;3:193–214.
25. Nashner LM, Peters JF. Dynamic posturography in the diagnosis and management of dizziness and balance disorders. Neurologic Clinics 1990;8:331–349.
26. Pollock AS, Durward BR, Rowe PJ, Paul JP. What is balance? Clinical Rehabilitation 2000;14:402–406.
27. Jacobs JV, Horak FB. Cortical control of postural responses. Journal of Neural Transmission 2007;114(10):1339–1348.
28. Hoffmann S, Tittgemeyer M, von Cramon DY. Cognitive impairment in multiple sclerosis. Current Opinion in Neurology 2007;20(3):275–280.
29. Hamilton F, Rochester L, Paul L, Rafferty D, O'Leary CP, Evans JJ. Walking and talking: An investigation of cognitive-motor dual tasking in multiple sclerosis. Multiple Sclerosis 2009;15(10):1215–1227.

30. Massion J. Movement, posture, and equilibrium: Interaction and coordination. Progress in Neurobiology 1992;38:35–56.

31. van der Kooij H, Jacobs R, Koopman B, van der Helm F. An adaptive model of sensory integration in a dynamic environment applied to human stance control. Biology 2001;84(2):103–115.

32. Peterson EW, Cho CC, Finlayson ML. Fear of falling and associated activity curtailment among middle aged and older adults with multiple sclerosis. Multiple Sclerosis 2007;13(9):1168–1175.

33. Brown TR, Kraft GH. Exercise and rehabilitation for individuals with multiple sclerosis. Physical Medicine and Rehabilitation Clinics of North America 2005;16(2):513–555.

34. Confavreux C, Vukusic S, Moreau T, Adeleine P. Relapses and progression of disability in multiple sclerosis. New England Journal of Medicine 2000;343(20):1430–1438.

35. Confavreux C, Adeleine P. Age at disability milestones in multiple sclerosis. Brain 2006;129:595–605.

36. Iezzoni LI, Rao SR, Kinkel RP. Patterns of mobility aid use among working age persons with multiple sclerosis living in the community in the United States. Disability and Health Journal 2009;2:67–76.

37. Finlayson M, Peterson E, Cho C. Risk factors for falling among people aged 45 to 90 years with multiple sclerosis. Archives of Physical Medicine and Rehabilitation 2006;87:1274–1279.

38. Dunn JE, Rudberg MA, Fumer SE, Cassel CK. Mortality, disability, and falls in older persons: The role of underlying disease and disability. American of Journal of Public Health 1992;82:395–400.

39. Stolze H, Klebe S, Zechlin C, Baecker C, Friege L, Deuschl G. Falls in frequent neurological diseases—Prevalence, risk factors and aetiology. Journal of Neurology 2004;251(1):79–84.

40. Peterson EW, Cho CC, von Koch L, Finlayson ML. Injurious falls among middle aged and older adults with multiple sclerosis. Archives of Physical Medicine and Rehabilitation 2008;89:1031–1037.

41. Tinetti M, Richman D, Powell L. Falls efficacy as a measure of fear of falling. Journal of Gerontology 1990;45(6):P239–P243.

42. Motl RW, Goldman M. Physical inactivity, neurological disability, and cardiorespiratory fitness in multiple sclerosis Acta Neurologica Scandinavica 2011;123:98–104.

43. Delbaere K, Crombez G, Vanderstraeten G, Willems T, Cambier D. Fear-related avoidance of activities, falls and physical frailty: A prospective community-based cohort study. Age and Ageing 2004;33:368–373.

44. Powell LE, Myers AM. The activities-specific balance confidence (ABC) scale. Journals of Gerontology Series A: Biological Sciences and Medical Sciences 1995;50A(1):M28–M34.

45. Cattaneo D, Regola A, Meotti M. Validity of six balance disorders scales in persons with multiple sclerosis. Disability and Rehabilitation 2006;28(12):789–795.

46. Cattaneo D, Jonsdottir J, Repetti S. Reliability of four scales on balance disorders in persons with multiple sclerosis. Disability & Rehabilitation 2007;29(24):1920–1925.

47. Mak MK, Pang MY. Balance confidence and functional mobility are independently associated with falls in people with Parkinson's disease. Journal of Neurology 2009;256(5):742–749.

48. Tennstedt S, Howland J, Lachman M, Peterson E, Kasten L, Jette A. A randomized, controlled trial of a group intervention to reduce fear of falling and associated activity restriction in older adults. Journals of Gerontology Series B: Psychological Sciences & Social Sciences 1998;53:384–392.

49. Tesio L, Perucca L, Franchignoni FP, Battaglia MA. A short measure of balance in multiple sclerosis: Validation through rasch analysis. Functional Neurology 1997;12(5):255–265.

50. Nieuwenhuis MM, Van Tongeren H, Sørensen PS, Ravnborg M. The six spot step test: A new measurement for walking ability in multiple sclerosis. Multiple Sclerosis 2006;12:495–500.

51. Berg KO, Maki BE, Williams JI, Holliday PJ, Wood-Dauphinee SL. Clinical and laboratory measures of postural balance in an elderly population. Archives of Physical Medicine and Rehabilitation 1992;73:1073–1080.

52. Whitney SL, Hudak MT, Marchetti GF. The dynamic gait index relates to self-reported fall history in individuals with vestibular dysfunction. Journal of Vestibular Research 2000;10:99–105.

53. McConvey J, Bennett SE. Reliability of the dynamic gait index in individuals with multiple sclerosis. Archives of Physical Medicine and Rehabilitation 2005;86:130–133.
54. Schoppen T, Boonstra A, Groothoff JW, de Vries J, Goeken LN, Eisma WH. The timed "up and go" test: Reliability and validity in persons with unilateral lower limb amputation. Archives of Physical Medicine and Rehabilitation 1999;80:825–828.
55. Goldman MD, Marrie RA, Cohen JA. Evaluation of the six-minute walk in multiple sclerosis subjects and healthy controls. Multiple Sclerosis 2008;3:383–390.
56. Verheyden G, Nuyens G, Niewboer A, Van Asch P, Ketalaer P, De Weerdt W. Reliability and validity of trunk assessment for people with multiple sclerosis. Physical Therapy 2006;86:66–76.
57. Nashner LM, Peters JF. Fixed patterns of rapid postural responses among leg muscles during stance. Experimental Brain Research 1977;30(1):13–24.
58. Shumway-Cook A, Horak FB. Assessing the influence of sensory interaction of balance: Suggestions from the field. Physical Therapy 1986;66:1548–1550.
59. Medical Research Council of the United Kingdom. Aids to examination of the peripheral nervous system: Memorandum no 45. Palo Alto, CA: Pedragon House; 1978.
60. Asano M, Dawes DJ, Arafah A, Moriello C, Mayo NE. What does a structured review of the effectiveness of exercise interventions for persons with multiple sclerosis tell us about the challenges of designing trials? Multiple Sclerosis 2009;15:412–421.
61. Rietberg MB, Brooks D, Uitdehaag BMJ, Kwakkel G. Exercise therapy for multiple sclerosis. Cochrane Database of Systematic Reviews 2004;(3). Art. No.: CD003980. DOI: 10.1002/14651858. CD003980.pub2.
62. White LJ, Castellano V. Exercise and brain health—Implications for multiple sclerosis: Part 1—Neuronal growth factors. Sports Medicine 2008;38(2):91–100.
63. Dalgas U, Stenager E, Jakobsen J, Petersen T, Hansen H, Knudsen C, Overgard K, Ingemann-Hansen T. Resistance training improves muscle strength and functional capacity in multiple sclerosis. Neurology 2009;73:1478–1484.
64. Dalgas U, Stenager E, Jakobsen J, Petersen T, Hansen HJ, Knudsen C, Overgaard K, Ingemann-Hansen T. Fatigue, mood and quality of life improve in MS patients after progressive resistance training. Multiple Sclerosis 2010;16(4):480–490.
65. Oken BS, Kishiyama S, Zajdel D, Bourdette D, Carlsen J, Haas M, Hugos C, Kraemer DF, Lawrence J, Mass M. Randomized controlled trial of yoga and exercise in multiple sclerosis. Neurology 2004;62(11):2058–2064.
66. Snook EM, Motl RW. Effect of exercise training on walking mobility in multiple sclerosis: A meta-analysis. Neurorehabilitation and Neural Repair 2009;23(2):108–116.
67. Freeman JA, Thompson AJ, Freeman JA. Building an evidence base for multiple sclerosis management: Support for physiotherapy. Journal of Neurology, Neurosurgery, and Psychiatry 2001;70(2):147–148.
68. DeBolt LS, McCubbin JA. The effects of home-based resistance exercise on balance, power, and mobility in adults with multiple sclerosis. Archives of Physical Medicine and Rehabilitation 2004;85(2):290–297.
69. Cantalloube S, Monteil I, Lamotte D, Mailhan L, Thoumie P. Évaluation préliminaire des effets de la rééducation sur les paramètres de force, d'équilibre et de marche dans la sclérose en plaques. Annales De Réadaptation Et De Médecine Physique 2006 5;49(4):143–149.
70. De Souza LH, Worthington JA. The effect of long-term physiotherapy on disability in multiple sclerosis patients. In: F. C. Rose, R. Jones, editors. Multiple sclerosis: Immunological, diagnostic and therapeutic aspects. London: John Libbey; 1987.
71. Finkelstein J, Lapshin O, Castro H, Cha E, Provance PG. Home-based physical telerehabilitation in patients with multiple sclerosis: A pilot study. Journal of Rehabilitation Research & Development 2008 12;45(9):1361–1373.
72. Prosperini L, Kouleridou A, Petsas N, Leonardi L, Tona F, Pantano P. The relationship between infratentorial lesions, balance deficit and accidental falls in multiple sclerosis. Journal of the Neurological Sciences 2011;304:55–60.

73. Prosperini L, Leonardi L, De Carli P, Mannocchi ML, Pozzilli C. Visuo-proprioceptive training reduces risk of falls in patients with multiple sclerosis. Multiple Sclerosis 2010;16(4):419–499.

74. Lord SE, Wade DT, Halligan PW. A comparison of two physiotherapy treatment approaches to improve walking in multiple sclerosis: A pilot randomized controlled study. Clinical Rehabilitation 1998;12(6):477–486.

75. Cattaneo D, Ferrarin M, Jonsdottir J, Montesano A, Bove M. The virtual time to contact in the evaluation of balance disorders and prediction of falls in people with multiple sclerosis. Disability Rehabilation 2012;34(6):470–477.

76. Cattaneo D, Jonsdottir J, Zocchi M, Regola A. Effects of balance exercises on people with multiple sclerosis: A pilot study. Clinical Rehabilitation 2007;21:771–781.

77. Burks J, Bigley G, Hill H. Rehabilitation challenges in multiple sclerosis. Annals of Indian Academy of Neurology 2009;12(4):296–306.

78. Bronson C, Brewerton K, Ong J, Palanca C, Sullivan SJ. Does hippotherapy improve balance in persons with multiple sclerosis: A systematic review. European Journal of Physical and Rehabilitation Medicine 2010;46(3):347–353.

79. Silkwood-Sherer D, Warmbier H. Effects of hippotherapy on postural stability, in persons with multiple sclerosis: A pilot study. Journal of Neurologic Physical Therapy 2007;31(2):77–84.

80. Cattaneo D, Marazzini F, Crippa A, Cardini R. Do static or dynamic AFOs improve balance? Clinical Rehabilitation 2002;16(8):894–899.

81. Finlayson M, Peterson EW, Cho C. Pilot study of a fall risk management program for middle aged and older adults with MS. NeuroRehabilitation 2009;25(2):107–115.

82. Dalgas U, Stenager E, Ingemann-Hansen T. Multiple sclerosis and physical exercise: Recommendations for the application of resistance-, endurance- and combined training. Multiple Sclerosis 2008;14(1):35–53.

83. Romberg A, Virtanen A, Ruutiainen J, Aunola S, Karppi S, Vaara M, Surakka J, Pohjolainen T, Seppanen A. Effects of a 6-month exercise program on patients with multiple sclerosis: A randomized study. Neurology 2004;63:2034–2038.

84. Surakka J, Romberg A, Ruutiainen J, Aunola S, Virtanen A, Karppi SL, Mäentaka K. Effects of aerobic and strength exercise on motor fatigue in men and women with multiple sclerosis: A randomized controlled trial. Clinical Rehabilitation 2004;18:737–746.

85. Wiles CM, Newcombe RG, Fuller KJ, Shaw S, Furnival-Doran J, Pickersgill TP, Morgan A. Controlled randomised crossover trial of the effects of physiotherapy on mobility in chronic multiple sclerosis. Journal of Neurology, Neurosurgery and Psychiatry 2001;70:174–179.

86. Shumway-Cook A, Woollacott M. Motor control: Theory and practical applications. 2nd ed. Philadelphia, PA: Lippincott Williams and Wilkins; 2001.

87. Berg KO. Measuring balance in the elderly: Preliminary development of an instrument. Physiotherapy Canada 1989;41(6):304–311.

7

Muscle Strength, Tone, and Coordination

Francois Bethoux and Matthew H. Sutliff

CONTENTS

Muscle strength, tone, and coordination are among the main determinants of a person's ability to perform voluntary movements, and to function in the environment. Those living with multiple sclerosis (MS) and their caregivers know how impairments of these bodily functions often coexist and can therefore be difficult to address. Fortunately, interventions and treatments are available to remediate, albeit partially, some of these impairments. Thus, it is essential to assess each impairment separately and to integrate impairment-specific interventions into individualized treatment and rehabilitation planning. After reading this chapter, you will be able to:

1. Define and describe changes in strength, tone, and coordination in relation to MS,
2. Explain how these impairments and their consequences can be assessed clinically, and

3. Give an overview of rehabilitation and treatment strategies, and illustrate with case studies.

The purpose of this chapter is to define and describe the assessment tools for the measurement of strength and muscle tone in MS, as well as describe the various interventions that are available to help patients with MS maximize their ability to perform daily tasks when they are confronted with these deficits. These interventions include medications, exercise approaches, and a variety of orthotics and devices to help compensate for deficits. It is intended that the reader will be able to apply these concepts as they establish a rehabilitation plan of care to help maximize the functional outcomes for people with MS.

7.1 Muscle Tone

The International Classification of Functioning, Disability and Health (ICF) defines muscle tone functions as follows:

Functions related to the tension present in the resting muscles and the resistance offered when trying to move the muscles passively.

Inclusions: functions associated with the tension of isolated muscles and muscle groups, muscles of one limb, one side of the body and the lower half of the body, muscles of all limbs, muscles of the trunk, and all muscles of the body; impairments such as hypotonia, hypertonia, and muscle spasticity.[1]

Owing to the heterogeneous and unpredictable nature of MS, a variety of muscle tone disorders can be encountered. Hypotonia (which can result from cerebellar dysfunction), extrapyramidal hypertonia (characterized by cogwheeling and rigidity), and dystonia (consisting of abnormal sustained or intermittent muscle contractions with twisting movements and abnormal postures) are not common. Spastic hypertonia is by far the most frequent disorder of muscle tone in MS and will be the focus of our discussion.

Spasticity is defined as a velocity-dependent increase in stretch reflex.[2] Spasticity is present in a wide variety of central nervous system disorders and is thought to result from decreased inhibitory input from supraspinal centers. Two forms of spasticity have been identified, each with distinct clinical features: spinal origin spasticity (characterized by enhanced excitability of polysynaptic pathways, slow rise in excitation, flexor and extensor spasms) and cerebral origin spasticity (characterized by hyperexcitability of monosynaptic pathways, rapid rise in excitation, stereotypical postures involving antigravity muscle groups). In spinal origin spasticity, response to an afferent stimulation (e.g., touching the skin, stretching the muscle) can lead to a response (increased tone, spasm) away from the original limb segment involved. Strong flexor muscles are frequently involved, but the stimulation can spread to extensor muscles (e.g., knee extensors). Cerebral origin spasticity often leads to what is known as the "hemiplegic posture" (shoulder adduction; elbow, wrist, and finger flexion; hip adduction; knee extension; and ankle plantarflexion). Although MS-related spasticity is usually identified as spinal, features of cerebral spasticity can be present, and sometimes predominant, since MS lesions can occur throughout the neuraxis.

Although the muscles affected by spasticity vary greatly from patient to patient, the following are more commonly affected or bothersome: in the upper extremities, shoulder adductors, elbow flexors/extensors, pronators, wrist flexors, and finger flexors; in the lower extremities, hip flexors and adductors, knee extensors/flexors, and ankle plantarflexors. Spasticity can also affect trunk muscles.

7.1.1 Prevalence and Impact

Prevalence estimates of spasticity in MS are high. For example, in a cross-sectional survey of over 20,000 individuals with MS in the North American Research Committee on Multiple Sclerosis (NARCOMS) registry, 80% of responders reported some spasticity (34% reported that spasticity interfered with daily activities).[3] Potential consequences of spasticity include:

- Discomfort: for example, painful spasms, pain with passive stretch of the limb, difficulty attaining or keeping a good sitting posture due to stiffness;
- Interference with care and hygiene leading to increased caregiver burden: for example, difficulty performing hygiene or catheterizing the bladder due to decreased ability to spread the legs out with hip adductor muscle spasticity, difficulty with dressing due to stiffness and deformity of the limbs;
- Activity limitations: for example, difficulty bending the legs can interfere with transferring and walking and often increases the risk of falling, difficulty extending the arm to reach for objects due to stiffness in the upper-extremity flexor muscles; and
- Medical complications (contractures, decubitus ulcers).

Overall spasticity can have a major impact on a patient's quality of life.

7.1.2 Identifying and Measuring Spasticity

Symptoms commonly associated with spasticity in MS include muscle stiffness or tightness, spasms, clonus, pain, and difficulty moving the affected limbs. Examination findings may include abnormal posture, decreased passive range of motion, hyperreflexia, involuntary resistance to passive movement with or without "clasp-knife phenomenon" (catch followed by release), observed spasms and clonus, and abnormal movement patterns (synergies, co-contraction of agonist and antagonist muscles, "spastic dystonia"). Electrodiagnostic tests (H reflex and vibration inhibitory index) and the pendulum test have been used to assess spasticity, mostly for research purposes, and do not always correlate with clinical measures.

Quantitative assessment of spasticity and its consequences are mostly clinical (see Table 7.1). The use of pictures and videotapes can be useful in tracking changes with treatment. For example, pictures of a limb deformity or of the patient's sitting posture over time, or videoclips of the patient walking or performing simple functional tasks, can help document the results of rehabilitation or the efficacy of assistive devices. Images give a more synthetic view than analytical scales, and they can be a great communication tool with patients and their families, with other health care providers, and with third-party payers. When using images one must always keep privacy regulations in mind (in the United States, the Health Insurance Portability and Accountability Act [HIPAA] Privacy Rule). Passive and active range of motion (ROM) should be measured. The Ashworth scale and its modified versions,[4] as well as the Tardieu scale,[5] assess resistance to passive movement. The Ashworth Scale has been widely used in the clinical and research settings and has been validated.[6] The rating scale ranges from 0 (no increase in muscle tone) to 4 (affected

TABLE 7.1

Scales and Tests Used to Evaluate Changes in Muscle Tone

Name	Purpose	Method	Measurement Scale	Pluses/Minuses
Modified Ashworth Scale (MAS)	To assess spasticity via resistance to passive movement (impairment level)	Clinical examination	Ordinal scale from 0 = no increase in muscle tone to 4 = affected part(s) rigid in flexion or extension	Easy to perform, not very sensitive to change, proper training required to ensure inter- and intrarater reliability. Does not measure all aspects of spasticity, and there can be resistance to passive movement without spasticity.
Tardieu Scale	To assess spasticity (impairment level)	Clinical examination	Continuous measurement (angles in degrees), within physiological limits. Testing is performed at three different speeds (as slow as possible = V1, at the speed of the limb falling under gravity = V2, and as fast as possible = V3); two angles are measured: first point of resistance (catch) to a quick stretch = R1, passive range of motion (ROM) with a slow stretch = R2	Similar concerns to MAS regarding reproducibility and sensitivity to change. The calculation of R2 minus R1 is conceived as a measurement of "dynamic tone," which is not provided in the MAS.
Spasm Frequency Scale (SFS)	To assess spasticity via spasm frequency (impairment level)	Self-report and direct observation	Ordinal scale from 0 = no spasm to 4 = >10 spontaneous spasms per hour	Not very sensitive to change, does not assess severity of spasms or spasm-related pain.
Spasticity Numeric Rating Scale (NRS)	To assess the overall severity of spasticity	Self-report	Ordinal scale from 0 = no spasticity to 10 = worst possible spasticity	Easy to use, recently introduced and validated; therefore, there is limited experience with this scale. The authors report good test–retest reproducibility, significant correlation with the SFS. A 30% change in score has been defined as a clinically important difference (CID).
MS Spasticity Scale-88 (MSSS-88)	To assess the impact of spasticity (impairment, activity, and participation levels)	Self-report	For each item, from 1 = not at all bothered to 4 = extremely bothered; total score from 88 to 352	Recently developed using qualitative and quantitative methods, and validated in various MS populations. The MSSS-88 gives a thorough picture of the impact of spasticity, but responder burden is a concern in a clinical setting.

part(s) rigid in flexion or extension), with intermediate grades reflecting increasing degrees of resistance to passive mobilization of the limb or limb segment tested.

Assessment of muscle tone at rest should, as much as possible, be complemented by an assessment of dynamic tone during the performance of simple movements and functional tasks. The Spasm Frequency Scale[7] is a useful complement, since spasms can be painful and often disrupt activities and sleep. Similar to pain severity measurement, a spasticity numeric rating scale (NRS) has been validated.[8] However, when planning spasticity management and evaluating treatment outcomes, it is equally important to assess the consequences of spasticity (pain, sleep disruption, activity limitations, etc.). Unfortunately, generic scales of activity, participation, and quality of life are often insensitive to the effects of interventions to relieve spasticity. The Multiple Sclerosis Spasticity Scale-88 items (MSSS-88) is, to our knowledge, the only MS-specific self-report scale assessing the impact of spasticity in persons with MS.[9]

7.1.3 Improving Spasticity

Severe untreated spasticity can have devastating consequences on an individual's quality of life and health. Therefore, it is important to recognize and address the problem when appropriate.

However, prior to designing a treatment plan, it is essential to determine realistic goals and expectations. This approach is commonly used in rehabilitation, but it is often not as rigorously applied in symptom management. Following the broad consequences of spasticity listed above, we identify four broad potential goals for spasticity management:

- To improve comfort (i.e., to reduce the discomfort or bother from spasticity-related symptoms). This goal is usually easier to attain than the others, provided that the treatments can be tolerated.

- To improve posture (standing, sitting, or in bed). Positive outcomes include decreasing the risk of pressure sores, facilitating positioning in the wheelchair or bed, and reducing caregiver burden. Achieving this goal often requires rehabilitation. The use of appropriate seating/cushioning is facilitated by increased flexibility. However, decreasing tone may unmask profound core weakness and further compromise posture.

- To improve passive movement (i.e., to improve range of motion and reduce deformity, to decrease resistance to passive mobilization). Positive consequences are similar to those observed with improved posture. This goal can be attained in many individuals but often requires daily stretching and the use of orthotics, for which treatment adherence can be a problem.

- To improve active function, usually by controlling spasms and reducing the co-contraction of antagonist muscles to a desired movement or function. Examples include reducing plantarflexor tone to improve foot clearance while walking, or reducing finger flexor tone to facilitate the release of objects. This proves to be the most challenging goal to attain because there is often significant loss of motor power "underneath the spasticity" and because many other impairments can contribute to the loss of function.

Fortunately, health care professionals can draw from a variety of interventions to formulate an individualized treatment plan (see Table 7.2). Depending on the country and health care system, pharmacologic treatment options may be formally indicated for the management

TABLE 7.2

Interventions to Relieve Spasticity

Treatment	Components	Comments
Rehabilitation/Exercise	Stretching/ROM Serial casting/Splinting Light pressure/Stroking Biofeedback Orthotics with gait training Weight bearing Vibration Positioning devices/Bed positioning Electrical stimulation/TENS Education Cooling	Rehabilitation can be useful across the spectrum of spasticity severity.
Oral medications	Baclofen Tizanidine Dantrolene sodium Benzodiazepines Other (gabapentin, levetiracetam, clonidin)	Side effects include sedation and weakness. Starting at a low dose and titrating slowly is advised. Medications can be combined, with close attention to cumulative side effects.
Local treatments	Botulinum toxin injections Phenol/Alcohol injections	Usually indicated for focal spasticity or to address a focal problem related to spasticity.
Orthopedic surgery	Tendon release Osteotomy	It is important to optimize spasticity control before performing orthopedic surgery.
Neuromodulation	Intrathecal baclofen therapy	Indicated for severe diffuse spasticity refractory to oral medications.
Neurosurgery	Neurotomy, selective dorsal rhizotomy	Rarely used in MS.

of spasticity or used off-label. In general, these treatments have been used for a long time, and, for most of them, there is evidence of their effectiveness on spasticity although data on functional outcomes is scarce.[10] The Consortium of MS Centers and Paralyzed Veterans of America have published clinical practice guidelines for the management of spasticity.[11] It is important to remember that spasticity can also be helpful (e.g., lower-extremity spasticity in the extensor muscles can help preserve a patient's ability to stand despite profound weakness). Therefore, both clinicians and patients should be aware of a possible adverse effect of decreasing tone on functional performance, and the treatment plan needs to be adjusted accordingly if this occurs. There are anecdotal reports of increased severity of ataxia with tone reduction. It is also suspected that spasticity helps prevent deep venous thrombosis and may help alleviate pressure on bony structures.

It is impossible to give a universal strategy for spasticity management that could be applied to any patient and any situation. However, a few simple guidelines can be helpful:

Always assess other components of motor control (e.g., strength and coordination) and functional performance along with spasticity. When spasticity is a significant problem, it is essential to document in detail the adverse consequences of spasticity (e.g., "severe painful spasms in the arms and legs disrupting sleep," "severe stiffness in the hand that prevents proper hygiene and causes pain with passive mobilization"). This will greatly facilitate setting goals setting and planning treatment.

As a general rule, daily stretching is recommended (unless medically contraindicated) in every patient with spasticity. As basic and "low-tech" as this recommendation appears, more sophisticated (and costly) interventions may not lead to the expected outcomes due to lack of stretching. If the patient is unable to perform stretching exercises, a reliable caregiver should be identified and trained.

The categories of interventions described in Table 7.2 are not mutually exclusive, but should be combined to achieve the desired goals. For example, if functional improvement is sought, rehabilitation should be combined with treatments to reduce tone.

Oral medications, combined with stretching, are often the first line of intervention when spasticity is bothersome. Due to a significant potential for side effects (particularly sedation, increased weakness, increased imbalance with falls), these medications should be started at a low dose and titrated slowly. Because MS patients may have cognitive problems, giving the titration instructions in writing is highly recommended. It makes sense to start with a monotherapy, but medications can be combined (e.g., a medication causing sedation can be used at bedtime to control night spasms and facilitate sleep, while a less sedating medication can be used during the day; or a second medication can be added when spasticity relief is still incomplete when the patient reaches the maximum tolerated dose of one medication). Baclofen and tizanidine have both been tested in MS patients. Dantrolene, although indicated for spasticity, is less frequently used in MS due to a higher risk of increased muscle weakness and to a known risk of liver toxicity.

When spasticity is focal, or focally bothersome, oral medications may not be the best choice. Focused rehabilitation interventions (e.g., orthotics) and local pharmacologic treatments such as botulinum toxin or phenol injections may allow the health care provider to address the problem with a lower risk of generalized decrease in tone and systemic side effects.

When spasticity is severe and diffuse and refractory to other treatment options, intrathecal baclofen (ITB) therapy can be considered. Since ITB therapy involves the surgical implantation of a pump and catheter, requires compliance with visits to adjust and refill the pump, and is associated with potential complications, candidates should be carefully evaluated (including a test injection of ITB) and thoroughly educated about the risks and benefits of this treatment option. Selective dorsal rhizotomy is a possible alternative to ITB therapy in patients with diffuse severe spasticity, but it is rarely used in practice.

Treatment options for decreased passive range of motion (including fixed contractures) include stretching, bracing/splinting (including adjustable splints), serial casting, and orthopedic surgery (particularly tendon release). Spastic hypertonia and spasms should be controlled as much as possible to implement these strategies since spastic muscles have a tendency to "fight back," potentially leading to discomfort, increased risk of pressure sores, and recurrence of contractures.

7.1.4 Key Points about Spasticity

- Spasticity rarely comes alone; it is usually associated with other components of the upper motor neuron syndrome (particularly weakness and loss of dexterity), other neurologic impairments (e.g., visual, sensory, cognitive), and musculoskeletal impairments.

- Spasticity can be helpful: hypertonia and spasms can help preserve function by compensating for underlying loss of motor control.

- Assessing the consequences of spasticity and predicting the results of treatments can be difficult due to the heterogeneity of clinical presentations and symptom fluctuations characterizing MS.
- Clearly defining goals and expectations and combining interventions involving multiple discipline increase the chances of achieving a positive outcome.

7.2 Muscle Strength

The ICF defines the functions related to muscle power as follows:

> Functions related to the force generated by the contraction of a muscle or muscle groups. Inclusions: functions associated with the power of specific muscles and muscle groups, muscles of one limb, one side of the body, the lower half of the body, all limbs, the trunk and the body as a whole; impairments such as weakness of small muscles in feet and hands, muscle paresis, muscle paralysis, monoplegia, hemiplegia, paraplegia, quadriplegia and akinetic mutism.[1]

> In MS, muscle weakness is mostly related to decreased central command due to a disruption of excitatory signal transmission along the neuraxis. Other factors include deconditioning, chronic elongation of muscles due to sustained nonphysiologic postures (e.g., chronic stretch of hip and knee extensors in patients who sit in a wheelchair), and rheological changes (i.e., changes in the physical properties of the muscle and tendon tissue such as plasticity) from chronic nonuse. A decrease in the proportion of Type I muscle fibers (slow twitch, fatigue resistant), a decrease in the overall size of muscle fibers, and an increase in anaerobic metabolism have been observed in the muscles of MS patients.[12,13]

> Clinically, in MS the pattern of muscle weakness varies from patient to patient. Examples of muscles more commonly affected by weakness, or for which weakness has more significant consequences, include in the upper extremities, shoulder flexors/abductors, elbow flexors/extensors, wrist extensors, finger extensors; in the lower extremities, they include hip flexors/extensors/abductors, knee flexors, and ankle dorsiflexors.

7.2.1 Prevalence and Impact

Although weakness is frequently encountered with MS, its prevalence is not well known. Clinically, loss of muscle power is considered one of the most important (if not the most important) factors of loss of function in MS. Weakness can affect most, if not all, activities of daily life. Few published studies directly evaluate the relationship between muscle weakness and loss of function. A cross-sectional study in 100 MS patients showed statistically significant correlations between walking speed (high-speed walk on 7 m), and hamstring/quadriceps strength (correlation coefficients 0.35–0.47, $p < 0.01$).[14]

7.2.2 Measuring Muscle Strength

Muscle strength is typically assessed via manual muscle testing (MMT) using the Medical Research Council rating scale (from 0 = no muscle contraction to 5 = normal strength)

TABLE 7.3

Rating Scale for Manual Muscle Testing (MMT)

5	WNL
4+	Good plus (negligible deficit)
4–	Good (mild deficit)
4–	Good minus (able to break with moderate resistance)
3+	Fair plus (able to break with little resistance)
3	Fair (hold against gravity)
3–	Fair minus (moves through almost full ROM against gravity)
2+	Poor plus (moves through partial ROM against gravity)
2	Poor (moves through ROM in gravity neutral position)
2–	Poor minus (moves through partial ROM in gravity neutral position)
1	Trace (palpable muscle contraction but no movement)
0	No palpable muscle activity

Note: Based on the Medical Research Council Oxford system.[17]

(see Table 7.3). The MMT is easy to administer, but standardization of maneuvers is essential to ensure interrater and intrarater reliability, particularly if "half-grades" are used (e.g., 3+ or 4–). Changes on the MMT are clinically significant, but its sensitivity to change is suboptimal compared to dynamometry. Muscle testing should be preceded by measurement of passive range of motion (ROM). Dynamometry provides a more precise and reliable quantification of muscle strength. Hand-held dynamometers require a careful setup to ensure reproducibility of measurements. Isometric and isokinetic dynamometers can provide measurements of maximal isometric torque and sustained contraction to quantify motor fatigue.[15] Although dynamometry is more sensitive to change and more reproducible than MMT, the clinical significance of measurements (particularly their functional significance) is not fully established. Electromyography (EMG) recordings have been used to quantify changes in muscle strength in research settings.[16] Examples of scales and tests used to evaluate changes in muscle strength are shown in Table 7.4.

Synergies and spasticity represent a challenge in the assessment of muscle strength. Spasticity in the agonist muscle may lead to an overestimation of muscle strength (commonly encountered when testing knee extensors), while spasticity in the antagonist muscle may lead to an underestimation of muscle strength (e.g., spasticity in the ankle plantarflexors overcomes activation of the dorsiflexors).

TABLE 7.4

Scales and Tests Used to Evaluate Changes in Muscle Strength

Name	Purpose	Method	Measurement Scale	Pluses/Minuses
Manual Muscle Testing (MMT)	To measure muscle strength (impairment level)	Clinical examination	Ordinal scale from 0 = no muscle contraction to 5 = normal strength	Easy to use, not very sensitive to change, proper training required to ensure good inter- and intrarater reproducibility
Dynamometry	To measure muscle strength (impairment level)	Performance based	Continuous measurement (force or torque) within physiological limits	More producible and sensitive to change than the MMT, but requires equipment and data processing

Functional testing of muscle strength involves performance of typical duties such as walking, balancing, stooping, bending, climbing stairs, lifting, and carrying objects. Objective tests such as the Nine Hole Peg Test (NHPT), Timed 25 Foot Walk (T25FW), the Timed Up and Go (TUG), the Berg Balance Scale (BBS), and the Six Minute Walk Test (6MWT) provide measurable outcomes for many daily tasks, although performance on these tests is not solely dependent on strength and can also be affected by impaired tone, coordination, sensation, postural control, balance (walking tests), and fine motor control (NHPT). The measurement of lifting, carrying, and overhead reaching is performed through task-specific assessment. These measures are needed for vocational assessments, recommendations for reasonable accommodations at work, or for recommendations regarding disability criteria. Baseline measurements of these specific tasks, when compared with job requirements, also provide firm goal-setting opportunities for rehabilitation programs that can allow for a return to work once the goals have been met.

Muscle strength output often varies over time with MS. One of the reasons for these fluctuations in performance is motor fatigue, which manifests itself as a more rapid decline in muscle strength during a prolonged maximal isometric contraction compared to healthy individuals. Motor fatigue is not necessarily correlated with the subjective sensation of fatigue.[15,18] It has been primarily linked to abnormal central activation, with decreased cortical inhibition[19] and more diffuse cortical activation.[20] An index can be calculated to quantify motor fatigue.[15,21] The Fatigue Index is based on a 30-s recording of maximum isometric strength on a dynamometer, which generates a force–time curve (with a peak force reached within the first 5 s, then a gradual decrease of force output over time). The area under the curve is calculated and divided by the hypothetical area under the curve (if the patient was able to sustain the initial maximal force throughout the 30-s period). This index is not routinely used in clinical practice to our knowledge, but it has proven to be reliable; has been used to evaluate the relationship between subjective fatigue and motor fatigue; and could be used to test the efficacy of various treatments, including rehabilitation, on motor fatigue. Metabolic changes in the muscles due to disuse and deconditioning, although commonly found, are thought to be a secondary cause of motor fatigue. Motor fatigue has been linked to decreased walking endurance.[21,22] It is important to remember that motor fatigue is not equivalent to the overall concept of MS fatigue, which is addressed in Chapter 5.

7.2.3 Improving Muscle Strength

Because loss of muscle power in MS is mainly related to decreased central command, due itself to the damage caused by MS lesions, opportunities to improve muscle strength appear limited. It is conceivable to reverse the weakness due to deconditioning and non-use (or under-use). There is also some evidence of CNS plasticity in MS.[23] CNS plasticity has been defined as "the tendency of synapses and neuronal circuits to change because of activity" (p. 310).[24] In other conditions, such as stroke, a fast component of plasticity (due to changes in neurotransmitters) and a slow component of plasticity (due to long-term functional and structural changes in neuronal circuits) have been demonstrated. The consistent demonstration of CNS plasticity in MS would further support the use of functional training with the hope of obtaining long-term benefits, not by altering the disease process but by allowing one to "work around" the damage caused by MS. Obvious obstacles to CNS plasticity in MS are the presence of multiple areas of CNS damage and the fact that the disease process is ongoing over time rather than a single occurrence, such as a stroke.

7.2.3.1 Medications

There is currently no FDA-approved medication to improve strength in persons with MS. Some symptomatic medications may improve motor performance indirectly. For example, spasticity or neuropathic pain may inhibit voluntary movement, and motor performance may improve when these symptoms are relieved.

Dalfampridine (extended release 4-aminopyridine [4-AP], also named fampridine) is an oral medication, indicated to improve walking in MS; 4-AP is a potassium channel blocker, which has been shown to facilitate the conduction of action potentials along demyelinated axons in animal models. The main finding in the clinical trials of dalfampridine was a sustained improvement in walking speed on the T25FW, based on a responder analysis. In two separate clinical trials, there was a significantly higher proportion of responders in the treatment group compared to the placebo group, and the average improvement of walking speed was 25%. It was interesting that subject characteristics at baseline (e.g., disease course, severity of disability) did not appear to have a significant impact on the proportion of responders. Statistically significant improvement of lower-extremity strength (based on MMT) with dalfampridine compared to placebo was also reported.[25,26] Dalfampridine is prescribed at the dose of 10 mg twice daily. Higher doses did not lead to greater average efficacy in one clinical trial but did lead to increased risk of serious adverse events. There is a risk of seizures with this medication, and it is contraindicated in patients with a history of seizure or moderate-to-severe renal impairment.

7.2.3.2 Strength Training

Strength training during periods of MS relapse should be done with caution. Overtraining during this period can be exhausting and lead to short-term functional decline if a client is pushed to a point of muscular fatigue. Instead, gentle progression of exercise is more beneficial and does not impair short-term function. The concept of "start slow, go slow" is most effective during this stage. This concept is best relayed through the eyes of a former patient. She was a "tri-athlete" and used to ride her bicycle 100 miles in races. When she presented to physical therapy, she stated that she could barely ride a half mile without fatigue, so she had stopped exercising altogether in disgust over her decline in function. With encouragement, she agreed to start riding again on a stationary bike for 5 min, 5 times a week. The next week she added 1 min so that she rode for 6 min, 5 times a week; then the next week she added another minute, and so on. The goal was to work up to 30–45 min a day. Using this approach of "starting slow and progressing slowly," she was able to build her endurance and return to riding, albeit not for 100 miles, but she regained an activity that had provided her with so much joy. In the summer heat, she learned to appreciate a cooling vest while riding, as this kept her body cool and helped avoid the heat-related fatigue that is so common in MS.

Outside the realm of a defined MS relapse, the typical progression of exercise in the presence of limited active range of motion (ROM) presented below[27] has been found to be an effective intervention strategy for improving walking and functional ability in moderately disabled persons with MS.[28]

1. Begin passive range of motion (PROM) until full ROM is achieved. Active assistive range of motion (AAROM) may be incorporated in the pain-free range of movement.

2. Once PROM is within expected limits, begin active range of motion (AROM) in addition to AAROM.

3. Once AROM can be completed through the expected range of movement without pain, isometric strengthening of surrounding muscle groups can begin. Isometric exercises are static exercises against stable resistance that offer strengthening of the muscle groups surrounding the joint while providing stabilization and protection to the joint because there is negligible joint movement while the exercise is being performed.

4. Once isometric strengthening can be performed without pain, isotonic and isokinetic strengthening can begin. These are typically done in a concentric (muscle contraction while shortening) manner. With isotonic exercises, the tension in the muscle remains constant despite a change in muscle length, such as when using hand weights while performing elbow flexion strengthening. Isotonic exercises are typically performed in concentric (muscle contraction while the muscle shortens) rather than in eccentric (muscle contraction while the muscle lengthens) fashion. These exercises involve dynamic muscle activity performed at a constant angular velocity while torque and tension remain constant as the muscles shorten or lengthen. Isokinetic strengthening exercises typically involve the use of a machine to isolate a specific joint movement.

5. Eccentric training (muscle contraction while lengthening) may begin for specific tasks that require eccentric control to be performed properly. For example, if a client has difficulty descending stairs, they may drop abruptly to the next lowest stair due to weakness in the knee extensors, hip extensors, or in the ankle plantarflexors of the stance leg. Repetitive eccentric training of these muscles, using hand rails for safety and control, is a task-specific method of using eccentric contractions to achieve a functional goal. Eccentric strengthening should be performed with caution in MS as excessive strain can quickly cause muscle fatigue. For this reason, fewer repetitions and more sets with short breaks in between can help reduce muscular fatigue (e.g., 3 sets of 10 would typically be preferable to 1 set of 30). Excess muscle soreness or evidence of weakening (such as declining quality of performance of the task) are indicators to stop.

6. Functional training, activity-specific training, or sport-specific training may then be initiated as the patient continues to improve with the above steps. For example, a patient initially came in presenting with falls, particularly when she bent over to pick up an object. Her treatment started with AROM exercise for the legs performed in a cool water pool (80–85°F), which allows the buoyancy of the water to *assist* movement rather than *resist* movement. As she improved, her treatment progressed to resistance exercises using elastic bands or cuff weights for light resistance, which were taught as part of a seated and lying home exercise program that was safe to perform because she would not fall while doing these exercises. In the clinic, she was taught proper sit-to-stand and performed these repetitively (concentric and eccentric strengthening). She then advanced to using isokinetic machines and free weights for resistance during strengthening, physioball exercises for core strengthening and balance, lunges and mini-squats, then to the activity-specific training of picking up objects. Once she mastered this, she performed lateral movements while bending to pick up objects, such as bending and moving laterally to pick up a soccer ball that was rolled to her. These activity-specific exercises were then repeated, simulating a home environment: picking up laundry, the TV remote, stooping to do gardening, and using varied surfaces (grass, gravel, etc.).

TABLE 7.5

Strength Training Guidelines in MS

- Begin strength training at 70% of a 10-repetition maximum. This is 70% of the weight that one can perform an exercise 10 times. When 25 repetitions at this weight can be performed for two consecutive sessions, increase the weight by 10%.
- Training should be performed two to three times per week, for three sets, 8–12 repetitions per set, 10–15 min per session.
- Do not strength-train the same muscle groups on consecutive days.

A variety of equipment can be used, depending on the levels of balance/coordination, plasticity/tremor, strength, and/or fatigue:

- Free weights
- Isokinetic machines
- Stretch band exercises
- Sandbag weights
- Water resistance exercises

Source: Reproduced from Strength training guidelines in multiple sclerosis [Internet]; c2007 [cited 2010 July 1]. Available from: http://www.ncpad.org/disability/fact_sheet.php?sheet=79&section=595. With permission.

Note: This copyrighted article is reproduced from the National Center on Physical Activity and Disability at www.ncpad.org. It may be freely distributed in its entirety as long as it includes this notice but cannot be edited, modified, or otherwise altered without the express written permission of NCPAD. Contact NCPAD at 1-800-900-8086 for additional details.

The challenge of strength training lies in carefully following exercise guidelines that avoid excessive muscular fatigue. Researchers at the University of Illinois of Chicago (UIC) were able to establish guidelines for strength training that help optimize the benefit of this treatment option (see Table 7.5). It should be noted that periods of exercise should be carefully timed to avoid the hotter periods of the day and to prevent excessive fatigue.

7.2.3.3 Assistive Technology

When traditional methods of strengthening and conditioning exercises are not effective, compensatory devices may be used to help optimize function (see Table 7.6). In particular, passive orthoses are frequently used to correct a nonphysiologic posture due to weakness (e.g., wrist/hand splint, ankle foot orthosis, or AFO). Although these orthoses are helpful in avoiding contractures and facilitating function (e.g., an AFO can improve foot clearance by preventing foot drop), the concern is that their passive nature will increase muscle weakness through decreased stimulation of the sensori-motor cortex and through nonuse atrophy. It has been reported that traditional AFOs improve static balance but have a negative impact on dynamic balance.[30] Furthermore, AFOs do not remediate hip flexor weakness, which is a significant contributor to decreased foot clearance.

The hip flexion assist device (HFAD) uses two elastic bands attached to a waist belt on one end, and to the shoe on the other end, to facilitate hip flexion and improve foot clearance for patients with weakness of the three main antigravity muscle groups of the lower extremities: the hip flexors, knee flexors, and ankle dorsiflexors (see Figure 7.1). A popliteal strap may be added for patients with more extensor tone who have difficulty initiating knee flexion as it creates a flexion moment at the knee (see Figure 7.2). An uncontrolled pilot study of the HFAD on 21 subjects showed significant improvement in performance on several walking tests for up to 12 weeks and significant improvement of strength output (tested without the device) in the limb wearing the HFAD.[31]

TABLE 7.6

Assistive Technology to Compensate for Weakness in the Lower Extremities

Problem	Device Options	Discussion
Exertional foot drop (mild)	Foot FlexR	Inexpensive, used for brisk walking or jogging, minimal support
	Foot-Up or Dictus Band	Inexpensive, minimal-to-moderate support
Foot drop (moderate to severe)	Ankle foot orthosis (AFO), carbon fiber with *posterior* shell	Fairly consistent result, good cosmesis, not custom molded
	AFO, carbon fiber with *anterior* shell	Less common but can help with "crouching gait" as it provides knee extension assist, good cosmesis, not customized
	AFO, plastic, can be posterior leaf-spring variety	Fairly consistent results, fair cosmesis, custom, accommodates orthopedic foot abnormalities well, usually needs larger shoes
	FES	"Active" device, can be more difficult to use (e.g., electrode positioning) but works well for some patients, good cosmesis
Foot drop with ankle plantar flexion contracture (Supine ankle DF <= 0 degrees)	AFO, plastic, solid or articulated with dorsal strap	Consistent results, fair cosmesis, custom fit, accommodates orthopedic foot abnormalities well, usually needs larger shoes, provides medial–lateral stability
Foot drop with mild genu recurvatum (hyperextension of the knee)	AFO, plastic, articulated ankle with adjustable plantarflexion stop	Fairly consistent results, fair cosmesis, custom fit, accommodates orthopedic foot abnormalities well, usually needs larger shoe, provides medial–lateral stability, adjustable stop point can help control recurvatum (usually set at 2–5° of dorsiflexion at its stop point) by encouraging flexion at knee
Foot drop with moderate-to-severe genu recurvatum (hyperextension of the knee)	Knee orthosis (KO) with separate AFO	Easier to don and doff than KAFO, protects knee joint and can reduce knee pain
	Knee ankle foot orthosis (KAFO)	One-piece device for control of severe genu recurvatum, fairly heavy and cumbersome, protects knee joint and can reduce pain
Hip flexor weakness	HFAD	Relatively inexpensive, fairly consistent results, improves walking speed and walking distance, can also use popliteal strap to control genu recurvatum and initiate knee flexion, fair cosmesis

Functional electrical stimulation (FES) is also used to compensate for decreased motor control. FES represents the use of neuromuscular electrical stimulation to activate weak muscles in a specific sequence and magnitude in order to enhance function. Several devices have been marketed for the upper and lower extremities. In the lower extremities, the efforts have been focused on correcting foot drop through peroneal nerve stimulation, but the options presented in Figure 7.3 do not compensate for hip and knee weakness. The stimulation is triggered by either a heel switch or a tilt sensor with either wired or wireless connectivity. The heel switch activates the FES at the heel-off segment of stance phase, stimulating the deep peroneal nerve and stimulating a sustained contraction of the anterior tibialis, therefore dorsiflexing the ankle. Once the foot completes its stride and heel strike occurs, the heel switch then deactivates the FES and allows the ankle to passively move into a foot-flat position. FES that is controlled by a tilt sensor does not require a heel

FIGURE 7.1
Hip flexion device. (Courtesy of Becker Orthopedic.)

FIGURE 7.2
Hip flexion device with popliteal strap. (Courtesy of Becker Orthopedic.)

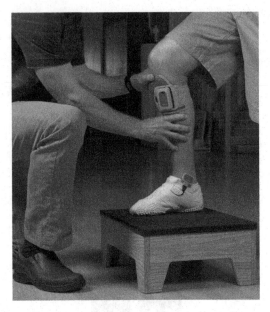

FIGURE 7.3
Bioness L300 FES unit. (Courtesy of Bioness, Inc.)

switch and may be worn barefoot, an appealing option for some users. The FES is activated when the proximal tibia is tilted anteriorly (at toe off) and deactivated when the proximal tibia is tilted posteriorly (at heel strike). A preliminary study on 21 patients with MS found an increase in walking speed and a decrease in the perceived effort of walking with an FES device for foot drop.[32] Another pilot study of the immediate effects of FES for foot drop found modest benefits on walking tests.[33] Beyond the immediate effects of FES, the hope is to promote cortical reorganization and to improve motor control. However, this has not been demonstrated in MS to our knowledge.

Cost is an issue for some of these devices (particularly for newer and more technologically advanced devices, such as FES), and reimbursement varies across countries and health care systems. These economic factors must be taken into account in the decision process, along with more traditional medical and functional parameters.

7.3 Coordination

In the ICF, coordination of voluntary movement is defined as "functions associated with coordination of simple and complex voluntary movements, performing movements in an orderly combination. Inclusions: right-left coordination, coordination of visually directed movements, such as eye-hand coordination and eye-foot coordination; impairments such as dysdiadochokinesia (http://apps.who.int/classifications/icfbrowser/)."[1]

Impaired coordination, in MS and other neurologic disorders, is usually referred to as ataxia. There are two types of ataxia: cerebellar ataxia and sensory ataxia. Cerebellar ataxia is caused by lesions that develop within the circuits that connect the three different parts of the cerebellum: the archicerebellum, paleocerebellum, and neocerebellum.[34] Cerebellar ataxia usually causes several types of movement abnormalities: asynergy (lack of coordination between muscle contractions), dysmetria (inability to reach a target position precisely), and dysdiadochokinesia (inability to perform rapid, alternating movements). Cerebellar ataxic gait is characterized by a widened base, unsteadiness, irregularity of steps, and lateral veering off of the intended path. Sensory ataxia is caused by a loss of sensory input that is needed to control movement. Sensory ataxia is distinguished from cerebellar ataxia by the presence of nearly normal (or markedly improved) coordination when the patient can watch the movement in question, but marked worsening of coordination when the eyes are closed.[35] Before diagnosing any movement abnormality as ataxia, however, weakness, spasticity, sensory disturbances, or vestibular dysfunction must be excluded as potential causes. This is often difficult since these impairments frequently coexist.

In addition to ataxia, we will discuss tremor, one of the most common movement disorders in MS. A tremor is a rhythmic, involuntary oscillating movement. Action tremor occurs during a targeted, voluntary movement (e.g., evident during finger-to-nose or heel-to-shin testing). Postural tremor is observed when a limb is maintained in a fixed position against gravity, with proximal large oscillations and/or distal, more rapid, oscillations. Axial tremor, with titubation of the head and trunk, may also be observed. MS tremor is most likely related to lesions in the cerebellar nuclei and cerebellar efferent pathways. The thalamus is also likely to play a role in MS tremor since the destruction or stimulation of thalamic nuclei has led to improved tremor in MS patients.

7.3.1 Prevalence and Impact

Tremor and ataxia affect many MS patients and, when severe, can be totally incapacitating, even in the absence of weakness or spasticity. Tremor of varying degrees was found to be present in 25% of MS patients from a specific geographical region. However, only 3% of those people with MS were found to have tremor that qualified as "severe."[36]

Tremor and ataxia are among the most difficult MS symptoms to treat,[37] as they are frequently nonresponsive to medications. The rehabilitation of disorders of coordination in general and, more specifically, cerebellar ataxia, is not well developed. It is rarely supported by published evidence, and anecdotal results are often disappointing.

7.3.2 Identifying and Measuring Incoordination

Upper-extremity coordination is measured with several types of clinical examination methods. Finger-to-nose testing is performed with the patient seated, eyes closed, with shoulders abducted to 90°, and elbows extended. The patient is asked to touch his or her index finger to the tip of the nose and hold for 5 s, alternating right and left side.[38] This allows the examiner to determine the presence of dysmetria and tremor. Rapid forearm pronation and supination is performed with one hand facing palm up, while the other forearm is rapidly pronating and supinating from a "palm-down" to "palm-up" position to assess dysdiadochokinesia. For the lower extremity, the heel-to-shin test consists of asking the patient to place one heel on the opposite knee, then slide the heel down the shin, looking for dysmetria and tremor. Heel tapping or foot tapping is used to demonstrate dysmetria and dysdiadochokinesia. The Romberg test can be helpful in distinguishing cerebellar ataxia from sensory ataxia, since an increased swaying with eyes closed compared to eyes open is expected when there is sensory loss.

Standardized methods for measuring coordination are used mostly for the upper extremity. Fahn's Tremor Rating Scale is a widely used ordinal scale for the assessment of tremor severity, with satisfactory interrater and intrarater reliability.[39,40] Quantitative measurement of tremor can be performed with EMG recordings, accelerometers, or both. Functional tests can be helpful in judging the impact of coordination impairments and, when repeated over time, in monitoring their evolution and the effect of treatments. These may include Archimedes spiral drawings, butterfly drawings, handwriting samples, or simple everyday tasks such as drinking water from a glass. The Nine Hole Peg Test,[41] which is one of the three components of the Multiple Sclerosis Functional Composite (MSFC), or for more severely impaired clients, the Box and Blocks Test,[42] may also be used to measure upper extremity coordination. Functional tests for lower-extremity ataxia are essentially tests of balance and gait, which are discussed elsewhere (see Chapters 6 and 13). Performance on these tests can be videotaped and in some cases quantified via digitized recordings, such as the recently proposed digitized spirography.[43] While digitized spirography appears as a valid method and was found to be sensitive to short-term changes, it is not a technology that is routinely available or as practical to administer in a clinical setting as the previously mentioned tools.

7.3.3 Improving Abnormal Coordination

7.3.3.1 Medications

A variety of medications have been proposed to treat tremor and ataxia in MS, all of them off label. Isoniazid, an antibiotic used to treat tuberculosis, has been tested in several randomized clinical trials with variable results.[44,45] Side effects of isoniazid include visual problems, abnormal liver function tests, sleepiness, fever, rash, nausea, dysphagia, increased bronchial secretions, and weakness. Other medications that have not been as extensively tested include levetiracetam, oral tetrahydrocannabinol, primidone, carbamazepine, clonazepam, gluthetimide, and L-tryptophan. Propranolol, frequently used to treat essential tremor, is usually not effective on intention tremor.

7.3.3.2 Thalamic Surgery

Surgical treatment is sometimes performed for severe MS tremor. Stereotactic thalamotomy was shown to produce immediate relief of MS tremor in most patients who underwent the procedure in a published case series,[46] but long-term results are not as satisfactory, particularly when looking at functional improvement. Complications may include new onset or worsening of neurologic deficits (e.g., hemiparesis, dysphagia, speech disturbance, cognitive deficits, bladder dysfunction), which are usually transient, but may persist in the long term. Speelman et al.[46] reported a worsening of EDSS scores in 4 of 11 patients following thalamotomy.

Deep brain stimulation (DBS) was initially approved by the Food and Drug Administration for the treatment of essential and Parkinson's tremor. DBS consists in implanting an electrode in specific areas of the brain using stereotactic and intraoperative recording methods. The other end of the electrode is connected to a pulse generator (pacemaker) inserted under the skin of the upper chest. This method is more attractive than thalamotomy because it is nondestructive and more flexible (the stimulation parameters can be adjusted with a programming device). Relief of MS tremor in the first few weeks has been reported in almost all patients treated with DBS. Long-term results are usually not as consistent as those observed in Parkinson's disease and essential tremor. While adjustments to the stimulation parameters generally produce an immediate improvement of tremor, the benefit seems to decrease rapidly with time, requiring further adjustments.[47]

7.3.3.3 Rehabilitation

Exercise programs for the treatment of ataxia typically focus on weight-bearing activities that provide distal stabilization while promoting proximal control. For example, quadruped exercises may be performed (patient on hands and knees), which stabilize the joints distally and allows the therapist to target proximal muscle groups (e.g., scapular stabilizers). Alternative positions include, but are not limited to, standing with arms stabilized against a wall, sitting with arms outstretched against a mat table, tall kneeling with arms stabilized against a physioball, or half kneeling with arms stabilized against a wall. It is important to understand that proximal control and stability are required to foster an improvement in distal coordination, but, in the case of the ataxic patient, the proximal control must be combined with distal stabilization. These closed-chain exercise examples that are listed above provide stabilization that helps dampen ataxic movements and are most beneficial for the treatment of ataxia in the extremities and the trunk. Open-chain exercises, in which the hands or feet are not stabilized distally, typically result only in worsening of tremor, poor outcomes, and a frustrated patient.

Balance-based torso weighting (BBTW) is a technique that uses a weighted vest in an effort to provide greater truncal stability and improve gait stability. The weights are placed within the vest and are intended to counteract directional loss of balance. The weights are typically very light (<1.5% of the patient's body weight) and are placed opposite the direction of loss of balance. Research by Gibson-Horn and Widener indicated that the imbalanced patients had an immediate improvement in upright mobility and walking speed in the T25FW and in the TUG.[48]

For upper-extremity incoordination, strengthening exercises can be very beneficial in maintaining strength and preventing disuse weakness; however, strengthening will not reduce the tremor. The most practical physical approach is to apply a several-pound weight to the upper limb, increasing the mass, which consequently decreases the excursion produced by any given force applied. Weakness can be an obstacle to the use of weights, especially for repetitive movements. Along the same line, weighted objects, such as weighted utensils, pens, or cups, may be used for mild tremors. For more severe tremors, the upper extremity may be stabilized distally to minimize the magnitude of the tremor. The use of Dycem® or Velcro® may help to stabilize the upper extremity at the wrist when performing functional tasks. For feeding, the wrists may be supported distally on a tray table or raised support, which provides stability and a dampening effect on the severity of the tremor. In addition, wrist cuffs or wrist rests provide distal stabilization during computer use. To provide stability on unstable objects such as tub seats, clamps or suction cups may be used.[49] Computer adaptations that may also reduce the effects of tremor include the use of Sticky Keys, which allow a user to press one key at a time rather than holding down two simultaneously (e.g., CTRL or ALT with another key), or the use of filter keys, which can ignore brief or repeated keystrokes or slow down the repeat rate. The sensitivity of the computer mouse may also be dampened for lower sensitivity, thereby reducing the effect of the tremor.

For lower-extremity loss of coordination, few rehabilitation techniques have been proposed. Assistive devices for mobility, such as canes and walkers, can be helpful in broadening the base of support.

The link between the oculomotor system and the locomotor system suggests that deficits in ocular movements may translate into worsening of intention tremor in affected patients with MS. Conversely, using gaze stabilization techniques may help reduce the effect of upper-extremity tremor and ataxic gait. Crowdy et al. (p. 244)[50] used a technique of "rehearsal by eye movement," in which patients made repeated saccadic eye movements to the first 6 footfall targets (in a sequence of 18) while standing stationary at the start of the walkway. This technique was employed immediately prior to ambulation through a walkway and showed a significant improvement in locomotor performance. Other studies have indicated that in patients who have MS with intention tremor, unsteady gaze fixation influences the severity of the intention tremor during eye–hand coordinated tasks.[51] These studies suggest a need to incorporate ocular exercise into a rehabilitation program aimed at reducing the effects of tremor on function in MS. Further research is needed to clarify these findings and offer more guidance on these rehabilitation techniques.

7.3.4 Key Points about Coordination and Tremor

- Disorders of coordination are common in MS and are often a source of frustration.
- The impact of ataxia on functional performance is often difficult to separate from the impact of other impairments, such as weakness and spasticity.
- Interventions to improve coordination are limited, but encouraging recent findings regarding specific rehabilitation techniques warrant further research.

7.4 Impact of Rehabilitation Setting

The information and case study presented in this chapter refer to outpatient rehabilitation settings, since most of the care of MS patients is delivered in outpatient offices or clinics. Inpatient rehabilitation is mostly indicated after severe acute exacerbations of MS, after acute health events that result in a worsening of MS-related impairments (e.g., surgery, sepsis), and when progression of complex MS symptoms threatens a patient's ability to manage at home. Home rehabilitation is the preferred option when patients have difficulty going to an outpatient rehabilitation facility (e.g., severely decreased mobility, lack of adequate transportation, long travel time to the nearest facility that has experience with MS rehabilitation). The principles for the rehabilitation of disorders of strength, tone, and coordination are identical in each setting, but the intensity, modalities, and focus will change. In the home, limited rehabilitation equipment is available, but the therapist has a unique opportunity to work on functional tasks in the patient's own environment, to better understand specific obstacles (including the need for modifications to the environment), and to adjust the goals and techniques accordingly. In an acute inpatient rehabilitation setting, the intensity and frequency of rehabilitation sessions is greater, and it is necessary to make sure that the patient will be able to tolerate this level of rehabilitation through a preliminary inpatient, outpatient, or home physical and occupational therapy evaluation. Inpatient rehabilitation can also be provided in a subacute setting (skilled nursing facility) when the patient requires a less intense level of rehabilitation or to ease the transition from acute inpatient rehabilitation to the home setting.

7.5 Conclusion and Directions for Future Research

Although clinicians can already draw from a substantial body of knowledge (both empirical and evidence based) to address changes in tone, strength, and coordination, significant gaps remain that represent opportunities for future research. For example, the pathophysiology of these impairments is not yet completely understood. There is also a need to better understand how various combinations of impairments affect functional performance and which treatment strategies are best suited, depending on which impairments are present.

CASE STUDY

"Kathy" is a 55-year-old woman who was diagnosed with relapsing–remitting MS six years prior to her physical therapy evaluation. She is a single mom with a 15-year-old son. She presented during an MS relapse with "more overwhelming fatigue than previous, feeling exhausted at night after working during the day." (Her job involves car travel, then standing for extended periods and offering presentations to customers.) Over the previous four to six weeks she had noticed worsening imbalance, more frequent falls, a need to use a cane all of the time, leg pain at night rated 5/10 on a numeric pain-rating scale (NPRS), and spasms in both legs at night. She has been walking with trekker poles in each hand on walking trails or grassy surfaces. She enjoyed yoga and felt that it was beneficial, but had "lost her motivation" to exercise despite acknowledging that it had been helpful to her.

Kathy has been under the care of a neurologist and nurse practitioner in the MS center. After obtaining a detailed history of her symptoms and performing a neurologic examination to assess underlying impairments, they decided against a course of i.v. steroids due to a prior history of severe side effects, but prescribed symptomatic medications for her leg symptoms at night, including baclofen and pramipexole. Unfortunately, she developed nausea and vomiting with both medications so she stopped them.

During her first evaluation, the Modified Fatigue Impact Scale score was 62/84, indicating a "maximal" level of fatigue. On examination she presented with proximal weakness in both lower extremities (MMT grade 4/5 in hips all planes as well as 4/5 in knee flexors and extensors bilaterally). All other muscle groups of the upper and lower extremities had normal strength. There was mild spasticity in the right lower extremity only (Modified Ashworth score 1/4 in the right knee extensors and 1+/4 in the right plantarflexors). Upper extremity coordination was normal, but she had mild impairment bilaterally with heel to shin sliding maneuver. Proprioceptive testing of the lower extremities was mildly impaired, but visual testing was normal. Her gait was wide based, moderately ataxic, taking wide turns that were markedly unstable. She had poor eccentric control when descending stairs, resulting in an abrupt "drop" down to the next step.

Performance on timed tests and assessment scales was as follows:

- Timed 25 Foot Walk (T25FW, a test of maximum but safe walking speed on 25 ft): 7.84 s with a cane. This corresponds to a walking speed of 97.2 cm/s, while the published average maximum walking speed for healthy women in their 50s is 201.0 mc/s.[52]
- Timed Up and Go (TUG): 10.37 s (mean TUG time in healthy, elderly aged 60–69 years, 8.1 s).[53]
- Berg Balance score 40/56 (normal score 56/56).

- Dynamic Gait Index (DGI) 12/24 (normal score 24/24), with particular imbalance during 180° turns and walking with horizontal and vertical head movements. She was unable to safely bend to the floor to pick up objects.

Her initial treatment focused on safety, so she received gait training on various surfaces using a rollator walker. Although this was difficult to accept, the rollator was meant to supply short-term gait safety while she recovered from her MS relapse. She continued to work and was busy at home; therefore, she could not return for physical therapy sessions and relied only on a home exercise program. Her nocturnal spasms were a primary concern, so she was instructed in gentle stretching exercises for the painful muscle groups, to be performed 3 times a day (when first waking, midday, and in the evening before bed). Supine positioning tended to increase her extensor spasms, so she was taught to sleep side lying using a full-length body pillow (the same type that pregnant women may choose to use) for upper and lower extremity support. In an effort to combat her fatigue, aerobic exercise was included in her home program. She had a stationary bike at home and started riding it 5 minutes per day, 4 days a week at a "moderate" level of perceived exertion. She was asked to add 1 minute each week as long as she had no adverse effects from this aerobic training.

She returned 3 weeks after her initial visit, publicly complaining about her rollator but privately acknowledging its safety and that the "seat is always available" when she was tired. She was proud that she had faithfully performed her exercises and reported that her nocturnal spasms had slowly decreased, now rating 1/10 on the NPRS.

A progression of her exercise program was now indicated to encourage strengthening of the hip and lower extremity musculature. Weight-bearing exercise was the treatment of choice as it provided distal stabilization and joint approximation, minimizing her ataxic incoordination. She performed quadruped exercises for trunk and hip strengthening, tall kneel-to-heel sitting to tall knee, prone hip extension and side-lying hip abduction, and abdominal strengthening. These were performed at a submaximal level with 30-s rests between exercises to avoid fatigue. Half of the exercises were performed on Monday, Wednesday, Friday and the other half on Tuesday, Thursday, Saturday, to avoid strengthening the same muscle groups on consecutive days. Her previously taught stretches were still completed 3 times a day.

One week later she expressed that her side-lying hip abduction exercise was "exhausting" so these were scaled back to the easier-to-perform "clamshells." This is performed in side lying, with knee flexed to 45°. The upper leg is abducted while the feet remain together, creating an appearance of a clamshell opening and closing as the patient performs each repetition. Her overall tolerance to exercise was very good and her strength was improving, so it was decided that she needed more frequent therapy sessions (2×/week) to challenge her balance and gait coordination. She was able to arrange her work schedule to accommodate this increased frequency of visits. Her goal was "to walk without a cane or rollator."

Seated balance and strengthening exercises with a physioball ensued. Standing balance exercises were initiated, such as reaching for varied targets, trunk rotation, side bends, and bending to the floor to pick up objects. Proprioceptive exercises such as soccer ball alternating foot "taps," side-to-side and forward/backward single leg "rolls" were added. She did these initially with eyes trained on her feet, then straight ahead, and then eyes closed—a progression of techniques meant to advance the proprioceptive and balance challenges. Her gait had then improved to a level where she wanted to stop using the rollator walker indoors (with agreement from her PT), but she needed to practice negotiating surfaces that she would typically encounter in her home. She had reviewed fall safety tips with her occupational therapist, such as removing throw rugs and clutter, adding night lights, and avoiding carrying laundry on stairs. She was able to successfully discontinue the rollator indoors without falls or stumbles. Dynamic gait exercises were then added in her PT sessions, such as side-stepping, braiding, and pivot turns. The areas of major deficit on the Dynamic Gait Index (horizontal and vertical head turns) were then incorporated into the balance training. She walked down hallways with exam rooms, but had to turn her head alternately to each side of the hallway and read room numbers aloud while maintaining a steady gait path. Once this was mastered, cognitive challenges were added to her dynamic balance routine ("initially count backwards from 100 by 3s," then adjust challenge by counting backwards by 6s, then 7s) while performing these tasks.

As her PT discharge neared, she was able to stand alone, pass and trap a soccer ball with her son from a 15-foot distance on carpet, and step laterally to bend to pick up his soft "shots" with the ball. She had continued her work, but was glad to no longer need her rollator. To help avoid fatigue, she did learn to use barstool-type chairs rather than standing nonstop during her training seminars. Her scores improved as follows: T25FW: 5.69 s (from 7.84 s), TUG 7.23 s (from 10.37 s), Berg Balance Score 46/56 (from 40/56), DGI 19/24 (from 12/24). Her MFIS score improved to 45/84 (from 62/84). Her pain was now 0/10, and she reported only very rare nonpainful spasms at night. She was now riding her stationary bike 20–25 min, 4 times a week, and had also returned to yoga classes. In total, she had received 20 PT visits over a nearly five-month period and had successfully maintained her home exercise program, with her son's encouragement, after her discharge.

References

1. World Health Organization. International classification of functioning, disability and health. Geneva: World Health Organization; 2001.
2. Lance J. Symposium synopsis. In: R. G. Feldman, R. R. Young, W. P. Koella, editors. Spasticity: Disordered motor control. Chicago, IL: Year Book Medical Publishers; 1980.
3. Rizzo M, Hadjimichael O, Preiningerova J, Vollmer T. Prevalence and treatment of spasticity reported by multiple sclerosis patients. Multiple Sclerosis 2004; 10:589–595.
4. Ashworth B. Preliminary trial of carisodoprol in multiple sclerosis. Practitioner 1964; 192:540–542.
5. Tardieu G, Shentoub S, Delarue R. A la recherche d'une technique de mesure de la spasticite. Rev Neurol 1954; 91:143–144.
6. Bohannon RW, Smith MB. Interrater reliability of a Modified Ashworth Scale of muscle spasticity. Physical Therapy 1987; 67:206–207.
7. Penn RD, Savoy SM, Corcos D, Latash M, Gottlieb G, Parke B, Kroin JS. Intrathecal baclofen for severe spinal spasticity. New England Journal of Medicine 1989; 320:1517–1521.
8. Farrar JT, Troxel AB, Stott C, Duncombe P, Jensen MP. Validity, reliability, and clinical importance of change in a 0–10 numeric rating scale measure of spasticity: A post hoc analysis of a randomized, double-blind, placebo-controlled trial. Clinical Therapeutics 2008; 30:974–985.

9. Hobart JC, Riazi A, Thompson AJ, Styles IM, Ingram W, Vickery PJ, Warner M, Fox PJ, Zajicek JP. Getting the measure of spasticity in multiple sclerosis: The multiple sclerosis spasticity scale (MSSS-88). Brain 2006; 129:224–234.

10. Bethoux F. Management of spasticity in multiple sclerosis. In: J. A. Cohen, R. A. Rudick, editors. Multiple sclerosis therapeutics. 3rd ed. London: Informa UK Ltd.; 2007.

11. Multiple Sclerosis Council for Clinical Practice Guidelines. Spasticity management in multiple sclerosis. Consortium of Multiple Sclerosis Centers; 2003.

12. Kent-Braun JA, Ng AV, Castro M, Weiner MW, Gelinas D, Dudley GA, Miller RG. Strength, skeletal muscle composition, and enzyme activity in multiple sclerosis. Journal of Applied Physiology 1997; 8(6):1998–2004.

13. Castro MJ, Kent-Braun JA, Ng AV, Miller RG, Dudley GA. Muscle fiber type-specific myofibrillar actomyosin Ca^{2+} ATPase activity in multiple sclerosis. Muscle & Nerve 1998; 21(4): 547–549.

14. Thoumie P, Lamotte D, Cantalloube S, Faucher M, Amarenco G. Motor determinants of gait in 100 ambulatory patients with multiple sclerosis. Multiple Sclerosis 2005; 11(4):485–491.

15. Surakka J, Romberg A, Ruutiainen J, Virtanen A, Aunola S, Maentaka K. Assessment of muscle strength and motor fatigue with a knee dynamometer in subjects with multiple sclerosis: A new fatigue index. Clinical Rehabilitation 2004; 18(6):652–659.

16. Milner-Brown HS, Mellenthin M, Miller RG. Quantifying human muscle strength, endurance and fatigue. Archives of Physical Medicine & Rehabilitation 1986; 67(8):530–535.

17. Medical Research Council. Aids to the examination of the peripheral nervous system. London: Her Majesty's Stationery Office; 1976.

18. Iriarte J, Subira ML, Castro P. Modalities of fatigue in multiple sclerosis: Correlation with clinical and biological factors. Multiple Sclerosis 2000; 6(2):124–130.

19. Leocani L, Colombo B, Magnani G, Martinelli-Boneschi F, Cursi M, Rossi P, Martinelli V, Comi G. Fatigue in multiple sclerosis is associated with abnormal cortical activation to voluntary movement—EEG evidence. NeuroImage 2001; 13(6 Pt 1):1186–1192.

20. Filippi M, Rocca MA, Colombo B, Falini A, Codella M, Scotti G, Comi G. Functional magnetic resonance imaging correlates of fatigue in multiple sclerosis. NeuroImage 2002; 15(3): 559–567.

21. Schwid SR, Thornton CA, Pandya S, Manzur KL, Sanjak M, Petrie MD, McDermott MP, Goodman AD. Quantitative assessment of motor fatigue and strength in MS. Neurology 1999; 53(4):743–750.

22. Goldman MD, Marrie RA, Cohen JA. Evaluation of the six-minute walk in multiple sclerosis subjects and healthy controls. Multiple Sclerosis 2008; 14:383–390.

23. Zeller D, Kampe K, Biller A, Stefan K, Gentner R, Schutz A, Bartsch A, Bendszus M, Toyka KV, Rieckmann P et al. Rapid-onset central motor plasticity in multiple sclerosis. Neurology 2010; 74(9):728–735.

24. Cauraugh JH, Summers JJ. Neural plasticity and bilateral movements: A rehabilitation approach for chronic stroke. Progress in Neurobiology 2005; 75(5):309–320.

25. Goodman AD, Brown TR, Cohen JA, Krupp LB, Schapiro R, Schwid SR, Cohen R, Marinucci LN, Blight AR. Dose comparison trial of sustained-release fampridine in multiple sclerosis. Neurology 2008; 71(15):1134–1141.

26. Goodman AD, Brown TR, Krupp LB, Schapiro RT, Schwid SR, Cohen R, Marinucci LN, Blight AR. Sustained-release oral fampridine in multiple sclerosis: A randomised, double-blind, controlled trial. Lancet 2009; 373(9665):732–738.

27. Dalgas U, Stenager E, Jakobsen J, Petersen Y, Hansen HJ, Knudsen C, Overgaard K, Ingemann-Hansen T. Resistance training improves muscle strength and functional capacity in multiple sclerosis. Neurology 2009; 73:1478–1484.

28. Gutierrez GM, Chow JW, Tillman MD, McCoy SC, Castellano V, White LJ. Resistance training improves gait kinematics in persons with multiple sclerosis. Archives of Physical Medicine & Rehabilitation 2005; 86(9):1824–1829.

29. Strength training guidelines in multiple sclerosis [Internet]; c2007 [cited 2010 July 1]. Available from: http://www.ncpad.org/disability/fact_sheet.php?sheet=79&section=595.

30. Cattaneo D, Marazzini F, Crippa A, Cardini R. Do static or dynamic AFOs improve balance? Clinical Rehabilitation 2002; 16(8):894–899.

31. Sutliff M, Naft J, Stough D, Lee JC, Arrigain S, Bethoux F. Efficacy and safety of a hip flexion assist orthosis in ambulatory multiple sclerosis patients. Archives of Physical Medicine & Rehabilitation 2008; 89(8):1611–1617.

32. Taylor PN, Burridge JH, Dunkerley AL, Wood DE, Norton JA, Singleton C, Swain ID. Clinical use of the Odstock dropped foot stimulator: Its effect on the speed and effort of walking. Archives of Physical Medicine & Rehabilitation 1999; 80(12):1577–1583.

33. Sheffler LR, Hennessey MT, Knutson JS, Chae J. Neuroprosthetic effect of peroneal nerve stimulation in multiple sclerosis: A preliminary study. Archives of Physical Medicine & Rehabilitation 2009; 90(2):362–365.

34. Mochizuki H, Ugawa Y. Cerebellar ataxic gait. Brain Nerve 2010; 62(11):1203–1210.

35. Moeller J, Macaulay R, Valdmanis P, Weston L, Rouleau G, Dupré N. Autosomal dominant sensory ataxia: A neuroaxonal dystrophy. Acta Neuropathologica 2008; 116(3):331–336.

36. Pittock SJ, McClelland RL, Mayr WT, Rodriguez M, Matsumoto JY. Prevalence of tremor in multiple sclerosis and associated disability in the Olmsted County population. Movement disorders: Official Journal of the Movement Disorder Society 2004; 19(12):1482–1485.

37. Alusi SH, Glickman S, Aziz TZ, Bain PG. Tremor in multiple sclerosis. Journal of Neurology, Neurosurgery, and Psychiatry 1999 Feb;66(2):131–134.

38. Bickerstaff ER. Chapter 26: Co-ordination. In: E. R. Bickerstaff, editor. Neurological examination in clinical practice. 3rd ed. Oxford: Blackwell Scientific Publications; 1976.

39. Fahn S, Tolosa E, Marin C. Clinical rating scale for tremor. In: J. Jankovic, E. Tolosa, editors. Parkinson's disease and movement disorders. 2nd ed. Edinburgh: Churchill Livingstone; 1991.

40. Hooper J, Taylor R, Pentland B, Whittle IR. Rater reliability of Fahn's Tremor Rating Scale in patients with multiple sclerosis. Archives of Physical Medicine and Rehabilitation 1998; 79:1076–1079.

41. Rushmore JR. The R-C pegboard test of finger dexterity. Journal of Applied Psychology 1942; 26:253–259.

42. Cromwell FS. Occupational therapist's manual for basic skill assessment; primary prevocational evaluation. Pasadena, CA: Fair Oaks Printing; 1976.

43. Feys P, Helsen W, Prinsmel A, Ilsbroukx S, Wang S, Liu X. Digitised spirography as an evaluation tool for intention tremor in multiple sclerosis. Journal of Neruoscience Methods 2007; 160:309–316.

44. Bozek CB, Kastruckoff LF, Wright JM, Perry TL, Larsen TA. A controlled trial of isoniazid therapy for action tremor in multiple sclerosis. Journal of Neurology 1987; 234:36–39.

45. Hallett M, Lindsey JW, Adelstein BD, Riley PO. Controlled trial of isoniazed therapy for severe postural cerebellar tremor in multiple sclerosis. Neurology 1985; 35:1374–1377.

46. Speelman JD, Van Manen J. Stereotactic thalamotomy for the relief of intention tremor of multiple sclerosis. Journal of Neurology, Neurosurgery, & Psychiatry 1984; 47:596–599.

47. Montgomery EBJ, Baker, K.B., Kinkel, R.P., Barnett, G. Chronic thalamic stimulation for the tremor of multiple sclerosis. Neurology 1999; 53(3):625–628.

48. Widener GL, Allen DD, Gibson-Horn C. Randomized clinical trial of balance-based torso weighting for improving upright mobility in people with multiple sclerosis. Neurorehabilitation and Neural Repair 2009; 23(8):784–791.

49. Bhasin C. Occupational therapy in the management of multiple sclerosis. Physical Disabilities Special Interest Section Newsletter, American Occupational Therapy Association 1989; 12(4):1–3.

50. Crowdy KA, Kaur-Mann D, Cooper HL, Mansfield AG, Offord JL, Marple-Horvat DE. Rehearsal by eye movement improves visuomotor performance in cerebellar patients. Experimental Brain Research 2002; 146(2):244–247.

51. Feys P, Helsen W, Nuttin B, Lavrysen A, Ketelaer P, Swinnen S, Liu X. Unsteady gaze fixation enhances the severity of MS intention tremor. Neurology 2008; 70(2):106–113.

52. Bohannon RW. Comfortable and maximum walking speed of adults aged 20–79 years: Reference values and determinants. Age and Ageing 1997; 26:15–19.

53. Bohannon RW. Reference values for the Timed Up and Go Test: A descriptive meta-analysis. Journal of Geriatric Physical Therapy 2006; 29(2):64–68.

8

Cognition

Eynat Ben Ari, Ralph H. B. Benedict, Nicholas G. LaRocca, and Lauren S. Caruso

CONTENTS

Multiple sclerosis (MS) is a disease of the central nervous system, commonly affecting the brain; therefore, it is expected that many people with MS will exhibit deficits in cognitive functioning. However, in contrast to this understanding—dating back to Charcot's observations of memory and concept formation dysfunction[1]—cognitive manifestations of MS were underestimated and seldom examined in the neurology literature for many decades. With the emergence of the comprehensive care model in the early 1980s, there was a resurgence of interest in the topic, resulting in increased understanding of the nature of cognitive changes in MS, the development of standardized tests for quantifying cognitive changes, and identification of intervention methods.

Cognition refers to the integrated functions of the human mind and involves the acquisition, processing, and application of information. Through cognition an individual can process information from the environment and utilize past experiences to form behaviors and generate adaptive strategies. Cognitive impairments as a result of brain damage or illness can lead to profound functional limitations.[2] When cognitive impairments occur, an individual can experience difficulties in tasks that require cognitive abilities such as memory, learning, and organization of thought. Cognitive impairments can affect almost all life domains and lead to difficulties performing daily activities (such as dressing and eating), vocational activities, and socialization. In addition, cognitive impairments can cause behavioral changes, such as aggression or impulsivity.[2]

The pervasive impact of cognitive impairments on general functioning implies that cognitive deficits, when present, can affect almost all areas of rehabilitation. Thus, all health care professionals involved in MS rehabilitation can benefit from a basic understanding of cognitive impairments in this population. For example, cognitive impairment can affect mobility and balance since reduced attention and distractibility force people with MS to actively think about their walking to reduce falls. Cognitive decline can cause people with MS to limit social interactions for fear of appearing forgetful or slow, consequently increasing depression. Cognitive decline can also cause people to forget to take their medications.

The overarching aim of this chapter is to provide health care professionals working in the field of MS a detailed description of cognitive symptoms in MS and ways to address them in rehabilitation. After reading this chapter, you will be able to:

1. Discuss the prevalence and impact of cognitive impairments in MS on functioning and rehabilitation,

2. Introduce assessment tools and procedures appropriate for cognitive evaluation of people with MS, and

3. Provide recommendations for cognitive interventions in various rehabilitation settings.

8.1 Prevalence of Cognitive Impairments

It is commonly accepted that roughly one-half of all people with MS will experience clinical problems related to cognitive impairment during the course of their illness. Prevalence estimates vary for many reasons, including the type of MS diagnosed and demographic characteristics of the sample being assessed. Some studies that focused on people with MS who are mildly affected and recently diagnosed reported lower prevalence rates of impairment.[3] Similarly, volunteer samples from the community at large are typically found to have lower rates of impairment.[4,5] In contrast, in studies on consecutive attendees at neurology clinics, where there is more a progressive disease course and greater degrees of neurological disability, the frequency of cognitive impairment tends to be higher, roughly 60%.[6,7] In some people with MS, the degree of cognitive impairment reaches a threshold for diagnosis of dementia, that is, very marked or severe impairment in memory plus one other domain of cognitive function, and subsequent decline in activities of daily living. Roughly 10% of clinical samples[6] experience cognitive impairments that reach the threshold for diagnosis of dementia. These 10% have a higher risk for neurogenic personality changes or euphoria sclerotic (i.e., a chronic state of perpetual cheerfulness with poor insight and little appreciation for disability).[8–12] Cognitive changes in MS can occur at the early stages of the disease and independently of disease duration or severity.[13,14]

8.2 Domains of Cognitive Impairments

There is good consensus about the domains of cognitive function that are most commonly affected in MS. In the seminal study of Rao et al.,[4] a battery of 22 neuropsychological tests was administered to 102 participants with MS. The battery yielded 31 test measures, spanning the domains of intelligence, attention, mental-processing speed, memory (e.g., working memory, episodic memory), executive function, new learning, language, and spatial perception. Frequency of impairment was most apparent on tests measuring processing speed and episodic memory. Since then, extensive research has been conducted investigating the performance of people with MS in different cognitive domains to identify cognitive abilities that are more susceptible to change in this population. Impairments in cognition and perception commonly diagnosed in MS can be found in Table 8.1 and in the description below.

Probably the most common cognitive impairment in MS is memory, with a prevalence rate of 40–60%.[15] Studies have shown that difficulties retrieving information from long-term memory and deficits in new learning result from encoding impairments during the acquisition phase.[16,17] Though more trials are needed, people with MS can retrieve information similar to the way healthy controls do when given sufficient time to acquire new information.[16,18] People with MS also have difficulties with tasks requiring prospective memory.[19] Short-term memory (particularly verbal), procedural memory, and general knowledge are typically preserved.[15]

When tested on tasks that call for visual or auditory attention and rapid information-processing speed, individuals with MS performed more poorly than healthy controls.[4,18] Studies have shown that reduced information-processing in MS is mostly a matter of speed and that provided sufficient time people with MS can perform information processing tasks as well as healthy controls.[20–23] Tasks that require sustained or divided attention

TABLE 8.1

Cognitive Domains Commonly Impaired in MS

Cognitive Domain (%)[a]	Explanation
Visuo-spatial perception (~20%)	The ability to register and accurately interpret information. Most often examined with visually mediated tasks that require right/left discrimination, figure/ground discrimination, and estimating spatial relations.
Mental-processing speed and working memory (~50%)	The ability to efficiently and quickly process information and respond correctly, manipulate and/or resist interference of incoming information.
Learning and episodic memory (~54%)	Ability to learn and retrieve new information with successive exposure, such as a word list or an array of figures.
Executive functions (~20%)	Mental operations needed to independently perform complex, nonroutine, goal-oriented tasks. Executive functions include initiation, organization, planning, problem solving, decision making, mental flexibility, abstraction, behavior monitoring, and awareness.

Source: Adapted from Sohlberg MM, Mateer CA. Cognitive rehabilitation: An integrative neuropsychological approach. New York: Guilford Press; 2001; Benedict RH, et al. Journal of the International Neuropsychological Society 2006;12(4):549–558; Gillen G. Cognitive and perceptual rehabilitation: Optimizing function. St. Louis, MO: Mosby; 2009; Winkelmann A, et al. Journal of Neurology 2007;254(Suppl 2):II/35–II/42; Zoltan B. Vision, perception, and cognition. 4th ed. Thorofare, NJ: Slack; 2007.

[a] Percentage of people with MS who may exhibit impairments in the cognitive domain.

seem to be more challenging for those with MS[24,25] whereas simple attention can remain intact.[4] Given the reciprocal relationship between attention, memory, and executive functioning,[26] reduced processing speed can influence the performance of people with MS on everyday tasks that require working memory and higher mental functions, such as planning, reasoning, and organization.[27]

Individuals with MS perform less effectively on tests of executive functions, which suggests poorer problem-solving strategies; compromised decision making; and difficulty in completing complex, nonroutine tasks.[28,29] Fluency or the spontaneous production of words can also be impaired in MS (especially phonemic and semantic fluency[30]). Executive function impairments also have behavioral implications, such as reduced flexibility,[31] and affect awareness.[8] Studies have shown that reduced awareness in MS is not consistent across different dimensions of cognitive functioning.[32,33] For example, people with MS may be aware of more concrete declines (such as forgetting) and less aware of abstract declines (such as difficulties problem solving).[34]

Impairments in perception (i.e., the ability to appreciate relevant information and then distinguish between the characteristics of that information) can also occur in MS. Approximately 20% of individuals with MS experience changes in visuospatial abilities due to visual processing impairments (e.g., optic neuritis) or visual perception deficits.[35] General intelligence is typically preserved.[36]

Some demographic characteristics can affect cognitive impairment and functioning by lending cognitive reserves. For example, increased years of education or high IQ scores prior to disease diagnosis may afford a person increased processing capacity and strategies that can overcome neuropathologic change to a certain extent. On the other hand, neurological changes to the brain as a result of aging may accelerate cognitive decline among older individuals.

Clinicians should be aware of additional MS symptoms that can further impede cognitive functioning. The most common mood disorder in MS is depression. Studies with sufficient power have shown a modest, but often significant, relationship between cognitive functioning and depression in MS.[37] Depressive symptoms may aggravate cognitive changes, especially in tasks that require greater mental effort.[38,39] In addition, depressive symptoms are better correlated with subjective perceptions of cognitive difficulties rather than actual functional performance.[34] Therefore, it is possible that cognitive impairments reported by people with MS are associated with their depressed state and not with true cognitive dysfunction.[40,41] Fatigue, another common MS symptom,[42] may affect cognitive functioning since cognitively demanding tasks can potentially increase fatigue.[43,44] However, studies show that cognitive performance is not correlated with subjective fatigue.[45] The relationship between fatigue and cognition in MS is poorly understood and requires further investigation. For further information on mood changes and fatigue in MS rehabilitation, refer to Chapters 10 and 5, respectively. Many people receive medications to treat other MS symptoms, some of which can have adverse effects on cognitive abilities. Examples include steroids during exacerbation of symptoms and medications for pain or spasticity.

8.3 Cognitive Impairments in Pediatric Populations with MS

MS has been found to affect pediatric populations, which account for approximately 3–4% of all MS cases.[46] Studies have shown that cognitive changes can also appear in children with MS; specifically, changes have been noted in memory and learning, attention, executive functioning, processing speed, visual–perceptual abilities, and verbal fluency.[47,48] Longer disease duration and an early disease onset seem to be risk factors for cognitive deficits in children with MS.[49] Experiencing cognitive changes during critical formative years can have detrimental effects on cognitive maturation (including loss of neural networks) and educational attainment.[49] Children with MS report that cognitive changes have negative effects on their ability to perform daily activities and learn in school.[50]

8.4 Pathogenesis

MS most often affects the integrity of the brain, and thus it is of no great surprise that many people with MS are impaired when examined neuropsychologically. Unlike diseases that affect well-circumscribed or known regions of the brain in a homogenous fashion (e.g., Alzheimer's disease, Huntington's disease), MS is very heterogeneous from the standpoint of cerebral pathology. The list of regions affected is long, and a full review of MRI/cognition relationships is beyond the scope of this chapter. Table 8.2 offers a summation of some of the more well-established correlations in the literature.

The cognitive profile of MS is variable and complex, which makes delineating the brain/ behavior relationships among people with MS difficult. For any given client presenting with cognitive complaints (or presenting due to reports of cognitive problems by doctors or family members), the clinician must not only determine if the person is impaired, but have at least some notion as to why or what aspect of the brain pathology may account for the cognitive disorder.

The hallmark of MS is inflammatory lesions of the cerebral white matter most often occurring in the periventricular area, thus affecting not only pyramidal tracts but also long inter-cortical tracts such as the longitudinal fasiculus. These lesions can be acutely active with breach of the blood–brain barrier, which is detected by gadolinium enhancement MRIs (see Chapter 2). Some white matter lesions shrink or disappear altogether on conventional MRI while others are seen chronically and become more apparent on T1 weighted sequences, the so-called black holes.[51] In recent years, MRI technology has improved and lesions surrounding axons are also reported in the cerebral cortex.[52,53]

TABLE 8.2

Recommended Literature on MRI/Cognition Relationships in MS

Tissue Compartment or Pathology	Example Citations
Subtle tissue destruction in the NABT	Filippi et al.,[167] Iannucci et al.[168]
White matter lesions	Comi et al.,[169] Rao et al.[87]
Frontal-lobe lesions	Arnett et al.,[170] Swirsky-Sacchetti et al.[171]
Whole-brain atrophy	Benedict et al.,[58] Christodoulou et al.,[99] Rao et al.[87]
Regional atrophy—Corpus callosum	Rao et al.[172]
Regional atrophy—Cerebral cortex	Amato et al.,[100] Benedict et al.[7]
Regional atrophy—Frontal cortex	Benedict et al.,[173] Tekok-Kilic et al.[174]
Regional atrophy—Ventricle enlargement	Benedict et al.,[8] Benedict et al.,[58] Christodoulou et al.[99]
Regional atrophy—Deep gray matter	Benedict et al.,[13] Houtchens et al.[59]
Regional atrophy—Hippocampus	Benedict et al.,[13] Sicotte et al.[175]

While MS was once regarded as an exclusively autoimmune, white matter disease, the last decade has witnessed an increasingly greater appreciation of neurodegeneration, most often recognized on MRI as cerebral atrophy. This atrophy is mainly seen in the gray matter, in both cortical and subcortical regions. Gray matter atrophy is frequently associated with progressive disease course, and it does not consistently correspond with the degree of lesion volume,[54] especially among individuals with a progressive course.[55] Thus, MS may consist of two interrelated diseases, one axonal/inflammatory, the other neurodegenerative.[56,57]

These theoretical considerations notwithstanding, it is now known that indicators of brain atrophy account for more variance in predicting cognitive impairment in individuals with MS than lesions per se. This conclusion is based on the recognition of more robust correlations between atrophy and cognitive impairment. In addition, multiple regression analyses show greater retention of atrophy metrics rather than lesion burden in models predicting cognitive outcomes. This literature is extensive, but by way of example, a few studies from one of the authors are summarized below (RHBB).

In 2003, Benedict and colleagues, interested in using MRI to predict cognitive disorders among people with MS, investigated which conventional MRI metrics were most strongly associated with cognition. The research found that third ventricle width (TVW) best correlated with the outcome measures, spanning the domains of processing speed, memory, and executive function.[58] The researchers later determined in post-hoc analysis that the TVW was also predictive of neuropsychiatric/personality disturbance in this same sample.[8,12] The third ventricle separates the thalamic hemispheres, and thus the dilatation of this ventricle may arise ex-vacuo, with atrophy of the thalamus. The researchers then conducted manual tracings of the thalami in a new, larger sample[59] and found marked atrophy in MS versus age-matched controls. In addition, thalamic volume was strongly correlated with the cognitive outcome measures. Subsequent research from this group replicated the robust correlation between TVW and cognitive impairment in MS.[7,60] Brain/behavior correlations were also identified.

Altogether, these data reveal MS to be far more than a white matter disease, at least from a neuropsychiatric perspective. MS cognitive disorder is strongly correlated with several compartments of gray matter atrophy as well as central and whole brain atrophy. One can therefore appreciate how cognitive impairment in MS is caused by numerous, different sources of cerebral disease.

8.5 Impact of Cognitive Impairments on Everyday Life

The relative impact of cognitive impairments on functioning depends on the severity of the symptoms. Some people with MS may experience relatively mild impairments (e.g., with executive functions on occasion during very complicated tasks) while others may experience difficulties that are more pervasive. However, even mild changes can have significant ramifications on functioning should the impairments prevent the ability to perform important and valuable tasks (such as meeting deadlines at work or providing parental guidance to a child).

It has been reported in both professional and consumer-based publications that cognitive impairments in MS can be disruptive with significant effects on everyday functioning

> **BOX 1**
>
> "The images and thoughts racing through my head might best be described as watching a speeding train passing by. While standing close at a railroad crossing one tries to focus on a single train car while unsuccessfully attempting to catch a word or phrase as it speeds by. That is what it is like for me when I become cognitively frustrated, lost, dizzy, and unable to regain control."
>
> **Dan***

(see, e.g., references 61–63). Although cognitive impairments in MS are far from rare, some health care professionals tend to underestimate the impact of these changes.

Regardless of the populations in which they occur, cognitive impairments can lead to profound functional limitations in everyday activities and routines. Activities of daily living such as self-care may be compromised along with instrumental activities of daily living, such as driving, shopping, and managing finances. Performance can change considerably under different conditions and activities that are unpredictable or conducted in unfamiliar environments are particularly vulnerable. Cognitive impairments can affect social functioning as a result of communication difficulties or inability to register subtle social nuances. In addition, cognitive impairments can affect work-related activities and compromise vocational status. They are not always physically visible and, therefore, may be unknown to others unless disclosed.[2]

> **BOX 2**
>
> "When I first started experiencing cognitive symptoms, I was working as an attorney. I started having difficulty concentrating, my reading comprehension declined drastically, and I was falling asleep whenever I was doing a lot of reading and writing. When I realized this was something real, and not just me going crazy, I started telling people that I "thought with a limp" because no one understood why I was tired all the time."
>
> **Liz**

People with MS who experience cognitive impairments tend to experience similar disruptions in everyday activities. They may have difficulty performing personal activities of daily living, such as dressing and grooming.[64] Particular difficulties are found in daily activities that have greater cognitive demands, such as meal preparation, home maintenance, telephone use, paying bills, and managing finances.[34,64] These instrumental activities of daily living require processing, organization, planning, and problem solving that are frequently impaired in MS. In a recent study, Kalmar et al.[65] examined the relationship between cognitive symptoms and performance in daily activities (particularly activities requiring executive functioning) using an objective, standardized assessment, the Executive Functions Performance Test.[66] The authors found that cognitive symptoms contributed to reduced functional capacity in that cognitively impaired participants had more

---

* The authors would like to thank the generous and candid contributions from Dan, Jeffrey, and Liz, whose words breathe life into the topic of cognition and MS.

difficulties completing daily tasks and required greater assistance than people with MS who did not have cognitive symptoms or healthy controls. Taken together, these studies indicate that cognitive impairments have a major negative influence on functioning, participation, and quality of life, and further contribute to the disability in MS.

People with MS are required to manage a complex disease typically treated with a variety of disease-modifying and symptomatic interventions. As a result, adherence to treatment regimens can become a significant challenge. Among the reasons cited in the literature for failure to maintain adherence was forgetting to administer the medication.[67,68] Although many individuals can sometimes forget to take medications, individuals with memory impairment or organizational problems are more likely to experience such lapses, undermining the potential efficacy of treatment.

Unemployment is common in MS,[69] with cognitive impairments strongly contributing to vocational status.[5,70,71] In a 10-year follow-up study by Amato et al.,[72] cognitive impairments, independent of physical abilities, were related to participants' workplace activities. As work environments become more demanding, changes in processing speed, memory, and executive functions make it very challenging to negotiate the work environment. Changes in work status have considerable implications not only to well-being and participation, but also to available income and medical insurance. The attainment of higher or vocational education is also affected by cognitive changes as it calls upon many cognitive abilities.[62] Since MS can be diagnosed in early adulthood,[73] difficulties obtaining education beyond high school can further impede the ability to gain employment.

> **BOX 3**
>
> "The [cognitive] test is horribly long and difficult for those of us who struggle with cognitive issues. At a certain point I began to get teary eyed as I was struggling in areas which (in the past) seemed so naturally simple. I recognize the mental decline—I'm painfully aware—it hurts! The test strained my mind and emotions and I needed a rest."
>
> **Dan**

Cognitive symptoms can compromise the ability of people with MS to drive. Specifically, executive functions, visual memory, information processing, concentration, and visuo-spatial abilities have been found to be related to diminished driving capabilities in MS.[74] In a cross-sectional study incorporating neuropsychological testing and assessments of driving skills, Schultheis et al.[75] found that people with MS who had cognitive impairments demonstrated significantly poorer performance on driving tasks when compared with cognitively intact individuals with MS and healthy controls. People with MS who are experiencing cognitive impairments have also been involved in more motor accidents than people with MS with no impairments.[76] The inability to drive can have negative implications, such as difficulty getting to work and limiting participation in social activities, particularly for those who live in suburban and rural communities, where public transportation can be scarce. This growing area of MS research supports the incorporation of driving evaluations among people with MS, particularly individuals showing signs of cognitive difficulties.

People with MS who are managing cognitive changes are less likely to participate in social activities with family members and friends because of their reduced communication abilities. They also report reduced engagement in leisure activities.[5,77,78] Cognitive changes, especially verbal memory deficits and slowed information processing, can affect the fulfillment of family roles and responsibilities. People with MS may become unreliable and unable to complete familiar tasks, such as coordinating the family schedule. As a result, family roles may need to shift and certain types of responsibilities transferred from the person with MS to other family members.[79,80]

Cognitive changes also have a significant impact on emotions within the family. Cognitive changes are rarely discussed at the time of diagnosis, and family members are often confused when the person with MS begins to exhibit cognitive symptoms. The initial reaction may be anger and resentment, especially if the cognitive changes result in the individual's failure to fulfill basic family roles. Frustration and doubt can follow if family members feel they can no longer rely on the person with MS.[79] Feelings of loss are also possible, especially if the cognitive changes are severe enough to compromise the individual's ability to interact with other family members and participate in family activities.[81] This strain may require cognitive assessment, family counseling, or both to clarify what is happening, sort out the associated feelings, and move forward with necessary changes in family roles and responsibilities.

Legal competency issues can emerge when a person with MS becomes so cognitively impaired that he or she is no longer capable of managing personal affairs. Fortunately, only a small percentage (5–10%) of individuals with MS progress to this point. However, other individuals may experience deficits in judgment or worsening of their condition that can compromise their welfare and that of their families. The best way to prepare for this issue is through careful planning and consultation with a financial planner and an attorney who specializes in wills and trusts.[82] The potential for conflict and strain within families when legal competency is at issue can be tremendous, and the services of a social worker and other professionals may prove valuable.

> **BOX 4**
>
> "Cognitive fatigue is a complete game changer—the MS "rules" constantly change your energy level and ability to focus as you try to move forward. My energy is better spent understanding the limits on my mental and physical fatigue rather than arguing with MS. I tend to second guess whether I will have the mental tenacity to deal with even the simplest tasks and conversations."
>
> **Jeffrey**

In summary, the impact of cognitive changes among people with MS and their social networks can be varied (depending on severity of symptoms), multifaceted, and at times devastating. Cognitive changes can affect a client's confidence in his or her ability to engage in desired and valued occupations that can result in reduced self-esteem and sense of purpose. Some cognitively demanding activities may be eliminated; however, more meaningful activities, such as valued leisure activities or work, can be more difficult to give up and doing so is likely to be at an emotional cost.[62] Understanding cognitive changes requires a solid understanding of the context in which activities are performed, including personal, social, and environmental implications. It is important for professionals working with people who have MS to understand which activities are more meaningful to their clients (that may be more resistant to change) in order to suggest alternatives and increase participation and satisfaction within desired roles.

8.6 Cognitive Assessment Strategies

An essential first step to building a successful treatment plan is an evidence-based assessment process proved appropriate for people with MS and conducted by a multidisciplinary care team. Common reasons for conducting cognitive evaluations in MS can be found in Table 8.3. The high prevalence of cognitive symptoms in MS and their potentially debilitating effects on rehabilitation outcomes and functioning in daily activities justify the need for a cognitive evaluation for every MS client. In order to increase the sensitivity and specificity of the evaluation process, it is important to apply evaluation procedures that are reliable and valid for this population. This is particularly important when evaluating mild cognitive impairments as these can be harder to detect.

Various health care professionals may conduct cognitive evaluations with people who have MS. The chosen professional depends on the country's health care system, the specific rehabilitation setting and its policies, and available resources. Neuropsychologists have particular expertise in cognition and psychometric analysis. Occupational therapists are trained to evaluate the impact of cognitive impairments on disability and levels of functioning in daily activities. Speech–language pathologists focus on the relationship between cognitive impairments and language/communication capabilities. Other professionals on the care team who may conduct cognitive screenings are neurologists and nurses. Although cognitive evaluations should only be conducted by personnel with specific training in the field of cognition, it is imperative that all members of the rehabilitation team be aware of warning signs that signal the need for a cognitive evaluation.

Increased awareness of cognitive issues in MS by professionals is vital as cognitive difficulties may only become apparent during particular activities (such as taking medications or during meals) and visible to only some members of the rehabilitation team. For example, the physical therapist may notice an individual having difficulties learning a new exercise routine. This difficulty could be a sign of decreased executive function that can affect other important areas of functioning, such as driving and maintaining a household. Common warning signs for cognitive decline include:

- Confusion
- Forgetfulness
- Distractibility

TABLE 8.3

Common Reasons for Conducting Cognitive Evaluations in MS

- Establish a baseline for future comparison.
- Evaluate treatment efficacy.
- Assess cognitive strengths and weaknesses, especially before beginning cognitive and/or vocational rehabilitation.
- Identify one or more treatable conditions (e.g., affective issues, sleep disturbances, fatigue) that may be compounding cognitive symptoms.
- Address any questions that may arise concerning job performance and/or job accommodations.
- Determine the need for disability benefits.
- Address an individual's need to know and understand perceived changes in ability.

- Difficulties following instructions
- Difficulties learning something new or problem solving
- Safety risks
- Slowed thinking
- Impulsivity
- Unawareness or denial of difficulties when pointed out
- Complaints about difficulties with daily activities that require cognitive functioning
- Reports of communication problems with others

BOX 5

"My neuropysch testing was grueling. The tester tried to break it up over two days so as not to be too taxing, but it left me feeling totally wiped out and exhausted. The testing was a little intimidating because I didn't know what facets were being tested, and I had no idea how I was doing. But, in the end, when the results were explained to me, I felt validated. With my weaknesses/deficits identified, I was able to work around them."

Liz

Since cognitive evaluations can be time and cost consuming, brief procedures might be utilized to identify those at higher risk for impairment. The pretreatment assessment can provide guidance that the therapist, client, and family can utilize throughout the intervention. In the MS literature one can find a variety of assessments used to detect cognitive changes. Recent reviews of the literature have borne out the general impression that neuropsychological tests involving processing speed, working memory, and episodic memory are most sensitive for the evaluation of people with MS.[27,83] This section will include assessments that are most frequently reported and have supporting evidence. Functional-based assessments of daily activities and the use of computer technology for cognitive testing will also be discussed.

8.6.1 Self-Report

One method of obtaining information regarding cognitive symptoms and their impact is self-report questionnaires that collect subjective perspectives. These questionnaires typically include a series of questions regarding cognitive difficulties and ask respondents to rank symptom severity or task performance on a low-to-high scale. Self-report questionnaires can be completed by the person with MS or a caregiver (informant).

The MS Neuropsychological Screening Questionnaire (MSNQ) is a brief (5–10 min) 15-item questionnaire assessing client and informant perceptions of cognitive and neuropsychiatric changes in MS.[40] Items used for screening cognitive impairments are scored on a 5-point rating scale (0–4) accounting for both frequency and severity of symptoms. Carone et al.[41] examined client/informant discrepancies on the MSNQ. They found that MSNQ discrepancy scores were normally distributed and that the absolute values of these discrepancies can be viewed as a measure of awareness. When the discrepancies are in the direction of the person with MS overreporting cognitive problems relative to the informant, depression is commonly seen. In contrast, when the person with MS underreports cognitive problems, very often there is evidence of euphoria and personality changes. In addition, informant reports are more strongly correlated with the actual performance of people with MS on cognitive tests. This finding is in line with the executive function deficits and reduced self-awareness associated with MS. Although other self-report questionnaires can be found in the MS literature, the MSNQ provides the strongest evidence of validity and reliability.

Individuals' self-report of cognitive difficulties are an insufficient indicator of cognitive impairment, and clinicians should refrain from using this method as the sole means of assessment. It is often beneficial to include a caregiver to provide additional information and personal insight, and it is highly recommended to include objective testing.[84]

8.6.2 Neuropsychological Assessment Batteries

A neuropsychological assessment of an individual's cognitive abilities and impairments is obtained by administrating a selected battery of objective tests and inventories. The length of the battery can vary from a half-hour screen to a 16-h comprehensive battery, depending on the depth of functional information required. Because the pattern of cognitive changes in MS varies from person to person and since the individual's intervention goals also vary, the evaluation needs to be comprehensive enough to cover a wide variety of cognitive domains as well as sensitive to the variations between individuals. Neuropsychological assessment batteries allow clinicians to determine which cognitive domains are most impaired and, thus, work to improve or compensate for these areas during treatment (e.g., memory, processing speed, attention, etc.).

Since neuropsychological evaluations tend to be expensive and time consuming, two approaches have been developed and validated that narrow the scope of neuropsychological testing: honing in on those domains most often compromised and using tests with known brevity, sensitivity, and reliability.

The first approach is widely known for its originator, Stephen Rao, who conducted much of the seminal early work in the neuropsychology of MS.[4,5,85–91] In 1991, Rao et al.[4,5] administered a comprehensive neuropsychological battery to measure the prevalence and psychosocial impact of cognitive impairment in MS. To this day, these are the most frequently cited works on the topic. Rao also set out to identify a subset of the most sensitive tests from the battery that could be used for screening purposes. The Neuropsychological Screening Battery for MS (NSBMS) was proposed. These tests included the Selective Reminding Test (SRT),[92] the 7/24 Spatial Recall Test,[93] the Controlled Oral Word Association Test (COWAT),[94,95] and the Paced Auditory Serial Addition Test (PASAT).[96]

Soon afterward there was an initiative to study the effects of new disease-modifying therapy in MS, and some investigators, especially in Europe, called for a repeatable, brief battery based on the original publications by Rao and colleagues. In answer to the call, Rao et al.[4] developed a similar battery called the Brief Repeatable Neuropsychological Battery (BRNB). For this battery, the researchers took the original NSBMS and made two modifications—they added the Symbol Digit Modalities Test (SDMT)[97] and changed the 7/24. The SDMT augmented the assessment of mental-processing speed. The 7/24 Visual Recall Test was replaced by the 10/36 Spatial Recall Test (10/36) in order to have a larger matrix from which to develop alternate test forms. Multiple test forms were created (to reduce practice effects), and the BRNB was translated into multiple European languages, enabling cross-cultural research. The BRNB takes 45–60 min to administer.

Subsequent studies demonstrated that the BRNB tests are sensitive[98] and correlate well with brain MRI measures, such as ventricle enlargement[99] and neocortical volume.[100] It is still a core test battery used throughout Europe,[101,102] and it reveals longitudinal changes in people with MS.[72,103]

While the Rao batteries were valid and very useful in many respects, it was recognized that the battery was perhaps too brief, with little appraisal of higher executive functions and visual/spatial processing, two domains that are often involved in cerebral disease. In addition, little was known about the psychometric properties of the NSBMS/BRNB. The following question was posed: If a clinician had more time, what would be an optimal, yet minimal, record of neuropsychological status for people with MS? The Consortium of MS Centers then hosted an expert panel in 2001 to address this question. The objective was to determine by examination of the scientific literature and expert consensus the most reliable and valid neuropsychological battery for routine clinical and research use in MS. This meeting culminated in what has come to be known as the Minimal Assessment of Cognitive Functioning in Multiple Sclerosis (MACFIMS).[104]

The MACFIMS includes much of Rao's original contribution but replaces the 10/36 with the Brief Visuospatial Memory Test-Revised (BVMTR)[105,106] and the SRT with the California Verbal Learning Test-Second Edition (CVLT2).[107] The SRT and the 7/24 or 10/36 were judged to have many merits, but little was known about their reliability, and the normative data were sparse. In contrast the BVMTR and CVLT2 had been extensively studied in this regard, and the available literature indicated that they were as sensitive as the BRNB tests. The MACFIMS battery also added the Delis–Kaplan Executive Functioning System Sorting Test (DKEFS),[108] also known as the California Card Sorting Test.[32] In addition, the Judgment of Line Orientation Test (JLO) was included to measure visual/spatial processing.[109] In total, the MACFIMS takes approximately 90 min to administer. More information on the MACFIMS subtests can be found in Table 8.4.

Much work has been done since the 2001 meeting regarding prospective evaluation of the reliability and validity of the MACFIMS. In one interesting study,[110] 34 persons with MS were examined at baseline and one week later. After baseline, the participants were randomly divided into alternate and same form conditions, with the primary objective

TABLE 8.4

MACFIMS Subtests

Test Name	Cognitive Domain	Time to Administer	Published
Controlled Oral Word Association Test	Language	5 min	Psychological Assessment Resources, Inc.
Judgment of Line Orientation	Spatial processing	15 min	Psychological Assessment Resources, Inc.
California Verbal Learning Test, 2nd edition	Auditory/Verbal learning	20 min	Pearson Assessment, Inc.
Brief Visuospatial Memory Test Revised	Visual/Spatial memory	10 min	Psychological Assessment Resources, Inc.
Symbol-digit Modalities Test	Visual-processing speed and working memory	5 min	Stephen Rao, PhD
Paced Auditory Serial Addition Test	Auditory-processing speed and working memory	15 min	Stephen Rao, PhD
Delis Kaplan Executive Function System (DKEFS) Sorting Test	Executive function	15 min	Pearson Assessment, Inc.

being to test for repeat testing effects in these conditions. Reliability for the CVLT2 and BVMTR was good or excellent, provided that the alternate forms were used at the second testing. Large sample cross-sectional studies have also shown that the MACFIMS is sensitive, has good construct validity, and predicts vocational outcomes well.[111,112] Research is also being conducted to provide new normative data that accounts for demographic characteristics.[113]

Neither the Rao nor the MACFIMS approach is intended to dictate precisely which cognitive tests should be used in clinical or research settings; however, clinicians who are conversant with the literature are able to determine which domains should be examined and to take into account psychometric standards. In one recent study, the batteries were compared, yielding similar results when discriminating people with MS from controls.[114]

In the past few years, computer-based alternatives have been proposed for the screening of cognitive impairments in MS. Wilken et al.[115] evaluated the Automated Neuropsychological Assessment Metrics (ANAM), an eight-subtest, computerized screening battery assessing cognitive domains most affected in MS. The researchers tested the ANAM's ability to differentiate between people with MS who have cognitive impairments and controls compared to a traditional set of paper–pencil cognitive tests. Moderate-to-high correlations were found between participants' performance on the ANAM and the traditional tests. Overall, 95.8% of participants were correctly classified as cognitively intact or impaired using the ANAM. In a more recent study by Hanly et al.,[116] impairment on at least one ANAM subtest was detected in 75% of participants with MS, with 20% failing more than four subtests.

Younes et al.[117] used the Cognitive Stability Index (CSI) to screen for cognitive impairment over the Internet. The CSI, which includes nine subtests covering attention, memory, processing speed, and response time, was compared to a standard neuropsychological battery. The results of the study indicated that the CSI was a valid screening tool for cognitive changes in MS and was superior to the PASAT. Further research is needed to weigh the CSI's ability to substitute for in-person testing.

Computer-based testing is another assessment option available for clinicians for testing cognitive abilities and impairments with MS. This option may be useful for clinical settings as well as evaluation of populations with limited access to MS clinics (e.g., rural dwelling). It is recommended that computer-based testing be conducted under clinician supervision as the testing context and environment can have significant implication on the results. Further research is needed in this area to support this method of evaluation, particularly its ability to detect changes in executive functioning that are best evaluated in novel, complex tasks.

8.6.3 Assessments of Performance in Daily Activities

Another important component of the cognitive assessment in MS is the evaluation of the impact of cognitive symptoms on performance in daily activities. Previous procedures mentioned in this chapter capture cognitive changes at the impairment level, allowing clinicians to identify the cognitive domains most affected in their clients (e.g., memory, attention, etc.). Assessments of performance in daily activities supplement impairment-based tests by clarifying the relationship between cognitive impairments and functional capacity. The evaluation of functional capacity allows clinicians to assess how people with MS utilize their cognitive skills and guided cues in simulated, real-world activities. Hence, these assessments have increased ecological validity due to their ability to predict functional impairment. Identifying activities that are particularly challenging for people with

MS can guide therapists in their choice of intervention activities and facilitate meeting the rehabilitation goals.

A number of assessments of daily activities have been tested among MS populations. The Executive Functions Performance Test (EFPT)[66] includes five activities: hand washing, simple cooking, telephone use, medication management, and bill payment, and takes approximately 60 min to administer. A hierarchy of cues (gestural, verbal, or physical) is used as assistance when difficulty with task execution is identified. Task completion is evaluated based on level of independent performance and need of assistance. A study by Goverover et al.[34] indicated that people with MS performed significantly worse than controls on the EFPT tasks, needing more assistance and cues.

Another objective measure of performance of daily activities is the Behavioral Assessment of the Dysexecutive Syndrome (BADS).[118] The BADS, which has been used in several MS studies (e.g., references 119 and120), takes 40 min to administer and includes six subtests examining different areas of executive functioning (e.g., problem solving, planning, and monitoring). The BADS has been validated among various neurologic populations, including people with MS.[121] It showed good concurrent validity as its subtests were significantly correlated with impairment-based assessment of executive functions. Construct validity was also established. The authors claim that the BADS holds stronger ecological validity in its ability to predict real-world functioning.[121]

The Rivermead Behavioral Memory Test (RBMT)[122] and the Test of Everyday Attention (TEA)[123] have also been used in MS research.[119,124] The 30 min RBMT includes tasks analogous to everyday situations that are difficult for people with MS exhibiting memory impairments, such as remembering names, recalling a series of actions, and new learning.[122] The TEA (45–60 min administration time) is comprised of eight subtests (such as searching a map and a telephone directory) requiring selective, sustained, divided, and shifting attention.[123] Higginson et al.[124] concluded that the RBMT and the TEA were better predictors of functional difficulties than standard tests or self-report questionnaires because they are more closely related to daily activities. Both the RBMT and the TEA include alternative versions, thus reducing learning effects when multiple administrations are needed (e.g., pre- and post intervention).

An alternative method to assess performance in daily activities is the use of virtual reality (VR). VR allows users to engage in simulated daily environments that have high ecological validity through the use of real-life events and objects.[125] VR has been found to facilitate performance among persons with cognitive changes (e.g., reference 126). Examples of VR environments include virtual apartments, supermarkets, and pedestrian street crossings. Using VR software, therapists can conduct initial cognitive evaluations in monitored and safe environments while receiving detailed reports of the client's performance in graded tasks and across multiple variables (e.g., incorrect vs. correct responses, time usage, organization, etc.).[127] Studies using VR to assess cognitive functioning among neurological populations, including those with MS, have shown promise, though more research is needed in the area.[128,129]

The use of the Internet as a platform to assess cognitive functioning has also been examined in MS. Goverover et al.[130] developed an assessment protocol using Actual Reality, instructing participants with MS to purchase airline tickets using an Internet site. The task called upon several cognitive abilities, such as decision making, problem solving, and organization of information and time. Correct and incorrect actions were calculated and need for cuing was monitored. People with MS performed potentially worse than controls and the Actual Reality task was significantly correlated with participants' cognitive impairments as measured by the MACFIMS battery.

8.6.4 Summary

A wide range of cognitive assessment tools are available to clinicians working with MS populations (a summary of the standardized assessment batteries can be found in Table 8.5). This variety can be burdensome, making the selection of assessment methods difficult. The evidence points to the wisdom of compiling several types of information in the evaluation process (i.e., clinical interview, self-report, impairment based, and assessment of daily activities)

TABLE 8.5

Summary of Assessments Batteries for Cognitive Impairments in MS

Assessment Tool	Clinical Utility	Psychometric Properties	Who Can Administer	Where to Obtain
MS Neuro-psychological Screening Questionnaire (MSNQ)[40]	Self-report questionnaire measuring frequency and severity of cognitive and neuropsychiatric changes in MS	The questionnaire includes 15 items with a reliability coefficient of $r = 0.93$. Informant reports are more strongly correlated with the actual performance on cognitive tests of people with MS.	Can be completed by person with MS or informant (e.g., caregiver)	Ralph H. B. Benedict, PhD
Brief Repeatable Neuro-psychological Battery (BRNB)[4]	A brief (45–60 min) neuro-psychological battery of selected tests measuring processing speed, memory, word retrieval, and working memory	BRNB tests are sensitive and correlate well with brain MRI measures. The battery reveals longitudinal changes in people with MS.	Neuropsychologist or health care professional with specific training and experience in neuropsychological assessment	Stephen Rao, PhD
Minimal Assessment of Cognitive Functioning in MS (MACFIMS)[104]	Neuropsychological assessment battery (90 min) developed by expert panel. Areas of assessment similar to BRNB + visual/spatial processing and executive functioning	Reliability and validity of each test is well established in the literature. Psychometric properties are well known and measures correlate well with brain MRI, vocational status, and other external validators	Neuro-psychologist or health care professional with specific training and experience in neuropsychological assessment	See Table 8.4
Automated Neuro-psychological Assessment Metrics (ANAM)[176]	A computerized screening battery including eight subtests (30 min) measuring reaction time, information-processing speed, memory, working memory, mathematical calculations, problem solving, and reasoning.	Moderate-to-high correlations were found between validated, conventional cognitive tests and ANAM subtests (among people with relapsing–remitting MS). ANAM scores are sensitive to cognitive changes in MS	Neuropsychologist or health care professional with specific training and experience in neuropsychological assessment	Center for the Study of Human Operator Performance, University of Oklahoma, Norman, OK

in order to link the impairments of those with MS to their functional capacity and to guide clinicians in the setting of intervention goals.

Clinicians should be aware that cognitive changes can exist even with mild physical impairment. In these cases, knowledge of functional difficulties identified through a clinical interview or a self-report questionnaire can raise red flags and signal the need for a more comprehensive, objective assessment. It is important to note that evaluation procedures should be conducted by professionals trained both in MS and in the specific assessment protocols to be utilized. In addition, clinicians need to be cognizant of the personal and emotional impact cognitive testing can have on their clients. Testing can indeed be a grim wake-up call for some. Others may feel a sense of relief at finally understanding what is wrong. Whatever the emotional reaction may be, clinicians are advised to offer support and validation.

8.7 Treatment Planning: Interventions for Cognitive Challenges

8.7.1 Pharmacological Interventions

Medication targeted at cognitive symptoms in MS is an area of burgeoning interest and activity. Different agents may be selective in their ability to effectively treat the two primary domains of cognitive dysfunction in MS, mental-processing speed and episodic memory. Potassium channel blockers could conceivably facilitate neuronal function in regions important for attention or processing efficiency. Bever et al.[131] studied the effects of oral 3,4 diaminopyridine compared to nicotinic acid control over 30 days. Significant gains were seen in the individuals treated on measures of lower-extremity function (leg strength), but there was no impact on neuropsychological testing. A larger double-blind study on 4-aminopyridine was carried out by Rossini et al.,[132] with cognition serving as a secondary outcome. Again there were no effects of the drug on neuropsychological testing.

> **BOX 6**
>
> "[Difficulties] during verbal communication invariably lead to anxiety, frustration, and "MS shutdown." By writing e-mails I think I am protecting my relationships from the frustration since I might otherwise be mistaken for an ornery and/or unhappy old man."
>
> **Dan**

Regarding the testing of disease-modifying agents on cognitive abilities, Wilken et al.[133] conducted a pilot study comparing people with MS interferon-β-1a with modafinil against those receiving only interferon-β-1a. While the authors did report beneficial effects on multiple tests emphasizing processing speed or executive function, there were some substantial methodological limitations to the study. Two studies examining the effects of methylphenidate[134] and L-amphetamine[135] showed positive effects on cognitive testing when the tests were administered shortly after drug administration. However, the effects were not replicated in a large sample study of daily 30 mg dose.[45]

Other studies have tested pharmaceuticals targeted at specific cognitive abilities. In the area of memory, Krupp et al.[136] examined the effects of donepezil over 24 weeks. While there were a few methodological shortcomings (e.g., small sample, treatment groups not matched on disease course), positive effects favoring the active arm of the study were seen on an adapted version of the Selective Reminding Test and a clinician impression of change. However, these findings were not replicated in a study examining the effects of a similar medication, rivastigmine[137] or in a larger multicenter study using donepezil.[138]

Although the above research shows little promise concerning the potential benefits of pharmaceutical agents on cognitive symptoms in MS, studies continue to explore new leads. However, until complete alleviation of symptoms or a cure can be achieved, people with MS will continue to struggle with cognitive changes and their implications. For this, cognitive rehabilitation has much to offer.

8.7.2 Cognitive Rehabilitation

The results of the cognitive evaluation coupled with the individual's psychosocial history, status, and goals are the primary factors that shape the planning and execution of a cognitive rehabilitation program. All functional difficulties should be considered in creating a cognitive rehabilitation program based on strengths and intact abilities. Individual personality style provides insight into the way the person confronts and manages difficulties. This understanding helps delineate maladaptive behaviors as reactions to cognitive changes and concerns and is important in setting the tone and communication style of the cognitive rehabilitation program.

> **BOX 7**
>
> On cognitive impairments and their implications: "The simple tasks, the ones that you could take for granted because you have done them a thousand times. Multitasking, even small details, seems overwhelming and difficult to follow and complete."
>
> **Jeffrey**

Cognitive rehabilitation is based on the use of techniques designed to improve cognitive functioning or compensate for deficiencies, with the goal of achieving the most independent or highest level of functioning.[139] Evidence shows that rehabilitation plans should address specific cognitive deficits as identified in the assessment procedures. Some people with MS may require the strengthening of basic abilities, such as attention and orientation, before more complex abilities can be addressed, such as memory and problem solving. As the rehabilitation protocol progresses, the goals and objectives are increasingly refined to address more higher-order functions as they pertain to the individual's life. Cognitive interventions should be provided by certified or licensed health care professionals with experience in cognitive rehabilitation (e.g., neuropsychologists, occupational therapists, psychologists, and speech–language pathologists).[140]

Cognitive interventions can incorporate different approaches depending on the nature and severity of the impairment. There are two distinctive approaches to cognitive rehabilitation: (1) restoring cognitive abilities (the remedial approach), or (2) compensating for cognitive deficits that cannot be restored (the compensatory approach). Below is a brief description of each approach followed by evidence supporting the use of intervention protocols that incorporate these approaches with people who have MS.

8.7.2.1 Remedial Approach

The remedial or restorative approach, often referred to as direct retraining, attempts to improve impaired abilities and strengthen previously learned behaviors through the use of exercises or practice drills (e.g., tabletop paper–pencil, computer, or virtual tasks). The underlying assumption of this approach is based on the concept of neuroplasticity in that the brain can partially recover from injury. This recovery is facilitated by using graded exercises of impaired functions to stimulate neural pathways. For example, verbal-learning difficulties can be addressed by practicing memorization of word lists of increasing length; improved perception of visual information can be achieved using computer-based drills or computer games. According to the remedial approach, skill transfer will occur naturally from the clinical setting to the client's natural environment.[141–144] Although particularly helpful in the recognition and practiced use of strategic concepts, such as chunking and semantic organization, the restorative approach generally requires continual practice and is best maintained if incorporated into the individual's lifestyle.[2]

8.7.2.2 Compensatory Approach

Unlike the remedial approach, a compensatory approach does not assume that neuroplasticity will occur and, therefore, aims to minimize cognitive functional limitations by establishing alternative compensatory activities. Compensation can be achieved through various learned strategies. A description of cognitive compensatory strategies can be found in Table 8.6. According to this approach, people with MS need to be taught how to transfer the strategies they learn during the intervention into their natural environments because transfer will not occur automatically.

The theoretical underpinnings of the compensatory approach can be found in the World Health Organization's *International Classification of Functioning, Disability and Health* (ICF).[145] In the parlance of the ICF, an important daily *activity* such as completing one's daily routine is accomplished through alternative means in spite of the fact that some crucial, underlying *function* is impaired. For example, if a *mental function*, such as long-term memory (ICF b1441) is impaired, shopping for groceries, part of the *activity* of completing one's daily routine (ICF d2302), can be accomplished through the compensatory strategy of using a shopping list. The single most important concept for both client and therapist to

TABLE 8.6

Compensatory Cognitive Strategies

I. Internal strategies
Recruiting one's own mental capacity and effort to achieve optimal performance in a cognitive task
Examples: Self-monitoring techniques, mental rehearsal and repetition, mnemonics (e.g., visual imagery), active organization of thoughts
II. External strategies
Changing the demands of an activity, making environmental modifications, or using assistive devices to help perform cognitive tasks
Examples: Incorporating daily routines with fewer tasks, use of pagers for external cuing, use of calendars/planners, working in quiet/organized spaces

Source: Adapted from Fleming JM, et al. Prospective memory rehabilitation for adults with traumatic brain injury: A compensatory training programme. Brain-Injury 2005;19(1):1–13; Kirsch NL, et al. Rehabilitation Psychology 2004;49(3):200–212; Wilson BA. Neuropsychology Review 2000;10(4): 233–243.

keep in mind is that impaired functions do not necessarily mean that important life activities are impossible.

8.7.2.3 Supporting Evidence for Cognitive Intervention in MS

The literature investigating the effects of cognitive rehabilitation interventions among MS populations is growing but still relatively in its infancy.[146] The studies currently available to guide clinicians vary considerably with respect to sample characteristics and size, type of intervention provided, and outcome measures used. Thus, results have been mixed and this variability makes it difficult to conclude whether cognitive interventions are beneficial for people with MS, and, if so, what elements within the intervention contributed to change. A full review is beyond the scope of this chapter, but a few highlights and recommended intervention protocols will be illustrated.

A number of studies have evaluated the use of computer-based training programs with people who have MS. These programs rely heavily on concepts derived from the remedial approach to cognitive intervention. In an earlier study Plohmann et al.[147] assessed the effects of a computerized, remedial 12-session program for attention deficits through the use of an attention test battery. Tasks included controlling the speed of driving a car in the face of obstacles, targeting stimuli on a monitor screen, and monitoring a radar screen for appearance of objects. Twenty-two people with MS who were not hospitalized participated in the individualized program. Participants showed improvement on selected attentional abilities at a 9-week follow-up; however, generalized improvement was not seen. Participants self-reported less distractibility and fatigue in their daily activities following the intervention and felt that their participation contributed to their self-esteem. Hildebrandt et al.[148] tested the benefits of a six-week, home-based computer training program for verbal and working memory. Seventeen individuals with relapsing–remitting MS were included in the controlled single-blinded study. Participants were evaluated on cognitive performance, quality of life, depression, and fatigue. Although training improved verbal learning and working memory, it did not have a significant effect on quality of life or fatigue.

In more recent studies, Vogt et al.[149] examined a home-based computer training program targeting working memory. Tasks included path finding on a map (spatial orientation), pairing overturned picture cards (visual memory), and memorization of numbers. Participants received a high-intensity program (45 min/4 times per week/4 weeks), a distributed program (45 min/2 times per week/8 weeks), or no intervention. Participants in the training programs were able to advance in task difficulty, reported significant decrease in fatigue, and showed improvements on tests measuring working memory and a visual measure of mental speed (face recognition). Treatment effects were not found for short-term memory, quality of life, or depression. Flavia et al.,[150] using a double-blind, controlled design, examined an intensive, three-month computer training program (1 h sessions/3 times per week) targeting attention, information processing, and executive functions in simulated real-life situations. Training tasks (such as planning and scheduling tasks on a city map and simulating a train conductor's control panel with distractions to practice divided attention) were catered to the individual's specific impairments (as measured by objective cognitive tests). People with MS who received the intervention showed significant improvements in their specific impairments and reduced depression as compared with a control group (no rehabilitation). Fink et al.[151] evaluated a training program to improve executive functioning with people with relapsing–remitting MS as compared to a placebo group (computer-controlled exercises measuring reaction time to visual stimuli) and control group (no intervention). The six-week training program involved textbook exercises completed four times per week (~30 min per day) and weekly meetings with a psychologist to discuss exercises and receive feedback. Improvements in executive functioning and verbal learning among people with MS in the training program were noted.

Although some studies were able to identify the benefits of computer training programs, others did not. For example, Solari et al.[152] assessed the efficacy of an individual 45 min, biweekly (8 weeks total), computer-aided remedial program of memory and attention compared to a computer-aided program consisting primarily of motor training. Eighty-two participants were randomly allocated to the groups. With 45% of those with MS showing improvement on only two BRBN subtests at 8 weeks, the authors concluded that the cognitive training was not effective as compared to the control group.

The results from the above studies highlight a few issues related to cognitive rehabilitation using a remedial approach. It seems that cognitive training (through the use of computer technologies) has the ability to improve cognitive functioning in the specific areas that were targeted (e.g., working memory or attention). Remedial interventions are also supported by imaging studies. For example, Mattioli et al.[153] noted recruitment of prefrontal and cingular cortices during tasks supporting brain plasticity theories and the notion that neural networks can be reestablished. On the other hand, remedial interventions are yet to prove their ability to affect daily functioning, quality of life, fatigue, and depression.

Some studies have begun to address these limitations, specifically by testing the benefits of learning and compensatory strategies to overcome cognitive difficulties. Tesar et al.[154] tested the use of a computer training program coupled with teaching compensatory strategies and relaxation exercises in order to test the impact of the intervention on everyday life. Nineteen individuals with MS were allocated into two groups. The study group received a weekly, 90 min training program and strategies training (e.g., external memory aids, building routines, visualization) in addition to the accepted outpatient rehabilitation. The control group received outpatient rehabilitation alone. Results showed some benefits of the training program for executive functioning, spatial abilities, and memory as compared with the control group. Participants reported the benefits of the program to their daily lives, including the use of compensatory strategies. Kardiasmenos et al.[19] examined prospective memory in people with MS and the benefits of using mnemonic strategies for impairments in this area using a table-top board game. The recommended strategy (i.e., using both visualization and verbal description) was shown to be helpful in remembering future tasks.

Chiaravalloti and colleagues conducted multiple studies investigating the benefits of internal compensatory strategies and self-generated learning to improve cognitive abilities in MS. Chiaravalloti and DeLuca[155] examined a generation effect as a means of maximizing learning by assisting in the encoding process. Generation effect is based on the assumption that items are better remembered if they are self-generated. Thirty-one community-based participants with MS were compared to 17 age- and education-matched healthy controls. The data showed that participants recalled self-generated stimuli better than provided stimuli. Chiaravalloti et al.[156] tested to see if a repetition effect increased learning in the form of recall among people with MS. Sixty-four outpatients were compared with 20 healthy controls. Participants showed decreased learning with additional repetition trials. In other words, people with MS can benefit from repetition; however, increasing the number of repetition trials after a certain point reduces recall. The authors concluded that repetition alone does not improve recall but rather learning in different modalities to create better encoding, such as organization, in-depth encoding procedures, reducing distractions, and providing more time to increase learning does. Chiaravalloti et al.[157] tested the effectiveness of visual imagery and contexts (using stories) to improve learning among people with MS. Twenty-eight participants were randomly allocated to the experimental group (story technique) or a control group (general memory exercises). Participants in the experimental group showed significant improvement both on objective cognitive tests and subjective reports. Goverover et al.[158] also showed the benefits of self-generation in functional tasks (meal preparation and financial management) among people with MS.

Thus, studies that have incorporated strategy formation or taught alternative ways to learn and conduct activities are showing that compensatory techniques have the potential to improve cognitive skills and self-reported evaluation of cognitive functioning. Research has also highlighted the importance of teaching people with MS how to transfer learned skills and strategies during rehabilitation into the real world. Difficulties in executive functioning hinder the ability to problem solve and relate learned strategies to new situations. Thus, transfer can become a significant rehabilitation challenge and must be directly addressed. An intervention technique that addresses transfer of learned skills has been introduced to the rehabilitation of people with MS (including cognitive rehabilitation): self-management (SM) programs.

SM programs are interventions that teach participants how to manage disease-related symptoms on a daily basis using cognitive-behavioral therapy and problem-solving techniques.[159] Typically delivered in a group format, SM programs are based on the needs and perceived problems of the participants. The programs increase clients' knowledge and resources about their conditions to promote decision making and empower them to take responsibility for their health. People with MS are taught specific skills and strategies to manage their conditions, thus increasing their sense of self-efficacy and confidence.[159] SM is particularly important for people with chronic diseases who are responsible for their own daily care.[160]

Since cognitive changes in MS are variable and can deteriorate as the disease progresses, people with MS can benefit from rehabilitation efforts that teach strategies that can be utilized as functional capacities change. The application of such strategies would require the individuals to reconsider how they perform daily activities and make certain behavioral changes. During SM programs, participants are taught compensatory strategies that are flexible and can be used over time. If taught how to appropriately translate learned strategies into natural environments, they can modify the strategies according to their personal needs and as their capabilities change. The earlier participants are taught strategies to cope with cognitive symptoms, the longer the time they can maintain functional independence in valued roles.

Recently an SM cognitive-intervention program designed specifically for people with MS was pilot tested.[161] The program, "Mind over Matter,"© included five 2-hour weekly sessions delivered in community settings by an occupational therapist who followed a facilitator manual. Due to the educational and problem-solving nature of the program, it is best suited for individuals with MS experiencing mild-to-moderate cognitive changes. The goals of the program were to increase participants' knowledge of cognitive impairments, increase levels of self-efficacy to manage cognitive difficulties, and increase use of management strategies. During the program, participants learned about the impact of common MS cognitive symptoms (including memory, attention, information processing, and executive functioning) and how they interacted with other MS symptoms (e.g., fatigue and depression).[161]

Teaching methods included oral instruction, evaluation of case examples, and demonstration. Participants were taught various cognitive compensatory strategies (e.g., mnemonics, incorporating a day planner or digital recorder, and organizing spaces) and practiced their implementation between sessions in order to increase treatment transfer. The program also addressed the social and emotional implications of cognitive impairments. Participants completed homework assignments in between sessions to apply the program content to their individual needs and challenges.

Results from the pilot testing of "Mind over Matter" indicated that participants increased their knowledge of cognitive symptoms and levels of self-efficacy in their ability to manage cognitive difficulties. Participants also reported that the program had a positive impact on their ability to manage cognitive symptoms by using strategies that they learned and were able to generalize strategies to other activities.[162] The development of "Mind over Matter" continues, and the program is currently being investigated in a randomized controlled trial.

The use of SM programs to address cognitive challenges in MS holds promise to promote independent and successful management of these symptoms. Therefore, further research is needed in the area. A recent review of evidence-based cognitive rehabilitation in MS[146] called for the investigation of programs that are contextualized, address generalization of learned materials, and can reduce activity limitations and increase participation. SM programs have the potential to meet these requirements.

8.7.2.4 Recommended Cognitive Strategies for Intervention Programs

The National Multiple Sclerosis Society in the United States published an expert opinion paper on the assessment and management of cognitive impairment in MS.[140] In this document, the task force recommends the use of compensatory strategies in cognitive interventions for people with MS as these have shown greater promise in circumventing cognitive difficulties. It is recommended that learned strategies be actively transferred into individuals' daily environment in order to increase generalization and continued strategy use.[140]

Most people with MS will arrive already equipped with two potential assets: their own home-grown compensatory strategies and their personal styles. These two elements provide pivotal information for the therapist in designing a treatment plan. Personal style is a key element to incorporate if treatment is going to be successful. Does the client like to organize things by color or is alphabetization more his or her style? Is the client a techie or more comfortable with a good old-fashioned notebook? Are you working with a lone wolf or with someone who is most successful when drawing upon a lot of social support? To what extent is it important that compensatory aids be stylish or is the utilitarian look acceptable? Adapting treatment to the individual's personal style will help to ensure that rehabilitation is a comfortable fit.

The workplace (if applicable) and the social network are important contextual elements that need to be incorporated in treatment planning and implementation. If the client is working, the issue of reasonable accommodations may be a central focus of the intervention. Dealing with the characteristics of the workplace, the nature of the client's duties, and the employer's understanding of the client's assets and limitations may be among the most helpful parts of treatment. In reference to the social network, the client's family is a significant factor in treatment. Families do not always know that MS can cause cognitive changes, and the therapist may need to help build understanding concerning the nature and causes of these changes. In addition, family members may need to become partners in the client's management strategies. Modifications in the family's routine, such as keeping a family calendar, storing objects in the same place, or providing a distraction-free environment, may provide important environmental supports. In addition, family members may need supportive counseling to help them manage their reactions to the cognitive changes affecting their family member with MS.

Overarching principles that apply to the compensatory and self-management approaches are provided in Table 8.7. Given the variety of patterns that MS-related cognitive impairment can take and the unique nature of individuals and their environments, the potential

TABLE 8.7

Overarching Principles That Apply to the Compensatory and Self-Management Approach

- Important life activities are possible even if the underlying body and/or mental functions associated with those activities are impaired.
- A good working knowledge of strengths and limitations is important to identify the specific problems that need to be addressed and the intact abilities that can be used to address them.
- It is OK to do things differently from before MS.
- Organization is critical and is the underpinning of many compensatory strategies.
- Each person is unique and what works for one might not work for another.
- Each problem may require its own individual strategy.
- Try to keep strategies as simple and uncomplicated as possible.
- Learn from trial and error what works best—think of failure as an opportunity to learn.
- Follow a self-management plan consistently.
- Try to work with another person, either a professional or a peer—self-management is enhanced by social support.

list of treatment goals and corresponding compensatory strategies could go on forever. Presented in Table 8.8 is an illustrative list drawn from the authors' clinical experience and research and enhanced by anecdotal reports from people with MS about strategies that have worked for them. This list is designed to whet the imagination of readers, who can then design personalized treatment plans for people with MS that draw upon some of the general and specific strategies provided. For a more complete discussion on these, see LaRocca and Caruso.[163]

TABLE 8.8

Examples of Compensatory Strategies for Cognitive Impairments

The family calendar	*Treatment goal*: Reduce missed appointments with family, friends, therapists, etc.
	Cognitive domain: Memory, organization of time
	Use of strategy: Maintain a highly visible family calendar to keep track of activities and appointments. Large calendars are preferable as they can store multiple entries. A family calendar can also be maintained electronically if all family members use the same calendar application and there is a way to tie each individual calendar together the way businesses do.
Packing list	*Treatment goal*: Improve ability to stay on-task while packing for a trip
	Cognitive domain: Attention, organization of thoughts
	Use of strategy: Packing can be simplified with fewer forgotten items by using a packing list. Compile a comprehensive list of items during a quiet moment when a trip is not imminent. This can be done on a computer and several copies printed for future reference. Use the list for each new trip and check off each item as it is added to the luggage. Items that are not applicable to a particular trip can just be skipped.
Electronic notes and reminders	*Treatment goal*: Reduce forgetfulness in everyday tasks (e.g., running errands, shopping)
	Cognitive domain: Memory
	Use of strategy: Technology has made it possible for reminders to be sent using previously scheduled text messages or e-mails. The calendar feature available on many smart phones has notifications built in. Ensure that the technology does not become too complicated. The therapist has an important role in advising individuals with MS how best to use technology as a tool in a self-management plan.

TABLE 8.8   (continued)

Examples of Compensatory Strategies for Cognitive Impairments

To-do list with priorities	*Treatment goal*: Reduce daily activity load and allow more rest periods
	Cognitive domain: Planning, cognitive fatigue
	Use of strategy: A to-do list can accomplish two things: it prevents forgetting to do important tasks and provides the ability to decide in what order things should be done by establishing priorities. To-do lists can be organized in different ways; for example, they can be divided into categories such as calls, errands, and tasks. Categories for deciding priority can be "must be done today," "would like to have it done today," and "do only if all the higher priority items are completed." To-do lists should not be too long, and small tasks are best done on the spot. Self-management of time can be aided by the use of a to-do list but requires a realistic assessment of the time required for tasks, one's energy level and endurance, and the probability of unplanned events interfering with daily activities.
Project/task templates	*Treatment goal*: Reduce errors in complex, multistep tasks (e.g., paying bills or refinancing a mortgage)
	Cognitive domain: Planning, organization, problem solving
	Use of strategy: Similar to a cookbook recipe, a project template or flowchart includes a list of needed resources, the tasks to be accomplished, a timeline, and the end goal. Project templates help to accomplish everyday projects at home or at work. Therapists can assist people with MS to become proficient in the use of templates and guide their use in a variety of situations.
Filing system	*Treatment goal*: Improve organization of office (at home or work) to reduce distractibility
	Cognitive domain: Categorization and organization of objects, attention
	Use of strategy: An assessment of a client's home environment can reveal organizational difficulties in the handling of bills, records, mail, and other forms of paper. People with MS should incorporate a well-organized filing system that includes a systematic way to deal with the processing, storing, retrieving, and disposing of paper. Some prefer filing systems organized alphabetically; others are more comfortable with categories; and many find color coding to be very effective. Only records that are absolutely necessary should be kept, and files should be purged regularly. Paper that arrives should be processed as soon as possible, and a decision should be made quickly whether the item needs attention or can be discarded. An effective technique is to open the mail standing or seated next to a wastepaper basket. The paper management and filing system will help to compensate for both the disorganization and indecisiveness that may affect those with MS and also memory deficits.
Character card and/or active reading	*Treatment goal*: Reduce difficulties reading
	Cognitive domain: Memory
	Use of strategy: Many people with MS have trouble reading books because they are not able to keep track of the story characters. A useful technique is the character card—an index card that includes two pieces of information for each character, the name of the character and who that character is, for example, "stepson of the murder victim." The card also serves as the bookmark.
	A more active approach can be helpful for readers who have trouble remembering what they read from one sitting to another. Here, the reader makes a conscious effort to pay attention to main points being covered and to jot down notes concerning the main points and supporting details. The reader can talk to a family member about what he or she has been reading. Although this strategy may make reading a bit choppy and lesson-like, it can make a big difference in one's ability to achieve continuity in recall of the narrative in spite of memory issues.

continued

TABLE 8.8   (continued)

Examples of Compensatory Strategies for Cognitive Impairments

Driving directions	*Treatment goal*: Increase independence in driving and reduce getting lost
	Cognitive domain: Orientation, spatial perception (way-finding)
	People with MS may experience problems finding their way to destinations with which they are familiar but do not visit regularly. GPS devices can assist in finding distant and obscure places. However, these devices at times require quick reactions and can provide sensory overload. Many people find reviewing written directions before heading out very helpful since they provide a refresher. Written directions can be consulted en route if needed. Some individuals with MS carry with them a folder or other container holding driving directions for all of the places they go. When they go to a new place, they will add those driving direction to the file
Name cards	*Treatment goal*: Increase confidence during social interactions with new people
	Cognitive domain: memory
	Use of strategy: When a client meets someone new, he or she should mentally repeat the name several times while looking for some feature of the person that stands out. For example, is the person unusually tall, particularly thin, wearing unusually shaped eyeglasses, etc. As soon as it is practical, the client should jot down the name of the person and one or more of the noted salient features. If the client has an opportunity to interact with the person some more during that first meeting, he or she should try to address that person by name, so long as doing so is appropriate. These names along with the associated information about the person and when they met can then be put into a notebook or computer file.

It is important to evaluate the progress of people with MS during the course of treatment. This can be done using subjective judgment of improvement based on the perceptions of people with MS or significant others and the therapist's thoughts concerning the extent to which goals have been achieved. Therapists also use the results of standardized testing of cognitive functions and standardized assessment of performance of daily activities. Each of these provides a different type of information and all are relevant. But perhaps the most important is the extent to which treatment is making a difference in the successful execution of important daily activities.

In order to decide when treatment should be terminated, the client and therapist, perhaps in conjunction with one or more significant others, needs to determine the extent to which the original goals of treatment have been achieved and whether or not additional treatment is likely to lead to further improvement. In many cases, a plateau in functioning will be reached, signaling that further treatment may not be productive and that it is time for the client to continue with his or her SM strategies. Other reasons for termination of treatment can be the end of insurance coverage or logistical difficulties. Following discharge, individuals may falter in their use of the SM strategies that they learned in therapy and may need to return in order to become re-energized to tackle the many challenges of MS. A component of any set of return visits could be an evaluation of the client's current cognitive and functional status, whether or not treatment follows.

8.8 Assessment and Interventions across Rehabilitation Settings

Addressing cognitive symptoms in the rehabilitation of people with MS can take different forms depending on the individual's medical status and the setting where rehabilitation is provided.

8.8.1 Inpatient Rehabilitation

Typically, people with MS require hospitalization or inpatient care due to an exacerbation of symptoms. In these cases, cognitive abilities may worsen (as do other MS symptoms) and may require time to stabilize. Under these circumstances, administering an elaborate cognitive evaluation is not recommended for several reasons: (1) assessment results may be skewed and vary greatly due to disease activity or medication side effects; (2) the assessment procedure, which requires significant mental effort, may cause unwanted fatigue; and (3) if performance on tests is dramatically less than previous capabilities, people with MS may experience elevated stress, anxiety, or depression. Therapists should be constantly aware of any dramatic changes in cognitive functioning as these can be a result of disease activity. A cognitive evaluation, preferably a brief one, may be conducted toward the end of the hospitalization in order to guide discharge planning and alert therapists in the following rehabilitation settings to the individual's status.

8.8.2 Outpatient Rehabilitation

During outpatient rehabilitation, the symptoms of people with MS are typically more stable and the therapist can assess cognitive abilities and impairments with greater accuracy. Since a large proportion of people with MS can experience cognitive impairments, it is important to conduct a cognitive screening to detect any possible changes and, if detected, follow up with a more comprehensive evaluation. The comprehensive evaluation should include both objective and subjective information. Objective evaluations of both cognitive abilities (impairments) and levels of functioning in daily tasks are advised. Subjective perceptions should be obtained from informants, such as caregivers, as well as the person with MS.

On the basis of evaluation results, the therapist can formulate a treatment plan that includes remedial exercises such as memory drills as well as compensatory and SM principles that fit the areas of weakness of the person with MS. Compensatory strategies (such as use of cue cards or a daily planner) should be taught across a variety of activities in order to promote transfer between activities since studies show that transfer will typically not occur automatically.[164] This implies that outpatient cognitive rehabilitation should be provided across multiple sessions. A visit to the person's home and work environment can provide valuable information. Therapists can note environmental barriers and facilitators (both physical and social) and thus shape intervention goals and activities to fit the person's lifestyle and demands. Outpatient rehabilitation is also an appropriate setting to address driving abilities. Individuals can be referred to a driving evaluation (e.g., by an occupational therapist specializing in this practice area). The rehabilitation process can include simulated and on-road training to improve capacities (such as spatial processing) as well as the prescription of assistive technologies (such as hand controls).

It is important to follow-up with people with MS after discharge to a community setting (e.g., home, independent living facility). Cognitive rehabilitation typically requires that people change their daily habits and incorporate new routines into daily life. Since habits are difficult to change, therapists can encourage continued use of strategies and help troubleshoot difficulties through follow-up sessions, phone calls, or a home visit. The therapist can communicate information regarding support for people with MS and their families for the psychosocial adjustment of cognitive changes (such as the availability of support groups or local MS organizations).

8.8.3 Community-Based Rehabilitation and Home Care

Rehabilitation provided in the community or in the homes of people with MS may be appropriate for two groups: (1) people with stable MS who are functioning relatively well but still struggling with some cognitive challenges that they want to address, or (2) people with MS who have difficulty getting out of the home (due to mobility limitations or physical barriers). High-functioning people with MS may be good candidates for community-based SM programs that teach specific skills for the ongoing, daily management of cognitive symptoms. SM programs are typically administered in groups, which allow participants to meet other individuals dealing with similar challenges and, thus, learn from each other as well as from the facilitator. Home-based rehabilitation has the advantage of allowing therapists to assess functioning within individuals' natural environments. Daily activities such as dressing, cooking, or bill paying can be assessed "in vivo" with direct input to treatment planning, such as detecting the need for assistive devices or environmental modifications. Community or home-based interventions can also be delivered via telephone or video communications. Telecommunications can be administered one-on-one or in groups.

8.9 Gaps in Knowledge and Directions for the Future

Cognition in MS has received increased attention over the past few decades as both researchers and clinicians become aware of the negative impact these symptoms can have on people with MS. Although the field has seen clarification on some issues, much more research is warranted.

Greater understanding is needed regarding the manifestation and course of cognitive decline among MS subgroups (e.g., pediatric; early onset; relapsing remitting, or progressive course). Studies that utilize imaging technologies are holding great promise. In addition, more research is needed to better understand the link between types of cognitive impairments commonly seen in MS and their impact on levels of functioning and participation in daily activities. For example, how do cognitive changes affect life roles, vocational abilities, education attainment, parenting, and more? It is recommended that future studies address the impairment–function–participation continuum in greater detail, utilizing imaging, impairment-based, and functional evaluations in daily tasks. The use of computer technology for cognitive evaluation is also warranted.

In comparison with other cognitive research in MS, studies testing the effectiveness and efficacy of rehabilitation methods have received relatively little attention. A proportion of the studies currently available include methodological problems that hinder the ability to use their results to drive evidence-based practice. Although emerging, more studies are needed to evaluate the benefits of remedial, compensatory, and self-management intervention strategies among people with MS. Studies should be methodologically sound with adequate sample sizes, include a comparison group, and follow participants over time to determine the continued use of cognitive strategies. Intervention studies should include detailed descriptions of intervention protocols so that these can be replicated by other researchers and in clinical settings.

Other complementary intervention areas are emerging. Researchers have already begun to investigate the relationship between physical exercise and cognitive abilities. This research is showing promise and pointing to the benefits of exercise on mental capacities.

Another area of inquiry is the use of computer games and their ability to enhance cognitive functioning. Research on the impact of pharmaceutical agents on cognition using better designs and larger samples is also warranted.

References

1. Charcot JM. Lectures on the diseases of the nervous system: Delivered at La Salpetriere. London: New Sydenham Society; 1877.
2. Katz N. Cognition and occupation across the life span: Models for intervention in occupational therapy. Baltimore, MD: American Occupational Therapy Association; 2005.
3. Patti F, Amato MP, Trojano M, Bastianello S, Tola MR, Goretti B, Caniatti L, Di Monte E, Ferrazza P, Brescia Morra V et al. Cognitive impairment and its relation with disease measures in mildly disabled patients with relapsing–remitting multiple sclerosis: Baseline results from the cognitive impairment in multiple sclerosis (COGIMUS) study. Multiple Sclerosis 2009;15(7):779–788.
4. Rao SM, Leo GJ, Bernardin L, Unverzagt F. Cognitive dysfunction in multiple sclerosis. I. Frequency, patterns, and prediction. Neurology 1991;41(12):685–691.
5. Rao SM, Leo GJ, Ellington L, Nauertz T, Bernardin L, Unveragt F. Cognitive dysfunction in multiple sclerosis. II. Impact on employment and social functioning. Neurology 1991;41(12):692–696.
6. Benedict R, Bobholz J. Multiple sclerosis. Seminars in Neurology 2007;27:78–86.
7. Benedict RH, Bruce JM, Dwyer MG, Abdelrahman N, Hussein S, Weinstock-Guttman B, Garg N, Munschauer F, Zivadinov R. Neocortical atrophy, third ventricular width, and cognitive dysfunction in multiple sclerosis. Archives of Neurology 2006;63(9):1301–1306.
8. Benedict RHB, Carone D, Bakshi R. Correlating brain atrophy with cognitive dysfunction, mood disturbances, and personality disorder in multiple sclerosis. Journal of Neuroimaging 2004;14(3 Suppl):36–45.
9. Benedict RHB, Priore RL, Miller C, Munschauer F, Jacobs L. Personality disorder in multiple sclerosis correlates with cognitive impairment. Journal of Neuropsychiatry & Clinical Neurosciences 2001;13(1):70–76.
10. Benedict RHB, Shapiro A, Priore RL, Miller C, Munschauer FE, Jacobs LD. Neuropsychological counseling improves social behavior in cognitively-impaired multiple sclerosis patients. Multiple Sclerosis 2001;6(6):391–396.
11. Figved N, Benedict R, Klevan G, Myhr KM, Nyland HI, Landrø NI, Larsen JP, Aarsland D. Relationship of cognitive impairment to psychiatric symptoms in multiple sclerosis. Multiple Sclerosis 2008;14(8):1084–1090.
12. Fishman I, Benedict RHB, Bakshi R, Priore R, Weinstock-Guttman B. Construct validity and frequency of euphoria sclerotica in multiple sclerosis. Journal of Neuropsychiatry & Clinical Neurosciences 2004;16(3):350–356.
13. Benedict RHB, Fazekas F. Benign or not benign MS: A role for routine neuropsychological assessment? Neurology 2009;73(7):494–495.
14. Rovaris M, Barkhof F, Calabrese M, De Stefano N, Fazekas F, Miller DH, Montalban X, Polman C, Rocca MA, Thompson AJ et al. MRI features of benign multiple sclerosis: Toward a new definition of this disease phenotype. Neurology 2009;72(19):1693–1701.
15. Rao SM, Grafman J, DiGiulio D, Mittenberg W, Bernardin L, Leo GJ, Luchetta T, Unverzagt F. Memory dysfunction in multiple sclerosis: Its relation to working memory, semantic encoding, and implicit learning. Neuropsychology 1993;7(3):364–374.
16. DeLuca J, Gaudino EA, Diamond BJ, Christodoulou C, Engel RA. Acquisition and storage deficits in multiple sclerosis. Journal of Clinical & Experimental Neuropsychology 1998;20(3):376–390.
17. Thornton AE, Raz N, Tucker KA. Memory in multiple sclerosis: Contextual encoding deficits. Journal of the International Neuropsychological Society 2002;8(3):395–409.

18. DeLuca J, Berbieri-Berger S, Johnson SK. The nature of memory impairments in multiple sclerosis: Acquisition versus retrieval. Journal of Clinical and Environmental Psychology 1994; 16(2):183–189.
19. Kardiasmenos KS, Clawson DM, Wilken JA, Wallin MT. Prospective memory and the efficacy of a memory strategy in multiple sclerosis. Neuropsychology 2008;22(6):746–754.
20. DeLuca J, Chelune GJ, Tulsky DS, Lengenfelder J, Chiaravalloti ND. Is speed of processing or working memory the primary information processing deficit in multiple sclerosis? Journal of Clinical & Experimental Neuropsychology 2004;26(4):550–562.
21. Demaree HA, Gaudino EA, DeLuca J, Ricker JH. Learning impairment is associated with recall ability in multiple sclerosis. Journal of Clinical and Experimental Neuropsychology 2000; 22(6):865–873.
22. Denney DR, Lynch SG, Parmenter BA, Horne N. Cognitive impairment in relapsing and primary progressive multiple sclerosis: Mostly a matter of speed. Journal of the International Neuropsychological Society 2004;10(7):948–956.
23. Denney DR, Lynch SG, Parmenter BA. A 3-year longitudinal study of cognitive impairment in patients with primary progressive multiple sclerosis: Speed matters. Journal of the Neurological Sciences 2008;267(1–2):129–136.
24. Kujala P, Portin R, Revonsuo A, Ruutiainen J. Attention related performance in two cognitively different subgroups of patients with multiple sclerosis. Journal of Neurology, Neurosurgery & Psychiatry 1995;59(1):77–82.
25. McCarthy M, Beaumont JG, Thompson R, Peacock S. Modality-specific aspects of sustained and divided attentional performance in multiple sclerosis. Archives of Clinical Neuropsychology 2005;20(6):705–718.
26. Sohlberg MM, Mateer CA. Cognitive rehabilitation: An integrative neuropsychological approach. New York: Guilford Press; 2001.
27. Chiaravalloti ND, DeLuca J. Cognitive impairment in multiple sclerosis. Lancet Neurology 2008;7(12):1139–1151.
28. Brassington JC, Marsh NV. Neuropsychological aspects of multiple sclerosis. Neuropsychology Review 1998;8(2):43–72.
29. Birnboim S, Miller A. Cognitive strategies application of multiple sclerosis patients. Multiple Sclerosis 2004;10(1):67–73.
30. Henry JD, Beatty WW. Verbal fluency deficits in multiple sclerosis. Neuropsychologia 2006;44:1166–1174.
31. Mahler ME. Behavioral manifestations associated with multiple sclerosis. Psychiatric Clinics of North America 1992;15(2):427–438.
32. Beatty WW, Monson N. Problem solving by patients with multiple sclerosis: Comparison on performance on the Wisconsin and California card sorting tests. Journal of the International Neuropsychological Society 1996;2(2):134–140.
33. Randolph JJ, Arnett PA, Higginson CI. Metamemory and tested cognitive functioning in multiple sclerosis. Clinical Neuropsychologist 2001;15(3):357–368.
34. Goverover Y, Chiaravalloti N, DeLuca J. The relationship between self-awareness of neurobehavioral symptoms, cognitive functioning, and emotional symptoms in multiple sclerosis. Multiple Sclerosis 2005;11(2):203–212.
35. Vleugels L, Lafosse C, van Nunen A, Nachtergaele S, Ketelaer P, Charlier M, Vandenbussche E. Visuospatial impairment in multiple sclerosis patients diagnosed with neuropsychological tasks. Multiple Sclerosis 2000;6(4):241–254.
36. Bobholz JA, Rao SM. Cognitive dysfunction in multiple sclerosis: A review of recent developments. Current Opinion in Neurology 2003;16(3):283–288.
37. Arnett PA, Barwick FH, Beeney JE. Depression in multiple sclerosis: Review and theoretical proposal. Journal of the International Neuropsychological Society 2008;14(5):691–724.
38. Arnett PA, Higgonson CI, Voss WD, Bender WI, Wurst JM, Tippin JM. Depression in multiple sclerosis: Relationship to working memory capacity. Neuropsychology 1999;13(4):546–556.

39. Arnett PA, Higgonson C, Voss WD, Wright B, Bender WI, Wurst JM, Tippen JM. Depressed mood in multiple sclerosis: Relationship to capacity-demanding memory and attentional functioning. Neuropsychology 1999;13(3):434–446.

40. Benedict RHB, Cox D, Thompson LL, Foley FW, Weinstock-Guttman B, Munschauer F. Reliable screening for neuropsychological impairment in MS. Multiple Sclerosis 2004;10:675–678.

41. Carone D, Benedict RHB, Munschauer FE, Fishman I, Weinstock-Guttman B. Interpreting patient/informant discrepancies of reported cognitive symptoms in MS. Journal of the International Neuropsychological Society 2005;11(5):574–583.

42. Bakshi R. Fatigue associated with multiple sclerosis: Diagnosis, impact and management. Multiple Sclerosis 2003;9(3):219–227.

43. Bryant D, Chiaravalloti N, DeLuca J. Objective measurement of cognitive fatigue in multiple sclerosis. Rehabilitation Psychology 2004;49(2):114–122.

44. Krupp LB, Elkins LE. Fatigue and declines in cognitive functioning in multiple sclerosis. Neurology 2000;55:934–939.

45. Morrow SA, Kaushik T, Zarevics P, Erlanger D, Bear MF, Munschauer FE, Benedict RH. The effects of L-amphetamine sulfate on cognition in MS patients: Results of a randomized controlled trial. Journal of Neurology 2009;256(7):1095–1102.

46. Yeh EA, Chitnis T, Krupp L, Ness J, Chabas D, Kuntz N, Waubant E. Pediatric multiple sclerosis. Nature Reviews Neurology 2009;5:621–631.

47. MacAllister WS, Belman AL, Milazzo M, Weisbrot DM, Christodoulou C, Scherl WF, Preston TE, Cianciulli C, Krupp LB. Cognitive functioning in children and adolescents with multiple sclerosis. Neurology 2005;64(8):1422–1425.

48. Montiel-Nava C, Peña JA, González-Pernía S, Mora-La Cruz E. Cognitive functioning in children with multiple sclerosis. Multiple Sclerosis 2009;15(2):266–268.

49. Banwell BL, Anderson PE. The cognitive burden of multiple sclerosis in children. Neurology 2005;64:891–894.

50. Amato MP, Goretti B, Ghezzi A, Lori S, Zipoli V, Portaccio E, Moiola L, Falautano M, De Caro MF, Lopez M et al. Cognitive and psychosocial features of childhood and juvenile MS. Neurology 2008;70:1891–1897.

51. Zivadinov R, Bakshi R. Role of MRI in multiple sclerosis I: Inflammation and lesions. Frontiers in Bioscience 2004;9:665–683.

52. Bo L, Geurts J, van der Valk P, Polman C, Barkhof F. Lack of correlation between cortical demyelination and white matter pathologic changes in multiple sclerosis. Archives of Neurology 2007;64(1):76–80.

53. Roosendaal SD, Moraal B, Pouwels PJ, Vrenken H, Castelijns JA, Barkhof F, Geurts JJ. Accumulation of cortical lesions in MS: Relation with cognitive impairment. Multiple Sclerosis 2009;15(6):708–714.

54. Fisher E, Lee JC, Nakamura K, Rudick RA. Gray matter atrophy in multiple sclerosis: A longitudinal study. Annals of Neurology 2008;64(3):255–265.

55. De Stefano N, Matthews PM, Filippi M, Agosta F, De Luca M, Bartolozzi ML, Guidi L, Ghezzi A, Montanari E, Cifelli A et al. Evidence of early cortical atrophy in MS: Relevance to white matter changes and disability. Neurology 2003;60(7):1157–1162.

56. Frohman EM, Filippi M, Stuve O, Waxman SG, Corboy J, Phillips JT, Lucchinetti C, Wilken J, Karandikar N, Hemmer B et al. Characterizing the mechanisms of progression in multiple sclerosis: Evidence and new hypotheses for future direction. Archives of Neurology 2005;62(9): 1345–1356.

57. Hauser SL, Oksenberg JR. The neurobiology of multiple sclerosis: Genes, inflammation, and neurodegeneration. Neuron 2006;52(1):61–76.

58. Benedict RHB, Weinstock-Guttman B, Fishman I, Sharma J, Tjoa CW, Bakshi R. Prediction of neuropsychological impairment in multiple sclerosis: Comparison of conventional magnetic resonance imaging measures of atrophy and lesion burden. Archives of Neurology 2004; 61:226–230.

59. Houtchens MK, Benedict RH, Killiany R, Sharma J, Jaisani Z, Singh B, Weinstock-Guttman B, Guttmann CR, Bakshi R. Thalamic atrophy and cognition in multiple sclerosis. Neurology 2007; 69(12):113–123.

60. Benedict RH, Ramasamy D, Munschauer F, Weinstock-Guttman B, Zivadinov R. Memory impairment in multiple sclerosis: Correlation with deep grey matter and mesial temporal atrophy. Journal of Neurology, Neurosurgery & Psychiatry 2009;80(2):201–206.

61. Gingold J. Mental sharpening stones: Manage the cognitive challenges in multiple sclerosis. New York: Demos Health; 2009.

62. Shevil E, Finlayson M. Perceptions of persons with multiple sclerosis on cognitive changes and their impact on daily life. Disability and Rehabilitation 2006;28(12):779–788.

63. Yorkston. K.M., Johnson K, Klasner ER, Amtmann D, Kuehn CM, Dudgeon B. Getting the work done: A qualitative study of individuals with multiple sclerosis. Disability & Rehabilitation 2003;25:369–379.

64. Mansson E, Lexell J. Performance of activities of daily living in multiple sclerosis. Disability and Rehabilitation 2004;26(10):576–585.

65. Kalmar JH, Gaudino EA, Moore NB, Halper J, DeLuca J. The relationship between cognitive deficits and everyday functional activities in multiple sclerosis. Neuropsychology 2008;22(4):442–449.

66. Baum CM, Morrison T, Hahn M, Edwards DF. Test manual for the executive function performance test. St. Louis, MO: Washington University; 2003.

67. Costello K, Kennedy P, Scanzillo J. Recognizing nonadherence in patients with multiple sclerosis and maintaining treatment adherence in the long term. Medscape Journal of Medicine 2008;10(9):225.

68. Treadaway K, Cutter G, Salter A, Lynch S, Simsarian J, Corboy J, Jeffery D, Cohen B, Mankowski K, Guarnaccia J et al. Factors that influence adherence with disease-modifying therapy in MS. Journal of Neurology 2009;256(4):568–576.

69. Busche KD, Fisk JD, Murray TJ, Metz LM. Short term predictors of unemployment in multiple sclerosis patients. Canadian Journal of Neurological Sciences 2003;30(2):137–142.

70. Edgley K, Sullivan MJL, Dehoux E. A survey of multiple sclerosis: Part 2. Canadian Journal of Rehabilitation 1991;4:127–132.

71. Benedict RHB, Wahlig E, Bakshi R, Fishman I, Munschauer F, Zivadinov R. Predicting quality of life in multiple sclerosis: Accounting for physical disability, fatigue, cognition, mood disorder, personality, and behavior change. Journal of the Neurological Sciences 2005;231:29–34.

72. Amato MP, Ponziani G, Siracusa G, Sorbi S. Cognitive dysfunction in early-onset multiple sclerosis: A reappraisal after 10 years. Archives of Neurology 2001;58(10):1602–1606.

73. Weinshenker BG. The natural history of multiple sclerosis: Update 1998. Seminars in Neurology 1998;18(3):301–307.

74. Lincoln NB, Radford KA. Cognitive abilities as predictors of safety to drive in people with multiple sclerosis. Multiple Sclerosis 2008;14(1):123–128.

75. Schultheis MT, Garay E, DeLuca J. The influence of cognitive impairment on driving performance in multiple sclerosis. Neurology 2001;56(8):1089–1094.

76. Schultheis MT, Garay E, Millis SR, DeLuca J. Motor vehicle crashes and violations among drivers with multiple sclerosis. Archives of Physical Medicine and Rehabilitation 2002;83(8): 1175–1178.

77. Hakim EA, Bakheit AM, Bryant TN, Roberts MW, McIntosh-Michaelis SA, Spackman AJ, Martin JP, McLellan DL. The social impact of multiple sclerosis—A study of 305 patients and their relatives. Disability and Rehabilitation 2000;22(6):288–293.

78. Stenager E, Knudsen L, Jensen K. Multiple sclerosis: The impact of physical impairment and cognitive dysfunction on social and spare time activities. Psychotherapy and Psychosomatics 1991;56(3):123–128.

79. Kalb RC. The emotional and social impact of cognitive changes. In: N. G. LaRocca, R. C. Kalb, editors. Multiple sclerosis: Understanding the cognitive challenges. New York: Demos Medical Publishing; 2006.

80. LaRocca NG. Emotional and cognitive issues. In: R. C. Kalb, editor. Multiple sclerosis: A guide for families. 3rd ed. New York: Demos Medical Publishing; 2006.

81. LaRocca NG, Miller PH. Cognitive challenges: Assessment and management. In: Multiple sclerosis: The questions you have; the answers you need. 4th ed. New York: Demos Medical Publishing; 2008.

82. Cooper L. Effective life planning begins now. In: R. Kalb, editor. The questions you have, the answers you need. 4th ed. New York: Demos Medical Publishing; 2008.

83. Amato MP, Zipoli V, Portaccio E. Multiple sclerosis-related cognitive changes: A review of cross-sectional and longitudinal studies. Journal of Neurological Sciences 2006;245:41–46.

84. Benedict RH, Munschauer F, Linn R, Miller C, Murphy E, Foley F, Jacobs L. Screening for multiple sclerosis cognitive impairment using a self-administered 15-item questionnaire. Multiple Sclerosis 2003;9(1):95–101.

85. Peyser JM, Rao SM, LaRocca NG, Kaplan E. Guidelines for neuropsychological research in multiple sclerosis. Archives of Neurology 1990;47(1):94–97.

86. Rao SM. Neuropsychology of multiple sclerosis: A critical review. Journal of Clinical & Experimental Neuropsychology 1986;8(5):503–542.

87. Rao SM. On the nature of memory disturbance in multiple sclerosis. Journal of Clinical & Experimental Neuropsychology 1989;11(5):699–712.

88. Rao SM, Glatt S, Hammeke TA, McQuillen MP, Khatri BO, Rhodes AM, Pollard S. Chronic progressive multiple sclerosis: Relationship between cerebral ventricular size and neuropsychological impairment. Archives of Neurology 1985;42(7):678–682.

89. Rao SM, Hammeke TA, McQuillen MP, Khatri BO, Lloyd D. Memory disturbance in chronic progressive multiple sclerosis. Archives of Neurology 1984;41(6):625–631.

90. Rao SM, Leo GJ, Haughton VM, St. Aubin-Faubert P, Bernardin L. Correlation of magnetic resonance imaging with neuropsychological testing in multiple sclerosis. Neurology 1989;39:161–166.

91. Rao SM, Leo GJ, St.Aubin-Farbert P. Information processing speed in patients with multiple sclerosis. Journal of Clinical & Experimental Neuropsychology 1989;11(4):471–477.

92. Buschke F, Fuld PA. Evaluating storage, retention, and retrieval in disordered memory and learning. Neurology 1974;24(11):1019–1025.

93. Barbizet J, Cany E. Clinical and psychometric study of a patient with memory disturbances. International Journal of Neurology 1968;7:44–54.

94. Benton AL, Hamsher K. Multilingual aphasia examination. Iowa City, IA: AJA Associates; 1989.

95. Borkowski JG, Benton AL, Spreen O. Word fluency and brain damage. Neuropsychologia 1967;5(2):135–140.

96. Gronwall DMA. Paced auditory serial addition task: A measure of recovery from concussion. Perceptual and Motor Skills 1977;44(2):367–373.

97. Smith A. Symbol digit modalities test: Manual. Los Angeles, CA: Western Psychological Services; 1982.

98. Huijbregts SC, Kalkers NF, de Sonneville LM, de Groot V, Polman CH. Cognitive impairment and decline in different MS subtypes. Journal of the Neurological Sciences 2006;245(1–2): 187–194.

99. Christodoulou C, Krupp LB, Liang Z, Huang W, Melville P, Roque C, Scherl WF, Morgan T, MacAllister WS, Li L et al. Cognitive performance and MR markers of cerebral injury in cognitively impaired MS patients. Neurology 2003;60:1793–1798.

100. Amato MP, Bartolozzi ML, Zipoli V, Portaccio E, Mortilla M, Guidi L, Siracusa G, Sorbi S, Federico A, De Stefano N. Neocortical volume decrease in relapsing–remitting MS patients with mild cognitive impairment. Neurology 2004;63:89–93.

101. Camp SJ, Stevenson VL, Thompson AJ, Ingle GT, Miller DH, Borras C, Brochet B, Dousset V, Falautano M, Filippi M et al. A longitudinal study of cognition in primary progressive multiple sclerosis. Brain 2005;128(Pt 12):2891–2898.

102. Sepulcre J, Vanotti S, Hernández R, Sandoval G, Cáceres F, Garcea O, Villoslada P. Cognitive impairment in patients with multiple sclerosis using the brief repeatable battery-neuropsychology test. Multiple Sclerosis 2006;12(2):187–195.

103. Amato MP, Ponziani G, Pracucci G, Bracco L, Siracusa G, Amaducci L. Cognitive impairment in early-onset multiple sclerosis: Pattern, predictors, and impact on everyday life in a 4-year follow-up. Archives of Neurology 1995;52(2):168–172.

104. Benedict RH, Fischer JS, Archibald CJ, Arnett PA, Beatty WW, Bobholz J, Chelune GJ, Fisk JD, Langdon DW, Caruso L et al. Minimal neuropsychological assessment of MS patients: A consensus approach. Clinical Neuropsychologist 2002;16(3):381–397.

105. Benedict RHB. Brief visuospatial memory test—Revised: Professional manual. Odessa, FL: Psychological Assessment Resources, Inc; 1997.

106. Benedict RHB, Schretlen D, Groninger L, Dobraski M, Shpritz B. Revision of the brief visuospatial memory test: Studies of normal performance, reliability, and validity. Psychological Assessment 1996;8(2):145–153.

107. Delis DC, Kramer JH, Kaplan E, Ober BA. California verbal learning test manual: Adult version. Second ed. San Antonio, TX: Psychological Corporation; 2000.

108. Delis DC, Kaplan E, Kramer JH. Delis–Kaplan Executive Function System (D-KEFS). San Antonio, TX: The Psychological Corporation; 2001.

109. Benton AL, Sivan AB, Hamsher K, Varney NR, Spreen O. Contributions to neuropsychological assessment. New York: Oxford University Press; 1994.

110. Benedict RHB. Effects of using same vs. alternate form memory tests in short-interval, repeated assessment in multiple sclerosis. Journal of the International Neuropsychological Society 2005;11(6):727–736.

111. Benedict RH, Cookfair D, Gavett R, Gunther M, Munschauer F, Garg N, Weinstock-Guttman B. Validity of the minimal assessment of cognitive function in multiple sclerosis (MACFIMS). Journal of the International Neuropsychological Society 2006;12(4):549–558.

112. Parmenter BA, Zivadinov R, Kerenyi L, Gavett R, Weinstock-Guttman B, Dwyer MG, Garg N, Munschauer F, Benedict RH. Validity of the Wisconsin card sorting and Delis–Kaplan Executive Function System (DKEFS) sorting tests in multiple sclerosis. Journal of Clinical & Experimental Neuropsychology 2007;29(2):215–223.

113. Parmenter BA, Testa SM, Schretlen DJ, Weinstock-Guttman B, Benedict RH. The utility of regression-based norms in interpreting the Minimal Assessment of Cognitive Function in Multiple Sclerosis (MACFIMS). Journal of the International Neuropsychological Society 2010; 16(1):6–16.

114. Strober L, Englert J, Munschauer F, Weinstock-Guttman B, Rao S, Benedict RHB. Sensitivity of conventional memory tests in multiple sclerosis: Comparing the Rao Brief Repeatable Neuropsychological Battery and the Minimal Assessment of Cognitive Function in MS. Multiple Sclerosis 2009;15(9):1077–1084.

115. Wilken JA, Kane R, Sullivan CL, Wallin M, Usiskin JB, Quig ME, Simsarian J, Saunders C, Crayton H, Mandler R et al. The utility of computerized neuropsychological assessment of cognitive dysfunction in patients with relapsing–remitting multiple sclerosis. Multiple Sclerosis 2003;9(2):119–127.

116. Hanly JG, Omisade A, Su L, Farewell V, Fisk JD. Assessment of cognitive function in systemic lupus erythematosus, rheumatoid arthritis, and multiple sclerosis by computerized neuropsychological tests. Arthritis & Rheumatism 2010;62(5):1478–1486.

117. Younes M, Hill J, Quinless J, Kulduff M, Peng B, Cook SD, Cadavid D. Internet-based cognitive testing in multiple sclerosis. Multiple Sclerosis 2007;13:1011–1019.

118. Wilson BA, Alderman N, Burgess PW, Emslie H, Evans JJ. Behavioral assessment of the dysexecutive syndrome. St Edmunds, UK: Thames Valley Test Company; 1996.

119. Lincoln NB, Dent A, Harding J, Weyman N, Nicholl C, Blumhardt LD, Playford ED. Evaluation of cognitive assessment and cognitive intervention for people with multiple sclerosis. Journal of Neurology, Neurosurgery & Psychiatry 2002;72(1):93–98.

120. Simioni S, Ruffieux C, Bruggimann L, Annonia JM, Schluep M. Cognition, mood and fatigue in patients in the early stage of multiple sclerosis. Swiss Medical Weekly 2007;137:496–501.

121. Norris G, Tate LT. The Behavioural Assessment of Dysexecutive Syndrome (BADS): Ecological, concurrent and construct validity. Neuropsychological Rehabilitation 2000;10(1):33–45.

122. Wilson BA, Cockburn J, Baddeley A. The Rivermead behavioral memory test. Reading, England/ Gaylord, MI: Thames Valley Test Co./National Rehabilitation Services; 1985.

123. Robertson IH, Ward T, Ridgeway V, Nimmo-Smith I. The structure of normal human attention: The test of everyday attention. Journal of the International Neuropsychological Society 1996;2(6):525–534.

124. Higginson CI, Arnett PA, Voss MD. The ecological validity of clinical tests of memory and attention in multiple sclerosis. Archives of Clinical Neuropsychology 2000;15(3):185–204.

125. Rizzo AA, Schultheis M, Kerns KA, Mateer C. Analysis of assets for virtual reality applications in neuropsychology. Neuropsychological Rehabilitation 2004;14:207–239.

126. Zalla T, Plassiart C, Pillon B, Grafman J, Sirigu A. Action planning in a virtual context after prefrontal cortex damage. Neuropsychological Rehabilitation 2001;8:759–770.

127. Schultheis MT, Rizzo AA. The application of virtual reality technology in rehabilitation. Rehabilitation Psychology 2001;46(3):296–311.

128. Josman N, Schenirderman-Elbaz A, Klinger E, Shevil E. Using virtual reality to evaluate executive functioning among persons with schizophrenia: A validity study. Schizophrenia Research 2009;115(2–3):270–277.

129. Pugnetti L, Mendozzi L, Motta A, Cattaneo A, Barbieri E, Brancotti A. Evaluation and retraining for adults' cognitive impairment: Which role for virtual reality technology? Computers in Biology and Medicine 1995;25(2):213–227.

130. Goverover Y, O'Brien AR, Moore NB, DeLuca J. Actual reality: A new approach to functional assessment in persons with multiple sclerosis. Archives of Physical Medicine & Rehabilitation 2010;91(2):252–260.

131. Bever CT, Anderson PA, Leslie J, Panitch HS, Dhib-Jalbut S, Khan OA, Milo R, Hebel JR, Conway KL, Katz E et al. Treatment with oral 3,4 diaminopyridine improves leg strength in multiple sclerosis patients: Results of a randomized, double-blind, placebo-controlled, cross-over trial. Neurology 1996;47(6):1457–1462.

132. Rossini PM, Pasqualetti P, Pozzilli C, Grasso MG, Millefiorini E, Graceffa A, Carlesimo GA, Zibellini G, Caltagirone C. Fatigue in progressive multiple sclerosis: Results of a randomized, double-blind, placebo-controlled, crossover trial of oral 4-aminopyridine. Multiple Sclerosis 2001; 7(6):354–358.

133. Wilken JA, Sullivan C, Wallin M, Rogers C, Kane RL, Rossman H, Lawson S, Simsarian J, Saunders C, Quig ME. Treatment of multiple sclerosis related cognitive problems with adjunctive modafinil: Rationale and preliminary supportive data. International Journal of MS Care 2008;10:1–10.

134. Harel Y, Appleboim N, Lavie M, Achiron A. Single dose of methylphenidate improves cognitive performance in multiple sclerosis patients with impaired attention process. Journal of the Neurological Sciences 2009;276(1–2):38–40.

135. Benedict RH, Munschauer F, Zarevics P, Erlanger D, Rowe V, Feaster T, Carpenter RL. Effects of l-amphetamine sulfate on cognitive function in multiple sclerosis patients. Journal of Neurology 2008;255(6):848–852.

136. Krupp LB, Christodoulou C, Melville P, Scherl WF, MacAllister WS, Elkins LE. Donepezil improves memory in multiple sclerosis in a randomized clinical trial. Neurology 2004;63: 1579–1585.

137. Shaygannejad V, Janghorbani M, Ashtari F, Zanjani HA, Zakizade N. Effects of rivastigmine on memory and cognition in multiple sclerosis. Canadian Journal of Neurological Sciences 2008;35(4):476–481.

138. Krupp LB, Christodoulou C, Melville P, Scherl WF, Pai LY, Muenz LHD, He D, Benedict RHB, Goodman AD, Rizvi S et al. A multi-center randomized clinical trial of donepezil for memory impairment in multiple sclerosis. Neurology 2011;76:1500–1507.

139. Gillen G. Cognitive and perceptual rehabilitation: Optimizing function. St. Louis, MO: Mosby; 2009.

140. National Multiple Sclerosis Society. Expert opinion paper: Assessment and management of cognitive impairment in multiple sclerosis. New York, NY: National Multiple Sclerosis Society; 2006.

141. Averbuch S, Katz N. Cognitive rehabilitation: A retraining model for patients with neurological disabilities. In: N. Katz, editor. Cognition and occupation across the life span: Models for intervention in occupational therapy. Baltimore, MD: American Occupational Therapy Association; 2005.

142. Dirette DK, Hinojosa J, Carnevale GJ. Comparison of remedial and compensatory interventions for adults with acquired brain injuries. Journal of Head Trauma Rehabilitation 1999;14(6): 595–601.

143. Fleming JM, Shum D, Strong J, Lightbody S. Prospective memory rehabilitation for adults with traumatic brain injury: A compensatory training programme. Brain Injury 2005;19(1):1–13.

144. Katz N, Hartman-Maeir A. Higher-level cognitive functions: Awareness and executive functions enabling engagement in occupation. In: N. Katz, editor. Cognition and occupation across the life span: Models for intervention in occupational therapy. Baltimore, MD: American Occupational Therapy Association; 2005.

145. World Health Organization. International classification of functioning. Geneva: World Health Organization; 2001.

146. O'Brien AR, Chiaravalloti N, Goverover Y, DeLuca J. Evidence-based cognitive rehabilitation for persons with multiple sclerosis: A review of the literature. Archives of Physical Medicine and Rehabilitation 2008;89(4):761–769.

147. Plohmann AM, Kappos L, Ammann W, Thordai A, Wittwer A, Huber S, Bellaiche Y, Lechner-Scott J. Computer assisted retraining of attentional impairments in patients with multiple sclerosis. Journal of Neurology, Neurosurgery & Psychiatry 1998;64(4):455–462.

148. Hildebrandt H, Lanz M, Hahn HK, Hoffmann E, Schwarze B, Schwendemann G, Kraus JA. Cognitive training in MS: Effects and relation to brain atrophy. Restorative Neurology & Neuroscience 2007;25(1):33–43.

149. Vogt A, Kappos L, Calabrese P, Stocklin M, Gschwind L, Opwis K, Penner IK. Working memory training in patients with multiple sclerosis—Comparison of two different training schedules. Restorative Neurology & Neuroscience 2009;27(3):225–235.

150. Flavia M, Stampatori C, Zanotti D, Parrinello G, Capra R. Efficacy and specificity of intensive cognitive rehabilitation of attention and executive functions in multiple sclerosis. Journal of the Neurological Sciences 2010;288:101–105.

151. Fink F, Rischkau E, Butt M, Klein J, Eling P, Hildebrandt H. Efficacy of an executive function intervention programme in MS: A placebo-controlled and pseudo-randomized trial. Multiple Sclerosis 2010;16:1148–1151.

152. Solari A, Motta A, Mendozzi L, Pucci E, Forni M, Mancardi G, Pozzilli C, CRIMS Trial. Computer-aided retraining of memory and attention in people with multiple sclerosis: A randomized, double-blind controlled trial. Journal of Neurological Sciences 2004;222:99–104.

153. Mattioli F, Stampatori C, Bellomi F, Capra R, Rocca M, Filippi M. Neuropsychological rehabilitation in adult multiple sclerosis. 2010;31:S271–S274.

154. Tesar N, Bandion K, Baumhack U. Efficacy of a neuropsychological training programme for patients with multiple sclerosis—A randomised controlled trial. Wiener Klinische Wochenschrift 2005;117(21–22):747–754.

155. Chiaravalloti ND, DeLuca J. Self-generation as a means of maximizing learning in multiple sclerosis: An application of the generation effect. Archives of Physical Medicine and Rehabilitation 2002;83:1070–1079.

156. Chiaravalloti N, Demaree H, Gaudino EA, DeLuca J. Can the repetition effect maximize learning in multiple sclerosis? Clinical Rehabilitation 2003;17:58–68.

157. Chiaravalloti ND, DeLuca J, Moore NB, Ricker JH. Treating learning impairments improves memory performance in multiple sclerosis: A randomized clinical trial. Multiple Sclerosis 2005;11:58–68.

158. Goverover Y, Chiaravalloti N, DeLuca J. Self-generation to improve learning an memory of functional activities in persons with multiple sclerosis: Meal preparation and managing finances. Archives of Physical Medicine & Rehabilitation 2008;89(8):1514–1521.

159. Lorig KR, Holman RH. Self-management education: History, definition, outcomes, and mechanisms. Annals of Behavioral Medicine 2003;26(1):1–7.

160. Holman H, Lorig K. Patient self-management: A key to effectiveness and efficacy in care of chronic disease. Public Health Reports 2004;119(3):239–243.

161. Shevil E, Finlayson M. Pilot study of a cognitive intervention program for persons with multiple sclerosis. Health Education Research 2010;25(1):41–53.

162. Shevil E, Finlayson M. Process evaluation of a self-management cognitive program for persons with multiple sclerosis. Patient Education and Counseling 2009;76(1):77–83.

163. LaRocca N, Caruso L. Assessment of cognitive changes. In: N. LaRocca, R. Kalb, editors. Multiple sclerosis: Understanding the cognitive challenges. New York: Demos Medical Publishing; 2006.

164. Toglia JP. A dynamic interactions approach to cognitive rehabilitation. In: N. Katz, editor. Cognition and occupation across the life span. Baltimore, MD: American Occupational Therapy Association; 2005.

165. Winkelmann A, Engel C, Apel A, Zettl UK. Cognitive impairment in multiple sclerosis. Journal of Neurology 2007;254(Suppl 2):II/35–II/42.

166. Zoltan B. Vision, perception, and cognition. 4th ed. Thorofare, NJ: Slack; 2007.

167. Filippi M, Tortorella C, Rovaris M, Bozzali M, Possa F, Sormani MP, Iannucci G, Comi G. Changes in the normal appearing brain tissue and cognitive impairment in multiple sclerosis. Journal of Neurology, Neurosurgery, & Psychiatry 2000;68(2):157–161.

168. Iannucci G, Rovaris M, Giacomotti L, Comi G, Filippi M. Correlation of multiple sclerosis measures derived from T2-weighted, T1-weighted, magnetization transfer, and diffusion tensor MR imaging. American Journal of Neuroradiology 2001;22(8):1462–1467.

169. Comi G, Filippi M, Martinelli V, Campi A, Rodegher M, Alberoni M, Sirabian G, Canal N. Brain MRI correlates of cognitive impairments in primary and secondary progressive multiple sclerosis. Journal of the Neurological Sciences 1995;132(2):222–227.

170. Arnett PA, Rao SM, Bernardin L, Grafman J, Yetkin FZ, Lobeck L. Relationship between frontal lobe lesions and Wisconsin card sorting test performance in patients with multiple sclerosis. Neurology 1994;44:420–425.

171. Swirsky-Sacchetti T, Mitchell DR, Seward J, Gonzales C, Lublin F, Knobler R, Field HL. Neuropsychological and structural brain lesions in multiple sclerosis: A regional analysis. Neurology 1992;42(7):1291–1295.

172. Rao SM, Bernardin L, Leo GJ, Ellington L, Ryan SB, Burg LS. Cerebral disconnection in multiple sclerosis: Relationship to atrophy of the corpus callosum. Archives of Neurology 1989;46(8):918–920.

173. Benedict RHB, Bakshi R, Simon JH, Priore R, Miller C, Munschauer F. Frontal cortex atrophy predicts cognitive impairment in multiple sclerosis. Journal of Neuropsychiatry & Clinical Neurosciences 2002;14:44–51.

174. Tekok-Kilic A, Benedict RH, Weinstock-Guttman B, Dwyer MG, Carone D, Srinivasaraghavan B, Yella V, Abdelrahman N, Munschauer F, Bakshi R et al. Independent contributions of cortical gray matter atrophy and ventricle enlargement for predicting neuropsychological impairment in multiple sclerosis. NeuroImage 2007;36(4):1294–1300.

175. Sicotte NL, Kern KC, Giesser BS, Arshanapalli A, Schultz A, Montag M, Wang H, Bookheimer SY. Regional hippocampal atrophy in multiple sclerosis. Brain 2008;131:1134–1141.

176. Reeves DL, Winter KP, LaCour S. Automated neuropsychological assessment metrics documentation: Test administration guide. Walter Reed Army Medical Institute of Research, Silver Spring, MD: Office of Military Performance Assessment Technology; 1992.

177. Kirsch NL, Shenton M, Spirl E, Rowan J, Simpson R, Schreckenghost D. Web-based assistive technology interventions for cognitive impairments after traumatic brain injury: A selective review and two case studies. Rehabilitation Psychology 2004;49(3):200–212.

178. Wilson BA. Compensating for cognitive deficits following brain injury. Neuropsychology Review 2000;10(4):233–243.

9

Chronic Pain

Dawn M. Ehde, Anna L. Kratz, James P. Robinson, and Mark P. Jensen

CONTENTS

Compared to the more visible symptoms related to the physically disabling nature of multiple sclerosis (MS), pain is one of the more hidden effects of the disease. This relative invisibility may have contributed to the lack of scientific attention to MS-related pain until the late 1990s, when researchers began to take a more serious look at the prevalence and impact of pain. Pain is now known to be a common problem for individuals with MS and has a widespread impact on the lives of many of those with MS. With the increased attention to the seriousness of pain in MS, there has been a corresponding increase on our understanding of the assessment and treatment of pain for individuals with MS. This chapter provides a general overview of the current state of knowledge about the prevalence, characteristics, impact, etiology, assessment, and treatment of chronic pain in MS. This chapter focuses on chronic pain, that is, pain that persists or is recurrent for 6 months or more. Although individuals with MS can also experience acute pain, chronic pain is more commonly the focus of rehabilitation interventions and, thus, the focus of this chapter.

9.1 Overview and Prevalence of Chronic Pain in MS

Pain can occur at any time during the course of MS, including early in the course of the disease. A large proportion of people with MS develop chronic pain. Prevalence rates of persistent pain in MS approximate 50–65% in community samples.[1–3] For those who do report chronic pain, it is generally described as a significant concern. For instance, studies have reported that more than half (57%) of those with pain report that it is constant,[3] one quarter report that pain is severe,[1] and 44% report that pain is "persistent and bothersome."[4] Though the evidence is not entirely consistent, the bulk of research indicates that an increased likelihood of having chronic pain is linked to older age, longer MS duration, MS subtypes other than relapsing–remitting, and greater disease severity.[5] Some studies have found that women report more severe pain than men with MS,[2,6] although evidence for sex differences in MS-related pain is equivocal.

The experience of pain varies among individuals with MS, and the pain condition of any one individual is often variable over time. Many individuals with MS have widespread pain across multiple body regions,[7] although surveys of individuals with pain indicate that pain is most commonly reported in the legs, followed by the hands and feet.[8]

9.2 The Biopsychosocial Model of Chronic Pain

Significant advances in the understanding and treatment of chronic pain in persons who present with pain as their primary problem (i.e., persons who were not otherwise disabled or have MS) occurred when biopsychosocial models of chronic pain were developed to replace exclusively biomedical models.[9] Biopsychosocial models of chronic pain acknowledge that pain has an underlying biological basis but that psychosocial factors, such as pain beliefs/thoughts, coping behavior, and the social environment, also have a significant and sometimes profound impact on the experience of pain and its effects on physical, psychological, and social functioning. Biopsychosocial models have been used to inform appropriate treatments for chronic pain.

A number of cognitions, beliefs, and coping strategies have been associated with functioning among persons with chronic pain as their primary disability.[10] One psychosocial factor in particular, catastrophizing, has been consistently and strongly associated with virtually all pain outcomes investigated.[11] Catastrophizing has been defined as the tendency to focus on pain and negatively evaluate one's ability to cope with it.[12] Examples of catastrophizing thoughts include "My pain is awful and I can't stand it," and "I can't deal with this pain." Numerous studies have shown catastrophizing to be associated with higher levels of pain intensity, pain-related interference with activities/participation, psychological distress, medical services utilization, and vocational dysfunction (for a review, see reference 10). In addition to catastrophizing, pain-related beliefs and coping efforts have also been associated with pain outcomes in samples in which pain is the primary problem.[10] These can be classified into those that are typically associated with poorer functioning ("maladaptive" beliefs and coping) and those that are typically associated with better functioning ("adaptive" beliefs and coping).[13] Associations between psychosocial variables and pain outcomes are typically strong, even when controlling for variables that may influence the relationships, such as demographics and pain level.

Research conducted in the last decade confirms the viability of a biopsychosocial model for understanding chronic pain in persons with MS.[13] As in the broader pain literature, pain catastrophizing is positively and strongly associated with greater pain, psychological distress, and pain interference in persons with MS.[14] Other biopsychosocial factors including MS-specific factors (e.g., disease severity, disease duration, number of pain sites), pain beliefs (especially belief in one's control over pain), pain coping behavior (especially guarding, resting, and task persistence), and social variables (particularly social support) are also strongly associated with pain intensity, physical functioning, and psychological functioning in MS samples.[5,14,15] Adaptive coping responses include the use of behavioral activities to distract oneself from pain,[15] task persistence, that is, carrying on with an activity despite pain and exercise.[14] Resting in response to pain has been associated with poorer outcomes.[14] These findings support the use of biopsychosocial models for conceptualizing chronic pain in persons with MS and provide an empirical guide for determining which psychosocial factors to target in treatment. This chapter therefore focuses not only on the "bio" aspects of MS pain but also on psychosocial aspects and treatments.

9.3 Pathophysiology and Classification of Pain Syndromes in MS

An important principle in understanding pain is the multiplicity of pathophysiologic sources for pain in MS disease. Given the multiple causes of MS pain, it is important for both researchers and clinicians to have some way to classify pain symptoms, in the hope that accurate classification will lead to treatment that is relevant to the pain problems of individual patients. As O'Connor and colleagues[5] point out, the absence of an accepted system for classifying MS pain has adverse effects on efforts to expand our understanding of MS pain.

Unfortunately, classifying pain in people with MS is difficult for several reasons. One problem is that neuropathic pain is inherently difficult to diagnose or characterize as there are no clear clinical decision-making rules to follow for its diagnosis.[16] Another is that the pain problems experienced by individuals with MS often do not fit neatly into one specific category; rather, they often reflect a combination of factors. For example, ectopic impulses from MS plaques, spasticity, and overuse might all contribute to pain around the elbow in a person with MS pain.

Ideally, a classification system of MS pain would provide insight into the mechanisms underlying an individual's pain complaint (pathophysiology), the settings and conditions that might have led to the pain (etiology), and the appropriate treatments. Tables 9.1 and 9.2

TABLE 9.1

Classification of Pain Conditions Associated with Multiple Sclerosis

Pain Condition	Examples
Continuous central neuropathic pain	Dysesthetic extremity pain
Intermittent central neuropathic pain	Lhermitte's sign
	Trigeminal neuralgia
Musculoskeletal pain	Painful tonic spasms
	Low-back pain
	Muscle spasms
Mixed neuropathic and nonneuropathic pain	Headache

Source: Proposed by O'Connor AB et al. Pain 2008;137(1):96–111.

TABLE 9.2

Classification of Pain Conditions Associated with Multiple Sclerosis

Pain Catogory	Subtype
Neuropathic pain due to MS	Pain due to exacerbations of MS
	Painful paroxysmal syndromes, including
	• Trigeminal neuralgia
	• Glossopharyngeal neuralgia
	• Paroxysmal sensory symptoms in the arms or legs
	• Paroxysmal painful tonic spasms
	• Painful Lhermitte's sign
	• Chronic pain (painful dysesthesias)
Pain as an indirect sequel of MS symptoms	Pain associated with spasticity
	Pain associated with malposition-induced burden on joints and muscles
	Pain or painful sensory symptoms due to pressure lesions
	Visceral pain
MS treatment-related pain	Adverse effects of interferon-ß therapy
	Adverse effects of glatiramer acetate therapy
	Adverse effects of intravenous immunoglobulins
Pain unrelated to MS	Back pain
	Painful peripheral neuropathies
	Headaches

Source: Proposed by Pollman W, Feneberg W. CNS Drugs 2008;22:291–324.

give classification systems proposed in two recent reviews[5,17] on pain in MS. Unfortunately, neither system has gained universal acceptance or (yet) been shown empirically to improve pain treatment in MS.

In the absence of any validated system for classifying pain in MS, the discussion in this chapter is organized around the following categories: neuropathic pain secondary to MS plaques, optic neuritis, pain associated with spasticity, musculoskeletal pain, headaches, and visceral pain syndromes.

9.3.1 Neuropathic Pain Secondary to MS Plaques

Neuropathic pain is thought to be the most common type of chronic pain in MS.[18] It is beyond the scope of this chapter to discuss how the neuronal demyelination in the central nervous system that occurs in MS leads to neuropathic pain. Briefly summarized, investigators have proposed that demyelination can lead to hyperexcitability in pain transmission pathways,[19] especially those pathways that involve the spinothalamic tracts and thalamus.[20] The central neuropathic pain (CNP) resulting from this hyperexcitability is appropriately viewed as a primary MS pain syndrome.

Although the symptoms described by patients with CNP syndromes are notoriously variable, two of the most common complaints are of burning, or dysesthetic, pain and sharp/stabbing, or lancinating, pain.[5] Dysesthetic pain, often described as a steady burning or electrical sensation, is thought to be the most common type of neuropathic pain in MS. The point prevalence of dysesthetic pain is estimated to be around 17%,[21] and approximately 25% of persons with MS report this type of pain at some point during the course of the disease.[18] Dysesthetic pain most commonly occurs in the extremities.

Although lancinating pain can occur essentially anywhere in the body of an individual with MS, two other presentations, Lhermitte's sign and trigeminal neuralgia, are common enough to warrant specific discussion. Lhermitte's sign, a sudden lancinating pain that runs down the neck, back, or limbs, is triggered by forward or backward head movement and is highly identified with MS, although it is also seen in other medical conditions. Studies have found that Lhermitte's sign is present in roughly 40% of individuals with MS at some point during the course of their disease, and point prevalence rates approximate 10%.[21,22] Patients with trigeminal neuralgia experience severe, lancinating pain in one of the divisions of the trigeminal nerve. It is experienced around the eyes, cheek, or jaw and can be triggered by light physical contact or even sound vibrations. Though estimates of the prevalence of trigeminal neuralgia in MS seem small at 1–2%,[23] they are approximately 20 times the prevalence rate of this condition in the general adult population.

9.3.2 Optic Neuritis

In MS, pain can also be associated with swelling/edema caused by the inflammatory processes that occur with this disease. In particular, painful optic neuritis is thought to be a product of swelling/edema of the optic nerve.[24] This pain is typically described as a steady aching pain, along with a sense of retro-orbital pressure. Unlike many other types of neuropathic pain, painful optic neuritis often responds well to pharmacotherapy, specifically, corticosteroids.

9.3.3 Pain Secondary to Spasticity

Pain secondary to spasticity can be viewed as a hybrid between pain of neurologic origin and pain of musculoskeletal origin.[25] Spasticity is an inherent cause of pain and can also cause pain due to the abnormal strains that it places on muscles, joints, and periarticular tissues. Lesions that damage the corticospinal tract lead to spasticity. In those with long tract signs, pain described as muscular is likely to reflect painful spasticity, particularly if it occurs in the context of observable indicators of obvious neuromuscular hyperexcitability.

9.3.4 Musculoskeletal Pain

MS lesions that cause loss of function in voluntary motor activity can produce secondary and tertiary pain syndromes. Voluntary motor function in people with MS is frequently compromised by a combination of weakness and cerebellar dysfunction that impairs the quality of movement. This combination of deficits sets the stage for a variety of musculoskeletal problems that can be broadly conceptualized as overuse syndromes. These can take several forms.

Patients who have lost voluntary control over some muscles that control motion of a joint may overuse the muscles that they can control. This can lead to aching pain in these muscles (comparable to the pain that patients with polio often report). In another overuse syndrome, the combination of weakness and poor coordination places strains on joints and periarticular tissues. Patients may thus develop traumatic arthritis or periarticular problems such as tendonitis of the shoulder. Individuals with severe incapacitation in voluntary motor function are at risk for other musculoskeletal problems, such as painful joint contractures.

Back pain is reported by 10–16% of individuals with MS.[2,21] For many, back pain has a musculoskeletal etiology and may be aggravated by physical posture, inactivity, or overexertion; however, in some cases back pain may have neurogenic origins.[18]

9.3.5 Headaches

Though certainly not specific to MS, headache pain has been found to be one of the more common types of pain reported in MS, with approximately 20% of people with MS reporting headache pain.[2] Gender predicts the report of migraine and tension headaches in MS with females at higher risk than males of experiencing both types.[26]

9.3.6 Visceral Pain Syndromes

MS can also lead to visceral pain, including urogenital pain syndromes. Some patients report bladder pain that can plausibly be related to bladder dysfunction, such as painful bladder distension secondary to a neurogenic bladder. Others report bladder pain in the absence of obvious loss of control of bladder function. Their symptoms are similar to those of patients with interstitial cystitis. Some patients report dyspareunia, or painful sexual intercourse. The pathophysiologic basis of this pain is generally obscure and probably multifactorial.

9.4 Impact of Pain on Everyday Life

For an individual whose life has been affected by an MS diagnosis, chronic pain may provide additional challenges and changes to one's life. A growing body of evidence suggests that individuals with MS and chronic pain report lower overall quality of life compared to those with MS and no pain.[3,7,27–30] Pain is associated with a negative impact on life quality across multiple life domains—occupational, social, recreational, and physical and mental health—even when controlling for the influence of clinical variables and demographics.[30]

Currently, the relation of pain to indicators of physical health, disability, and functioning are the most thoroughly studied areas of the impact of pain on people with MS. The influence of pain appears to be quite significant, with nearly 42% of participants in one study reporting that pain interfered with their daily lives most or all of the time.[31] The association between pain severity and greater physical disability in MS is well established.[1,2,28,32] In terms of physical health, pain is related to perceptions of poorer health and physical vitality[1,6] as well as to concerns about one's health and future abilities.[2] Those with MS and pain report greater overall health care utilization and consume more analgesic medications compared to those without pain.[2,3] The impact of pain on occupational functioning is not well studied, although some evidence suggests that those with MS and pain are less able to work, either full or part time, as a result of pain.[1,7] Reports also indicate that pain in MS significantly interferes with sleep and is related to greater fatigue.[1,32]

Though many individuals with MS show evidence of healthy psychosocial functioning despite having chronic pain, a substantial number are more likely than those without pain to suffer from lower levels of psychological functioning and greater social isolation or discord. Greater pain severity is associated with poorer general mental health and functioning.[1,6,7] Most studies have found that pain is related to more severe depressive symptoms.[2,32] Furthermore, there is some evidence that the association of pain to depressive and anxiety symptoms is greater for women than for men with MS.[33] Pain has also been associated with less vitality and enjoyment of life.[2] The impact of pain on social functioning is largely unknown, although those with MS and pain have reported lower levels of marital satisfaction and reduced ability to function in important social roles as spouse/partner, parent, and friend.[7,34] Focus groups of individuals with MS have described their pain as a very socially isolating experience due to the perception that others do not understand their private and unseen pain experience.[34] Individuals may also be hesitant to communicate their pain to others out of concern about burdening family members or friends with their pain.[35]

9.5 Pain Assessment Strategies

Given the multidimensionality of pain, assessment of several dimensions of pain is needed for the clinician and researcher to fully understand the chronic pain experience. What follows is a description of self-report measures tapping the key components of the pain experience based on recommendations made by a consensus panel.[36] These self-report measures may be integrated into the clinical assessment of the person with MS and pain; additional recommendations for the clinical assessment of pain are then offered. For guidance on selecting measures for clinical research on pain, see reference 36.

9.5.1 Self-Report Measures

Because pain is a multidimensional phenomenon, its assessment requires that clinicians and researchers consider its different domains. While pain intensity is the pain domain most often assessed, pain also has affective/emotional, quality, spatial, and temporal characteristics. Pain may also interfere with functioning and quality of life. This section summarizes the measures that can be used to assess each of these domains. The measures are not specific to MS but are drawn from the general pain literature, where they have proven useful across a wide range of painful conditions.

9.5.1.1 Pain Intensity

Although a large variety of measures may be used to assess pain intensity—that is, the magnitude of felt pain—three measures are most commonly used: the visual analog scale (VAS), the numerical rating scale (NRS), and the verbal rating scale (VRS). The VAS consists of a line, usually 100 mm in length, with the endpoints "no pain" on the left and some descriptor indicating extreme pain (e.g., "pain as bad as you can imagine") on the right. Respondents indicate the severity of their pain (worst, least, or average pain over a specified time period, or current pain) by making a mark on the line that represents its intensity. The NRS is similar except that respondents are presented with a series of numbers (usually 0 through 10) with the intensity endpoint descriptors below the "0" and "10," respectively. Respondents are asked to circle the single number that best represents their pain intensity along the continuum. VRSs consist of a list of words or phrases indicating different levels of pain severity (e.g., no pain, mild, moderate, severe), and respondents are asked to select the single word or phrase that best represents their pain intensity.

A large body of research supports the reliability and validity of each of these measures across a wide spectrum of patient populations.[13] All else being equal, the 0–10 NRS is recommended over the VAS or VRS because it has the most strengths and fewest weaknesses of the three. Specifically, it allows for more response options than the categorical VRS and is less confusing for individuals with mild to moderate cognitive difficulties than the VAS. Use of the 0–10 NRS also allows for more comparisons with the findings from current and future pain research given that consensus groups are increasingly recommending the 0–10 NRS over other measures.[36]

9.5.1.2 Pain Affect

The affective domain of pain represents its general unpleasantness or bothersomeness as well as the different kinds of emotional responses (e.g., fear, sadness, anger) that can accompany sensations of pain. Often, measures of pain affect are strongly related to measures of pain intensity, as pain tends to bother people more when it is more severe. When used as an outcome domain, the effects of treatment on measures of intensity are very similar to the effects on measures of pain affect. In this case, measuring affect may be seen as redundant. It is most appropriate to measure pain affect when the clinician or researchers expect there to be a discrepancy between affect and intensity, for example, if there is the possibility that a treatment might alter how much pain bothers the person, even if it has minimal effects on intensity. Examples of such treatments are acceptance-based or "mindfulness"-based treatments (see the next section).

VASs, NRSs, and VRSs have been developed for assessing the general unpleasantness of pain (for VASs' and NRSs' using endpoints such as "not bad at all" and "the most unpleasant feeling possible for me," see reference 37). The most common measure used to assess pain's broader emotional aspect is the Affect subscale of the McGill Pain Questionnaire[38] and its associated short form.[39] All three of these scales have demonstrated reliability and validity for assessing the affective domain of pain.[40]

9.5.1.3 Pain Quality

Pain quality refers to the how the pain feels. For example, two pain problems might have equal overall intensities, but one might feel more "dull" and "achy" and the other might feel more "electric" and "piercing." A large number of pain quality measures have been developed, and no two assess the same qualities. In general, pain quality measures have two primary purposes: (1) to discriminate neuropathic pain (pain due to nerve damage or dysfunction) from nonneuropathic (or nociceptive pain, pain due to the stimulation of nociceptors in bodily tissue, often skin and muscle), and (2) to measure changes in pain qualities over time.[41,42]

The measure with the largest amount of evidence supporting its reliability and validity for discriminating neuropathic from nociceptive pain is the Leeds Assessment of Neuropathic Signs and Symptoms (LANSS).[43,44] The pain quality measure with the largest amount of evidence supporting its validity as an outcome measure for detecting changes in pain quality is the Neuropathic Pain Scale (NPS).[45] However, a revised version of the NPS, the Pain Quality Assessment Scale (PQAS),[46] includes all of the NPS items (so evidence supporting the validity of the NPS also supports the validity of the new measure), but the PQAS adds other pain quality domains commonly seen in both neuropathic and nonneuropathic pain conditions. The additional items give the PQAS more content validity than the NPS, especially for assessing nociceptive pain.

9.5.1.4 Pain's Spatial Characteristics

Pain can be experienced as being in different body areas (e.g., low back, joints) as well as at different depths (surface, deep). Assessing pain location along either of these continuums is fairly straightforward. Body area location can be assessed using a pain diagram, in which the respondent indicates on a line drawing of a body the location(s) of any pain.[47] An alternative is a simple pain-site checklist in which the respondent places a checkmark next to one or more specific body areas in a list (e.g., head, neck, upper back, midback, low back) (e.g., reference 1) or indicates the relative magnitude of both "surface" and "deep" pain.[46]

9.5.1.5 Pain's Temporal Characteristics

Pain can, and almost always does, change over time. This temporal change can be reflected by different temporal domains: pain variability, frequency, and duration. The most common method for assessing pain variability is to ask the respondent to rate the least and most pain intensity over a specified time period (e.g., previous week), using an NRS, VAS, or VRS. Pain frequency can be assessed by asking the respondent to indicate the number of pain "episodes" or number of pain "flare-ups" over a specified period of time; duration of these can be rated in minutes and hours.

The PQAS,[46] mentioned above, also includes a checklist that allows respondents to describe the usual temporal pattern of their pain by asking them to indicate which of the following best describes their pain:

- I have intermittent pain (I feel pain sometimes but I am pain free at other times).
- I have variable pain ("background" pain all the time, but also moments of more pain, or even severe "breakthrough" pain, or varying types of pain).
- I have stable pain (constant pain that does not change very much from one moment to another and no pain-free periods).

It is also be possible to ask respondents to rate their current pain intensity on an intensity rating scale, such as the NRS, on multiple occasions over the course of a specified period of time (e.g., 4 times/day for 7 days). Variability (range of scores, as well as score standard deviation), frequency (rate at which pain intensity rises above a prespecified level), and duration (amount of time that pain intensity stays above a prespecified level) can then be computed from such diary scores. However, the technological effort needed both by the respondent to complete the ratings and by the assessor to download or input the ratings and compute the temporal variables preclude the practicality of this strategy in most clinical settings.

9.5.1.6 *Pain Interference*

To understand fully an individual's pain problem, it is helpful to assess how pain affects or interferes with activities and functioning. After pain intensity, pain interference is the next most common outcome measure in controlled trials of chronic pain interventions.[36] One pain interference measure that has been used in samples of individuals with MS is the Pain Interference Scale of the Brief Pain Inventory.[48] Using a 0–10 NRS, this scale has individuals rate how much pain has interfered with a variety of activities, such as sleep, self-care, recreational activities, mood, and relationships. This scale has demonstrated psychometric superiority to other measures of pain interference in disability samples, including persons with MS and pain.[14]

9.5.1.7 *Depression*

Given the prevalence of depressive disorders and symptoms among people with MS and the high rates of co-occurrence of depression and chronic pain, assessment of depression is recommended. Although a number of depression self-report measures are available, we recommend the Patient Health Questionnaire-Depression Scale (PHQ-9)[49] due to its brevity and ability to indicate the probability of a depressive disorder. It is a nine-item self-report module that assesses depressive symptom severity and screens for major and minor depressive episodes. Respondents rate the degree to which they have experienced symptoms required for a diagnosis of a major depressive episode within the last 2 weeks on a 4-point scale from 0 (not at all) to 3 (nearly everyday). Major depressive disorder (MDD) is coded as present if the respondent endorsed five or more symptoms as occurring "more days than not," and one of those items included depressed mood or anhedonia. Using these scoring procedures, Spitzer and colleagues[49] reported a strong agreement between depressive diagnoses based on the PHQ-9 and on structured clinical interviews.

9.5.2 Clinical Evaluation

The clinical evaluation of a pain problem in the context of MS may vary depending on the clinical setting, severity of the pain problem, and available resources. Given that pain is multidimensional, it follows that multiple rehabilitation team members play a role in its assessment, such as physicians, nurses, physical therapists, occupational therapists, psychologists, and vocational rehabilitation counselors. Ideally, clinical assessment is multimodal and interdisciplinary, particularly when the pain problem is unmanageable and significantly interfering with one or more areas of the patient's life and functioning. A fundamental strategy in pain assessment is acknowledging that the patient's pain problem is valid, even if it cannot be directly observed. What follows are other general suggestions for the clinical evaluation of pain in MS. For more information on clinical pain assessment, the reader is referred to the *Handbook of Pain Assessment*.[50]

9.5.2.1 Classification of Pain and Caveats

Clinical pain assessment is complicated by the multiplicity of pain syndromes in MS. To approach pain in MS in a systematic way, it is important for the clinician to consider the type or types of pain the patient has; the classification systems in Tables 9.1 and 9.2 may be helpful in this regard. However, a few cautions are warranted. First, as noted previously, there is no classification system for MS pain that is widely accepted or empirically validated. Second, diagnostic labels can often be misleading in patients with chronic pain. The processes underlying pain become blurred when the pain becomes chronic. Furthermore, it can be misleading for both the clinician and the patient to assume that they fully understand the pain problem simply because a diagnostic label has been attached to it. One problem is that diagnoses for some chronic pain problems—for example, the diagnosis of lumbago for chronic back pain—are nonspecific, and attempts to make more specific diagnoses based on presumed structural lesions have been controversial.[51] A related issue is that diagnoses in many chronic pain problems do not provide much insight into the degree of suffering or disability experienced nor the variability in functioning seen within the same pain condition.[52] For example, some individuals with a diagnosis of chronic headaches are productive in all major areas of their lives, whereas others are virtually bedridden.

9.5.2.2 Physical Assessment

9.5.2.2.1 Neuropathic Pain Assessment

There are no unequivocal clinical criteria that can be used to diagnose CNP; what is provided here is an overview of some of the considerations when clinically assessing it. Details regarding the assessment of neuropathic pain are provided elsewhere.[5,53] The first step in making the diagnosis is to decide whether a patient has a disorder that could plausibly cause CNP, and MS certainly qualifies as one. The next step is to determine whether the patient has neuropathic pain of any kind. There are four cardinal criteria of neuropathic pain:

1. Vague localization: Neuropathic pain is often described as existing over a wide area of the body, such as an entire limb.
2. Characteristic qualities: Although neuropathic pain can have many qualities,[54] it should be suspected if the pain is described as either burning or shooting (lancinating) in quality.
3. Associated symptoms: The patient with neuropathic pain will often describe paresthesias (pins and needles) and or numbness in the same area where pain is experienced.
4. Associated physical findings: There may be evidence of loss of neurologic function (sensory or motor loss) in the area where the patient is symptomatic.

Once a clinician has decided that a patient has neuropathic pain, the next question is whether this reflects central, as opposed to peripheral, neuropathic pain. Often, specific clinical features and diagnostic tests will determine the presence or absence of a lesion or disorder of the peripheral nervous system. Sometimes the location of symptoms is indicative; in other instances, electrodiagnostic testing can determine the presence or absence of a disorder of the peripheral nervous system. If the person with MS has neuropathic symptoms that cannot be attributed to dysfunction in the peripheral nervous system, a diagnosis of CNP is appropriate. It should be clear from this discussion, though, that several complexities are involved in the identification of CNP. For an example of a complex algorithm used to determine whether people with known MS were suffering from CNP, see reference 18.

Assessment of painful Lhermitte's sign warrants a brief description given that it is common in MS. Lhermitte's sign is a sign elicited during physical examination. To assess the sign, the physician has the patient flex his or her neck and then exerts a gentle push to increase neck flexion. The patient is said to have a positive Lhermitte's sign if he or she reports lancinating pain in the trunk or lower extremities in response to this maneuver. The presence of Lhermitte's sign in a patient with MS strongly suggests neuropathic pain secondary to CNS dysfunction.

9.5.2.2.2 Musculoskeletal Pain Assessment

It is often difficult to determine whether a patient's pain represents neurologic dysfunction or dysfunction of the musculoskeletal system. Although in theory the diffuse burning or lancinating pain of neurologic origin is quite different from the more localized, aching pain of musculoskeletal origin, the boundaries between neurologic pain and musculoskeletal pain are disputed in relation to many patients. For example, some physicians claim that irritation or dysfunction of muscles or ligaments can produce widespread symptoms that bear at least a superficial resemblance to neurologic pain.[55,56] As a practical matter, it is frequently necessary to work MS patients up for musculoskeletal disorders that could be contributing significantly to their reports of, for example, shoulder or knee pain.

Given the importance of physical activity in managing pain, as described below, physical assessment of the patient with MS and pain should include consideration of physical capabilities with respect to exercise and activity. Preliminary physical assessment would ideally consider the individual's premorbid level of physical activity and the impact that MS has had on exercise and overall activity levels. Information on the individual's current mobility will help inform what exercise activities are indicated. Evaluation should also include an assessment of the person's cardiopulmonary functioning, muscle strength, and active and passive range of motion in order to determine baseline physical fitness and to prioritize fitness goals.

9.5.2.3 Assessment of Pain's Impact and Other Factors

It is crucial to consider the ways in which the individual is affected by pain. This analysis should encompass all aspects of quality of life, including the impact of pain on physical health, emotional impact, the person's ability to engage in daily activities or to be independent, and the person's ability to participate in important social roles, such as family roles or employment. It should also include an assessment of how pain affects other MS symptoms, as they often interact. The treatment plan needs to be anchored in an understanding of how the patient is affected by his or her pain. For example, if the patient is unable to participate in valued family activities, such as attending a child's sporting events, the treatment plan is unlikely to be successful unless it addresses the issue of attending sporting events. It is important to note that a strictly medical evaluation of chronic pain rarely reveals how severely a patient is affected by the pain.

TABLE 9.3

Example Questions for Assessing Pain Impact, Treatment Goals, and Treatment Expectations

Domain	Example Questions
Pain impact	How is your pain affecting your life (such as your health, emotions, sleep, activities, exercise/activity level, relationships)? What are you doing more of because of your pain? What are you doing less of because of your pain? What does your pain keep you from doing?
Treatment goals and expectations	What, if anything, would you like to learn, or try, to help you manage your pain? What is one thing you would like to do more of if you could better manage your pain? What do you hope to get out of your pain rehabilitation program? What do you expect from pain treatment?
Pacing problem	Many people with MS pain describe problems with pacing in which they overdo it on a day they feel good, only to pay for it later with an increase in pain (and fatigue). Does this describe you? How?
Current pain self-management strategies	What do you do to manage your pain, if anything? Besides medications, do you use other strategies for managing or coping with your pain? What? What do you do that eases or relieves your pain?

Assessment should include a variety of factors other than the patient's MS that may be contributing to his or her pain and consequent disability/dysfunction. In particular, there is a need to assess other factors that are potentially modifiable in a pain-treatment program. For example, many patients become severely deconditioned as their MS progresses. In general, patients feel and function better when the effects of deconditioning have been reversed. Pacing problems should be assessed, given how common they are among individuals with chronic pain. Anxiety and depression should also be assessed and, if present, aggressively treated via pharmacologic and/or psychotherapeutic interventions. Assessing how the patient responds to and copes with pain is also important to guide recommendations for facilitating adaptive coping and functioning. Finally, inquiring about the patient's treatment goals and expectations is important in devising a realistic treatment plan. Table 9.3 lists example questions for assessing pain impact, treatment expectations, and other domains during a clinical assessment.

9.6 Setting Rehabilitation Goals and Expectations

An important step between pain assessment and treatment is the setting of treatment goals and expectations. Several issues are important regarding the goals of pain treatment. Given that complete pain relief is rarely achieved with currently available treatments in the context of MS, it is important to focus treatment not only on pain relief. It is also important to educate the patient that although pain relief is a goal, it is not always achievable. Equally important targets for pain treatment are increased functioning, activity, and participation in valued life activities. Improvements in other areas of physical, emotional, and behavioral health may also be targets. For example, goals of pain treatment might be to increase physical activity, decrease pain's impact on mood, or to decrease the reliance on medications that may have significant adverse consequences, such as high-dose opioid pain medications. The assessment of pain impact and patient-identified goals will be useful in setting these goals.

A crucial step in setting up a collaborative pain rehabilitation plan is to set realistic expectations, not only about the treatment goals but also about the course and nature of treatment. Providers are encouraged to use a biopsychosocial model for educating patients about the nature of chronic pain, particularly the need for patients to take an active role in self-managing their responses to their pain. The emphasis should be on optimal management of pain and its impact so that the patient can participate in valued life activities and roles. They should be informed that pain treatments often do not provide immediate relief but rather take time, particularly if the interventions involve exercise, increasing activity, or psychological pain-management skills.

As described below, exercise is often an important goal of rehabilitation interventions because it benefits not only pain relief but also overall health. However, it is common for those who are experiencing pain to be hesitant to exercise out of fear that it will exacerbate their pain. These individuals should be reminded that the evidence suggests that though there may be short-term increases in pain, as there often are in nondiseased populations, engagement in a regular exercise routine could result in long-term pain relief and other health benefits.

Some patients and providers have the potential to ignore improvements that fall short of complete pain relief. Therefore, it is crucial to establish measurable markers of progress toward goals so that both the provider and the patient can be aware of small improvements as they occur. In addition, while self-reported pain reduction should be counted as an indicator of improvement, you should also look for indicators that pain is interfering less with activities (e.g., that the patient is resuming community outings), that are objective (such as an increase in physical functioning), or that rely on the perceptions of people other than the patient (e.g., input from the patient's significant other).

It is also important to set ground rules for treatment. These should spell out the obligations and responsibilities of both the provider and the patient. This is particularly important if opioids will be prescribed as part of the treatment plan. Many pain specialists strongly recommend that any patient being treated with long-term opiate therapy sign a treatment agreement that outlines ground rules and consequences if the ground rules are violated.[57]

9.7 Pain Rehabilitation Interventions

Patients with MS often have refractory pain problems for which curative therapy is rarely available. Moreover, their pain problems may well wax and wane in conjunction with flares of their disease and may well gradually progress as their disease progresses, despite optimal treatment. In this difficult situation, a rehabilitative approach to MS pain is the most appropriate one. A few principles govern this approach. Treatment is often multimodal. Involvement of a treatment team is often necessary in order to optimize specific modalities, since no single practitioner is likely to be well versed or have the time to engage in all of them. It is important for this care to be coordinated such that team members regularly communicate with one another and work on common goals.

Within a rehabilitation context, some combination of four broad treatment approaches should be used:

1. Palliation—reduction of pain via medications, surgery, and so forth
2. Rehabilitation—helping the patient optimize function through utilization of residual capabilities, development of new skills, and practice of good behavior
3. Optimal coping—promoting emotional well-being and coping and developing strategies to manage the psychological and behavioral demands of pain and limits to activity
4. Environmental support—adapting the environment so that the patient can carry out activities of daily living even in the face of reduced capacity

Some treatments are relatively specific to particular kinds of problems, while others are applicable to a wide range of pain problems. An example of the former would be an ankle-foot orthosis for a patient suffering from pain because of repeated ankle sprains; an example of the latter would be cognitive behavioral therapy (CBT) to promote optimal psychological adaptation to and coping with pain.

9.7.1 Pharmacological Intervention

Once it has been determined that a patient has CNP, the therapies described below should be considered, along with more general rehabilitative treatments such as exercise and psychological interventions. A few caveats are in order, however. The goal of pain treatment is to address as many of these problems as possible rather than to focus exclusively on CNP and ignore other contributors to a patient's pain. When research identifies a treatment approach as helpful in MS, it is often unclear what "kinds" of pain have benefitted. Furthermore, the factors that influence neuropathic pain are often elusive. As a practical consequence, treatments that are reliably effective for one MS patient may exacerbate symptoms or cause unacceptable side effects in another, even though the two may appear to have very similar pain problems. Although there is a reasonably robust literature on the treatment of pain in MS, many interventions that have proven helpful in other pain-patient populations have not been adequately studied in MS.

In the face of these uncertainties, clinicians should take an eclectic approach when they treat MS pain. They should target multiple possible sources of pain at once rather than focusing strictly on NP. For example, they should address MS pain and spasticity. They should borrow liberally from knowledge of treatment of chronic pain in general. It is important to recognize the difficulty of predicting how a particular patient will respond to treatment for MS and how willing he or she will be to experiment with a variety of treatments.

Since there is only a modest amount of research on the effectiveness of pharmacologic agents on CNP in MS, it is reasonable to make inferences based on similarities between central pain in MS and other kinds of CNP (e.g., in stroke or spinal cord injury), and similarities between central pain and peripheral neuropathic pain.

9.7.1.1 Anticonvulsants

Research on the effectiveness of anticonvulsants in the treatment of central pain in MS is minimal. One small randomized controlled trial found evidence of benefit from levetiracetam[58]; two studies concluded that lamotrigine was not effective.[59,60] However, there is evidence that some anticonvulsants, including gabapentin[61–66] and pregabalin[67–71] are beneficial in the treatment of various kinds of neuropathic pain, including CNP. Thus, there is reason to believe that these two anticonvulsants, which both act by binding the alpha-2-delta subunit of L-type calcium channels, are reasonable drugs for the treatment of neuropathic pain in MS.

In the use of anticonvulsants, it may be crucial to distinguish between burning steady neuropathic pain and lancinating pain. A recent study supported the efficacy of oxcarbazepine, an anticonvulsant that acts on sodium channels of neurons, in the treatment of trigeminal neuralgia.[72] There is more substantial evidence that carbamazepine, which has the same mechanism of action, is effective in the treatment of trigeminal neuralgia.[19,73] Since carbamazepine is no longer under patent, there has been no recent research on its effectiveness in treating pain. Based largely on the positive results with carbamazepine, it is reasonable to use it or oxcarbazepine preferentially in the treatment of CNP that is lancinating in quality—for example, in patients with positive Lhermitte's signs as well as those with trigeminal neuralgia.

9.7.1.2 Antidepressants

As with anticonvulsants, there is a dearth of literature on the effectiveness of antidepressants in the treatment of CNP in MS, but there is support for some of these medications in the treatment of other neuropathic pain disorders. Most reviews have concluded that tricyclic antidepressants are effective in the treatment of neuropathic pain.[36,73–75] Medications in this group are often helpful at doses lower than those used for the treatment of depression. Among newer antidepressants, a useful distinction can be made between medications that selectively inhibit the reuptake of serotonin (SSRIs) and ones that inhibit the reuptake of both norepinephrine and serotonin (SNRIs). There is evidence that the SNRIs duloxetine and venlafaxine are effective in the treatment of neuropathic pain. The efficacy of SSRIs is less clear.

It should be noted that both antidepressants and anticonvulsants might be prescribed for MS patients with pain in part because of ancillary benefits from the drugs. Antidepressants can help with emotional dysfunction among MS patients, and gabapentin has demonstrated some efficacy in treating spasticity.[76]

9.7.1.3 Opioids

Opioids have received very little research attention in the treatment of pain associated with MS.[77] However, there is evidence that opioids reduce neuropathic pain in other patient groups, so it is reasonable to consider them in the treatment of CNP in MS. However, it is important to be aware that the use of opioids in any chronic pain disorder is controversial for many reasons, including the high frequency of aberrant behavior among chronic pain patients on long-term opioid therapy and the risk of tolerance or opioid-induced hyperalgesia.[78]

9.7.1.4 Cannabinoids

Cannabinoids are chemicals that can be extracted from the plant *Cannabis sativa*. Approximately 460 of them have been identified.[79] A significant amount of research has been done on the basic biology of cannabinoids, including the characterization of cannabinoid receptors[80] and the discovery of endogenous agents with cannabinoid activity.[81] The literature provides some evidence that cannabinoids reduce pain, presumably CNP[79,82] and spasticity[82,83] in MS patients. However, because marijuana is classified as a Schedule I drug by the Food and Drug Administration, physicians have limited access to cannabinoids for therapeutic use and take some risk when they prescribe cannabinoids or recommend their use for the treatment of pain. If the legal status of marijuana changes in the future, it will be important for researchers to identify which, if any, specific cannabinoids have analgesic properties and to explore different routes of administration (e.g., nasal spray) for the cannabinoids. Additional research is currently underway on the role of cannabinoids on pain, spasticity, and other MS symptoms and processes.[82]

9.7.1.5 Miscellaneous Other Agents

Anticonvulsants, antidepressants, and opioids are credible agents in a wide variety of painful conditions, and thus should be considered in the majority of patients with MS who need pharmacologic management of their pain. Other agents are more targeted in their effects. Nonsteroidal anti-inflammatory medications should be considered when a treating physician suspects that an MS patient's pain reflects a musculoskeletal problem with inflammation. Agents such as baclofen, diazepam, tizanidine, and dantrolene should be considered if an individual's pain is thought to be secondary to spasticity.

9.7.2 Exercise

Exercise is one of the most commonly prescribed self-management strategies for a variety of chronic pain conditions other than MS. However, historically, individuals with MS were counseled against physical exertion as a means of avoiding increased pain, fatigue, and thermosensitivity. Current thought has shifted to reflect an understanding that, for people with MS, exercise holds benefits that outweigh possible risks.[84] For example, among persons with MS, physical activity has been related to lower pain as well as other benefits, such as lower disability, fatigue, and depression and better social support, self-efficacy, and quality of life.[85–87] Regular exercise is also related to reduced risk for falls, thereby ostensibly preventing further injury and pain.[84] Nonetheless, most individuals with MS report that they do not engage in regular exercise.[88]

A recent Cochrane review[89] found that multidisciplinary rehabilitation that includes exercise therapy is helpful for patients with MS. The review contained nine studies, three of which showed that patients undergoing rehabilitative treatment experienced improvements in pain. Although research supports the conclusion that exercise therapy enhances physical function, mood, and pain control for individuals with MS, the optimal use of this type of treatment for pain in MS is unclear. In the absence of any convincing research otherwise, it is appropriate to have patients with pain engage in activities that promote progress across multiple exercise domains, including improved aerobic fitness, flexibility, strength, and skill in motor performance.

One key issue in using exercise for pain is the timing and progression of exercises. Too frequently, individuals with chronic pain engage in too much activity at the start of an exercise program. This predictably leads to a pain flare-up and a failure experience. A better strategy is for rehabilitation professionals to employ behavioral methods for gradually increasing physical activity as outlined by Fordyce[90]; see Table 9.4 for specific strategies. In particular, it is important for the physical therapist or other rehabilitation professional to establish baselines for patients in relevant exercises: that is, the level of performance that a

TABLE 9.4

Behavioral Suggestions for Increasing Activity Despite Pain Based on the Work of Fordyce

Strategy	Examples
Select an activity that can be quantified.	Number of minutes spent in the activity, number of repetitions, walking distance.
Establish a baseline for the targeted activity by measuring it for several days (e.g., 3–5 days). Ask patient to perform the behavior until pain or fatigue causes the patient to want to stop. Compute an average for the behavior.	Over 5 days, Kelly was able to walk an average of 10 min a day before she needed to stop due to pain.
Set an initial quota for the behavior that will provide success early in treatment. This is typically a value that is lower (e.g., 50–80%) than the average and was typically exceeded during baseline.	On the first day of the plan, Kelly's goal was to walk for 8 min.
Systematically increase the behavior by setting quotas that increase gradually, typically 5–10% or less each time.	Kelly was instructed to gradually increase her walking by 30 s per day.
Encourage patient to perform the quotas and to pace his or her activity (not overdo or underdo the activity). Do not allow pain to guide the amount of activity once using the quotas.	Even though Kelly wanted to walk further on day 10, she remembered not to in order to avoid overdoing her activity.

Source: Adapted from Fordyce WE. St. Louis: C.V. Mosby; 1976.

Note: For additional discussion on establishing baselines, setting initial values and increments, and managing quota failures, see reference 90.

patient can handle without undue pain. Once baselines have been established, the patient can be encouraged to build from these to progressively more demanding levels of exercise. These behavioral strategies for gradually increasing exercise are also helpful for the person with pain who is reluctant to participate in exercise, as it allows him or her to gradually build tolerance for activity.

9.7.3 Passive Physical Therapy and Other Rehabilitation Modalities

Rehabilitation providers, including physical and occupational therapists, routinely use a variety of passive pain-relieving modalities in the course of pain treatment. Commonly used modalities include electric stimulation, various heat treatments (e.g., ultrasound, hot packs), and ice. Instruction in proper body mechanics, posture, activity pacing, and energy management is also commonly provided. Clinical observations suggest that a number of these strategies can be useful in managing pain (as well as fatigue). However, it is beyond the scope of this chapter to describe the myriad of pain-relief modalities used in specific rehabilitation therapies, particularly given the absence of research examining their efficacy in MS.

9.7.4 Psychological Interventions

The shift in thinking from an exclusively biomedical to a biopsychosocial perspective has had an important effect on how researchers and clinicians understand and treat chronic pain. This shift in perspective has provided a theoretical rationale for psychological pain interventions, such as operant conditioning, CBT, self-hypnosis, relaxation training, family therapy, and education, all of which have been found to be effective for decreasing reported pain severity and decreasing the negative impact of pain on people's lives when pain is the primary disability.[91]

In contrast, research evaluating the efficacy of psychological interventions for pain is sparse in the MS literature. What follows is a summary of hypnosis, the only intervention to have an evidence base for use in MS, and a description of other potential psychological interventions worth consideration and further research, including CBT.

9.7.4.1 Hypnosis

There are descriptions of hypnosis and hypnotic-like treatments throughout recorded history, although the first scientific documentation describing the efficacy of hypnotic analgesia—in this case, as an anesthetic for surgery—was in the mid-1800s.[92] Since that time, interest in hypnosis as a treatment for pain has waxed and waned. Currently, interest in hypnosis and hypnotic analgesia appears to be again on the rise, perhaps because of three related conclusions from recent research in the area.[93] First, as a result of recent research on the neurophysiology of pain from imaging studies, we have confirmed that the experience of pain is associated much more closely with activity in the brain than with activity in the periphery—it is not the input so much as how the brain processes input that influences the experience of pain. Second, recent research, also taking advantage of advances in imaging technology, has demonstrated that hypnotic analgesia has direct effects on the areas of the brain that are involved in the experience of pain, and that these effects vary as a function of the hypnotic suggestions used. For example, suggestions for decreases in pain sensations are associated with decreases in activity in the sensory cortex (which processes sensory information), while suggestions for decreases in the bothersomeness of pain are associated with decreases in activity in the anterior cingulate cortex (which processes information about the affective response to pain).[94,95] Finally, controlled trials during the past decade have confirmed that hypnosis is effective beyond any effects produced by patient expectancy or the passage of time alone.[96]

At its most basic, hypnosis can be defined as an induction followed by a suggestion.[97] The induction can take anywhere from a few seconds to many minutes and usually involves asking the hypnotic subject to focus his or her awareness on a single stimulus (voice, point in a wall, breathing, etc.). When used for chronic pain management, the suggestions that follow can range from suggestions for experiencing decreases in pain severity and increases in comfort or relaxation to improvements in one or more of the many domains associated with chronic pain, such as beliefs about pain controllability, sleep quality, and activity levels.[93,98] Additional suggestions that any benefits of the session will become "permanent" and that the individual can use self-hypnosis for obtaining pain relief are also usually added. Audio recordings of the sessions are often made and given to the person for home use and practice.

Over the past few years, our group has been evaluating the efficacy of this approach in a number of disability populations, including individuals with MS.[99–102] The findings from these studies indicate that the intervention is effective for reducing daily pain intensity relative to control conditions, and that the benefits of treatment are maintained for at least 12 months. Defining a clinically meaningful decrease in pain intensity as one that is 30% or more, the number of individuals with MS and pain who reported clinically meaningful decreases in pain in these studies has ranged from 33%[100] to 47%,[99] although many more obtain at least some pain relief. Moreover, the "side effects" of the treatment are also overwhelmingly positive, with the participants who receive this treatment reporting an increased sense of general well-being, improved sleep, and increased energy, among other benefits. For information on how to apply hypnosis to individuals with chronic pain, see references 103 and 104.

9.7.4.2 *Cognitive Therapy and Cognitive Behavioral Therapy*

Cognitive therapy (CT) was originally developed as a treatment for depression and anxiety, and is based on a cognitive model in which irrational thoughts and thought processes are at the core of emotional disorders.[105] In CT for depression and anxiety, individuals are taught to monitor their thoughts, evaluate them with respect to their effects on mood, and replace any irrational and unhelpful thoughts with more rational and reassuring ones. A substantial body of evidence supports the efficacy of CT for reducing symptoms of depression and anxiety.[106,107]

Due perhaps in large part to the demonstrated efficacy of CT in reducing depression and anxiety symptoms, clinicians began combining it with behavioral approaches for chronic-pain management (i.e., CBT) starting in the late 1970s. The CT component of CBT for pain management focuses on reducing the frequency of *pain-related* unhelpful thoughts, the most common of which are catastrophizing thoughts. Based on a cognitive model of pain, CT and CBT treatments are hypothesized to be effective, at least in part, because they reduce the frequency of catastrophizing cognitions, which are thought to contribute to hypervigilance and amplification of pain signals.[108] The behavioral component of CBT for pain management typically includes training in relaxation skills, such as progressive muscle relaxation, guided imagery, and diaphragmatic breathing. It also commonly includes promoting behavioral activation, including graduated increases in activity despite the presence of pain, along with strategies for pacing of activities (to avoid over- or underdoing activity based on pain). Meta-analyses of CBT treatments support their efficacy for treating a variety of painful conditions.[109–111]

Given the evidence linking catastrophizing and other unhelpful thoughts and coping behavior to pain severity, pain-related disability, and psychological distress in MS,[14,15] it is logical to apply CBT to pain in MS. However, PubMed and PsychInfo searches did not yield any published randomized, controlled trials evaluating cognitive behavioral interventions for chronic pain in MS. Our pilot work[112] supports the feasibility of CT for reducing daily pain intensity in individuals with disabilities and chronic pain. We are also aware of three studies currently underway that are evaluating the efficacy of CBT for MS pain, two from our group and a third by researchers at the VA Connecticut Healthcare System in West Haven (R. D. Kerns, personal communication, March 28, 2011; unreferenced).

In the absence of empirically validated treatments to guide implementation of CBT for MS pain, clinicians may find it useful to consult the general pain literature, including Otis,[113] Thorn,[108] and Turk and Gatchel.[91]

9.7.4.3 *Other Psychosocial Interventions*

In recent years, several additional psychosocial interventions for pain have generated empirical support in the broader pain literature. Although they have not been tested in MS via randomized controlled trials, they merit mention given their growing empirical support for managing pain. Mindfulness-based interventions are increasingly being applied to a range of psychological and health problems, including chronic pain. Although mindfulness interventions vary in terms of their techniques, at their core, they include a focus on nonjudgmental awareness and acceptance of the present moment and any feelings, sensations, or thoughts that arise (mindfulness). Such interventions often include prescribing regular practice of mindfulness, often through mindfulness meditation. A recent randomized controlled trial of an 8-week structured, mindfulness-based intervention for individuals with MS[114] found the intervention to significantly improve health-related quality of life and well-being (including reduced depressive symptoms) relative to a usual care condition. However, this study did not include pain as an outcome.

Acceptance and commitment therapy (ACT),[115] which often includes a mindfulness component, is an intervention with empirical support for use in a range of psychological conditions. One of the main ways that ACT differs from other psychosocial interventions is that it considers suffering, including the experience of pain, to be an unavoidable part of the human experience. As such, rather than aiming to control, avoid, change, or cure pain, ACT interventions teach individuals skills for living more fully in the present, developing self-compassion and flexibility, and engaging in and committing to value-centered goals and activities in daily life. In contrast to CT, which aims to change unhelpful thoughts and feelings about pain, in ACT individuals are taught to recognize that uncomfortable thoughts and feelings are not necessarily true reflections of reality but simply momentary experiences that will change over time. A small study[116] evaluated the effects of a half-day ACT workshop on MS-related impairment, emotional distress, and quality of life in a sample of individuals ($N = 15$) with MS. They saw significant improvements in depression and pain impact, but no other changes. A recent meta-analysis[117] that included nine randomized, controlled trials, found that acceptance-based therapies, defined to include both ACT and mindfulness-based stress reduction, had small to medium effects on physical and mental health in patients with chronic pain (but not MS) that are comparable to the effects seen in meta-analyses of CBT for pain.

A small, but increasing, body of randomized, controlled trials, primarily in samples with chronic low-back pain, suggests that yoga may have benefits in treating chronic pain conditions.[118] Although it might be considered an exercise intervention, yoga, particularly as applied to chronic pain, often includes not only physical activity (poses, movements) but also relaxation and mindfulness techniques, body awareness, postures, and home practice. In MS, a randomized trial comparing yoga (weekly Iyengar yoga class plus home practice), exercise (weekly stationary bicycling class plus home exercise), and usual care (wait-list control) found that both yoga and exercise significantly improved fatigue relative to the usual care condition; mood and cognitive functioning were not affected, nor was pain reported as an outcome.[119]

9.8 Gaps in Knowledge and Future Directions

A number of gaps are highlighted by this review of the status of chronic pain treatment in people with MS. Although a few have been suggested,[5,17] empirical investigations of systems for classifying pain in MS are needed. Such research would provide insight into the pathophysiology and etiology of MS-related pain and also inform treatment. Given the role psychosocial factors appear to play, biopsychosocial models should be included in research on pain in MS.

As discussed above, many of the pharmacologic approaches currently recommended are based on research performed on patient groups other than those with MS. It will be important to examine the efficacy of various agents (e.g., tricyclic antidepressants) in people with MS. Another potentially important area of research concerns the possibility that cannabinoids may ease pain in MS.

The area of exercise in the context of MS is an exciting one given the potential for exercise to not only address pain but also other areas of functioning, including mood.[120] Although overall, exercise is thought to be beneficial, there is much to be learned about the specifics of exercise, including the types of exercise that are beneficial, dosing, and how to promote exercise and motivation to exercise in individuals with MS. There is some evidence that motivational interviewing techniques may be beneficial in this regard.[121]

It is well established that psychological interventions benefit individuals with a variety of painful conditions other than MS. Given the concordance of psychosocial factors in MS pain with psychosocial factors in other painful conditions, it is suspected that psychosocial interventions will likely benefit individuals with MS pain as well. However, this hypothesis awaits confirmation by randomized controlled trials in samples with MS and pain. Even if psychological treatments for pain are proven beneficial in MS, many questions remain regarding their efficacy, delivery, dosing, and mechanisms. For example, we do not know whether specific psychological treatments are more effective than others, nor do we know how to enhance the factors that produce the beneficial effects. Research on the mechanisms underlying any benefits is also needed, not only in MS pain but also in the broader field of pain.[122] We also have much to learn about the biopsychosocial variables that can influence an individual's response to treatment, which would allow better matching of patients to specific psychosocial interventions.

Another gap and area for future research and consideration is the matter of access to pain rehabilitation interventions. It is not known if individuals with MS are able to access the broader complement of pain-management strategies that may be used adjunctive to medications. Studies examining utilization of services for pain have found that most persons with disabilities do not obtain or participate in psychosocial interventions for chronic pain.[123,124] In our research, we have found that fewer than 20% of persons with MS pain[1] reported ever using a psychological intervention such as relaxation, hypnosis, or CBT for pain. Our observation is that few individuals with MS and chronic pain are offered, or have access to, psychological or other rehabilitation pain-management interventions. When offered, they are usually seen as a last resort rather than as an integral aspect of chronic pain self-management.

In conclusion, the chronic pain problems that persons with MS experience are inadequately treated, with many gaps in the scientific and clinical literatures. Given that pain itself can be quite disabling, chronic pain as a co-occurring condition of MS has the potential to contribute to disability and poor quality of life over and above the effects of MS itself. Pain in MS is an area warranting further rehabilitation efforts, both scientifically and clinically.

Acknowledgments

This chapter was supported in part by grants from the National Multiple Sclerosis Society (#MB 0008), the National Center for Medical Rehabilitation Research, National Institute of Child Health and Human Development (#5R01HD057916), and the National Institute of Disability and Rehabilitation Research, Department of Education (# H133B080025).

References

1. Ehde DM, Osborne TL, Hanley MA, Jensen MP, Kraft GH. The scope and nature of pain in persons with multiple sclerosis. Multiple Sclerosis 2006;12(5):629–638.
2. Hadjimichael O, Kerns RD, Rizzo MA, Cutter G, Vollmer T. Persistent pain and uncomfortable sensations in persons with multiple sclerosis. Pain 2007;127(1–2):35–41.
3. Khan F, Pallant J. Chronic pain in multiple sclerosis: Prevalence, characteristics, and impact on quality of life in an Australian community cohort. Journal of Pain 2007;8(8):614–623.
4. Ehde DM, Gibbons LE, Chwastiak L, Bombardier CH, Sullivan MD, Kraft GH. Chronic pain in a large community sample of persons with multiple sclerosis. Multiple Sclerosis 2003;9(6):605–611.
5. O'Connor AB, Schwid SR, Herrmann DN, Markman JD, Dworkin RH. Pain associated with multiple sclerosis: Systematic review and proposed classification. Pain 2008;137(1):96–111.
6. Grasso MG, Clemenzi A, Tonini A, Pace L, Casillo P, Cuccaro A, Pompa A, Troisi E. Pain in multiple sclerosis: A clinical and instrumental approach. Multiple Sclerosis 2008;14(4):506–513.
7. Archibald CJ, McGrath PJ, Ritvo PG, Fisk JD, Bhan V, Maxner CE, Murray TJ. Pain prevalence, severity and impact in a clinic sample of multiple sclerosis patients. Pain 1994;58(1):89–93.
8. Rae-Grant AD, Eckert NJ, Bartz S, Reed JF. Sensory symptoms of multiple sclerosis: A hidden reservoir of morbidity. Multiple Sclerosis 1999;5(3):179–183.
9. Novy DM, Nelson CV, Francis DJ, Turk DC. Perspectives of chronic pain: An evaluative comparison of restrictive and comprehensive models. Psychological Bulletin 1995;118:238–247.
10. Boothby JL, Thorn BE, Stroud MW, Jensen MP. Coping with pain. In: D. C. Turk, R. J. Gatchel, editors. Psychosocial factors in pain. New York, NY: Guilford Press; 1999.
11. Sullivan MJ, Thorn B, Haythornthwaite JA, Keefe FJ, Martin M, Bradley LA, Lefebvre JC. Theoretical perspectives on the relation between catastrophizing and pain. Clinical Journal of Pain 2001;17:52–64.
12. Keefe FJ, Rumble ME, Scipio CD, Giordano LA, Perri LM. Psychological aspects of persistent pain: Current state of the science. The Journal of Pain 2004;5:195–211.
13. Jensen MP, Moore MR, Bockow TB, Ehde DM, Engel JM. Psychosocial factors and adjustment to chronic pain in persons with physical disabilities: A systematic review. Archives of Physical Medicine and Rehabilitation 2011;92:146–160.
14. Osborne TL, Jensen MP, Ehde DM, Hanley MA, Kraft G. Psychosocial factors associated with pain intensity, pain-related interference, and psychological functioning in persons with multiple sclerosis and pain. Pain 2007;127(1–2):52–62.
15. Douglas C, Wollin JA, Windsor C. Biopsychosocial correlates of adjustment to pain among people with multiple sclerosis. The Clinical Journal of Pain 2008;24:559–567.
16. Attal N, Bouhassira D. Can pain be more or less neuropathic? Pain 2004;110:510–511.
17. Pollman W, Feneberg W. Current management of pain associated with multiple sclerosis. CNS Drugs 2008;22:291–324.
18. Osterberg A, Boivie J, Thuomas KA. Central pain in multiple sclerosis—Prevalence and clinical characteristics. European Journal of Pain 2005;9(5):531–542.
19. Finnerup NB. A review of central neuropathic pain states. Current Opinion in Anaesthesiology 2008;21:586–589.
20. Murray PD, Masri R, Keller A. Abnormal anterior pretectal nucleus activity contributes to central pain syndrome. Journal of Neurophysiology 2010;103:3044–3053.
21. Solaro C, Brichetto G, Amato MP, Cocco E, Colombo B, D'Aleo G, Gasperini C, Ghezzi A, Martinelli V, Quessy S. The prevalence of pain in multiple sclerosis: A multicenter cross-sectional study. Neurology 2004;63(5):919–921.
22. Al-Araji AH, Oger J. Reappraisal of Lhermitte's sign in multiple sclerosis. Multiple Sclerosis 2005;11(4):398–402.
23. Hooge JP, Redekop WK. Trigeminal neuralgia in multiple sclerosis. Neurology 1995;45(7):1294–1296.

24. Chen L, Gordon LK. Ocular manifestations of multiple sclerosis. Current Opinion in Ophthalmology 2005;16:315–320.
25. Beard S, Hunn A, Wight J. Treatments for spasticity and pain in multiple sclerosis: A systematic review. Health Technology Assessment 2003;7:iii, ix–x, 1–111.
26. Boneschi FM, Colombo B, Annovazzi P, Martinelli V, Bernasconi L, Solaro C, Comi G. Lifetime and actual prevalence of pain and headache in multiple sclerosis. Multiple Sclerosis 2008;14(4):514–521.
27. Warnell P. The pain experience of a multiple sclerosis population: A descriptive study. Axone 1991;13(1):26–28.
28. Brochet B, Deloire MS, Ouallet JC, Salort E, Bonnet M, Jové J, Petry KG. Pain and quality of life in the early stages after multiple sclerosis diagnosis: A 2-year longitudinal study. Clinical Journal of Pain 2009;25(3):211–217.
29. Svendsen KB, Jensen TS, Hansen HJ, Bach FW. Sensory function and quality of life in patients with multiple sclerosis and pain. Pain 2005;114(3):473–481.
30. Douglas C, Wollin J, Windsor C. The impact of pain on the quality of life of people with multiple sclerosis: A community survey. International Journal of Multiple Sclerosis Care 2009;11:127–136.
31. Svendsen KB, Jensen TS, Overvad K, Hansen HJ, Koch-Henriksen N, Bach FW. Pain in patients with multiple sclerosis: A population-based study. Archives of Neurology 2003;60(8):1089–1094.
32. Newland PK, Naismith RT, Ullione M. The impact of pain and other symptoms on quality of life in women with relapsing-remitting multiple sclerosis. Journal of Neuroscience Nursing 2009;41(6):322–328.
33. Kalia LV, O'Connor PW. Severity of chronic pain and its relationship to quality of life in multiple sclerosis. Multiple Sclerosis 2005;11(3):322–327.
34. Murray TJ. The psychosocial aspects of multiple sclerosis. Neurology Clinics 1995;13(1):197–223.
35. Douglas C, Windsor C, Wollin J. Understanding chronic pain complicating disability: Finding meaning through focus group methodology. Journal of Neuroscience Nursing 2008;40(3):158–168.
36. Dworkin RH, Turk DC, Farrar JT, Haythornthwaite JA, Jensen MP, Katz NP, Kerns RD, Stucki G, Allen RR, Bellamy N et al. Core outcome measures for chronic pain clinical trials: IMMPACT recommendations. Pain 2005;113:9–19.
37. Price DD, Harkins SW, Baker C. Sensory-affective relationships among different types of clinical and experimental pain. Pain 1987;28(3):297–307.
38. Melzack R. The McGill Pain Questionnaire: Major properties and scoring methods. Pain 1975;1:277–299.
39. Melzack R. The short-form McGill Pain Questionnaire. Pain 1987;30:191–197.
40. Jensen MP, Karoly P. Self-report scales and procedures for assessing pain in adults. In: D. C. Turk, R. Melack, editors. Handbook of pain assessment. 3rd ed. New York, NY: Guilford Press; 2011.
41. Jensen MP. Using pain quality assessment measures for selecting analgesic agents. The Clinical Journal of Pain 2006;22 Suppl:S9–S13.
42. Jensen MP. Measurement of pain. In: S. M. Fishman, J. C. Ballantyne, J. P. Rathmell, editors. Bonica's management of pain. 4th ed. Media, PA: Williams & Wilkins; 2010.
43. Bennett M. The LANSS pain scale: The Leeds Assessment of Neuropathic Symptoms and Signs. Pain 2001;92:147–157.
44. Bennett MI, Smith BH, Torrance N, Potter J. The S-LANSS score for identifying pain of predominantly neuropathic origin: Validation for the use in clinical and postal research. Journal of Pain 2005;6:149–158.
45. Galer BS, Jensen MP. Development and preliminary validation of a pain measure specific to neuropathic pain: The Neuropathic Pain Scale. Neurology 1997;48:332–338.
46. Jensen MP, Gammaitoni AR, Olaleye DO, Oleka N, Nalamachu SR, Galer BS. The Pain Quality Assessment Scale: Assessment of pain quality in carpal tunnel syndrome. Journal of Pain 2006;7:823–832.

47. Margolis RB, Tait RC, Krause SJ. A rating system for use with patient pain drawings. Pain 1986;24:57–65.

48. Cleeland CS, Ryan KM. Pain assessment: Global use of the brief pain inventory. Annals of the Academy of Medicine 1994;23:129–138.

49. Spitzer RL, Kroenke K, Williams JBW. Validation and utility of a self-report version of the PRIME-MD. Journal of the American Medical Association 1999;282:1737–1744.

50. Turk DC, Melack R. Handbook of pain assessment. 3rd ed. New York: Guilford Press; 2011.

51. Robinson JP, Apkarian AV. Low back pain. In: E. A. Mayer, M. C. Bushnell, editors. Functional pain syndromes: Presentation and pathophysiology. Seattle, WA: IASP Press; 2009.

52. Wasiak R, Young AE, Dunn KM, Côté P, Gross DP, Heymans MW, von Korff M. Back pain recurrence: An evaluation of existing indicators and direction for future research. Spine 2009;34(9):970–977.

53. Bouhassira D, Attal N, Fermanian J, Alchaar H, Gautron M, Masquelier E, Rostaing S, Lanteri-Minet M, Collin E, Grisart J et al. Development and validation of the neuropathic pain symptom inventory. Pain 2004;108:248–257.

54. Asbury AK, Fields HJ. Pain due to peripheral nerve damage: An hypothesis. Neurology 1984;34:1587–1590.

55. Hackett GS, Hemwall GA, Montgomery GA. Ligament and tendon relaxation treated by prolotherapy. Oak Park, IL: Hemwall; 1993.

56. Simons DG, Travell J, Simons LS. Travell and Simons' myofascial pain dysfunction: The trigger point manual. Baltimore, MD: Williams & Wilkins; 1999.

57. Agency Medical Directors' Group. Interagency guideline on opioid dosing for chronic non-cancer pain: An educational aid to improve care and safety with opioid therapy. Olympia, WA: Washington State Agency Medical Directors Group; 2010.

58. Rossi S, Mataluni G, Codecà C, Fiore S, Buttari F, Musella A, Castelli M, Bernardi G, Centonze D. Effects of levetiracetam on chronic pain in multiple sclerosis: Results of a pilot, randomized, placebo-controlled study. European Journal of Neurology 2009;16:360–366.

59. Silver M, Blum D, Grainger J, Hammer AE, Quessy S. Double-blind, placebo-controlled trial of lamotrigine in combination with other medications for neuropathic pain. Journal of Pain and Symptom Management 2007;34:446–454.

60. Breuer B, Pappagallo M, Knotkova H, Guleyupoglu N, Wallenstein S, Portenoy RK. A randomized, double-blind, placebo-controlled, two-period, crossover, pilot trial of lamotrigine in patients with central pain due to multiple sclerosis. Clinical Therapeutics 2007;29:2022–2030.

61. Keskinbora K, Pekel AF, Aydinli I. Gabapentin and an opioid combination versus opiod alone for the management of neuropathic cancer pain: A randomized open trial. Journal of Pain and Symptom Management 2007;34:183–189.

62. Yaksi A, Ozgönenel L, Ozgönenel B. The efficacy of gabapentin therapy in patients with lubar spinal stenosis. Pain 2011;152:939–942.

63. van de Vusse, AC, Stomp-van den Berg, SG, Kessels AH, Weber WE. Randomised controlled trial of gabapentin in complex regional pain syndrome type 1. BMC Neurology 2004 September;4:13.

64. Caraceni A, Zecca E, Bonezzi C, Arcuri E, Yaya Tur R, Maltoni M, Visentin M, Gorni G, Martini C, Tirelli W et al. Gabapentin for neuropathic cancer pain: A randomized controlled trial from the gabapentin cancer pain study group. Journal of Clinical Oncology 2004;22:2909–2917.

65. Levendoglu F, Ogün CO, Ozerbil O, Ogün TC, Ugurlu H. Gabapentin is a first line drug for the treatment of neuropathic pain in spinal cord injury. Spine 2004;29:743–751.

66. Serpell, MG & Neuropathic Pain Study Group. Gabapentin in neuropathic pain syndromes: A randomised, double-blind, placebo-controlled trial. Pain 2002;99:557–566.

67. Arezzo JC, Rosenstock J, Lamoreaux L, Pauer L. Efficacy and safety of pregabalin 600 mg/d for treating painful diabetic peripheral neuropathy: A double-blind placebo-controlled trial. BMC Neurology 2008 September;8:33.

68. Tölle T, Freynhagen R, Versavel M, Trostmann U, Young JPJ. Pregabalin for relief of neuropathic pain associated with diabetic neuropathy: A randomized, double-blind study. European Journal of Pain 2008;12:203–213.

69. Siddall PJ, Cousins MJ, Otte A, Griesing T, Chambers R, Murphy TK. Pregabaline in central neuropathic pain associated with spinal cord injury: A placebo-controlled trial. Neurology 2006;67:1792–1800.
70. Freynhagen R, Strojek K, Griesing T, Whalen E, Balkenohi M. Efficacy of pregabalin in neuropathic pain evaluated in a 12-week, randomised, double-blind, multicentre, placebo-controlled trial of flexible- and fixed-dose regimens. Pain 2005;115:254–263.
71. Vranken JH, Dijkgraaf MG, Kruis MR, van der Vegt, MH, Hollmann MW, Heesen M. Pregabalin in patients with entral neuropathic pain: A randomized, double-blind, placebo-controlled trial of a flexible-dose regimen. Pain 2008;136:150–157.
72. Solaro C, Restivo D, Mancardi GL, Tanganelli P. Oxcarbazepine for treating paroxysmal painful symptoms in multiple sclerosis: A pilot study. Neurological Sciences 2007;28:156–158.
73. Attal N, Cruccu G, Haanpaa M, Hansson P, Jensesn TS, Nurmikko T, Sampaio C, Sindrup S, Wiffen P. EFNS guidelines on pharmacological treatment of neuropathic pain. European Journal of Neurology 2006;13:1153–1169.
74. Dworkin RH, O'Connor AB, Backonja M, Farrar JT, Finnerup NB, Jensen TS, Kalso EA, Loeser JD, Miaskowski C, Nurmikko TJ et al. Pharmacologic management of neuropathic pain: Evidence-based recommendations. Pain 2007;132(3):237–251.
75. Saarto T, Wiffen PJ. Antidepressants for neuropathic pain: A Cochrane review. Journal of Neurology, Neurosurgery, and Psychiatry 2010;81(12):1372–1373.
76. Cutter NC, Scott DD, Johnson JC, Whiteneck G. Gabapentin effect on spasticity in multiple sclerosis: A placebo-controlled, randomized trial. Archives of Physical Medicine & Rehabilitation 2000;81:164–169.
77. Kalman S, Osterberg A, Sorensen J, Boivie J, Bertler A. Morphine responsiveness in a group of well-defined multiple sclerosis patients: A study with i.v. morphine. European Journal of Pain 2002;6:69–80.
78. Angst, MS, & Clark, JD. Opioid-induced hyperalgesia: A qualitative systematic review. Anesthesiology 2006;104:570–587.
79. Rahn EJ, Hohmann AG. Cannabinoids as pharmacotherapies for neuropathic pain: From the bench to the bedside. Neurotherapeutics 2009;6:713–737.
80. Howlett AC. Pharmacology of cannabinoid receptors. Annual Review of Pharmacology and Toxicology 1995;35:607–634.
81. Freund TF, Katona I, Piomelli D. Role of endogenous cannabinoids in synaptic signaling. Physiological Reviews 2003;83:1017–1066.
82. Zajicek JP, Apostu VI. Role of cannabinoids in multiple sclerosis. CNS Drugs 2011 2011;25(3):187–201.
83. Wade DT, Makela P, Robson P, House H, Bateman C. Do cannabis-based medicinal extracts have general or specific effects on symptoms in multiple sclerosis? A double-blind, randomized, placebo-controlled study on 160 patients. Multiple Sclerosis 2004;10:434–441.
84. White LJ, Dressendorfer RH. Exercise and multiple sclerosis. Sports Medicine 2004;34(15):1077–1100.
85. Motl RW, Gosney JL. Effect of exercise training on quality of life in multiple sclerosis: A meta-analysis. Multiple Sclerosis 2008;14(1):129–135.
86. Motl RW, McAuley E. Pathways between physical and quality of life in adults with multiple sclerosis. Health Psychology 2009;28(6):682–689.
87. Motl RW, McAuley E, Snook EM, Gliottoni RC. Physical activity and quality of life in multiple sclerosis: Intermediary roles of disability, fatigue, mood, pain, self-efficacy and social support. Psychology, Health & Medicine 2009;14(1):111–124.
88. Motl RW, McAuley E, Snook EM. Physical activity and multiple sclerosis: A meta-analysis. Multiple Sclerosis 2005;11(4):459–463.
89. Rietberg MB, Brooks D, Uitdehaag BM, Kwakkel G. Exercise therapy for multiple sclerosis. Cochrane Database of Systematic Reviews 2004 09(3).
90. Fordyce WE. Behavioral methods for chronic pain and illness. St. Louis: C.V. Mosby; 1976.

91. Turk DC, Gatchel RJ. Psychological approaches to pain management: A practitioner's handbook. 2nd ed. New York, NY: Guilford Press; 2002.

92. Esdaile J. Hypnosis in medicine and surgery. New York: Julian Press; 1957.

93. Jensen MP. Chronic pain. In: A. Barabasz, K. Olness, R. Boland, S. Kahn, editors. Evidence-based medical hypnosis: A primer for health care professionals. New York, NY: Routledge; 2009.

94. Hofbauer RK, Rainville P, Duncan GH, Bushnell MC. Cortical representation of the sensory dimension of pain. Journal of Neurophysiology 2001;86:402–411.

95. Rainville P, Duncan GH, Price DD, Carrier B, Bushnell MC. Pain affect encoded in human anterior cingulate but not somatosensory cortex. Science 1997;277:968–971.

96. Jensen MP, Patterson DR. Hypnotic treatment of chronic pain. Journal of Behavioral Medicine 2006;29:95–124.

97. Green JP, Barabasz AF, Barrett D, Montgomery GH. Forging ahead: The 2003 APA division 30 definition of hypnosis. International Journal of Clinical and Experimental Hypnosis 2005;53:259–264.

98. Jensen MP, Patterson DR. Hypnosis in the relief of pain and pain disorders. In: M. R. Nash, A. Barnier, editors. Contemporary hypnosis research. 2nd ed. Oxford, UK: Oxford University Press; 2008.

99. Jensen MP, Barber J, Romano JM, Molton IR, Raichle KA, Osborne TL, Engel JM, Stoelb BL, Kraft GH, Patterson DR. A comparison of self-hypnosis versus progressive muscle relaxation in patients with multiple sclerosis and chronic pain. International Journal of Clinical and Experimental Hypnosis 2009;57(2):198–221.

100. Jensen MP, Hanley MA, Engel JM, Romano JM, Barber J, Cardenas DD, Kraft GH, Hoffman AJ, Patterson DR. Hypnotic analgesia for chronic pain in persons with disabilities: A case series. International Journal of Clinical and Experimental Hypnosis 2005;53:198–228.

101. Jensen MP, McArthur KD, Barber J, Hanley MA, Engel JM, Romano JM, Cardenas DD, Kraft GH, Hoffman AJ, Patterson DR. Satisfaction with, and the beneficial side effects of, hypnotic analgesia. International Journal of Clinical and Experimental Hypnosis 2006;54:432–447.

102. Jensen MP, Ehde DM, Gertz KJ, Stoelb BL, Dillworth TM, Hirsh AT, Molton IR, Kraft GH. Effects of self-hypnosis training and cognitive restructuring on daily pain intensity and catastrophizing in individuals with multiple sclerosis and chronic pain. International Journal of Clinical and Experimental Hypnosis 2011;59(1):45–63.

103. Jensen MP. Hypnosis for chronic pain management: Therapist guide (treatments that work). USA: Oxford University Press; 2011.

104. Patterson DR. Clinical hypnosis for pain control. Washington, D.C.: American Psychological Association; 2010.

105. Beck AT. Cognitive therapy and the emotional disorders. New York: International Universities Press; 1976.

106. Norton PJ, Price EC. A meta-analytic review of adult cognitive-behavioral treatment outcome across the anxiety disorders. Journal of Nervous and Mental Disorders 2007;195(6):521–531.

107. Tolin DF. Is cognitive-behavioral therapy more effective than other therapies? A meta-analytic review. Clinical Psychology Review 2010;30:710–720.

108. Thorn BE. Cognitive therapy for chronic pain: A step-by-step guide. New York, NY: Guilford Press; 2004.

109. Butler AC, Chapman JE, Forman EM, Beck AT. The empirical status of cognitive-behavioral therapy: A review of meta-analyses. Clinical Psychology Review 2006;26(1):17–31.

110. Hoffman BM, Papas RK, Chatkoff DK, Kerns RD. Meta-analysis of psychological interventions for chronic low back pain. Health Psychology 2007;26(1):1–9.

111. Henschke N, Ostelo RW, van Tulder M, Vlaeyen JW, Morley S, Assendelft WJ, Main CJ. Behavioural treatment for chronic low-back pain. Cochrane Database of Systematic Reviews 2010 07(7).

112. Ehde DM, Jensen MP. Feasibility of a cognitive restructuring intervention for treatment of chronic pain in persons with disabilities. Rehabilitation Psychology 2004;49:254–258.

113. Otis JD. Managing chronic pain: A cognitive-behavioral therapy approach (treatments that work). Oxford, UK: Oxford University Press; 2007.

114. Grossman P, Kappos L, Gensicke H, D'Souza M, Mohr DC, Penner IK, Steiner C. MS quality of life, depression, and fatigue improve after mindfulness training: A randomized trial. Neurology 2010;75:1141–1149.

115. Hayes SC, Strosahl KD, Wilson KG. Acceptance and commitment therapy: An experimental approach to behavior change. New York, NY: Guilford Press; 1999.

116. Sheppard SC, Forsyth JP, Hickling EJ, Bianchi J. A novel application of acceptance and commitment therapy for psychosocial problems associated with multiple sclerosis. International Journal of MS Care 2010;12:200–206.

117. Veehof MM, Oskam MJ, Schreurs KM, Bohlmeijer ET. Acceptance-based intervention for the treatment of chronic pain: A systematic review and meta-analysis. Pain 2011;152:533–542.

118. Wren AA, Wright MA, Carson JW, Keefe FJ. Yoga for persistent pain: New findings and directions for an ancient practice. Pain 2011;152:477–480.

119. Oken BS, Kishiyama S, Zajdel D, Bourdette D, Carlsen J, Haas M, Hugos C, Kraemer DF, Lawrence J, Mass M. Randomized controlled trial of yoga and exercise in multiple sclerosis. Neurology 2004;62:2058–2064.

120. Patti F, Ciancio MR, Reggio E, Lopes R, Palermo F, Cacopardo M, Reggio A. The impact of outpatient rehabilitation on quality of life in multiple sclerosis. Journal of Neurology 2002;249:1027–1033.

121. Bombardier CH, Cunniffe M, Wadhwani R, Gibbons LE, Blake KD, Kraft GH. The efficacy of telephone counseling for health promotion in people with multiple sclerosis: A randomized controlled trial. Archives of Physical Medicine and Rehabilitation 2008;89:1849–1856.

122. Thorn BE, Burns JW. Common and specific treatment mechanisms in psychosocial pain interventions: The need for a new research agenda. Pain 2011;152(4):705–706.

123. Turner JA, Cardenas DD, Warms CA, McClellan CB. Chronic pain associated with spinal cord injuries: A community survey. Archives of Physical Medicine and Rehabilitation 2001 04;82(4):501–508.

124. Warms CA, Turner JA, Marshall HM, Cardenas DD. Treatments for chronic pain associated with spinal cord injuries: Many are tried, few are helpful. Clinical Journal of Pain 2002;18:154–163.

10

Depression

David C. Mohr

CONTENTS

Depression is common among people with multiple sclerosis (MS). The 12-month prevalence of major depressive disorder is twice that seen in the general population. This chapter reviews potential explanations for this increased prevalence, including MS pathogenic factors such as inflammation; lesion volume in specific frontal, temporal, and/or parietal brain regions; genetic predisposition; and psychosocial factors. This chapter also discusses issues in diagnosis and screening of depression in MS. Finally, trials examining the efficacy of pharmacotherapy and psychotherapy for the treatment of depression in MS are reviewed. After reading this chapter, you will be able to:

1. Identify potential reasons for the increased prevalence of depression among people with MS,

2. Describe the impact of depression on people with MS,

3. Implement depression screening strategies, and

4. Understand treatment methods for depression.

10.1 What Is Depression?

There are many different forms or diagnoses for depression. The two most common diagnoses in the *Fourth Edition of the Diagnostic and Statistics Manual (DSM-IV)* are major depressive disorder (MDD) and dysthymic disorder.[1] To be diagnosed MDD, a person must have five of the following symptoms for at least two weeks: (1) depressed mood most of the day nearly everyday, (2) anhedonia (loss of interest or pleasure in activities), (3) significant change in weight or appetite, (4) change in sleep (insomnia or hypersomnia), (5) psychomotor retardation or agitation, (6) fatigue, (7) feelings of worthlessness, (8) diminished ability to think or concentrate most of the day, (9) recurrent thoughts of death. Among the five or more symptoms, depressed mood or anhedonia must be present. It is important to note that some people can have a major depressive disorder without being in a depressed mood. Finally, these symptoms must cause significant levels of distress or impairment in social or occupational functioning. While 40–50% of people with MDD only have a single episode, most people have multiple episodes, which can cluster or become more frequent with age.

In contrast to MDD, dysthymic disorder is characterized by more consistent symptoms over a longer period of time. It is diagnosed when depressed mood is present most of the day, most days for more than two years, along with two more of the following: (1) poor appetite or overeating, (2) insomnia or hypersomnia, (3) low energy or fatigue, (4) low self-esteem, (5) trouble concentrating or making decisions, or (5) hopelessness. A diagnosis that is not frequently used, but that may be relevant, is mood disorder due to a medical condition (such as MS). This diagnosis requires depressed mood, anhedonia, or irritable mood that is a consequence of the medical disease and that causes significant distress or impairment in social or occupational functioning. While it is beyond the scope of this chapter, it should be mentioned that bipolar disorder II includes major depressive episodes along with hypomanic episodes. There are several other diagnoses of depressive disorders, such as substance-induced mood disorder and a catch-all diagnosis of mood disorder not otherwise specified.

Much of the research in MS has focused on depressive symptoms and severity of symptoms rather than a diagnosable disorder. Frequently, "depression" in these studies is defined using a cutoff score. This may be a reasonable approach, since even depressive symptoms that do not meet criteria for MDD or dysthymia can have significant effects on functioning. Indeed, persons with subthreshold levels of depression are frequently recommended for treatment.[2–4]

To summarize, there are a number of different depressive diagnoses. Research in MS has tended to focus more generally on depressive symptoms, although some research has also focused on MDD. While rehabilitation practitioners do not necessarily need to be able to differentiate the specific mood disorders, they should be familiar with and alert to the symptoms associated with these disorders. Screening is discussed in the following section.

10.2 Epidemiology of Depression in MS

Depression is common among people with MS. A population-based epidemiological study found that the 12-month prevalence of MDD was 15.7% among people with MS compared to 7.4% in the general population.[5] Among people with MS aged 18–45, MDD was particularly high, affecting 25.7% in a 12-month period. Another widely cited study found the lifetime risk of MDD to be 50.3%,[6] which is more than twice the rate seen in the general population. Studies evaluating the frequency of depressive symptoms reveal that far more people with MS are likely to experience significant symptoms of depression. A survey of people with MS found clinically significant levels of depressive symptoms in 42% of the sample.[7] Thus, depression will likely be common among people with MS who are treated by rehabilitation practitioners. A brief review of what is known about the potential etiology of depression in MS follows.

10.3 Pathogenesis of Depression in MS

The elevated levels of depressive symptoms and high rates of MDD indicate that people with MS face a unique vulnerability to depression. This section will focus on the biopsychosocial mechanisms of the co-morbidity between MS and depression. Included is a review of the literature on the role of genetics, MS pathogenic factors, MS pathology, MS symptoms, MS treatments, and psychosocial factors.

10.3.1 Genetic Factors

There has been very little work examining the genetic underpinnings of depression among people with MS. One study found that close relatives of depressed people with MS were more likely to be depressed than relatives of a depressed control group, suggesting some shared genetic etiology between MS and depression.[8] However, a second study found that the frequency of depression among first-degree relatives of depressed people with MS was lower than among those in the control group, suggesting that either the increased rates of depression among people with MS do not have a genetic basis, or, at least, that the causes of depression in MS are different from those in the general population.[6] The one study examining depression and apolipoprotein (Apo) E alleles found that the ApoE ε2 allele was associated with greater positive affect and potentially decreased depression, suggesting a protective effect.[9] While consistent with the literature outside of MS, this was a small study and as such should be reproduced. Thus, the literature on the genetics of depression in MS is small and inconclusive.

10.3.2 MS Pathogenesis

Inflammation is a major pathogenic component of MS, particularly during the relapsing phases of the disease. A growing body of literature outside of MS has implicated such inflammation as an etiologic factor in depression.[10] Increased proinflammatory cytokine production, including interleukin (IL)-1, IL-6, tumor necrosis factor (TNF)-α, and interferon (IFN)-γ have been shown to be associated with depression. Proinflammatory cytokines have been shown to produce "sickness behaviors" in animals and humans that are similar to symptoms of depression, including decreased appetite, weight loss, sleep disturbance, psychomotor retardation, reduced interest in the physical and social environment, and loss of libido. In humans, proinflammatory cytokines also produce reduced mood, anxiety, and memory impairment.

There is circumstantial support for cytokine-driven depression among people with MS. More than 90% of people with MS in exacerbation (the clinical manifestation of an inflammatory process) experience significant levels of distress, compared with 39% of people with MS who are not in exacerbation and 12% of people with spinal cord injuries.[11] While the symptoms of an exacerbation may be distressing, the consistency of distress in the presence of exacerbation suggests that these symptoms are not simply a psychosocial reaction, but likely have some biological etiology related to the exacerbation. Depression is more common among people who are younger and in earlier stages of the disease, when it is more likely to be inflammatory,[5] which is also consistent with the cytokine hypothesis. Indeed, proinflammatory cytokines such as TNF-α and IFN-γ are associated with both exacerbation and depression.[12] Furthermore, high levels of TNF-α six months after exacerbation continued to be significantly associated with continuing depression. Certainly, the data to date are not conclusive and much work remains. However, emerging evidence supports the hypothesis that the increased rates of depression among people with MS compared to medically healthy populations is likely due in part to underlying inflammatory processes.[13] Thus, depression should be seen not just as a reaction to MS, but as a symptom of MS that should be managed as part of MS care.[14]

10.3.3 MS Pathology

The pathological sequelae of MS pathogenic factors, particularly in terms of loss and damage to brain tissue, have long been a focus of research. Early studies of global brain atrophy did not show a relationship to psychiatric disturbance.[15] However, more recent studies that have located lesions in specific brain regions have found that depression is associated with focal lesions in the arcuate fasciculus[16] as well as areas in the frontal, temporal, and parietal areas.[17,18] Indeed, using conventional imaging, lesion volume in the frontal, prefrontal, and anterior temporal regions has been reported to account for as much as 40% of depression.[19,20] Using a newer neuroimaging technique, diffusion tensor imaging, changes in the left anterior temporal region were seen where no changes were evident with conventional imaging.[19] Thus, there is emerging evidence that damage to specific brain regions can increase risk of depression symptoms. Advances in neuroimaging technology, such as diffusion tensor imaging, magnetic transfer imaging, and spectroscopy, will likely provide greater insights into neuroanatomical involvement in depression.

While it is increasingly clear that lesions in these specific brain regions can increase the risk of depressive symptoms, it remains unclear whether the depressive symptoms associated with these lesions are similar in etiology to depression more generally. It is also unknown as to whether these symptoms respond to treatments in a manner similar to symptoms in medically healthy people, or if these symptoms represent a substantially different etiology and pathogenesis and would not respond to antidepressant treatment.[21,22] Studies examining the moderating effects of lesions in specific brain regions in response to treatments will be important for determining their prognostic value as well as for characterizing the nature of depressive symptoms resulting from these lesions.

10.3.4 Physical and Cognitive Symptom Severity

A common assumption is that greater symptom severity will result in greater distress and depression. In general, research has found that the association between severity of physical impairment and depression ranges from small to nonexistent.[18,23] A relatively large portion of this literature has focused on the relationship between fatigue and depression, where some investigators have found a relationship.[24,25] However, clinical trials of pharmacologic treatments for fatigue have shown no effect on depression[26] while treatments for depression consistently show improvements in fatigue.[27,28] This suggests that to the degree that fatigue and depression are associated, it is more likely that depression affects fatigue than the reverse. Thus, rehabilitation practitioners should not expect that treating fatigue or other symptoms related to depression will necessarily improve depression itself.

In the general population, people with chronic illnesses are more likely to experience depression.[29,30] People with MS are also vulnerable to the same illnesses experienced by the general population, and, indeed, these same comorbidities are associated with increased risk of depression among people with MS.[31] It is unclear at this point as to why depression is greater among people with other comorbidities, although some literature might implicate overall burden of disease or specific symptoms such as pain.[32,33]

A number of reviews in the literature have concluded that there is no evidence of a relationship between neuropsychological impairment and depression.[34,35] However, some recent research examining more refined areas of information processing suggest that specific areas of cognitive functioning may be related to depression and anxiety. It is hypothesized, based on work in geriatric populations, that frontosubcortical brain pathology can produce both impairments in executive function and symptoms that mimic depression.[21] Executive functioning is a cluster of high-order capacities, which include selective attention, behavioral planning and response inhibition, and the manipulation of information in problem-solving tasks. Preliminary work in this area has found an association between impairment in executive functioning and depression in MS.[22,36]

A similar but separate line of work has suggested that while performance on routine cognitive tasks may not be affected by depression, more effortful tasks that place a demand on attentional resources, such as executive functioning or speeded tasks, are more impaired among people with MS who are depressed.[37,38] Thus, while there is a growing literature supporting the relationship between depression and specific neuropsychological impairment, principally executive function, there is little information regarding the nature of this relationship. To what degree the relationship between depression and executive functioning is the result of a common neuropathological etiology and to what degree depression increases neuropsychological deficits in effortful tasks are unclear.

At this point, rehabilitation practitioners should not assume that people with low levels of physical or cognitive disability are in any way resilient with respect to depression. Likewise, people with greater physical or cognitive disability are not necessarily more vulnerable. And most important, treating or reducing the impact of disability on activity will not necessarily have any impact on depression.

10.3.5 MS Treatments

In rare cases, depression may be an iatrogenic effect of medications. Oral or i.v. glucocorticoids are used to treat exacerbations, and can produce changes in mood and cognition.[39]

There has also been some speculation that new-onset or increased depression may be an iatrogenic effect associated with some of the disease-modifying treatments (DMTs) commonly used to treat MS. Specifically, early uncontrolled studies suggested that the interferon-β's may be associated with increased risk of depression.[40–42] More recent studies have consistently shown that interferon-β's do not cause increases in depressive symptoms.[43,44] However, given the initial concerns, many of these clinical trials excluded people with severe, active depression. Thus, while the preponderance of evidence suggests that DMTs do not cause depression, one cannot rule out the possibility that they may activate depression among vulnerable people.

10.3.6 Psychosocial Factors

A variety of psychosocial factors have been identified that are associated with depression, anxiety, and emotional adjustment.[45] While these factors are similar to those found to be associated with increased risk of depression among people without medical illness, many of these factors can be augmented by the difficulties and burden posed by MS. Psychosocial factors are particularly important because they are potentially modifiable through psychological and behavioral intervention.

Stress and coping. A large literature has examined adjustment in terms of a stress and coping model, which defines coping as "constantly changing cognitive and behavioral efforts to manage specific external and/or internal demands that are appraised as taxing or exceeding the resources of the person" (p. 141).[46] A stressor, typically an external event, must be appraised as having two qualities to be stressful. First, the event must threaten something valued. Certainly threats to health or financial stability can be stressful. However, a growing body of evidence indicates that socially evaluative threats—threats from the judgment of others—can cause significant stress responses.[47] Second, an event must be perceived as uncontrollable.

A fairly large literature indicates that stressful life events as well as perceived stress are associated with increased risk of depression.[48–51] The appraisal of threat, a potential mediator of the effects of stress on well-being, has also been linked to depression and emotional well-being.[50,52,53]

Many studies have examined the relationship between coping and a variety of mental health outcomes, including depression, anxiety, and adjustment. In general, avoidant coping is associated with greater levels of depression. More active, problem-focused coping tends to be less related to depression or not related at all.[48,54–56] This literature, however, has not demonstrated causality. Many of these studies suggest that poor coping skills lead to greater depression. However, it is just as possible that depression causes greater avoidant coping and decreases an individual's capacity to engage in more active, problem-focused coping.

Social support. A large literature on social support suggests that it can moderate the effects of stressful events by providing active, tangible help as well as emotional support. Most studies do not find a significant independent relationship between social support and emotional well-being or depression. Studies examining the relationship between social support and depression have been mixed.[50,54] More focused studies suggest that perceived lack of support is associated with greater depression, while perceived positive support is unrelated to depression.[57] However, as with coping, these studies do not suggest causality. While negative social support may increase distress and depression, a substantial literature outside of MS also has found that depressed individuals can elicit negative interactions from people in their social networks.[58]

10.4 Impact of Depression

10.4.1 Quality of Life

A fairly large and growing literature has consistently shown that depression significantly and substantially diminishes quality of life (QOL) in people with MS.[59] This effect of depression on QOL is independent of disability and of the severity of other symptoms, such as fatigue,[60,61] and in some cases depression has been shown to be the main determinant of QOL.[62] Many of these studies have been cross sectional, which could raise questions regarding causality. However, trials of treatments for depression indicate that when depression is successfully treated, QOL improves dramatically, and improvements in QOL are associated with the improvements in depression.[63,64] Thus, overall, the literature supports the notion that depression can have a substantial impact on QOL.

10.4.2 Suicide

Early clinic-based studies suggested that suicide was very high among people with MS, with some lifetime prevalence rates higher than 28% and 7.5 times higher than the general population.[65] More recent population-based estimates conducted in Denmark and Sweden place the risk of suicide among people with MS at somewhat more than 2 times the rate seen in age-matched controls.[66–68] While this rate is higher in the year after diagnosis, particularly among women, it remains fairly consistent throughout the disease. Mental illness, depression, and social isolation significantly increase suicidal ideation and completed suicide.[69,70] Given that depression is at least twice as prevalent among people with MS, a large portion of the greater suicide risk among people with MS may be driven by depression.

Given the high risk of suicide, any comments related to suicidality, not wanting to live, or hopelessness should be taken seriously and indicate the need for further evaluation. A rehabilitation provider has an ethical and legal duty to ensure that a suicidal individual is safe. The following questions should be asked:

- What thoughts does the person have about harming or killing him- or herself?
- Does the person have a plan, and, if so, what is the lethality of that plan?
- Does the person have an intent to carry out that plan?

Often mental health professionals also assess factors that increase the risk of suicide by diminishing judgment capacity (e.g., substance or alcohol abuse, cognitive impairment, etc.) to cope with distress and suicidal ideation, history of self-destructive behavior, and other risk factors. However, if a rehabilitation provider encounters a person who expresses suicidal ideation, has a plan, and has intent to carry out that plan, the provider must ensure that the person obtains adequate mental health care. If the person is not safe to leave (e.g., intends to carry out the suicide plan in the near future), hospitalization is required and should be arranged. If the provider is unable to arrange for hospitalization of an acutely suicidal individual or if the person is unwilling to comply, emergency services should be contacted, such as calling 911 in the United States.

10.4.3 Participation Restriction and Implications for Rehabilitation

Depression and emotional distress are associated with participation restriction.[14,59] People who have higher levels of depression are less likely to engage in social activities[71] and may have more difficulty fulfilling work-related roles, and are less likely to be adherent to medical regimens.[72] While this author is unaware of any work specifically examining the effects of depression on rehabilitation in MS, depression has been found to have a negative impact on rehabilitation outcomes across many other conditions.[73]

10.4.4 Neuropsychological Functioning

Estimates of the prevalence of cognitive impairment ranges from 43% to 70%, depending on the assessment methods and cutoffs used.[74] While it has long been believed that depression affects neuropsychological performance among people with MS, this evidence has been largely mixed, leading some to conclude that there is no empirical support for such a relationship.[75] Others have suggested that the cognitive deficits may appear only at higher levels of depression.[76] Arnett has suggested a more complex, bidirectional relationship between depression and cognitive function in which depression may affect effortful cognition but not automatic cognition.[37] However, these conclusions are largely based on selected cross-sectional studies.

In contrast to objectively measured cognitive impairment, self-reported cognitive functioning tends to be very inaccurate, greater depression being associated with reports of greater impairment.[77,78] Indeed, subjective cognitive complaints are more strongly correlated with depression than objective neuropsychological functioning.[79] Successfully treating depression not only reduces self-reported severity of cognitive impairment, but it also leads to greater accuracy of those self-reports.[80] However, to date, trials examining neuropsychological outcomes among depressed people with MS who have been treated for depression do not find evidence of objective neuropsychological changes associated with improvements in depression.[80] Thus, while there is some debate regarding the effects of depression on cognitive performance, the largest effects appear to be on perceptions of cognitive functioning rather than on objective cognitive functioning.

10.4.5 Immunomodulatory Effects of Depression

There has been growing interest in the hypothesis that depression may have adverse immunomodulatory effects on MS.[13] This notion is not new. Charcot, who first characterized MS in the nineteenth century, wrote that grief, vexation, and adverse changes in social circumstance were related to the onset of MS.[81] There has also been a large number of studies that have consistently shown an association between stressful life events and an increased risk of MS exacerbations.[71] In addition, stressful life events, primarily in the form of family and work conflicts, are associated with a significant increase in the risk of the occurrence of new gadolinium-enhancing (Gd+) MRI brain lesions 8 weeks later.[82]

In spite of the interest in this area, the evidence remains quite sparse. Cross-sectional studies have found correlations between depression and markers of immune function associated with MS disease activity, including the gene expression of TNF-α and IFN-γ[12] as well as the number of lymphocytes.[83] Only one study has examined immune function among people with MS involved in a trial comparing treatments for depression.[35] People treated for depression showed significant reductions in T-cell production of IFN-γ. While the reductions in IFN-γ were related to improvements in depression, there was no evidence that these effects were related to clinical outcomes. Additional evidence of a common pathway comes from a small randomized, controlled trial of the serotonin-specific reuptake inhibitor (SSRI) fluoxetine among nondepressed people with relapsing forms of MS.[84] People receiving the SSRI had significantly fewer new gadolinium-enhancing MRI brain lesions over a 24-week period compared to people who received a placebo. Thus, the evidence to date is far from supporting any sort of meaningful effect of depression on MS, but it does suggest that this is an area worthy of further research.

10.5 Depression Assessment Strategies

While diagnosis of a mood disorder such as MDD requires a clinical interview, the severity of depression can be evaluated via interview as well as by self-report. Depression is underdiagnosed in MS, with as many as 67% of people with MS who have depression going undetected.[85-87] In response, a task force on depression in MS has recommended screening for depression.[88] The first screening study, using the 21-item Beck Depression Inventory (BDI)[89] recommended a cutoff of 13.[90] However, this was a small study, and, also, it may not be practical to give and score such a long measure at every visit. The most commonly used screening tool in general internal medicine is the Patient Health Questionaire-2 (PHQ-2),[91] which asks about mood and anhedonia (loss of interest). Using these two questions with people who have MS has the potential of identifying almost all people with MDD, and even one question has reasonable sensitivity and specificity.[92] Thus, screening for depression can be very efficient. However, it should be noted that outside MS, a growing number of studies indicated that without organizational enhancements and accessible systems of care, such screening tools will have little or no impact.[93] In other words, once a person is identified as depressed, systems must be in place that ensure people receive referrals, follow-up on referrals, access services, and receive adequate treatment. Without such systems, screening may not improve overall outcomes.

In summary, given the prevalence and the effects of depression, it is recommended that all people with MS be screened and monitored for depression. This can be done simply by asking the two screening questions in Table 10.1.[91,92] A score of 3 or greater indicates the need for referral for further assessment and possible treatment.[91] However, given that most people referred for mental health treatment never follow up,[87,94] it is critical that the rehabilitation practitioner monitor whether or not the person has, in fact, followed up and provide assistance if necessary.

10.5.1 Confounded Symptoms

There have been suggestions that some symptoms of depression are confounded with symptoms of MS.[72] For example, fatigue, cognitive impairment, insomnia, and change in weight could all potentially be symptoms of either MDD or MS. However, more recent evidence indicates that these symptoms may not have only one source but may have multiple sources. Indeed, a recent study has shown that all symptoms of depression improve with treatment for depression, including those that might be confounded with MS.[95] Thus,

TABLE 10.1

PHQ-2 Screen for Depression

1. Over the last two weeks, how often have you been bothered by feeling down, depressed, or hopeless?
2. Over the last two weeks, how often have you been bothered by little interest or pleasure in doing things?

Rating scale: 0 = Not at all; 1 = Several days; 2 = More than half the days; 3 = Nearly every day.

in diagnosing depression, symptoms used in the diagnosis of MDD should not be excluded just because they may also be related to MS. Any symptom of depression may be considered to contribute to a diagnosis of MDD if it occurs in the presence of either depressed mood or anhedonia.

10.6 Treatment

Evidence suggests that depression in MS, left untreated, will not improve.[56] While the etiology of depression in MS may include disease-specific factors, such as brain lesions or inflammation, the treatment literature suggests that, as a group, people with MS respond to treatment for depression about as well as people without MS.

10.6.1 Antidepressant Medications

Open-label studies have suggested treatment with serotonin-specific reuptake inhibitors, tricyclics, and other classes of antidepressant medications are effective at reducing depression.[96,97] The first small placebo-controlled trial of the tricyclic desimpramine with 28 depressed people with MS showed mixed outcomes.[98] While five weeks of treatment showed significant improvements on an interviewer-rated measure, compared to a placebo, there was no significant improvement in self-reported severity. At the end of the trial, most of the 28 individuals were receiving the minimal clinical dosage of 150 mg per day or less. A more recent trial compared 12 weeks of a flexible dose of paroxetine (10–40 mg) to a placebo in 42 people with MS diagnosed with MDD. Both groups improved, and there were no significant differences between the treatment arms.[99] However, effect sizes were similar to those seen in larger trials. It is possible that the failure to find significant effects was due both to the short treatment duration and the small sample sizes.

A somewhat longer trial of 16-week treatments compared sertraline to two forms of psychotherapy for major depressive disorder among people with MS.[100] Sertraline produced significant reductions in depressive symptoms on both objective and subjective measures of depression that were significantly larger than the group psychotherapy treatment arm and equivalent to individual cognitive behavioral therapy (CBT). Physicians met every four weeks with the individuals to evaluate progress and side effects and to adjust dosage accordingly. The median daily dosage by the end of treatment was 150 mg, or three times the minimal clinical dosage. To the best of our knowledge, there is no indication in the literature that there is any difference in efficacy between sertraline, paroxetine, or desimpramine. Thus, the difference between the trials is likely to be that those in the sertraline trial received more potent dosing over a longer period of time compared to the other two trials. This suggests that symptom reduction and remission requires regular monitoring and dose adjustments over longer periods of time.

10.6.2 Physician Care for Depression

The vast majority of treatment for depression is administered by non-mental-health physicians[101] and there is no reason to think this is not true for persons with MS.[87] A study of 260 individuals with MS cared for by neurologists in a large health maintenance organization in the United States found that 26% of them met MDD criteria as judged by an evaluation independent from the physician.[87] Of those individuals meeting MDD criteria, 67% received no antidepressant, 30% received antidepressant treatment that was at, or below, minimal clinical dosage, and only 3% received doses that exceeded minimal clinical dosages. A Danish national hospital registry reported that nearly 25% of those discharged filled prescriptions for antidepressants.[102] This was significantly greater than the rates for people in the control group, who had osteoarthritis. However, there was no information on the dosages. Overall, the literature suggests that many people go untreated and that, even when depression is identified, physicians may not be prescribing antidepressants at doses adequate to produce symptom remission. Thus, if depressed people with MS are receiving treatment for depression from a non-mental-health provider, the rehabilitation provider should not assume that the treatment is adequate. Indeed, the continuing symptoms should serve as an indicator of insufficient treatment and a referral for further treatment.

10.6.3 Psychotherapy for Depression

Two early trials comparing group-administered CBT to a waitlist or treatment as usual control conditions showed strong improvements in depressive symptoms.[103,104] Early trials comparing insight-oriented group treatments to no treatment controls also demonstrated reductions in depression;[105,106] however, these effect sizes were significantly smaller than outcomes for group CBT.[56]

The first comparative outcome trial randomized participants received 16 weeks of a manualized individual CBT,[107,108] a manualized supportive-expressive group therapy,[109] or pharmacotherapy with sertraline with a standardized follow-up protocol. The CBT and pharmacotherapy produced significant improvement in depression compared to the group supportive-expressive treatment, which produced no significant improvement,[100] further supporting the notion that a more problem-focused approach is more likely to be effective in treating depression.

While most people with depression want psychological treatments,[110] most people report barriers that prevent access, such as cost, transportation, and time constraints.[94,111] This is particularly true of people with MS and depression.[85] In an effort to overcome structural barriers to care, we investigated the use of the telephone to extend care. An initial pilot of an eight-week telephone-administered CBT (T-CBT) intervention produced significant improvement compared to treatment as usual in a neurology clinic.[112] A follow-up study compared a 16-week manualized individual T-CBT intervention with a 16-week manualized individual telephone-administered supportive-emotion focused therapy (T-SEFT).[113] While people showed significant improvements in both treatments on most measures of depression, T-CBT produced significantly greater improvements than T-SEFT.

The evidence indicates that psychological interventions can be effective for the treatment of depression among people with MS. Cognitive behavioral interventions have also been shown across several studies to be somewhat more effective than other emotion-focused treatments. This is notable, as such treatment differences have not been found among depressed but medically healthy people.[114] It may be that concerns related to the disease, management of MS symptoms, and the unique stresses associated with MS are more amenable to a cognitive behavioral focus. However, it is also true that many people improved in emotion-focused treatments, and the differences between the treatments disappeared rapidly after the cessation of treatment.[113] Thus, at this point the data would suggest that the individual's treatment preferences should play a significant role in the selection of treatment modality, but that absent strong preferences, CBT should be the default treatment of choice.

10.6.4 Predictions of Response and Relapse

In general, research has found no evidence that any global measures of physical or cognitive impairment predict initial responses to pharmaco- and psychotherapies. However, people with greater lesion load and greater levels of cognitive impairment are significantly less likely to maintain the treatment gains at six-month follow-up.[115] This suggests that people with MS and cognitive impairment as well as a high lesion load who are treated for depression should be followed closely to ensure maintenance of treatment gains.

While psychotherapy and antidepressant medications produce similar results in populations of depressed people, there may be some instances when one is preferable to the other. There is an emerging literature suggesting that antidepressant medications may be less effective than psychotherapy among people with executive dysfunction (e.g., deficits in planning, organizing, sequencing, and problem solving), which is associated with frontal and fronto-subcortical brain lesions. This effect was first seen in the treatment of depression among the elderly.[116–118] Similarly in MS, executive dysfunction is significantly associated with response to pharmacotherapy, with greater executive dysfunction predicting less response to pharmacotherapy. However, executive dysfunction is unrelated to response to psychotherapy.[22] This suggests that for depressed people with executive dysfunction, psychotherapy may be the treatment of choice.

If, indeed, psychological treatments are superior to pharmacotherapy for people with executive dysfunction, it will be important to understand why this is the case. One hypothesis is that depression resulting from fronto-subcortical brain lesions has a very different etiology from standard depression. It is marked by greater levels of anhedonia and difficulties initiating behavior.[21] Thus, psychological and behavioral treatments that focus on behavior activation and engagement with the environment may better address pathology underlying this form of depression than would pharmacotherapy. While intriguing, it should be emphasized that this is an emerging area of research, which requires more replication before executive dysfunction can be used clinically as a differential treatment outcome predictor.

10.6.5 Secondary Outcomes

Treatments for depression can have a variety of ancillary benefits for people with MS. Improvements in depression result in significant improvements in QOL and well-being.[63,64] Successful treatment for depression has been shown to result in significant improvements in disability.[27] Treating depression can also improve some MS symptoms, most notably fatigue.[28] Successful treatment for depression may also improve a person's adherence to DMTs.[112]

10.7 Summary and Implications for Rehabilitation Treatment

In summary, depression is common among people with MS compared to healthy populations. The increased prevalence of depression may be due to inflammation and lesion volume in specific frontal, temporal, and/or parietal brain regions. As in depression generally, genetic and psychosocial factors also contribute to depression, but it is unclear as to whether they account for the greater prevalence of psychiatric disorders and distress among people with MS compared to the general public. While depression and possibly other psychiatric disorders may result from etiologies that differ from etiologies of these disorders in healthy populations, evidence suggests that people with MS can respond to pharmacotherapeutic and psychotherapeutic interventions.

Overall, there is little research that specifically addresses the impact of depression on rehabilitation in MS or on how rehabilitation treatments should be modified for people with comorbid MS and depression. Given the prevalence of depression in MS, research in this area is needed. Nevertheless, based on existing research, the following recommendations can be made:

- All people with MS should be assessed at intake for depression.
- Individuals screening positive for depression should be screened for suicidality. The provider is responsible for ensuring the safety of suicidal individuals. Anyone who has a suicide plan and intent to carry out that plan should be hospitalized.
- People who screen positive for depression should be referred for further evaluation and treatment. Antidepressant medications and psychotherapy are both valid treatments and may be used together.
- The rehabilitation provider should follow up with the person on any recommendation or referral for evaluation and treatment to ensure that he or she has accessed and received appropriate treatment.
- The referring physician and primary care provider should be notified of the person's depression and the referral for evaluation and treatment, with the person's consent.
- Because depression can reduce motivation and engagement in treatment, rehabilitation providers should be prepared to accommodate these symptoms if necessary. These accommodations may include the following:
  - Evaluate the motivation and confidence that an individual has in his or her ability to complete recommended treatment. Modify the treatment plan so that the person is reasonably confident that he or she can adhere to treatment. This may involve beginning with more modest expectations, which can then be increased over time.
  - Offer more frequent appointments to people who cannot adhere to treatment exercises and recommendations.
  - Engage family members to support treatment efforts.
  - Engage in outreach to people who drop out of treatment. This may include making clear, emphatic recommendations to people who wish to discontinue treatment prematurely and calling people who miss appointments to encourage them to continue treatment.
- Continue to monitor depressed people for depressive symptoms, lack of response to treatment, and relapse.

References

1. American Psychiatric Association. Diagnostic and statistical manual of mental disorders. 4th ed. Washington, DC: American Psychiatric Association; 1994.
2. Pincus HA, Davis WW, McQueen LE. "Subthreshold" mental disorders. A review and synthesis of studies on minor depression and other "brand names". British Journal of Psychiatry 1999;174:288–296.

3. Wells KB, Burnam MA, Rogers W, Hays R, Camp P. The course of depression in adult outpatients: Results from the medical outcomes study. Archives of General Psychiatry 1992; 49(10):788–794.
4. The MacArthur Foundation's Initiative on Depression and Primary Care. The MacArthur initiative on depression and primary care at Dartmouth and Duke: Depression management toolkit. Hanover, NH: Trustees of Dartmouth College; 2004.
5. Patten SB, Beck CA, Williams JV, Barbui C, Metz LM. Major depression in multiple sclerosis: A population-based perspective. Neurology 2003;61(11):1524–1527.
6. Sadovnick AD, Remick RA, Allen J, Swartz E, Yee IM, Eisen K, Farquhar R, Hashimoto SA, Hooge J, Kastrukoff LF, et al. Depression in multiple sclerosis. Neurology 1996;46(3):628–632.
7. Chwastiak L, Ehde DM, Gibbons LE, Sullivan M, Bowen JD, Kraft GH. Depressive symptoms and severity of illness in multiple sclerosis: Epidemiologic study of a large community sample. American Journal of Psychiatry 2002;159(11):1862–1868.
8. Salmaggi A, Palumbo R, Fontanillas L, Eoli M, La Mantia L, Solari A, Pareyson D, Milanese C. Affective disorders and multiple sclerosis: A controlled study on 65 Italian patients. Italian Journal of Neurological Sciences 1998;19(3):171–175.
9. Julian LJ, Vella L, Frankel D, Minden SL, Oksenberg JR, Mohr DC. ApoE alleles, depression and positive affect in multiple sclerosis. Multiple Sclerosis 2009;15(3):311–315.
10. Capuron L, Dantzer R. Cytokines and depression: The need for a new paradigm. Brain, Behavior, and Immunity 2003;17(S1):S119–S124.
11. Dalos NP, Rabins PV, Brooks BR, O'Donnell P. Disease activity and emotional state in multiple sclerosis. Annals of Neurology 1983;13(5):573–577.
12. Kahl KG, Kruse N, Faller H, Weiss H, Rieckmann P. Expression of tumor necrosis factor-alpha and interferon-gamma mRNA in blood cells correlates with depression scores during an acute attack in patients with multiple sclerosis. Psychoneuroendocrinology 2002;27(6):671–681.
13. Gold SM, Irwin MR. Depression and immunity: Inflammation and depressive symptoms in multiple sclerosis. Immunology and Allergy Clinics of North America 2009;29(2):309–320.
14. Mohr DC. Stress and multiple sclerosis. Journal of Neurology 2007;254(S2):II65–II68.
15. Ron MA, Logsdail SJ. Psychiatric morbidity in multiple sclerosis: A clinical and MRI study. Psychological Medicine 1989;19(4):887–895.
16. Pujol J, Bello J, Deus J, Marti-Vilalta JL, Capdevila A. Lesions in the left arcuate fasciculus region and depressive symptoms in multiple sclerosis. Neurology 1997;49:1105–1110.
17. Bakshi R, Czarnecki D, Shaikh ZA, Priore RL, Janardhan V, Kaliszky Z, Kinkel PR. Brain MRI lesions and atrophy are related to depression in multiple sclerosis. Neuroreport 2000;11(6): 1153–1158.
18. Zorzon M, Zivadinov R, Nasuelli D, Ukmar M, Bratina A, Tommasi MA, Mucelli RP, Bmabic-Razmilic O, Grop A, Bonfigli L, et al. Depressive symptoms and MRI changes in multiple sclerosis. European Journal of Neurology 2002;9(5):491–496.
19. Feinstein A, O'Connor P, Akbar N, Moradzadeh L, Scott CJ, Lobaugh NJ. Diffusion tensor imaging abnormalities in depressed multiple sclerosis patients. Multiple Sclerosis 2010;16(2):189–196.
20. Feinstein A, Roy P, Lobaugh N, Feinstein K, O'Connor P, Black S. Structural brain abnormalities in multiple sclerosis patients with major depression. Neurology 2004;62(4):586–590.
21. Alexopoulos GS. The vascular depression hypothesis: 10 years later. Biological Psychology 2006;60(12):1304–1305.
22. Julian LJ, Mohr DC. Cognitive predictors of response to treatment for depression in multiple sclerosis. Journal of Neuropsychology and Clinical Neurosciences 2006;18:356–363.
23. Stenager E, Knudsen L, Jensen K. Multiple sclerosis: Correlation of anxiety, physical impairment and cognitive dysfunction. Italian Journal of Neurological Sciences 1994;15(2):97–101.
24. Ford H, Trigwell P, Johnson M. The nature of fatigue in multiple sclerosis. Journal of Psychosomatic Research 1998;45:33–38.
25. Schwartz CE, Coulthard-Morris L, Zeng Q. Psychosocial correlates of fatigue in multiple sclerosis. Archives of Physical Medicine and Rehabilitation 1996;77(2):165–170.

26. Krupp LB, Coyle PK, Doscher C, Miller A, Cross AH, Jandorf L, Halper J, Johnson B, Morgante L, Grimson R. Fatigue therapy in multiple sclerosis: Results of a double-blinded, randomized, parallel trial of amantadine, pemoline, and placebo. Neurology 1995;45(11):1956–1961.

27. Mohr DC, Hart S, Vella L. Reduction in disability in a randomized controlled trial of telephone-administered cognitive-behavioral therapy. Health Psychology 2007;26(5):554–563.

28. Mohr DC, Hart SL, Goldberg A. Effects of treatment for depression on fatigue in multiple sclerosis. Psychosomatic Medicine 2003;65(4):542–547.

29. Coulehan JL, Schulberg HC, Block MR, Janosky JE, Arena VC. Medical comorbidity of major depressive disorder in a primary medical practice. Archives of Internal Medicine 1990;150(11): 2363–2367.

30. Scott KM, Bruffaerts R, Tsang A, Ormel J, Alonso J, Angermeyer MC, Benjet C, Bromet E, de Girolamo G, de Graaf R, et al. Depression–anxiety relationships with chronic physical conditions: Results from the world mental health surveys. Journal of Affective Disorders 2007; 103(1–3):113–120.

31. Marrie RA, Cutter G, Tyry T, Campagnolo D, Vollmer T. Effect of physical comorbidities on risk of depression in multiple sclerosis. International Journal of MS Care 2009;11:161–165.

32. Arnow BA, Hunkeler EM, Blasey CM, Lee J, Constantino MJ, Fireman B, Kraemer HC, Dea R, Robinson R, Hayward C. Comorbid depression, chronic pain, and disability in primary care. Psychosomatic Medicine 2006;68(2):262–268.

33. O'Connor AB, Schwid SR, Herrmann DN, Markman JD, Dworkin RH. Pain associated with multiple sclerosis: Systematic review and proposed classification. Pain 2008;137(1):96–111.

34. Brassington JC, Marsh NV. Neuropsychological aspects of multiple sclerosis. Neuropsychology Review 1998;8(2):43–77.

35. Mohr DC, Cox D. Multiple sclerosis: Empirical literature for the clinical health psychologist. Journal of Clinical Psychology 2001;57(4):479–499.

36. Julian LJ, Arnett PA. Relationships among anxiety, depression, and executive functioning in multiple sclerosis. Clinical Neuropsychologist 2009;23(5):794–804.

37. Arnett PA, Barwick FH, Beeney JE. Depression in multiple sclerosis: Review and theoretical proposal. Journal of the International Neuropsychological Society 2008;14(5):691–724.

38. Arnett PA, Higginson CI, Voss WD, Wright B, Bender WI, Wurst JM, Tippin JM. Depressed mood in multiple sclerosis: Relationship to capacity-demanding memory and attentional functioning. Neuropsychology 1999;13:434–446.

39. Medical Economics. Coping and psychological adjustment among people with multiple sclerosis. Physicians' Desk Reference, 52nd ed. Montvale, NJ: Medical Economics Data Production Company; 1998.

40. Mohr DC, Goodkin DE, Likosky W, Gatto N, Neilley LK, Griffin C, Stiebling B. Therapeutic expectations of patients with multiple sclerosis upon initiating interferon beta-1b: Relationship to adherence to treatment. Multiple Sclerosis 1996;2(5):222–226.

41. Mohr DC, Likosky W, Boudewyn AC, Marietta P, Dwyer P, Van der Wende J, Goodkin DE. Side effect profile and adherence to in the treatment of multiple sclerosis with interferon beta-1a. Multiple Sclerosis 1998;4(6):487–489.

42. Neilley LK, Goodin DS, Goodkin DE, Hauser SL. Side effect profile of interferon beta-1b in MS: Results of an open label trial. Neurology 1996;46:552–554.

43. Borràs C, Río J, Porcel J, Barrios M, Tintoré M, Montalbon X. Emotional state of patients with relapsing–remitting MS treated with interferon beta-1b. Neurology 1999;52:1636–1639.

44. Feinstein A, O'Connor P, Feinstein K. Multiple sclerosis, interferon beta-1b and depression: A prospective investigation. Journal of Neurology 2002;249(7):815–820.

45. Dennison L, Moss-Morris R, Chalder T. A review of psychological correlates of adjustment in patients with multiple sclerosis. Clinical Psychology Review 2009;29(2):141–153.

46. Lazarus RS, Folkman S. Stress, appraisal, and coping. New York, NY: Springer Publishing Inc.; 1984.

47. Dickerson SS, Kemeny ME. Acute stressors and cortisol responses: A theoretical integration and synthesis of laboratory research. Psychology Bulletin 2004;130(3):355–391.

48. Aikens JE, Fischer JS, Namey M, Rudick RA. A replicated prospective investigation of life stress, coping, and depressive symptoms in multiple sclerosis. Journal of Behavior Medicine 1997;20(5):433–445.
49. Gilchrist AC, Creed FH. Depression, cognitive impairment and social stress in multiple sclerosis. Journal of Psychosomatic Medicine 1994;38:193–201.
50. Pakenham KI. Adjustment to multiple sclerosis: Application of a stress and coping model. Health Psychology 1999;18(4):383–392.
51. Patten SB, Metz LM, Reimer MA. Biopsychosocial correlates of lifetime major depression in a multiple sclerosis population. Multiple Sclerosis 2000;6(2):115–120.
52. Pakenham KI, Stewart CA, Rogers A. The role of coping in adjustment to multiple sclerosis-related adaptive demands. Psychology, Health and Medicine 1997;2:197–211.
53. Wineman NM, Durand EJ, Steiner RP. A comparative analysis of coping behaviors in persons with multiple sclerosis or a spinal cord injury. Research in Nursing and Health 1994; 17:185–194.
54. McCabe MP, McKern S, McDonald E. Coping and psychological adjustment among people with multiple sclerosis. Journal of Psychosomatic Research 2004;56(3):355–361.
55. Mohr DC, Goodkin DE, Gatto N, Van der Wende J. Depression, coping and level of neurological impairment in multiple sclerosis. Multiple Sclerosis 1997;3(4):254–258.
56. Mohr DC, Goodkin DE. Treatment of depression in multiple sclerosis: Review and meta-analysis. Clinical Psychology: Science and Practice 1999;6:1–9.
57. Wineman NM. Adaptation to multiple sclerosis: The role of social support, functional disability, and perceived uncertainty. Nursing Research 1990;39(5):294–299.
58. Coyne JC, Downey G. Social factors and psychopathology: Stress, social support, and coping processes. Annual Review of Psychology 1991;42:410–425.
59. Benito-Leon J, Morales JM, Rivera-Navarro J, Mitchell A. A review about the impact of multiple sclerosis on health-related quality of life. Disability Rehabilitation 2003;25(23):1291–1303.
60. Amato MP, Ponziani G, Rossi F, Liedl CL, Stefanile C, Rossi L. Quality of life in multiple sclerosis: The impact of depression, fatigue and disability. Multiple Sclerosis 2001;7(5): 340–344.
61. Janardhan V, Bakshi R. Quality of life in patients with multiple sclerosis: The impact of fatigue and depression. Journal of Neurological Sciences 2002;205(1):51–58.
62. D'Alisa S, Miscio G, Baudo S, Simone A, Tesio L, Mauro A. Depression is the main determinant of quality of life in multiple sclerosis: A classification-regression (CART) study. Disability Rehabilitation 2006;28(5):307–314.
63. Cosio D, Siddique J, Jin L, Mohr DC. The effect of telephone-administered cognitive-behavioral therapy on quality of life among patients with multiple sclerosis. Annals of Behavioral Medicine 2011;41(2):227–234.
64. Hart S, Fonareva I, Merluzzi N, Mohr DC. Treatment for depression and its relationship to improvement in quality of life and psychological well-being in multiple sclerosis patients. Quality of Life Research 2005;14(3):695–703.
65. Sadovnick AD, Eisen K, Ebers GC, Paty DW. Cause of death in patients attending multiple sclerosis clinics. Neurology 1991;41(8):1193–1196.
66. Bronnum-Hansen H, Stenager E, Nylev Stenager E, Koch-Henriksen N. Suicide among Danes with multiple sclerosis. Journal of Neurology, Neurosurgery and Psychiatry 2006; 76(10):1457–1459.
67. Stenager EN, Stenager E. Suicide and patients with neurologic diseases. Archives of Neurology 1992;49(12):1296–1303.
68. Fredrikson S, Cheng Q, Jiang GX, Wasserman D. Elevated suicide risk among patients with multiple sclerosis in Sweden. Neuroepidemiology 2003;22(2):146–152.
69. Feinstein A. An examination of suicidal intent in patients with multiple sclerosis. Neurology 2002;59(5):674–678.
70. Stenager EN, Koch-Henriksen N, Stenager E. Risk factors for suicide in multiple sclerosis. Psychotherapy and Psychosomatics 1996;65(2):86–90.

71. Mohr DC, Classen C, Barrera M, Jr. The relationship between social support, depression and treatment for depression in people with multiple sclerosis. Psychological Medicine 2004; 34(3):533–541.

72. Mohr DC, Goodkin DE, Likosky W, Beutler L, Gatto N, Langan MK. Identification of Beck depression inventory items related to multiple sclerosis. Journal of Behavioral Medicine 1997;20(4):407–414.

73. Thomas SA, Lincoln NB. Assessment and treatments of depression and anxiety after stroke. In: K. Laidlaw, B. Knight, editors. Handbook of emotional disorders in later life: Assessment and treatment. New York: Oxford Press; 2008.

74. Chiaravalloti ND, DeLuca J. Assessing the behavioral consequences of multiple sclerosis: An application of the frontal systems behavior scale (FrBe). Cognitive and Behavioral Neurology 2003;16(1):54–67.

75. Rao SM. Neuropsychology of multiple sclerosis. Current Opinion in Neurology 1995;8(3): 216–220.

76. Demaree HA, Gaudino E, DeLuca J. The relationship between depressive symptoms and cognitive dysfunction in multiple sclerosis. Cognitive Neuropsychiatry 2003;8(3):161–171.

77. Benedict RH, Fishman I, McClellan MM, Bakshi R, Weinstock-Guttman B. Validity of the Beck depression inventory-fast screen in multiple sclerosis. Multiple Sclerosis 2003;9(4):393–396.

78. Carone DA, Benedict RH, Munschauer FE, 3rd, Fishman I, Weinstock-Guttman B. Interpreting patient/informant discrepancies of reported cognitive symptoms in MS. Journal of the International Neuropsychological Society 2005;11(5):574–583.

79. Julian L, Merluzzi NM, Mohr DC. The relationship among depression, subjective cognitive impairment, and neuropsychological performance in multiple sclerosis. Multiple Sclerosis 2007;13(1):81–86.

80. Kinsinger SW, Latte E, Mohr DC. Relationship between depression, fatigue, subjective cognitive impairment, and objective neuropsychological functioning in patients with multiple sclerosis. Neuropsychology 2010;24(5):573–580.

81. Charcot JM. Lectures on diseases of the nervous system (G. Sigerson, trans.). London: New Sydenham Society; 1877.

82. Mohr DC, Goodkin DE, Bacchetti P, Boudewyn AC, Huang L, Marrietta P, Cheuk W, Dee B. Psychological stress and the subsequent appearance of new brain MRI lesions in MS. Neurology 2000;55(1):55–61.

83. Foley FW, Miller AH, Traugott U, LaRocca NG, Scheinberg LC, Bedell JR, Lennox SS. Psychoimmunological dysregulation in multiple sclerosis. Psychosomatics 1988;29:398–403.

84. Mostert JP, Admiraal-Behloul F, Hoogduin JM, Luyendijk J, Heersema DJ, van Buchem MA, De Keyser J. Effects of fluoxetine on disease activity in relapsing multiple sclerosis: A double-blind, placebo-controlled, exploratory study. Journal of Neurology, Neurosurgery and Psychiatry 2008;79(9):1027–1031.

85. Marrie RA, Horwitz R, Cutter G, Tyry T, Campagnolo D, Vollmer T. The burden of mental comorbidity in multiple sclerosis: Frequent, underdiagnosed, and undertreated. Multiple Sclerosis 2009;15(3):385–392.

86. Minden SL, Orav J, Reich P. Depression in multiple sclerosis. General Hospital Psychiatry 1987;9(6):426–434.

87. Mohr DC, Hart SL, Fonareva I, Tasch ES. Treatment of depression for patients with multiple sclerosis in neurology clinics. Multiple Sclerosis 2006;12(2):204–208.

88. Goldman Consensus Group. The Goldman Consensus statement on depression in multiple sclerosis. Multiple Sclerosis 2005;11(3):328–337.

89. Beck AT, Ward CH, Medelson M, Mock J, Erbaugh J. An inventory for measuring depression. Archives of General Psychiatry 1961;4:561–571.

90. Sullivan MJ, Weinshenker B, Mikail S, Bishop SR. Screening for major depression in the early stages of multiple sclerosis. Canadian Journal of Neurological Sciences 1995;22(3):228–231.

91. Kroenke K, Spitzer RL, Williams JB. The patient health questionnaire-2: Validity of a two-item depression screener. Medical Care 2003;41(11):1284–1292.

92. Mohr DC, Hart SL, Julian L, Tasch ES. Screening for depression among patients with multiple sclerosis: Two questions may be enough. Multiple Sclerosis 2007;13(2):215–219.
93. Gilbody S, Sheldon T, House A. Screening and case-finding instruments for depression: A meta-analysis. Canadian Medical Association Journal 2008;178(8):997–1003.
94. Mohr DC, Ho J, Duffecy J, Baron KG, Lehman KA, Jin L, Reifler D. Perceived barriers to psychological treatments and their relationship to depression. Journal of Clinical Psychology 2010;66(4):394–409.
95. Moran PJ, Mohr DC. The validity of the Beck depression inventory and Hamilton rating scale for depression items in the assessment of depression among patients with multiple sclerosis. Journal of Behavioral Medicine 2005;28(1):35–41.
96. Flax JW, Gray J, Herbert J. Effect of fluoxetine on patients with multiple sclerosis. American Journal of Psychiatry 1991;148(11):1603.
97. Scott TF, Allen D, Price TR, McConnell H, Lang D. Characterization of major depression symptoms in multiple sclerosis patients. Journal of Neuropsychiatry and Clinical Neuroscience 1996;8(3):318–323.
98. Schiffer RB, Wineman NM. Antidepressant pharmacotherapy of depression associated with multiple sclerosis. American Journal of Psychiatry 1990;147:1493–1497.
99. Ehde DM, Kraft GH, Chwastiak L, Sullivan MD, Gibbons LE, Bombardier CH, Wadhwani R. Efficacy of paroxetine in treating major depressive disorder in persons with multiple sclerosis. General Hospital Psychiatry 2008;30(1):40–48.
100. Mohr DC, Boudewyn AC, Goodkin DE, Bostrom A, Epstein L. Comparative outcomes for individual cognitive-behavior therapy, supportive–expressive group psychotherapy, and sertraline for the treatment of depression in multiple sclerosis. Journal of Consulting and Clinical Psychology 2001;69(6):942–949.
101. Regier DA, Narrow WE, Rae DS, Manderscheid RW, Locke BZ, Goodwin FK. The de facto US mental and addictive disorders service system: Epidemiologic catchment area prospective 1-year prevalence rates of disorders and services. Archives of General Psychiatry 1993;50(2):85–94.
102. Kessing LV, Harhoff M, Andersen PK. Increased rate of treatment with antidepressants in patients with multiple sclerosis. International Clinical Psychopharmacology 2008;23(1):54–59.
103. Foley FW, Bedell JR, LaRocca NG, Scheinberg LC, Reznikoff M. Efficacy of stress-inoculation training in coping with multiple sclerosis. Journal of Consulting and Clinical Psychology 1987;55(6):919–922.
104. Larcombe NA, Wilson PH. An evaluation of cognitive-behaviour therapy for depression in patients with multiple sclerosis. The British Journal of Psychiatry 1984;145:366–371.
105. Crawford JD, McIvor GP. Group psychotherapy: Benefits in multiple sclerosis. Archives of Physical Medicine and Rehabilitation 1985;66(12):810–813.
106. Crawford JD, McIvor GP. Stress management for multiple sclerosis patients. Psychological Reports 1987;61(2):423–429.
107. Mohr DC. The stress and mood management program for individuals with multiple sclerosis: Workbook. New York, NY: Oxford Press; 2010.
108. Mohr DC. The stress and mood management program for individuals with multiple sclerosis: Therapist guide. New York, NY: Oxford Press; 2010.
109. Spiegel D, Classen C. Group therapy for cancer patients: A research-based handbook of psychosocial care. New York, NY: Basic Books; 2000.
110. Dwight-Johnson M, Sherbourne CD, Liao D, Wells KB. Treatment preferences among depressed primary care patients. Journal of General Internal Medicine 2000;15(8):527–534.
111. Mohr DC, Hart SL, Howard I, Julian L, Vella L, Catledge C, Feldman MD. Barriers to psychotherapy among depressed and nondepressed primary care patients. Annals of Behavioral Medicine 2006;32(3):254–258.
112. Mohr DC, Likosky W, Bertagnolli A, Goodkin DE, Van Der Wende J, Dwyer P, Dick LP. Telephone-administered cognitive-behavioral therapy for treatment of depressive symptoms in multiple sclerosis. Journal of Consulting and Clinical Psychology 2000;68(2):356–361.

113. Mohr DC, Hart SL, Julian L, Catledge C, Honos-Webb L, Vella L, Tasch ET. Telephone-administered psychotherapy for depression. Archives of General Psychiatry 2005;62(9): 1007–1014.

114. Watson JC, Gordon LB, Stermac L, Kalogerakos F, Steckley P. Comparing the effectiveness of process-experiential with cognitive-behavioral psychotherapy in the treatment of depression. Journal of Consulting and Clinical Psychology 2003;71(4):773–781.

115. Mohr DC, Epstein L, Luks TL, Goodkin D, Cox D, Goldberg A, Chin C, Nelson S. Brain lesion volume and neuropsychological function predict efficacy of treatment for depression in multiple sclerosis. Journal of Consulting and Clinical Psychology 2003;71(6):1017–1024.

116. Alexopoulos GS, Kiosses DN, Choi SJ, Murphy CF, Lim KO. Frontal white matter microstructure and treatment response of late-life depression: A preliminary study. American Journal of Psychiatry 2002;159(11):1929–1932.

117. Alexopoulos GS, Meyers BS, Young RC, Kalayam B, Kakuma T, Gabrielle M, Sirey JA, Hull J. Executive dysfunction and long-term outcomes of geriatric depression. Archives of General Psychiatry 2000;57(3):285–290.

118. Kalayam B, Alexopoulos GS. Prefrontal dysfunction and treatment response in geriatric depression. Archives of General Psychiatry 1999;56(8):713–718.

11

Visual Impairments

Julie Ann Nastasi, Stephen Krieger, and Janet C. Rucker

CONTENTS

The human visual system relies on the constant and rapid transmission of information along myelinated nerve pathways and, therefore, is exquisitely vulnerable to the effects of a demyelinating disease such as multiple sclerosis (MS). Visual symptoms are very common in MS. They may be transient, lasting for several weeks at a time, or they may be more permanent or chronic and result in persistent visual impairment. In the latter situation, they may be a source of significant disability among clients of rehabilitation services.[1] Symptoms can be roughly categorized into two groups: symptoms that suggest a visual problem in one or both eyes (e.g., blurred, foggy, or cloudy vision in one or both eyes) and symptoms that suggest a problem with eye movement (e.g., double vision that goes away with either eye closed or a sense that whatever is being looked at is moving). MS can affect both the afferent visual stream, by which information from the eyes is transmitted to the brain and processed as visual information, and the efferent pathways that coordinate the movement of the eyes. In this way, MS can cause both visual blurring or loss in one or both eyes, as well as impairment of moving the eyes together that often manifests as diplopia (double vision). Few disease conditions have the potential to influence both the afferent and efferent aspects of the visual system, and evidence of both in an individual is highly suggestive of MS. After reading this chapter, you will be able to:

1. Review the underlying neuroanatomy and pathophysiology of both afferent and efferent visual impairment in MS,

2. Identify the impact of visual symptoms on the everyday lives of people with MS,

3. Provide an overview of the neuro-ophthalmologic history and assessment of an MS client with visual symptoms,

4. Discuss the roles of the different members of the vision rehabilitation team, and

5. Discuss the assessment methods and intervention strategies employed by the members of the rehabilitation team to address visual symptoms and reduce their impact on clients with MS.

11.1 Clinical Neuroanatomy of the Afferent Visual System

11.1.1 Optic Nerve and Visual Pathways

For normal vision to occur, light falls upon the retina, where it is first converted into electrical signals by rods and cones. Rods and cones are specialized cells in the retina in the back of the eye that respond to light. These signals are then routed to the afferent nerve of the visual system, the optic nerve. The optic nerves exit posteriorly from each eye and travel retro-orbitally (behind the eyeball through a bony cavity called the orbit) until they merge together to form a structure called the optic chiasm. In the chiasm, visual information deriving from the temporal (lateral) fields of each eye cross to the other side to join visual information from the nasal (medial) field of the other eye. These optic radiations, containing visual information from both eyes, continue toward the thalamus and, ultimately, to the primary visual cortex in the occipital lobes. Thus, anterior to (in front of) the optic chiasm, the optic nerves convey visual information from each eye individually, while posterior to (behind) the optic chiasm, the optic radiations contain bilateral visual information. The optic nerves and chiasm are the regions of the afferent visual system that are most vulnerable to MS. A substantial portion of the fibers of the optic nerve are devoted to conveying the detailed, color information of the central field of vision that is best detected by the macular area of the retina.

11.1.2 Pupillary Reflex

The pupillary reflex causes the pupils to constrict in presence of bright light in order to adjust the amount of light being conveyed into the afferent visual system. An important principle in considering the visual system is that the eyes are designed to work in concert—to move synchronously and to respond to light in a coordinated fashion. Thus, the pupillary reflex operates to bring about a consensual response of the pupils of both eyes in response to a light source shown to either one. This is accomplished by a nerve pathway that branches off from the main pathway that carries visual information to the brain. This pupil pathway (pupillary reflex) travels to the midbrain, where the parasympathetic nerves that make the pupil constrict begin in two structures called the Edinger–Westphal nuclei. The command to make the pupils constrict when a light is shined in the eye then travel from the Edinger–Westphal nuclei via cranial nerve III to the muscle that constricts the eye. It is important to note that the light input information from one eye goes to both Edinger–Westphal nuclei (see Figure 11.1). The bilaterality of this reflex is crucial to understanding the relative afferent pupillary defect (RAPD), as will be discussed subsequently.

11.1.3 Pathogenesis and Epidemiology of Afferent Visual Impairment in MS

Optic neuritis (ON) is the hallmark manifestation of MS on the afferent visual system. It is the initial manifestation of MS in 20% of cases, and may occur in 66% of MS clients during the course of their disease.[2] Inflammatory demyelination of the retro-bulbar optic nerve impairs the conduction of visual information from one eye and often presents rather acutely over the course of a day to a couple of days (Note: the retro-bulbar optic nerve is the portion of the nerve behind the eye rather than the part that can be seen when an examiner looks into the pupil to see the back of the eye). Clients will often report "blurred vision" from the symptomatic eye, which they describe as "hazy," like "seeing through a cloud," or "looking through a steamed-up window."[3] An area of dense visual obscuration known as a scotoma may develop in or near the central field of vision (see Figure 11.2). Since central vision is the region where color vision is perceived, clients will often report color desaturation, which is frequently depicted as a "washed out" or "gray and faded" appearance to vision from the affected eye. Optic neuritis is painful in up to 90% of cases, with the pain brought out by eye movements as the inflamed nerve is forced to flex in the retro-orbital space. Visual acuity is decreased, although in MS it is typically not worse than 20/200 at the worst point. Acute optic neuritis is usually unilateral in MS; in the presence of bilateral vision loss, optic neuritis affecting both optic nerves or the optic chiasm should be considered.

Using a swinging light, an RAPD may be detected, whereby the pupil of the symptomatic eye constricts consensually to light shown to the contralateral pupil, but when the light is swung to directly illuminate the symptomatic eye, paradoxically the pupil appears to dilate. Recall that the pupillary reflex is designed to cause the constriction of *both* pupils when light is shown to *either*; when the light is swung from the unaffected eye to the eye with the impaired optic nerve, relatively less light is transmitted into the visual system, and the pupils behave as though the light has been dimmed, that is, they dilate. This will be discussed further in the section on neuro-ophthalmic examination.

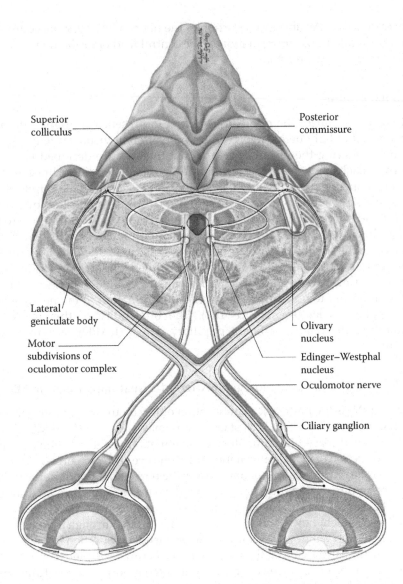

FIGURE 11.1
Schematic diagram of the pupillary light reflex. (Courtesy of *Journal of Comparative Neurology* by Wistar Institute of Anatomy and Biology. Reproduced with permission of John/Wiley & Sons, Inc. in the format Journal via Copyright Clearance Center.)

11.2 Clinical Neuroanatomy of the Efferent Visual System

The efferent visual system is designed to generate conjugate eye movements, that is, the matched rotation of the eyes within the orbit to stabilize the visual world, counter the movements of the head and body, and visually track objects in motion. The efferent visual system works in tight conjunction with the afferent system in that a primary purpose of extra-ocular movements is to keep the target of visual fixation on the macula, the area of

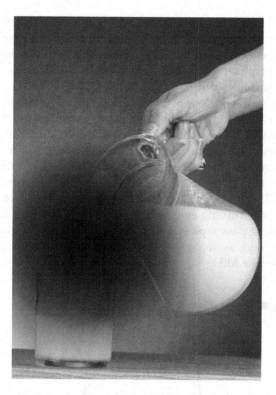

FIGURE 11.2
Central scotoma may impact functional tasks.

the retina responsible for high-acuity vision. Impairments of conjugate eye movements thus disrupt the afferent function of the visual system and can lead to significant, and varied, visual impairments seen in MS.

The extra-ocular movements are controlled by three cranial nerves, cranial nerves III, IV, and VI, which in turn drive the three paired muscle groups responsible for the movements of the eyes, the lateral and medial recti, the superior and inferior recti, and the superior and inferior oblique. The neural mechanisms of the eye movements involve complex inputs from voluntary gaze centers, the vestibular system, and the cerebellum, and separate systems are responsible for horizontal gaze, vertical gaze, and convergence. This discussion will focus on horizontal conjugate gaze, as it is through deficits in these pathways that the impact of MS is often seen. Cranial nerve III (the oculomotor nerve) innervates the medial rectus, and damage to this nerve (for instance, by an aneurysm or infarct) results in a "down and out" position of the affected eye with a failure of medial gaze. Cranial nerve VI (the abducens nerve) innervates the lateral rectus, and thus damage to this nerve causes a failure of lateral gaze. Since the eyes move horizontally in tandem, horizontal eye movements require reciprocal innervation of the eyes; to look to the right, the right eye must move laterally, and the left eye must move medially.

The signal for conjugate horizontal gaze (coordinated movement of both eyes together side to side) begins in the frontal lobe, where it descends to the pons in the brainstem to important brain centers for side-to-side eye movement. The control centers presented here represent a significant simplification of all of the machinery the brain uses to move the eyes. Once it receives a command to move the eyes to the side, CN VI causes the ipsilateral (on the same side) eye to abduct and look laterally, towards the ear (e.g., the left CN VI makes the left eye move toward the left ear) (see Figure 11.3). However, for the two eyes to move together, the contralateral (opposite) eye must move medially, toward the nose. To accomplish this, the nucleus (command center) of CN VI sends a command across the brainstem and back up to the nucleus of CN III in a pathway called the medial longitudinal fasciculus (MLF) that causes the eye to move medially, and thus the two eyes move in concert in the same direction. The eye that moves laterally, via direct command from CN VI, is the one that receives the gaze instruction from the cortex and can be considered the "leading eye." The signal that passes up through the MLF can be thought of as reaching over to the other eye and pulling it along medially. The eye moving medially can be thought of as the "trailing eye." The MLF can be conceptualized like a rope that the leading eye uses to "pull" the trailing eye along in the same direction. For conjugate gaze to occur, these signals must pass rapidly through these myelinated tracts, such as the MLF, within the brainstem, and, as such, they are prime targets for the pathophysiology of a demyelinating disease like MS.

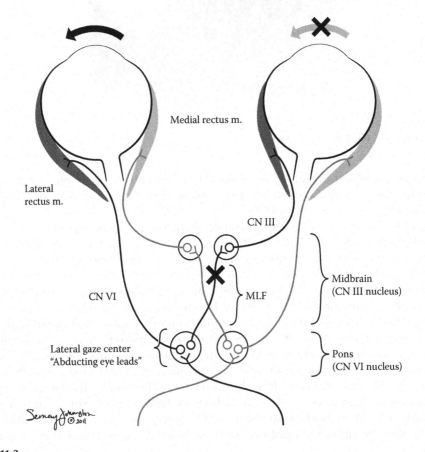

FIGURE 11.3

The lateral gaze center (PPRF) activates the ipsilateral eye to look laterally (toward the ear) and the contralateral eye to look medially (toward the nose). The signal travels from the nucleus of cranial nerve VI up to the nucleus of CN III on the contralateral side via the medial longitudinal fasciculus (MLF). The black "X" indicates a lesion in the MLF, which would be expected to cause an intranuclear ophthalmoplegia (INO). (Illustrated by Semay Johnston.)

11.2.1 Pathogenesis and Epidemiology of Efferent Visual Impairment in MS

As with acute optic neuritis, abnormal eye movements may be the initial, clinical, isolated demyelinating event, or they may arise during the chronic course of disease.[4] Many eye-movement abnormalities can occur as a consequence of MS, but particular ocular motility deficits are common. MS typically does not cause individual cranial nerve deficits of ocular motility, but rather it damages the myelinated pathways that connect these nuclei and allow these nerves to communicate to make the eyes move together. This leads to the failure of tightly coordinated, conjugate eye movements rather than the failure of a single movement in one eye. Perhaps the most common characteristic and highly localizing ocular movement abnormality is a defect in conduction through the MLF, which causes an internuclear ophthalmoplegia (INO).[5] With an MS lesion disrupting the "rope" connecting the abducting, leading eye to the adducting, trailing eye, the clinical findings of an INO are an impairment of adduction of the eye ipsilateral to the impaired MLF, with a dissociated nystagmus of the leading, abducting eye. Nystagmus refers to a bouncing, to-and-fro motion of the eyes due to dysfunction in the brainstem or cerebellar regions that hold the eyes steady. In an INO, the abduction nystagmus can be understood as a compensatory, overshooting mechanism on the part of the abducting eye's repeated attempts to "tug" on the contralateral eye medially via the impaired MLF. While the abduction nystagmus is often the more clinically obvious finding, the impairment of adduction is the crucial feature of an INO and may manifest only as a subtle adduction lag of the trailing eye.

True to its name, the MLF runs medially through the dorsal surface of the pons, and thus the MLFs originating from each side of the brainstem run together in close juxtaposition. A single large MS lesion in the midline of the brainstem can therefore affect bilateral MLFs, causing a bilateral INO. With a bilateral INO, there is impairment of adduction (medial movement) of both eyes, often accompanied by dissociative nystagmus of the abducting eye bilaterally evoked by gaze in either horizontal direction. (Convergence, the medial eye movements produced by a different brainstem pathway, is spared.) INO occurs in up to 30% of clients with MS[4] and may be the initial demyelinating event. Although no clinical sign is absolutely diagnostic of MS, the recognition of an INO in a young adult is extremely suggestive of the diagnosis, and it may be the only objective sign in a client with otherwise subjective symptoms, such as tingling or fatigue.[5] In chronic bilateral INO, an exotropia (outward deviation) of both eyes may develop, leading to what has been termed as a wall-eyed bilateral INO (WEBINO).[6]

The clinical symptoms of conjugate gaze impairment can vary considerably but will often involve a blurring or "smearing" of vision, which can evolve into true binocular diplopia (double vision) in directions of gaze that enhance the uncoordinated movement of the eyes. Normally, although the world is seen with two eyes, they are lined up together and move together with the image of visual interest falling on the same spot in the retina in each eye so that only one image is seen. If one or both eyes are not moving properly due a nerve problem from brainstem demyelination, binocular diplopia occurs due to misalignment of the eyes. Diplopia is often horizontal, but vertical diplopia may result from skew deviations if the vertical gaze pathways are affected. Nystagmus is often itself asymptomatic, although clients may report a shaking of the visual world termed oscillopsia, particularly if the nystagmus is coarse and sustained. If cerebellar function is impaired, particularly cerebellar inputs into the extra-ocular movements, ocular dysmetria commonly results. This incoordination of the eye movements, analogous to limb dysmetria often associated with cerebellar dysfunction, leads to a failure of visual stabilization. Clients may report difficulty focusing on a visual target with a perceived loss of visual acuity, particularly when either the client or the target is in motion.

11.3 Impact of Visual Impairments on Everyday Life

The World Health Organization (WHO) reports that 314 million people worldwide live with visual impairment.[7] Although the WHO does not identify MS as one of the top diseases causing visual impairment, it is a common problem for people with the disease. Up to 50% of individuals with MS report vision loss as an initial presenting symptom of their disease,[8,9] and over time 80% develop a visual impairment.[8-10] As a result, individuals with MS at various points along the disease course report self-perceived impairment in visual function.[10] Many report symptoms even though they present with normal visual acuity and no evidence of ocular misalignment. Vision is commonly reported as "washed-out" or "not right".[9] Deficits in contrast sensitivity function, "the ability to reliably detect the borders of objects as they degrade in contrast from their backgrounds" (pp. 2–4),[11] provides an explanation for the visual deficits reported. Mowry et al.[12] found that low-contrast letters correlate with health-related quality of life in MS. Participants in their study demonstrated two line differences in visual acuity when measured with a low-contrast visual-acuity chart.

Recent studies conducted on vision-related quality of life among individuals with MS, using the Visual Functioning Questionnaire–25 (VFQ-25)[8-10,12]; 10 Item Neuro-Ophthalmic Supplement to the National Eye Institute—Visual Functioning Questionnaire–25 (NEI-VFQ-25)[12]; Impact of Visual Impairment Scale and Short Form 36 Health Survey (SF-36)[12]; and the modified version of the Optic Neuritis Treatment Trial (ONTT) Patient Questionnaire,[9] reported that participants had significant difficulties in several areas of visual function. Findings included difficulty in:

- General vision[12]
- Near activities[12]
- Reading or completing close work[9]
- Using a computer[9,12]
- Distance activities[12]
- Looking at objects far away[9]
- Focusing on and following moving objects[12]
- Performing tasks when eyes were tired[9,12]
- Performing tasks in bright light[9,12]
- Seeing at night[9]
- Driving a car[12]
- Parking a car[12]

Additional significant findings included feeling both eyes were seeing differently[9,12] and blurry vision.[12] Impairments in these areas affect everyday life. Throughout the day, vision is needed to complete activities, seeing near and far as well as focusing on people and objects in the environment that move. The conditions of the outdoor environment influence function when it is bright and sunny or when it is dimly lit at night. Consequently, eye fatigue influences the ability of an individual with MS to perform tasks.

11.4 Vision Rehabilitation Team

In order to maximize participation in everyday life, providers of vision and rehabilitation services need to accurately identify the visual impairments and the activities that are being negatively affected. Once the impairments and activities are identified, members of the multidisciplinary treatment team can identify goals and interventions to enhance participation in everyday life. A number of professions may participate in this effort. These professions include, but are not limited to, ophthalmologists or neuro-opthalmologists, optometrists, occupational therapists, certified orientation and mobility specialists, and certified low-vision therapists.[13,14] Each of these professionals has his or her own unique role in the rehabilitation process and strives to allow the individual full participation in life. The ophthalmologist, neuro-ophthalmologist, or optometrist typically refers the individual with MS for occupational therapy services. Occupational therapists have a medical background that provides an understanding of MS. Within the field of occupational therapy, there are therapists that specialize in low vision, which allows the client to be seen by someone with a medical and visual background. The certified orientation and mobility specialist and the certified low-vision therapist have visual backgrounds but do not necessarily have medical backgrounds since it is not a requirement for certification. The ophthalmologist, neuro-opthalmologist, optometrist, or occupational therapist who refers a client for orientation and mobility services or low-vision services from a certified low-vision therapist need to be aware of this. Education on precautions and contraindications should be provided to ensure that the individual's symptoms will not be exacerbated during interventions provided. As members of the multidisciplinary team complete their evaluations and interventions, results should be communicated to the ordering physician and members of the team to ensure the individual's needs are met, which allows for optimal function. Additional referrals may be made during this process to address the needs of the individual.

11.4.1 Ophthalmologists and Neuro-Opthalmologists

The International Council of Ophthalmology defines ophthalmologists as "physicians who specialize in the medical and surgical care of the eyes and visual system and in the prevention of eye disease and injury" (p. 256).[15] Ophthalmologists are one of the members of the multidisciplinary team. Ophthalmologists are medical doctors (MDs) who have specialized in the visual system. As medical doctors they diagnose and medically manage the eye and visual system. This includes surgery, medications, and optical devices. Ophthalmologists are the only members of the team who perform surgery. The ophthalmologists provide referrals to other members of the multidisciplinary team. A neuro-ophthalmologist is a medical doctor who has has been trained in either neurology or ophthalmology and then completed specialized training in neuro-ophthalmology. This specialist, as opposed to a general ophthalmologist, does not take care of problems that affect the eye itself, but rather the vision nerves behind the eyes and problems of eye movements resulting from the brain. Clients are generally asked to see a neuro-ophthalmologist when their neurologists or ophthalmologists recognize a need for this specialized care.

11.4.2 Optometrists

The World Council of Optometry, representing 49 countries, defines optometry as "a healthcare profession that is autonomous, educated, and regulated (licensed/registered), and optometrists are the primary health care practitioners of the eye and visual system who provide comprehensive eye and vision care, which includes refraction and dispensing, detection/diagnosis and management of disease in the eye, and the rehabilitation of conditions of the visual system" (p. 263).[16] Within the field of optometry, there are optometrists who specialize in low-vision rehabilitation who are called low-vision optometrists. Low-vision optometrists evaluate the individual's vision as well as how well the individual is coping with the vision loss. Optical devices, including eyeglasses, contact lenses, prisms, telescopes, magnifiers, and such, and nonoptical devices, including large-print materials, 20/20 pens, and similar devices, and referrals may be prescribed by low-vision optometrists.

11.4.3 Occupational Therapists

The primary goal of occupational therapy is to enable people to participate in the activities of everyday life. Occupational therapists achieve this outcome by working with people and communities to enhance their ability to engage in the occupations they want to, need to, or are expected to do, or by modifying the occupation or the environment to better support their occupational engagement.[17]

Internationally, occupational therapists that specialize in low vision are able to earn the title certified low-vision therapist (CLVT) through the Academy for Certification of Vision Rehabilitation and Education Professions.[18] This certification body is outside the field of occupational therapy. Therapists with this certification have the credentials CLVT at the end of their names. In the United States, there is specialty certification in low vision through the American Occupational Therapy Association (AOTA), which is indicated with the credentials SCLV at the end of a therapist's name. This certification is only available to graduates of accredited programs through the Accreditation Council for Occupational Therapy Education (ACOTE).

11.4.4 Certified Orientation and Mobility Specialists

The Academy for Certification of Vision Rehabilitation and Education Professions, which resides in the United States, provides national and international certification in orientation and mobility. There are three ways to become a certified orientation and mobility specialist. All require passing certification through the Association for Education and Rehabilitation (AER) of the Blind and Visually Impaired. Certified orientation and mobility specialists may earn a bachelor's degree in orientation and mobility, a post-bachelor's certification, or master's degree in orientation and mobility.[19]

11.4.5 Certified Low-Vision Therapists

The Academy for Certification of Vision Rehabilitation and Education Professions also provides national and international certification in low-vision therapy. There are two ways to become a certified low-vision therapist (CLVT). Both require passing a certification examination through the Association for Education and Rehabilitation of the Blind and Visually Impaired. The certified low-vision therapist may earn a bachelor's degree from an accredited college with a major in low-vision therapy or earn a bachelor's degree in any field from an accredited college. Both ways require 350 hours of discipline-specific supervised training by a certified low-vision therapist and ophthalmologist or optometrist who has specialized in low vision.[18]

11.5 Neuro-Ophthalmologic Assessment: Clinical History

The first question for any client with visual complaints is "does it affect one eye or two?" If the client is uncertain, the examiner might interrupt the history and ask the client to cover each eye in turn and report the effect on visual symptoms. A complaint of visual change in one eye helps the examiner to focus the history and examination on the possibility of optic nerve involvement in MS or another problem in other eye structures (cornea, intraocular lens, retina). A complaint of visual change in each eye alone may suggest simultaneous involvement of both optic nerves or of visual pathways in the brain. A complaint of a visual change that is present only with both eyes open, but resolves with covering either eye, suggests that there is a problem with eye movement.

11.5.1 Clinical History Suggesting Optic Nerve Involvement

Optic nerve involvement is very common is MS, most often with an acute presentation of vision changes in one eye and eye pain that is typically worse with movement of the eye. This history in any young client with or without an existing diagnosis of MS suggests acute optic nerve inflammation, or optic neuritis, as a possible cause of the symptoms. The visual changes the client might describe include dimming, blurring, or fogging of vision; loss of color vision; or a spot in the vision. The client may report that initially he or she thought there might be a "smudge on my glasses," but the smudge failed to resolve, even when the glasses were off. While it is possible for acute optic neuritis to present simultaneously in both eyes, it most commonly affects only one eye at a time. With optic neuritis, the natural course of symptom progression is vision loss over a few days with eye pain, followed by a period of stable vision loss for a few weeks, and subsequent spontaneous resolution of pain and improvement in vision. After resolution of acute optic neuritis, the client may report 100% improvement in vision and pain, or the client may report near resolution but feelings that "tell something happened in that eye" and that slight residual visual symptoms of blur or color abnormalities remain. Very infrequently, vision fails to recover well after acute optic neuritis. Such a client is left with substantial complaints of vision loss in the affected eye. Any MS client sent for evaluation related to visual symptoms should be questioned about current or past episodes of sudden vision loss with painful eye movements.

A second possible mechanism of optic nerve involvement in MS is chronic optic nerve demyelination, which may be stable or progressive at subsequent visits. This is most often seen in clients with secondary progressive MS. The client may have vague complaints of difficulty with visual tasks, such as visual blurring or abnormal color vision; however, the optic nerve demyelination may also be subclinical, meaning that it is detected on examination only and the client has no visual symptoms. It is important to keep in mind that people with MS can have non-MS-related causes of visual symptoms also. Examples of this include presbyopia (age-related difficulty with focusing for near tasks such as reading), glaucoma, early cataracts in clients who have received multiple courses of corticosteroids for acute MS exacerbations, and macular edema (swelling of the specialized retinal region called the macula) as a complication of MS treatment with fingolimod.[20] These causes of visual symptoms are not typically associated with eye pain.

11.5.2 Clinical History Suggesting Abnormal Eye Movements

The second major category of common visual symptoms in MS are those due to abnormal eye movements.[5,21,22] Symptoms of this type of problem include diplopia (double vision, seeing two of everything) and oscillopsia (the client reports that the world seems to be bouncing or jumping). Double vision due to abnormal eye movements causes double vision that is present only with both eyes open and resolves with covering either eye. This type of double vision is called binocular diplopia. Once binocular diplopia is established as the symptom, it is helpful to ask if the diplopia is horizontal or vertical, to direct the examiner to assess for abnormal eye motion in a certain direction. If a client has double vision that is still present with either eye closed (monocular diplopia), it is more likely due to a need for glasses or a change in current glasses than to an eye-movement problem. In contrast to double vision from eye incoordination, oscillopsia is due to abnormal, spontaneous motion of the eyes when the client is looking straight ahead. This is most often a direct result of nystagmus.

11.6 Neuro-Ophthalmologic Examination

The basic neuro-ophthalmologic examination in any client includes assessment of visual acuity, color vision, pupillary function, visual fields, ophthalmoscopic examination of the optic nerves and retina, and examination of eye movements.[23] In MS, additional helpful components include assessment of contrast sensitivity and the thickness of the nerve fiber layer around the optic nerve in the back of the eye with a test called ocular coherence tomography (OCT). Visual-evoked potentials are also occasionally helpful when there remains a question of whether or not the optic nerves are functioning normally upon exam completion.

11.6.1 Examining Optic Nerve Involvement

The first step in neuro-ophthalmic examination is careful assessment of visual acuity for distance and near vision for each eye separately. The term "visual acuity" refers to the relationship between the distance at which testing is performed (usually 20 ft) and the minimal object size the client is able to discern at that distance. For example, 20/20 is normal acuity, meaning that the client's eye can see at 20 ft what a normal eye can see at 20 ft, and 20/50 means that the client's eye can see at 20 ft what a normal eye can see at 50 ft. Distance visual acuity is typically measured with a standard Snellen eye chart, and near acuity, with a near card. An explanation is required for any client whose vision cannot be corrected with glasses to 20/20. One helpful tool for assessing to what degree glasses might improve vision when glasses are not available is a small pinhole for a client to look through (this can be created by punching holes with a pin in an index card), which will correct for refractive error due to a need for glasses. Visual acuity is typically reduced in acute optic neuritis, often in the range between 20/25 (only one line of loss) and 20/400 (the large E on a distance Snellen eye chart), and typically recovers to at least 20/40 (often to 20/20) after resolution of the acute optic neuritis.[24] Loss of visual acuity is nonspecific and may be due to an optic nerve problem, but also to corneal, intraocular lens, and retinal dysfunction.

Color vision is often measured in the office clinical setting with a book of color test-plate numbers called the Ishihara Color Vision Test. A non-color-blind normal eye should be able to identify all of the numbers in this book. In acute optic neuritis, color vision is typically decreased to roughly the same extent as visual acuity—an eye with mild acute optic neuritis and visual acuity of 20/30 may miss just a few of the colored test plates, whereas an eye with severe acute optic neuritis and visual acuity of 20/400 may see few or none of the colored test plates. Color vision usually improves after optic neuritis, although it may do so more slowly than visual acuity. A simple bedside test of color vision can be performed by having the client compare the color of a red object between the two eyes. While seen as red in a normal eye, it will often be seen as pink or orange in an acute optic neuritis eye.

Contrast sensitivity is a very useful test for people with MS as it is a very sensitive measurement of optic nerve function and it may capture an aspect of visual dysfunction when other testing techniques are normal.[25] Although other contrast charts (such as low-contrast Sloan letter charts) are typically used for MS visual research,[26] a chart that can be quickly used in the office clinical setting is the Pelli–Robson chart, which presents letters of the same size but with progressively decreasing contrast toward the bottom of the chart (see Figure 11.4). Most normal subjects are able to read to line 12 or 14 on the chart (with each three letters on the chart representing one line). In acute optic neuritis, contrast sensitivity may be more severely reduced in the affected eye than visual acuity or color vision, and it may fail to recover well after improvement in visual acuity and color vision occur. Residual loss of contrast sensitivity likely explains continued complaints of abnormal vision in many clients after improvement of acute optic neuritis.[25,27]

Assessment of the pupils is critical in establishing the optic nerve as the source of loss of vision in a client with MS. Direct reaction of each pupil alone to a light stimulus is often normal, but a pupillary change called an afferent pupillary defect (APD, also called Marcus Gunn pupil) defines the optic nerve as the cause of vision loss. Corneal, intraocular lens, and retinal causes of vision loss are not associated with an APD. An APD is looked for with the swinging flashlight test, during which the client stares ahead at a distant target and the examiner shines a bright light into one eye and then swings the light to the other eye and continues swinging between the two. The normal response is for each pupil to constrict when the light is shined upon it. When an optic nerve problem is present and causing an APD, the pupil of the affected eye dilates when the light is shined upon it, as relatively less light is entering the pathway on the diseased optic nerve side compared to the normal side.

Visual-field testing provides an assessment of the peripheral and paracentral vision and should be tested with each eye separately. Confrontation visual-field testing is performed by having the client identify an object, such as the number of fingers the examiner is holding up, in each quadrant of vision as the client maintains fixation (holds the eye steady on

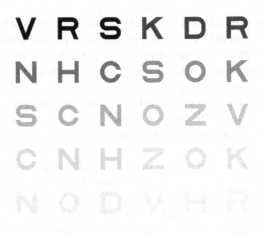

FIGURE 11.4
Pelli–Robson contrast sensitivity chart. (Courtesy of Pelli, D. G., Robson, J. G., and Wilkins, A. J. (1988) The design of a new letter chart for measuring contrast sensitivity. Clinical Vision Sciences 2, 187–199.)

the examiner's nose or eye); however, a more sensitive measure of visual fields is automated perimetry, a computerized method of detecting visual-field loss. This is the standard for visual-field testing in the neuro-ophthalmology office setting. As in confrontation testing, each eye is examined separately; the client fixates on a central visual target as visual stimuli are flashed in the periphery and indicates each time he or she sees a flash. With acute optic neuritis, visual-field deficits are typically seen in the affected eye, and they tend to improve or recover with time.

A final component of the basic optic nerve assessment is ophthalmoscopic examination. This assessment is rarely sufficient without pharmacologic dilation of the pupils with phenylephrine and tropicamide. Important structures to visualize during this component of examination include the optic nerve and retina. Optic nerve appearance is assessed, with specific notation of whether it is normal, swollen, or pale in appearance. The normal optic nerve is orange-red in color and flat with no evidence of swelling. Although the optic nerve appearance is often normal in acute optic neuritis since the inflammation tends to occur in the portion of the optic nerve behind the eye, a minority of clients with acute optic neuritis may have mild optic disc swelling at the time of vision loss. All optic nerves affected by acute optic neuritis develop pallor (the normal orange-red color changes to a light yellow) over a couple of months as residual evidence of optic nerve damage, even if visual recovery is otherwise complete.

OCT is a high-resolution, noninvasive imaging tool that generates retinal images via the use of infrared light, thereby allowing quantitative measurement of the retinal nerve fiber layer (RNFL) around the intraocular optic nerve. RNFL thinning in MS has been shown to correlate with measures of visual function (such as visual acuity and contrast sensitivity), client disability, brain atrophy, and disease subtype.[28–33] In addition, progressive RNFL thinning is seen over time in some MS clients.[34] Most clients with MS undergoing neuro-ophthalmologic evaluation are likely to get an OCT; however, correct utilization of the OCT information as part of clinical care for an individual client is not clearly defined at present.

11.6.2 Examining Abnormal Eye Movements

Neuro-ophthalmologic evaluation without a systematic evaluation of eye movements is incomplete. The most common eye movement abnormalities in MS include difficulty with smooth tracking of a smoothly moving target (called impaired smooth pursuit) and inaccurate placement of the eye on a visual target when fast jumps between two targets are made (saccadic dysmetria);[35,36] however, these are rarely a cause of visual complaints for the client, and they are nonspecific as they are caused by many different neurological diseases. Thus, this chapter will focus on two eye movement abnormalities: internuclear ophthalmoplegia (INO) and one type of nystagmus called pendular nystagmus. INO is one of the most common and characteristic eye movement abnormalities in MS, and pendular nystagmus is one of the most visually disabling.

The classic examination findings in INO are impaired adduction (medial motion of an eye toward the nose) in one eye on the same side as the demyelinating lesion in the brainstem and abducting nystagmus in the opposite eye (to and fro spontaneous motion of the opposite eye when it is placed in a lateral or outward position toward the ear; see Figure 11.5). In other words, a client with a right-sided INO will have trouble moving the right eye toward the nose when he or she tries to look to the left, and the left eye will jump or bounce when it moves toward the ear to look left. INO can have acute onset as an MS exacerbation, in which scenario it might be a cause of diplopia (since one eye does not move with the other), or it may be chronically present in a client with MS, in which scenario the client may not have visual complaints associated with it. INO may also be simultaneously present on both sides at the same time—a bilateral INO—in which case both eyes have trouble moving in toward the nose and both eyes jump when placed outward toward the ears.

There are many different forms of nystagmus, most of which can occur in MS clients; however, acquired pendular nystagmus (APN) is one of the most visually disabling. Nystagmus may only be present when the client moves the eyes to the far right, left, or upward. This is called gaze-evoked nystagmus and is rarely a cause of visual complaints in the client. In contrast, APN is a to-and-fro, continuous, spontaneous motion of both eyes that the client cannot voluntarily control. APN is present and constantly moving the eyes, even when the client looks straight ahead. Thus, to the client, vision is degraded and everything is in constant motion (the symptom of oscillopsia). This constant motion of the eyes may be a large-enough motion that it can be seen just by looking at the client's eyes, but it may also be a very tiny motion and only detectable when a magnified view of the eye is seen during the ophthalmoscopic examination. Any MS client with a complaint of

FIGURE 11.5
Right internuclear ophthalmoplegia. The patient is attempting to look to the left, but cannot move the right eye past the midline toward the nose. The left eye moves outward normally.

oscillopsia should have a dilated eye examination to look for subtle nystagmus if none is visible with the examiner's naked eye.

11.6.3 Visual Assessments by Other Members of the Team

11.6.3.1 Optometric Low-Vision Evaluation

During the optometric low-vision evaluation, the low-vision optometrist typically gathers information on the individual's case history, eye health, near and far visual acuities, central visual field, color vision, visual/mobility field, and contrast sensitivity. The low-vision optometrist refracts the individual's eyes and evaluates the need for magnification. Refraction refers to the process of determining the individual's refractive error, which occurs from myopia, hyperopia, and astigmatism, and then correcting it through the lenses (eyeglasses or contact lenses) with + or − diopters.[13] Refraction adjusts the focus of the images being seen by allowing the images to land on the retina, which allows for the images to come into focus. Once the client's eyes are refracted, the optometrist will determine if magnification is needed. If magnification is needed, the low-vision optometrist prescribes the magnification for the individual and then refers the client to the occupational therapist, who provides training in the magnification.[37]

11.6.3.2 Occupational Therapy Low-Vision Evaluation

The occupational therapy low-vision evaluation is a comprehensive evaluation that includes evaluation of the visual system as well as of the individual. During the evaluation the occupational therapist gathers information about the individual's medical and visual history. The occupational therapist evaluates visual acuity for near and far, eye dominance, contrast sensitivity function, the central and peripheral visual fields, and neurological function of the eyes.[38] See Table 11.1 for assessments that are part of the Brain Injury Visual Assessment Battery for Adults (biVABA) by Mary Warren, which addresses these areas.[11]

The biVABA assists occupational therapists in communicating with ophthalmologists, neuro-ophthalmologists, and optometrists. Occupational therapists using this battery of assessments should demonstrate competence in the assessments prior to administering and interpreting their results. The assessments are used to screen the areas of visual function and make appropriate referrals to ophthalmologists, neuro-ophthalmologists, and optometrists. The information gathered also assists in effectively addressing functional limitations due to visual impairment.[11] While the assessments were developed for adults with brain injury, they also address visual function as related to the areas where lesions from MS occur.

In addition to the visual assessments, the occupational therapist completes an occupational profile of the individual and analyzes his or her occupational performance. The profession of occupational therapy uses the term occupation to capture the breadth and meaning of "everyday activity."[39] The occupational profile consists of gathering information on the individual, why the individual is seeking services, what occupations and activities are problematic due to vision, the environments and context that support and hinder participation in the occupations and activities, the individual's occupational history (every activity), occupations and activities that are the individual's priorities, and the outcome the individual seeks to achieve.[39] During the analysis of occupational performance, that is, the

TABLE 11.1

Visual Assessments from the biVABA

Assessment	Purpose	Administration	Results	Clinical Application
Eye dominance Flower design card and 8 mm hole card[11]	To determine which eye is the dominant eye	Glasses on Therapist instructs the individual to pick up the card with a hole in it and to look through the hole to see the flower design	The card is held to the dominant eye	Vision is processed through the dominant eye The dominant eye should be tested first during assessments that test eyes separately Accommodations will be made based on the dominant eye
Visual acuity— Intermediate acuity Low-Vision LeaNumbers Chart[11]	To determine intermediate/ distance visual acuity The chart extends to 20/1000	Glasses on Therapist tests dominant eye, then nondominant, then both eyes together	Determines the individual's intermediate/ distance visual acuity.	Assists in making recommendations for the appropriate type of magnification and determining environmental adaptations[a]
Visual acuity—Near Warren Text Card[11]	To determine near visual acuity to 20/400	Glasses on Therapist tests both eyes together	Determines the individual's near visual acuity	Assists in making recommendations for the appropriate type of magnification and determining environmental adaptations[a]
Contrast-sensitivity function LeaNumbers Low Contrast Screener[11]	To determine the level of contrast sensitivity	Glasses on Therapist tests both eyes together at three different distances	Determines the individual's contrast sensitivity level	Assists in determining the individual's functional limitations for seeing facial expressions, water, black-and-white photographs, and reading materials; detecting curbs; and assessing driving conditions
Central-visual-field testing The Damato 30-point Multifixation Campimeter[11]	To measure the central visual field	Glasses on Therapist tests the dominant eye and then the nondominant eye	The test is scored in one direction and then rotated 180° to interpret the results Identifies areas of the central visual field affected	Provides a clear picture of the areas of the central visual field that are affected by the visual impairment Information will assist in developing compensatory strategies
Central-visual-field testing American Academy of Ophthalmology Red Dot Confrontation Test[11,55]	To measure the central visual field	Glasses off Therapist tests the dominant eye and then the nondominant eye	Identifies areas of central visual field affected as well as deficits in contrast sensitivity	Assists in identifying problematic areas in central vision as well as deficits in contrast sensitivity Information will assist in developing compensatory strategies

continued

TABLE 11.1   (continued)

Visual Assessments from the biVABA

Assessment	Purpose	Administration	Results	Clinical Application
Peripheral-visual-field testing The Kinetic Two Person Confrontation Test[11]	To measure the peripheral visual field	Glasses off Therapist tests the dominant eye and then the nondominant eye	Identifies areas of the peripheral visual field affected	Provides a clear picture of the areas of the peripheral visual field that are affected by the visual impairment Information will assist in developing compensatory strategies
Neurological function of the eyes Pupillary function Visual attention Oculomotor function Clinical observation[11]	To measure the neurological function of the eyes	Glasses on Therapist tests the dominant and then the nondominant eye for pupillary and oculomotor function Eyes are tested together for visual attention and clinical observation	Identifies areas of neurological impairment	Neurological function is based on a visual–perceptual hierarchy Deficits identified in lower areas of the hierarchy will result in deficits at higher levels Information will assist in developing compensatory strategies

[a] The ophthalmologist or optometrist determines the power of magnification.

ability to successfully perform everyday activities, the occupational therapist synthesizes the occupational profile, observes the individual performing the desired activity or occupation, notes the effectiveness of the individual and the areas that influence effectiveness, interprets the data to identify what facilitates and prevents performance, develops hypotheses, collaborates with the individual to establish goals, and identifies intervention based on best practice.[39]

During the evaluation, the occupational therapist will also identify if other referrals are necessary. The occupational therapist may refer the individual to the ophthalmologist, optometrist, or certified orientation and mobility specialist. If the individual was referred to the occupational therapist from his or her primary care physician, the occupational therapist will refer the individual to the ophthalmologist, neuro-ophthalmologist, or the optometrist to ensure the individual is working with his or her best-corrected vision. Best-corrected vision refers to the individual's visual acuity when wearing prescribed eyeglasses or contact lenses. The eyeglasses or contact lenses maximize the individual's visual acuity by correcting any refractive errors or astigmatism. The ophthalmologist and optometrist both prescribe eyeglasses and contact lenses to correct refractive errors as well as magnification. The occupational therapist needs to work with the best-corrected vision in order to maximize the individual's ability to participate in everyday life. Magnification may be necessary to maximize the individual's vision. The ophthalmologist or optometrist prescribes the amount of magnification necessary for the individual, and then the occupational therapist trains the individual in the magnification device. Magnification devices include, but are not limited to, handheld magnifiers, stand magnifiers, telescopes, and closed-circuit televisions. The occupational therapist will also refer the individual to a certified orientation and mobility specialist if the individual is having difficulty navigating and moving outside of the home.

11.6.3.3 Orientation and Mobility Specialists and Low-Vision Therapists

The certified orientation and mobility specialist evaluates the individual's spatial orientation, that is, the ability to sense position and its relationship to other objects in the environment.[40] Spatial-orientation skills are necessary to maintain orientation and navigate the environment. This includes crossing streets, traveling to different locations, and traveling safely.

The certified low-vision therapist (CLVT) evaluates the individual's visual acuity, visual fields, contrast-sensitivity function, color vision, stereopsis, visual perceptual and visual motor functioning, work history, performance of activities of daily living related to vision loss, and use of technology.[18] The certified low-vision therapist evaluates the individual's abilities based only on visual impairment.

11.6.3.4 Self-Report: Getting the Client's Perspective

In addition to using clinical histories to obtain the client's perspective on visual impairment, vision and rehabilitation providers can use the National Eye Institute's Visual Function Questionnaire (VFQ-25). This instrument measures self-reported vision-targeted health status for persons with chronic conditions.[41] While the VFQ-25 was not developed for individuals with MS, it has been found to be an effective measure of vision loss in MS[9] with reliability and validity.[8] The VFQ-25 contains the following subscales: global vision rating, difficulty with near-vision activities, difficulty with distant-vision activities, limitations in social functioning due to vision, role limitations due to vision, dependency on others due to vision, mental health symptoms due to vision, driving difficulties, limitations with peripheral and color vision, and ocular pain. It may be administered in the form of an interview or it may be self-administered. The interview format typically takes 10 min to complete. No psychometric testing has been completed on the self-administered version of the questions.[41] The instructions for the questionnaire instruct the individual who is answering the questions to answer the questions based on his or her vision when wearing glasses or contact lenses. The individual is also instructed to take as much time as needed to answer each of the questions.[42] After the individual answers the questions, the interviewer scores the responses. The original numeric values are recoded from 0 to 100, with higher scores representing better functioning. Then items in each of the subscales are averaged and a total composite score is calculated by averaging the subscale scores. This information assists the multidisciplinary team that works with the individual to identify the areas of greatest impairment. The ophthalmologist, optometrist, or occupational therapist who administers the questionnaire will share the findings with the multidisciplinary team. The findings will assist each of the members of the team in selecting the appropriate visual assessments to be completed during the initial evaluations.

11.7 Interventions

In addition to the above examinations to assess for visual problems directly related to MS, all clients with MS and visual complaints should have a careful refraction and receive a glasses prescription for the appropriate lenses to correct vision to the best extent possible for distance and near vision. That said, acute and chronic optic nerve injury in MS are not directly treatable with glasses.

11.7.1 Intervention for Optic Nerve Involvement

The natural course of acute optic neuritis is spontaneous improvement in the majority of cases. Rapid improvement in the first four to six weeks is the norm, but additional gradual improvement may occur over the course of the next year.[24] Corticosteroid treatment may be used to hasten recovery, but it has not been shown to change the overall long-term visual outcome. When prescribed, the standard of care is i.v. corticosteroids for three to five days. This may or may not be followed by a short course of oral corticosteroids. Oral corticosteroids without preceding i.v. corticosteroids are not a treatment option, as this may increase risk of recurrent optic neuritis.[43] There is little evidence that acute treatments such as corticosteroids have a role in chronic optic nerve involvement. As with other chronic MS deficits, the goal would be maintenance of stability via immunomodulatory therapy.

11.7.2 Intervention for Abnormal Eye Movements

If a client is symptomatic of diplopia from INO or other causes, prisms may be placed on the client's glasses to alleviate it. The type of prism used is a series of half cylinders that change the angle of entry of a visual stimulus into the eye in an attempt to align the visual object on the same retinal location in each eye. Rarely, eye muscle surgery to straighten the eyes may be helpful for MS clients troubled by chronic diplopia who have chronic INO.

Medications are the mainstay of therapy to treat the visual blurring and oscillopsia created by APN. Medication options include gabapentin, memantine, valproate, and clonazepam. A double-blind, placebo-controlled, crossover study comparing gabapentin and baclofen in APN revealed gabapentin to be highly effective in minimizing oscillopsia,[44] and subsequent masked, crossover studies of gabapentin and memantine found both medications to be effective.[45,46] Gabapentin is the preferred initial medication, as memantine at doses used to treat nystagmus may worsen underlying MS symptoms.[47]

11.7.3 Setting Goals and Planning Treatment for Vision Rehabilitation

The multidisciplinary team develops goals based on the results of the visual assessments. Goals will vary based on the discipline. The individual with visual impairment from MS requires an individualized treatment approach in order to maximize full participation in life. The results from the multidisciplinary evaluations guide the specific interventions that will be selected for the individual.

11.7.3.1 Compensatory Strategies

Visual impairment from MS results in permanent changes to the individual's visual function due to damage to the optic nerve. As a result, a compensatory approach is used to maximize the individual's participation in life. Through the use of magnification, lighting, and low-vision aids that incorporate concepts of contrast, the individual learns how to maximize his or her current vision and regains full participation in life.

11.7.3.1.1 Magnification

The ophthalmologist or optometrist will determine the amount of magnification required for the individual to complete tasks that are near or far. Options for near vision are as follows:

1. *Eyeglasses* are "the prescription of choice . . . if the magnification required is achieved by plus lenses up to approximately 10 diopters" (p. 304).[13] Plus lenses are specifically used to magnify near objects. When magnification requires more than 10 diopters, handheld magnifiers, stand magnifiers, or closed-circuit televisions (CCTV) are used.[13]

2. *Handheld magnifiers* are often used for tasks that do not require a lot of time and require portability. The individual needs to have good use of the upper extremity in order to use the handheld magnifier correctly. The individual using the magnifier needs to maintain the proper distance of the magnifier to the object being read.[48] Handheld magnifiers are helpful for checking prices of items in the supermarket or store. Since proper distance must be maintained when using the magnifier, it is best to incorporate this type of magnifier into tasks that do not require a lot of time. Handheld magnifiers are available with and without built-in illumination.

3. *Stand magnifiers* are most useful for reading. The stand magnifier is placed on the material being read. The individual using the magnifier moves the magnifier across the materials being read. Advantages to the stand magnifier include reduced fatigue since the magnifier sits on the surface of the material being read and the proper distance is maintained between the lens of the magnifier and the reading material.[48] Stand magnifiers are also available with and without built-in illumination.

4. *Closed-circuit televisions (CCTVs) or electronic magnification* are commonly used when the individual requires higher power magnification. Electronic magnification refers to CCTVs and adaptive computer hardware and software. CCTVs are used for reading and writing tasks that require more than 12 diopters of magnification.[13] The CCTV uses a mounted camera that magnifies images in the defined area onto a monitor. Models vary and should be selected based on the tasks the individual desires to complete. They may be used for reading, writing, and viewing pictures. Adaptive computer hardware and software include modified keyboards, larger monitors, scanners, and specialized software for screen magnification.[13] Different options should be trialed with the individual prior to investing in the CCTV and adaptive computer hardware and software since these forms of magnification are more costly. Obtaining the correct CCTV and adaptive computer hardware and software facilitates greater independence in tasks that require vision.

Options for far vision are as follows:

5. *Telescopes* are used for magnification of objects that are at a distance.[48] Telescopes may be handheld or mounted on eyeglasses. An individual may use a telescope to watch television or even to drive. Not all states allow for driving with telescopes, and state laws should be checked prior to prescribing and training an individual to use telescopes to drive.[48] Since objects at a distance are being magnified, the individual should be instructed in the proper use and circumstances for using the telescope in order to maintain safety.

11.7.3.1.2 Lighting

Research has found that performing tasks in bright light[9,12] and seeing at night[9] are difficult for individuals with visual impairment due to MS. As a result, it is necessary to look at the type of lighting and whether there is too much or too little light for the activities the individual desires to complete. Task lighting, with which light is directed to the activity being completed, provides increased illumination for the activity without the light being directed into the eyes. Gooseneck lamps are commonly used as task lights to direct light to the activity being completed.[49] By directing the light to the task, the individual gains the benefits of increased lighting without the problems of performing tasks in bright light. In addition, if the individual uses a magnifier, one of the options available is to obtain a magnifier with or without a light in it. This option assists those who need additional lighting but also provides the option of not having additional light if it is not needed.

11.7.3.1.3 Low-Vision Aids

Low-vision aids provide increased contrast and large print to commonly used household items. Research has found that decreased contrast sensitivity in individuals with MS affects visual acuity.[12] By increasing contrast, the individual with visual impairment from MS has increased functional vision. Low-vision aids include bold-lined paper and 20/20 pens, which provide increased contrast when writing on the paper since the dark lines orient the individual and the dark ink provides increased contrast. Using black-and-white cutting boards, plates, and cups also increases contrast. Pouring milk into a dark cup or cutting a steak on a white plate increases contrast. In addition to contrast, large print also assists in increasing visual function. Large-print items include clocks, telephones, checkbooks and check registers, magazines and books, and keyboards. If visual impairment prevents the individual's use of visual input, tactile indicators may be used to identify key buttons on appliances, telephones, and other objects. The individual would use his or her tactile abilities to feel for the indicator and then make the appropriate selections.

11.7.3.2 Self-Management

Self-management has been described as a particular approach to providing services, a set of skills, as well as an outcome of service delivery.[50] As a service approach, self-management is a particular type of client education that focuses on building skills and empowering clients to make decisions and take charge of their own care. As a set of skills, self-management encompasses the knowledge and confidence to solve problems, make decisions, set and follow-through on goals, find and use resources, build relationships with providers, and self-monitor one's situation as it changes.[51] As an outcome, a person who is a self-manager is one who has achieved the knowledge, skills, and confidence to manage his or her health condition and is actively engaged in behaviors that support health, well-being, and disease or symptom management.

In the context of vision rehabilitation in MS, individuals need to develop skills necessary to manage the magnification devices, lighting options, and low-vision devices that are prescribed or issued. Learning to manage these devices will require an educational process that enables the client to solve problems and make decisions about the use of these devices across life situations, as well as if and when visual symptoms change. Finding ways to effectively cope with the emotions associated with visual changes and loss may also be necessary. Without these types of skills, it may be difficult for a person with MS to minimize the impact of visual symptoms on participation in activities. While there are models for self-management programs related to vision loss,[52-54] existing programs target older adults with age-related vision changes. To date, there have not been self-management programs developed or tested specifically for people with MS. This is an approach to intervention that will need exploration, development, and testing in the future.

11.7.4 How Assessment and Intervention May Look Different across Rehabilitation Settings

Assessments and interventions may vary from setting to setting depending on which disciplines are represented on the multidisciplinary team. It is essential to have an ophthalmologist or optometrist as a member of the team. If the individual is experiencing problems completing activities due to vision problems, then an occupational therapist or other health care professional with specialization in low vision should also be included or a referral should be made. If the individual has problems with mobility due to his or her vision, then an orientation and mobility specialist should also be a member of the team or a referral should be made. A team approach is necessary in order to maximize full participation in life.

11.7.5 Case Study

To illustrate the process of setting goals and planning treatment in low-vision rehabilitation, the case of Mary Ann will be presented. Mary Ann was diagnosed with MS about five years ago. Her primarily symptoms are fatigue, numbness and tingling in her extremities, and problems concentrating. She has not had difficulty with mobility to date and does not use any form of walking aid. When she was initially diagnosed, Mary Ann reported problems with her vision. Over the last year, she experienced an acute exacerbation that included optic neuritis that failed to resolve. She reported a decline in vision in both eyes over a one-week period that eventually stabilized. Mary Ann's visual acuity is now 20/200.

Mary Ann recently made an appointment and went to see her ophthalmologist because she was encountering more problems with her vision. She reported to the doctor that she was having problems completing activities that she used to do and that her vision was making it harder and harder for her to complete the activities. The ophthalmologist identified that Mary Ann had changes in her visual acuity, central field of vision, peripheral field of vision, and contrast sensitivity. The ophthalmologist provided Mary Ann with a new prescription for eyeglasses, but explained to her that the glasses would not compensate for the areas of her central and peripheral field that were affected nor would they help with her decreased contrast sensitivity. The ophthalmologist recommended that Mary Ann be seen by other members of the rehabilitation team to address the activities that were becoming harder and harder for her to complete.

Mary Ann made appointments with an occupational therapist specializing in low vision and with an orientation and mobility specialist. During the occupational therapy low-vision evaluation, the occupational therapist identified that Mary Ann's visual impairments were negatively influencing her ability to read, write, cut food, and use her computer for work. Mary Ann reported that she needed to complete these activities. During the orientation and mobility session, the orientation and mobility specialist identified that Mary Ann's visual deficits were affecting her ability to travel safely. Mary Ann reported that she wanted to be able to travel independently to and from work as well as go out to run errands. At the end of the sessions with the occupational therapist and orientation and mobility specialist, Mary Ann set rehabilitation goals with the guidance of the multidisciplinary team. She had three short-term goals:

Goal 1: Mary Ann will write independently using a 20/20 pen in two weeks.

Goal 2: Mary Ann will read writing with a 20/20 pen independently in two weeks.

Goal 3: Mary Ann will cut food independently and safely using a contrasting surface in two weeks.

She set the following long-term goals:

Goal 1: Mary Ann will use magnification software to complete computer activities independently in one month.

Goal 2: Mary Ann will use a closed-circuit television to read mail independently in one month.

Goal 3: Mary Ann will use a sight cane to travel to and from work safely and independently in one month.

Goal 4: Mary Ann will use a sight cane to travel safely and independently when running errands in one month.

Based on the findings of Mary Ann's evaluation and her goals, it was determined that Mary Ann would benefit from magnification, low-vision devices, and training in orientation and mobility. Mary Ann's visual acuity made it difficult for her to see her computer. Her treatment plan included trialing different screen magnification software to determine which one provided her with the options that allowed her to complete her work activities on the computer. In addition, intervention involved training Mary Ann to use a CCTV for reading and writing. One of the desirable options on her CCTV was the white on black option, which converts black print to white and the white background of the paper to black. This maximized contrast and allowed Mary Ann to read the print. Mary Ann was also taught principles of contrast and learned to prepare and cut her food on contrasting cutting boards and plates. Finally, Mary Ann was seen by an orientation and mobility specialist who taught her how to use a sight cane for safe travel in and around the work environment as well as how to navigate the public transportation system.

Treatment plans need to be created based on the individual's current visual function and current needs. If vision worsens, the individual will need to be re-evaluated to identify the current visual status and appropriate compensatory measures to meet the individual's needs. Proper evaluation allows the ophthalmologist or optometrist to prescribe the appropriate magnification, which will facilitate the best possible vision and visual function.

11.8 Gaps in Knowledge and Directions for the Future

Currently, there is limited rehabilitation literature specific to assessment and intervention strategies for people with MS who are experiencing visual symptoms. Although research on the VFQ-25 is useful for rehabilitation providers, development of other functionally oriented self-report tools may be warranted. Development and validation of functionally oriented performance-based assessments are also needed. For example, evaluation of the biVABA in a MS population would provide valuable information. Research also needs to be completed on the effectiveness of magnification, lighting, contrast, and adaptive equipment commonly used to compensate for visual impairments that people with MS experience. Examination of how these interventions may need to be modified to accommodate other MS symptoms (e.g., fatigue, cognition, weakness) would inform clinical reasoning of rehabilitation professionals working in this area. Since visual impairments vary across individuals with MS, studies using single-subject, multiple baseline designs may be one way to begin evaluating the rehabilitation assessments and intervention approaches for clients affected by visual symptoms and related functional restrictions.

References

1. McDonald WI, Barnes D. The ocular manifestations of multiple sclerosis: 1: Abnormalities of the afferent visual system. Journal of Neurology, Neurosurgery and Psychiatry 1992;55:747–752.

2. Rodriguez M, Siva A, Cross SA. Optic neuritis: A population-based study in Olmsted County, Minnesota. Neurology 1995;45:244–250.

3. Patten J. Conjugate eye movements and nystagmus. In: Neurological differential diagnosis. London: Springer-Verlag; 1996.

4. Muri RM, Meinenberg O. The clinical spectrum of internuclear ophthalmoplegia in multiple sclerosis. Archives of Neurology 1985;42:851–855.

5. Rucker JC. Efferent visual dysfunction from multiple sclerosis. International Ophthalmology Clinics 2007;47:1–13.

6. Chen CM, Lin SH. Wall-eyed bilateral internuclear ophthalmoplegia from lesions at different levels in the brainstem. Journal of Neuro-Ophthalmology 2007;27:9–15.

7. World Health Assembly. Action plan for the prevention of avoidable blindness and visual impairment. Geneva, Switzerland: WHO; 2009.

8. Balcer LJ, Baier ML, Kunkle AM, Rudick RA, Weinstock-Guttman B, Simonian N. Self-reported visual dysfunction in multiple sclerosis: Results from 25-item National Eye Institute visual function questionnaire (VFQ-25). Multiple Sclerosis 2000;6:382–385.

9. Ma SL, Shea J, Galetta S, Jacobs D, Markowitcz C, Maguire M. Self-reported visual dysfunction in multiple sclerosis: New data from the VFQ-25 and development of an MS-specific vision questionnaire. American Journal of Ophthalmology 2002;133(3):686–692.

10. Noble J, Rorooghian F, Sproule M, Westall C, O'Connor P. Utility of the National Eye Institute VFQ-25 questionnaire in a heterogeneous group of multiple sclerosis patients. American Journal of Ophthalmology 2006;142(3):464–468.

11. Warren M. Brain injury assessment battery for adults. Birmingham, AL: visAbilities Rehab Services, Inc.; 2005.

12. Mowry EM, Loguidice MJ, Daniels AB, Jacobs EA, Markowitcz CE, Galetta SL. Vision-related quality of life in multiple sclerosis: Correlation with new measures of low and high contrast letter acuity. Journal of Neurology, Neurosurgery and Psychiatry 2009;80(7):767–772.

13. Markowitcz C. Principles of modern low vision rehabilitation. Canadian Journal of Ophthalmology 2006;41(3):289–312.

14. Warren M. An overview of low vision rehabilitation and the role of occupational therapy. In: M. Warren, E. A. Barstow, editors. Occupational therapy interventions for adults with low vision. Bethesda, MD: AOTA Press; 2011.

15. What are ophthalmologists? [Internet]; c2011 [cited 2011 July/11]. Available from: http://www.icoph.org/about/what_are_ophthalmologists.html.

16. About us. [Internet]; c2011 [cited 2011 July/11]. Available from: http://www.worldoptometry. org/en/about-wco/index.cfm.
17. What is occupational therapy? [Internet]; c2004 [cited 2011 July/11]. Available from: http:// www.wfot.org/information.asp.
18. ACVREP. Low vision therapist certification handbook. ACVREP; 2007.
19. Vision professional careers: Orientation and mobility specialists [Internet]; c2011 [cited 2011 June/24]. Available from: http://www.aerbvi.org/modules.php?name = News&file = article& sid = 1220.
20. Simao LM. Ophthalmologic manifestations commonly misdiagnosed as demyelinating events in multiple sclerosis patients. Current Opinion in Ophthalmology 2010;21:436–441.
21. Niestroy A, Rucker JC, Leigh RJ. Neuro-ophthalmologic aspects of multiple sclerosis: Using eye movements as a clinical and experimental tool Clinical Ophthalmology 2007;1:267–272.
22. Prasad S, Galetta SL. Eye movement abnormalities in multiple sclerosis. Neurologic Clinics 2010;28:641–655.
23. Rucker JC, Kennard C, Leigh RJ. The neuro-ophthalmological examination. Hand 2011;102:71–94.
24. Beck RW. Optic neuritis treatment trial. One-year follow-up results. Archives of Ophthalmology 1993;111:773–775.
25. Baier ML, Cutter GR, Rudick RA. Low-contrast letter acuity testing captures visual dysfunction in patients with multiple sclerosis. Neurology 2005;64:992–995.
26. Balcer LJ, Baier ML, Bohen JA. Contrast letter acuity as a visual component for the multiple sclerosis functional composite. Neurology 2003;61:1367–1373.
27. Beck RW, Cleary PA, Backlund JC. The course of visual recovery after optic neuritis: Experience of the optic neuritis treatment trial. Ophthalmology 1994;101:1771–1778.
28. Fisher JB, Jacobs DA, Markowitz CE. Relation of visual function to retinal nerve fiber layer thickness in multiple sclerosis. Ophthalmology 2006;113:324–332.
29. Gordon-Lipkin E, Chodkowski B, Reich DS. Retinal nerve fiber layer is associated with brain atrophy in multiple sclerosis. Neurology 2007;69:1603–1609.
30. Pulicken M, Gordon-Lipkin E, Balcer LJ. Optical coherence tomography and disease subtype in multiple sclerosis. Neurology 2007;69:2085–2092.
31. Sepulcre J, Murie-Fernandez M, Salinas-Alaman A. Diagnostic accuracy of retinal abnormalities in predicting disease activity in MS. Neurology 2007;68:1488–1494.
32. Siger M, Dziegielewski K, Jasek L. Optical coherence tomography in multiple sclerosis: Thickness of the retinal nerve fiber layer as a potential measure of axonal loss and brain atrophy. Journal of Neurology 2008;255:1555–1560.
33. Henderson AP, Trip SA, Schlottmann PG. An investigation of the retinal nerve fiber layer in progressive multiple sclerosis using optical coherence tomography. Brain 2008;131:277–287.
34. Talman LS, Bisker ER, Sackel DJ. Longitudinal study of vision and retinal nerve fiber layer thickness in multiple sclerosis. Annals of Neurology 2010;67:749–760.
35. Downey DL, Stahl JS, Bhidayasiri R. Saccadic and vestibular abnormalities in multiple sclerosis: Sensitive clinical signs of brainstem and cerebellar involvement. Annals of the New York Academy of Sciences 2002;956:438–440.
36. Derwenskus J, Rucker JC, Serra A. Abnormal eye movements predict disability in MS: Two-year follow-up. Annals of the New York Academy of Sciences 2005;1039:521–523.
37. Scheiman M, Scheiman M, Whitaker SG. Low vision rehabilitation: A practical guide for occupational therapists. Thorofare, NJ: Slack Inc.; 2007.
38. Meyers JR, Wilcox DT. Low vision evaluation. In: M. Warren, E. A. Barstow, editors. Occupational therapy interventions for adults with low vision. Bethesda, MD: AOTA Press; 2011.
39. American Occupational Therapy Association. Occupational therapy practice framework: Domain and process. American Journal of Occupational Therapy 2008;62:625–683.
40. Long RG, Hill EW. Establishing and maintaining orientation for mobility. In: B. B. Blasch, W. R. Wiener, R. L. Welsh, editors. Foundations of orientation and mobility. 2nd ed. New York, NY: AFB Press; 1997.

41. Mangione CM. Version 2000 the National Eye Institute 25-item visual function questionnaire (VFQ-25). National Eye Institute 2000.

42. National Eye Institute. National Eye Institute visual functioning questionnaire 25 (VFQ-25) version 2000. RAND; 2000.

43. Beck RW, Cleary PA, Anderson MM. A randomized, controlled trial of corticosteroids in the treatment of acute optic neuritis: The optic neuritis study group. New England Journal of Medicine 1992;326:581–588.

44. Averbuch-Heller L, Tusa RJ, Fuhry L. A double-blind controlled study of gabapentin and baclofen as treatment for acquired nystagmus. Annals of Neurology 1997;41:818–825.

45. Starck M, Albrecht H, Pollmann W. Acquired pendular nystagmus in multiple sclerosis: An examiner-blind cross-over treatment study of memantine and gabapentin. Journal of Neurology 2010;257:322–327.

46. Thurtell MJ, Joshi AC, Leone AC. Cross-over trail of gabapentin and memantine as treatment for acquired nystagmus. Annals of Neurology 2010;67:676–680.

47. Villoslada P, Arrondo G, Sepulcre J. Mermantine induces reversible neurologic impairment in patients with MS. Neurology 2009;72:1630–1603.

48. Nowakowski RW. Basic optics and optical devices In: M. Warren, E. A. Barstow, editors. Occupational therapy interventions for adults with low vision. Bethesda, MD: AOTA; 2011.

49. Steelman F, editor. Assistive technology interventions to improve occupational performance. Bethesda: AOTA Press; 2011.

50. McGowan P. A background paper for the new perspectives: International conference on patient self-management. Victoria, BC: Centre on Aging–University of Victoria; 2005.

51. Lorig KR, Holman H. Self-management education: History, definition, outcomes, and mechanisms. Annals of Behavioral Medicine 2003;26:1–7.

52. Rees G, Keeffe JE, Hassell J, Larizza M, Lamoureux E. A self-management program for low vision: Program overview and pilot evaluation. Disability & Rehabilitation 2010;32:808–815.

53. Rees G, Saw CL, Lamoureux EL, Keeffe JE. Self-management programs for adults with low vision: Needs and challenges. Patient Education and Counseling 2007;69:39–46.

54. Packer TL, Girdler S, Boldy DP, Dhaliwal SS, Crowley M. Vision self-management for older adults: A pilot study. Disability & Rehabilitation 2009;31:1353–1361.

55. Pandit RJ, McOptom KG, Griffiths PG. Effectiveness of testing visual fields by confrontation. The Lancet 2001;358(9290):1339–1340.

Section III

Multiple Sclerosis Rehabilitation for Activity Limitations and Participation Restrictions

12

Communication

Kathryn M. Yorkston and Carolyn R. Baylor

CONTENTS

The ability to communicate is a prerequisite for participation in many everyday activities. In fact, it is hard to imagine life without communication. It is necessary in order to establish and maintain interpersonal relationships, to raise children, to manage household business, to get appropriate health care services, and to take part in work or educational activities. When multiple sclerosis (MS) interferes with communication, it also restricts participation in life roles. As a result, many negative consequences arise, for example, loss of employment, social isolation, and difficulty pursuing services, including access to health care. After reading this chapter, you will be able to:

1. Recognize the communication problems that people with MS typically experience and how a speech–language pathologist (SLP) assesses them,

2. Explain how communication problems interact with other symptoms of MS and interfere with participation in everyday activities,

3. Describe intervention for communication disorders in MS, including reasons for referral and approaches to treatment,

4. List suggestions for facilitating communication in MS, and

5. Identify areas in need of future research.

12.1 Type and Prevalence of Communication Disorders

Communication disorders, especially speech disorders, have long been associated with MS. In fact, scanning speech, described as a drawling with words produced as if measured or scanned, was described as one of the hallmark features of MS in Charcot's original description.[1] Speech disorders are no longer considered a universal feature of MS.[2] Estimates of the prevalence of dysarthria (a disorder of motor-speech production due to neurologic impairment) in MS vary depending on how the data are obtained. Prevalence estimates range from 23% when the data were obtained by self-report in a community-based survey to 51% in a cohort study in which people with MS were followed for decades and data were obtained during professional examination.[3] The most common speech disorder associated with MS is a mixed dysarthria with components of both ataxia and spasticity.[4] The most common characteristics of this dysarthria are impaired loudness control, vocal harshness, and problems with articulation.[5]

Recently, there has been a growing appreciation of the effect that cognitive limitations may have on communication as well.[6,7] These cognitive-communication problems may be associated with slowness of information processing, word-finding problems, or fatigue. A more complete discussion of the cognitive problems seen in MS can be found in Chapter 8. Although the prevalence of language impairment in MS has not been examined as extensively as speech or cognitive disorders, in a study of 60 participants with MS, approximately half (53%) had essentially normal language ability, while one third (32%) had mild to moderate language impairment, and only 15% exhibited moderate to severe impairment.[6]

12.2 Association of Communication Disorders with Other Problems

In MS, communication disorders are typically seen in the presence of other symptoms. In the words of one man with MS, speech changes are "part of the soup." Just like carrots in a vegetable soup, he knew when they entered the mix, but after simmering for a while, it is hard to distinguish them from the rest of the soup. Thus, it was difficult for him to distinguish speech changes from the effects of cognitive changes, fatigue, mobility limitations, stressful situations, and so on because all of these variables affected his interactions with other people. In a community-based survey of over 700 people with MS, 40% reported some level of speech changes.[2] Those reporting moderate to severe changes were more likely to experience physical problems (arm strength/coordination or mobility changes). Vision and hearing problems, fatigue, and depression were also more likely in people with moderate-to-severe speech problems. More than half of those with moderate or severe speech problems also reported frequent problems with thinking, memory, reading, and writing.

Although most common communication problems are mild in MS, when combined with other problems, such as changes in mobility, vision, cognition, and fatigue, they restrict participation in important ways.[8] In a series of semistructured qualitative interviews, participants were asked, "What is communication like for you?"[9] Concerns about communication focused mainly on limited participation. Participants spoke of social interactions they could no longer engage in and roles they could not longer play because of the consequences of MS. They were not able to take part in valued life roles, including employment and leisure activities, because of fatigue, cognitive changes, speech or language changes, and visual changes. Fatigue was reported to be a particularly pervasive part of life, decreasing the opportunities for communication in work and family settings. They also spoke of the unpredictability of communication. Their communication ability and capacity to participate varied from day to day. Participants with MS frequently report that their social circle has shrunk as a consequence of MS.[10] This "depletion of social opportunity"[10] arises partly from MS-related issues, such as fatigue, and partly from the barriers created by the attitudes of those around them.

Themes that emerge from analysis of qualitative interviews are confirmed with quantitative methods. The association between self-reported communicative participation and a variety of other symptoms were examined in a sample of 498 community-dwelling adults with MS.[11] Regression analysis indicated that fatigue, slurred speech, depression, problems with thinking, employment, and social support all were associated with communicative participation, in that order. Thus, communicative participation is significantly associated with multiple variables, only some of which reflect communication disorders.

12.3 Assessment Strategies

Referrals for a speech evaluation can be prompted by a number of issues. Some questions that might spur referrals to an SLP are listed in Table 12.1. Communication disorders associated with MS can take many forms; therefore, assessment of these disorders typically includes evaluation of a variety of components. The *International Classification of Functioning, Disability and Health* (ICF)[12] provides a useful framework for organizing the assessment. The framework incorporates elements related to the physiologic consequences of the health condition and also provides a means of addressing psychosocial consequences of MS. The following section describes assessments at the levels of the impairment, activity limitation, and participation restriction.

12.3.1 Impairment of Structural or Functional Integrity

With dysarthria in MS, the anatomy of the speech production mechanism is normal, but the physiology of speech production may not be. To put this in ICF terminology, the speech structures are intact, but their function is not. Assessment of the impairment involves a

TABLE 12.1

Considerations for Referral to a Speech–Language Pathologist

- Are there complaints of recent changes in speech, language, or swallowing?
- Is speech difficult to understand?
- Do communication changes result in difficulty performing important roles at home or work?
- Does fatigue substantially change communication ability?
- Are important transitions (such as a new job) being considered?

physical examination of the components of the speech production mechanism. This involves assessment of respiratory, laryngeal, velopharyngeal, and oral articulatory function.[13] The goal of the *respiratory system* during speech production is to achieve a steady supply of air that is the energy source for speech. When evaluating the respiratory aspects of speech, SLPs might ask questions about the overall loudness of speech, the speaker's ability to alter the loudness of speech, and fatigue associated with extended periods of speech. The goal of the *laryngeal system* during speech is to serve as the sound source for speech by providing voicing. Laryngeal, or vocal fold, function is highly coordinated with the respiratory system. Clinicians might ask questions about the quality of the voice (is it harsh or breathy?), the coordination of vocal-fold movement with breathing and other aspects of articulation, and the ability to produce changes in pitch and loudness that signal stress patterning when speaking. The goal of the *velopharyngeal system* is to close off the nose from the mouth in order to produce clear speech sounds. When this system is impaired, speakers are unable to produce clear consonants and speech sounds nasal. Clinicians might ask questions about the ability to produce a group of sounds called "pressure consonants," such as /b/ or /p/, which require build-up of air pressure in the mouth. Speakers with poor velopharyngeal function will also complain of fatigue with extended periods of speech because they lose air through the nose as they speak. Finally, the goal of the *oral articulatory system* is to shape speech sounds by changing the flow of air in the mouth. When this system is impaired, vowels are distorted and consonants are difficult to understand. Clinicians might ask questions about the movement of the tongue and lips. The physical examination includes assessment of strength, range of motion, rate of movements, and muscle tone in these structures.

Because many language impairments associated with MS are mild, typical aphasia (an acquired disorder of receptive and expressive language processing) tests may not reveal impairment. However, clinicians may use tests of higher-order language processes to assess either language or cognitive-communication function. Examples of assessments include tests of word fluency (e.g., the Word Test[14] or the Controlled Oral Word Association Test [COWAT]),[15] as well as auditory comprehension and reasoning from the Test of Language Competence.[16] These tests are not designed specifically for MS. Rather, they are commonly used in educational or rehabilitation settings caring for those with high-level aphasia or cognitive-communication problems associated with traumatic brain injury. More details about tools to assess higher-order language process can be found elsewhere.[17]

12.3.2 Activity Limitations

In the ICF framework, activity is defined as the execution of a task or action. When this term is applied to dysarthria in MS, the activity is speaking. The adequacy of speech is typically measured with tests of speech intelligibility, speaking rate, and naturalness.[13] When testing intelligibility, speakers are recorded as they read a series of sentences of various lengths. These sentences are then played to listeners who transcribe what they hear. Intelligibility scores represent the proportion of words that listeners correctly transcribe. Measures of intelligibility are often used as an overall indicator of the severity of dysarthria. Dysarthria associated with MS may range from mild (dysarthria is present but intelligibility is intact) to severe (intelligibility is compromised) or profound (natural speech is not a functional means of communication). A more complete discussion on the levels of severity of dysarthria can be found elsewhere.[18]

The ability to engage in conversation may be used as the overall index of language and cognitive-communication skills. Scales or checklists that rate various behaviors found in conversation are available, for example, the Conversational Rating Scale[19] or the Checklist of Listening Behaviors.[20] Using these instruments, clinicians can rate processes such as sentence formulation, coherence of narrative, and initiation and maintenance of topics.

12.3.3 Participation Restrictions

Participation refers to a person's involvement in life situations. Communicative participation includes engagement in life situations in which the sharing of knowledge, information, ideas, or feelings is key.[21] The level of social engagement is dependent on the individual's preferences, opportunities, and capabilities. Assessment is typically carried out as an interview with topics including the identification of frequent communication environments and their characteristics, and frequent communication partners at home, in the community, and at work. Communication situations fall into several categories, from casual situations in which communication is typically one-on-one, to physically demanding situations in which loud speech is required for extended periods, to complex situations in which leadership is required and speech demands are high in terms of speed, accuracy, and complexity. Patterns of speech use vary considerably across speakers depending on age, education level, and work status. Simple scales are available to categorize speech usage patterns as undemanding (e.g., quiet for long periods every day), intermittent (e.g., quiet for long periods many days), routine (e.g., most talking is typical conversational speech), extensive (e.g., regularly talk for long periods), and extraordinary (e.g., high speech demands, participation depends on the quality of speech).[22] Knowing preferred speech usage patterns allows the clinician to tailor interventions to the goals of the client. This is particularly important when client needs and preferred roles extend beyond those of basic communication.

The definition of satisfactory communicative participation was explored in people with MS.[23] The results showed that the definition of satisfactory participation differed depending not only on how well the communication task was accomplished, but also on how comfortable the communication situation was and the personal importance or meaning of the communication situation. An item bank is currently under development for measurement of self-reported interference in typical communication situations.[24,25] Participants are asked to rate the level of interference they experience in a series of situations ranging from relatively easy situations (little interference), such as "letting your family know what you need at the store," to relatively difficult (considerable interference) situations, such as "making a telephone call to get information." Because items are placed on a continuum of difficulty, responses give an indication of overall level of interference that is experienced by a respondent. The following section reviews approaches to intervention at the level of activity limitation and restriction in participation.

12.4 Treatment Planning

Speech treatment should be delivered within the context of a comprehensive rehabilitation team because speech changes do not occur as an isolated symptom; rather they are more likely to occur with a complex constellation of other physical, cognitive, and psychosocial changes. Issues faced by people with MS cannot be resolved by a single discipline. Therefore, multidisciplinary teams may include physicians, physical and occupational therapists, vocational counselors, psychologists, and SLPs. Service delivery typically addresses different issues in different health care settings. For example, inpatient care typically focuses on the identification and management of acute exacerbations in which speech, language, and swallowing symptoms worsen temporarily. For an SLP, this may involve evaluation of the safety of swallowing and establishment of basic communication. In the outpatient setting, the SLP typically becomes involved during periods of transition during which a client is considering changes in employment or living situations. Treatment typically focuses on strategies to enhance functioning in particular everyday situations. Home care often involves management of people with more severe disabilities and may focus on maintenance of communication for self-care and social contact.

Unfortunately, a substantial proportion of the MS population does not receive the appropriate intervention for communication problems. Previous estimates suggested that only 2–3% of people with MS had been seen by an SLP.[3] Because of the chronic nature of the underlying pathophysiology in MS, restoring the function of the components of speech production is rarely the exclusive focus of treatment. Rather, management of communication disorders in MS typically focuses on the levels of activity and participation although information from the assessment at the impairment level informs treatment planning.

12.4.1 Dealing with Activity Limitations

The speech disorder or dysarthria associated with MS varies in terms of severity from mild to profound. Thus, management approaches also vary depending on the level of limitations. The following section summarizes treatment approaches for mild, moderate, and severe dysarthria,[18,23] focusing on managing dysarthria in MS. Information about the management of cognitive issues can be found in Chapter 8.

12.4.1.1 Mild Dysarthria

Speakers with mild dysarthria typically experience changes in their voices, including vocal tremor and a harsh voice. These symptoms worsen in the presence of fatigue. Although the dysarthria is noticeable, it does not interfere with speech intelligibility. Treatment approaches may include teaching how to conserve energy or techniques for regulating loudness. Techniques to increase vocal efficiency have also been reported.[26] Exercises may include the following: attention to posture, general relaxation, relaxed abdominal breathing, and controlled exhalation during speech. Individuals are also instructed on how to coordinate respiration with voicing and articulation most efficiently. Often, this requires close attention to how much an individual tries to say in one breath with the recommendation to pause more frequently at logical boundaries between phrases and sentences to take a breath. If the problems are related primarily to vocal loudness and endurance, other options may include having the individual with MS use a small portable voice amplifier. Amplifiers can be small enough to be worn on a belt at the waist, and can provide extra volume that lessens the need for the individual to project his or her voice. Amplifiers can be particularly helpful for people who frequently need to speak in situations such as in groups or in noisy areas.

12.4.1.2 Moderate Dysarthria

In addition to having the voice changes seen in mild dysarthria, those with moderate dysarthria experience a reduced speaking rate and compromised naturalness of speech.

Training in rate control techniques may be appropriate, especially for those with a sudden increase in ataxia caused by a recent exacerbation. Training focused on appropriate phrasing and breathing patterns during stages of mild dysarthria may also serve to pace speech. Speakers can also be taught to "chunk" utterances into meaningful units, thus improving the listener's ability to understand distorted speech.

12.4.1.3 Severe Dysarthria

In the late stages of MS, people may experience severe involvement of multiple neural systems. If natural speech can no longer be a functional means of communication, a variety of augmentative and alternative means of communication (AAC) are available.[27] Use of such systems may be a challenge because of decreased ability to use hands for writing and keyboard access, increasing visual problems, and changes in cognitive function. Multidisciplinary teams are often required to help an individual establish successful AAC communication. These teams often include specialists in seating/mobility, computer access, environmental control, and communication. A more comprehensive description of the composition and functioning of AAC teams can be found elsewhere.[28]

12.4.2 Reducing Restrictions in Participation

While one target of intervention may be to achieve the best function possible on a number of speech-related activities, it is also important to ensure that those activities can be carried out within the context of everyday life. People with MS suggest that participation is satisfactory when activities can be carried out easily and confidently without excess stress or effort.[23] The personal meaning of the activity appears also to be critically related to satisfaction. Targeting communication intervention to those activities that have personal meaning or importance to the client is critical. It may be more satisfying for a person with MS to focus on difficult, yet important, tasks as opposed to easy tasks that are trivial to them.

Communication serves many roles for people with MS. The following section focuses on two important roles of communication: providing access to health care and maintaining social connections.

12.4.2.1 Accommodating for Communication Problems in the Health Care System

The ability to communicate successfully is a critical factor in obtaining health care. Consider how many communication activities are required for a typical outpatient medical encounter. A telephone call is needed to make the appointment; the clinic visit involves a series of questions that must be answered and questions that must be asked; supplemental reading material is often provided; and subsequent visits or follow-up with other providers need to be planned. Research focused on the Medicare population suggests a significant relationship between the presence of communication problems and dissatisfaction with health care, including overall quality, accessibility, and receipt of information.[29] Communication problems also interfere in inpatient health services. The presence of a communication problem was significantly associated with an increased risk of experiencing a preventable adverse event.[30] In fact, hospital patients with communication problems were three times more likely to experience adverse events than those without such problems.

Health literacy is a term referring to the ability to understand the basic health information and services necessary for making appropriate decisions about health services.[31,32] It

TABLE 12.2

Suggestions for Health Care Providers Who Are Communicating with Those Who Have Communication Problems

- Know the pattern of communication problems because it will inform you of the patient's strengths and weaknesses. For example, dysarthria, an isolated motor execution problem, does not interfere with understanding language. So, health care providers do not need to adjust their language level when communicating.
- If the patient has communication aids (e.g., eye glasses, hearing aids, communication devices, memory aids), make sure they are available and used.
- Make sure the environment works in the patient's favor, that is, quiet, well lit, furniture arranged for face-to-face interactions.
- Make sure to allow extra time for communication.
- Limit the amount of information given in one session.
- Use "everyday" language, not medical jargon.
- Speak slowly, using short, simple sentences.
- Accompany verbal descriptions with diagrams, pictures, and writing.
- Confirm the patient's understanding with "teach back" approaches in which the patient tells you what has been said.
- Give take-home educational material in the preferred format and at the appropriate reading level. Simple pictures or drawings can help as well.

is a critical factor in health care, especially when communication disorders are present. Accommodations such as wheelchair ramps to allow physical access to health care facilities are now taken for granted. Unfortunately, similar accommodations to improve access to health care settings for those with health literacy issues are often not acknowledged as part of standard clinical practice.[32] Three types of accommodations in health care settings are needed to improve access to health care for people with communication disorders.[33] These include "brick and mortar" (e.g., a quiet room with furniture that allows eye-to-eye contact), tools (e.g., reading material at appropriate literacy levels), and policy changes (e.g., longer appointments). Training for health care providers is critical. Table 12.2 contains a list of suggestions for interacting with speakers with communication disorders. While many of these strategies can be adapted for anyone to use in interacting with someone with communication disorders, this list is tailored for health care providers.

12.4.2.2 Strategies for Maintaining Social Connections in the Face of Communication Disorders

Communication is unique because it serves a critical social need. It is not just about transactions such as exchanging information and transferring messages; it also serves an important role in establishing and maintaining social affiliation. Communication allows people to exert influence and to help others by listening, reflecting, and offering advice. If communication is compromised, social life will be affected. Communication is also critical in maintaining many preexisting roles, including parent, spouse, friend, employee, or homemaker. Table 12.3 contains a list of general tips for individuals with MS to consider when interacting with family and friends. These tips go beyond basic communication strategies; they address how strategies might be applied in the larger context of relationships with other people.

Another important challenge for those with MS is the premature transition out of valued roles that were expected to continue into older age. For example, they may need to retire long before their nondisabled peers. In an examination of how couples with MS work

TABLE 12.3

Suggestions for Interacting with Family and Friends

- Surround yourself with supportive friends and family. These should be people you are comfortable with and with whom you can be yourself.
- Allow time for adjustment. It may take time for people in your life to adjust to your diagnosis. It may also take time for them to adjust to new roles and responsibilities.
- Focus on maintaining your original relationships. Often it is easy to forget that your caregivers were originally your spouse, son, daughter, or best friend.
- Make an effort to participate in events or activities important to family and friends.
- Be clear about roles you want family and friends to play in your life, including their original roles and any new roles you may need them to assume, such as caregiver.
- See the lighter side of things. Having a positive attitude and sense of humor can help your family and friends be more comfortable during difficult situations.
- Be open to using different strategies. You may need to use different strategies to interact effectively with different people.
- Identify people in your life whom you can talk to about your MS. Identify at least one key person from work, home, and social settings.
- Ask your family and friends for help when you need it. But be specific about what you need.
- Do not let family and friends take over. They will often want to complete tasks for you. Do what you can first, and then ask for help.

collaboratively to manage these and other challenges, results have indicated that these transitions are rapid or undesirable, and psychosocial stress may occur.[34] Couples counseling may be appropriate during these periods of rapid change. See Chapter 22 for more information on social support and caregiving.

In treating communication disorders, partners play a critical role. By definition, communication is a social interaction involving an exchange between at least two people. Thus, the focus of intervention to address communication disorders, particularly at the level of participation restrictions, must be placed on both the person with the disorder and communication partners. Successful communication exchanges are especially difficult for speakers with severe dysarthria and their partners. Table 12.4 lists some strategies for partners of speakers who are difficult to understand. Note that many of the suggestions involve acknowledgment of the difficulty and working together to "costructure" the message. A more detailed description of intervention for dysarthria in MS can be found elsewhere.[18]

12.5 Gaps in Our Knowledge

Although much is known about the management of communication disorders in MS, there are important gaps in our current knowledge. The existence of these gaps dictates the direction for future research. This chapter concludes with a discussion of five areas in which work is needed to address gaps in our current knowledge. First, there is a growing appreciation that intervention should focus on reducing restrictions in participation. Unfortunately, there are few well-designed intervention programs that target participation within the environmental and social context. Such programs need to be developed and tested. Second, clinical measurement tools to track outcomes of participation-focused interventions are largely lacking. Although we have a long tradition of measuring the communication impairments associated with MS, the development of psychometrically

TABLE 12.4

Suggestions for Communicating with Speakers Who Are Difficult to Understand

Create a favorable environment

- Make sure the setting is as free of distractions and as quiet as possible (e.g., turn off the TV or radio to get rid of background noise).
- Watch the speaker.
- Pay attention and do not try to accomplish other tasks at the same time as you are communicating.

Set the stage for communication

- Ask for the topic of the message so you can use context cues to help you understand.
- Be honest by admitting that you are having difficulty understanding.
- Indicate a genuine interest in understanding the message.
- Ask permission to interrupt by asking questions or by guessing what the client is trying to say.

Resolve communication breakdown

- Signal when you have not understood. You can do this verbally or more subtly using a gesture.
- Repeat the part of the message that you have understood so that the speaker does not have to say the whole message again.
- Ask the speaker to repeat the message using exactly the same words.
- If this fails, give some prompts, such as "Say it slower next time" or "Say just a couple of words at a time."
- If this fails, ask the speaker to use an alternative system for the part of the message that you cannot understand. Examples include spelling the words out loud, pointing to an alphabet board, drawing, or using gestures.
- If this fails, ask yes/no questions to try to complete the message.
- If this fails, acknowledge the failure and say that you will try again later.

sound instruments for measuring participation are needed. Measurement tools based on item response theory are currently under development.[35] Third, because communication never takes place in a vacuum, the role of communication partners is a critical yet poorly understood topic. The role of social support in the development of resilience to change is an important topic for future study. Next, although MS is a chronic condition often spanning decades of life, few longitudinal studies are available. Tracking people with MS over time is imperative if we are to be able to identify critical periods of change where intervention is required. Finally, it is important to understand the issues faced by people aging with MS. In addition to changes in communication related directly to MS, a number of other conditions can arise as part of the process of aging. For example, hearing loss is ranked as the third most prevalent chronic condition in older adults.[36] Thus, the burden of communication disorders is cumulative over time, with disorders arising from the primary conditions and from other sources. Programs and approaches to support the needs of people aging with MS are critical.

Acknowledgments

The contents of this chapter were developed under a grant from the Department of Education, NIDRR grant number H133B031129 & H133B080025. However, these contents do not necessarily represent the policy of the Department of Education, and one should not assume endorsement by the U.S. federal government.

References

1. Darley F, Aronson A, Brown J. Motor speech disorders. Philadelphia, PA: W.B. Saunders; 1975.
2. Yorkston KM, Klasner ER, Bowen J, Ehde DM, Gibbons L, Johnson K, Kraft G. Characteristics of multiple sclerosis as a function of the severity of speech disorders. Journal of Medical Speech-Language Pathology 2003;11(2):73–85.
3. Hartelius L, Svensson P. Speech and swallowing symptoms associated with Parkinson's disease and multiple sclerosis: A survey. Folia Phoniatrica et Logopaedica 1994;46:9–17.
4. Duffy JR. Motor speech disorders: Substrates, differential diagnosis, and management. St. Louis, MO: Mosby; 2005.
5. Darley F, Brown JR, Goldstein NP. Dysarthria in multiple sclerosis. Journal of Speech Language and Hearing Research 1972;15(2):229–245.
6. Lethlean JB, Murdoch BE. Subgroups of multiple sclerosis patients based on language function. In: B. Murdoch, D. G. Theodoros, editors. Speech and language disorder in multiple sclerosis. London: Whurr Publishers; 2000.
7. Murdoch B, Lethlean JB. High-level language, naming and discourse abilities in multiple sclerosis. In: B. Murdoch, D. G. Theodoros, editors. Speech and language disorders in multiple sclerosis. London: Whurr Publishers; 2000.
8. Bringfelt P, Hartelius L, Runmarker B. Communication problems in multiple sclerosis: Nine-year follow-up. International Journal of Multiple Sclerosis Care 2006;8:130–140.
9. Yorkston KM, Klasner ER, Swanson KM. Communication in context: A qualitative study of the experiences of individuals with multiple sclerosis. American Journal of Speech-Language Pathology 2001;10(2):126–137.
10. Blaney BE, Lowe-Strong A. The impact of fatigue on communication in multiple sclerosis. The insider's perspective. Disability and Rehabilitation 2009;31(3):170–180.
11. Baylor CR, Yorkston K, Bamer A, Britton D, Amtmann D. Variables associated with communicative participation in people with multiple sclerosis: A regression analysis. American Journal of Speech-Language Pathology 2010;19:143–153.
12. World Health Organization (WHO). International Classification of Functioning, Disability and Health. Geneva: World Health Organization; 2001.
13. Yorkston KM, Beukelman DR, Strand E, Hakel M. Management of motor speech disorders in children and adults. 3rd ed. Austin, TX: Pro-Ed; 2010.
14. Huisingh R, Barrett M, Bagcen C, Ormna O. The word test–revised. East Moline, IL: Linguis Systems; 1990.
15. Benton AL, Hamsher K. Multilingual aphasia examination. Iowa City, IA: AJA Associates; 1983.
16. Wiig EH, Second E. Test of language competence–expanded. Columbus, OH: Charles E. Merrill Publishing Company; 1989.
17. Hinchliffe FJ, Murdoch BE, Theodoros DG. Treatment of language disorders in multiple sclerosis. In: B. Murdoch, D. G. Theodoros, editors. Speech and language disorders in multiple sclerosis. London: Whurr Publishers; 2000.
18. Yorkston KM, Miller RM, Strand EA. Multiple sclerosis. In: K. M. Yorkston, R. M. Miller, E. A. Strand, editors. Management of speech and swallowing disorders in degenerative disease. 2nd ed. Austin, TX: Pro-Ed; 2004.
19. Ehrlich JS, Barry P. Rating communication behaviours in the head-injured adult. Brain Injury 1989;3(2):193–198.
20. Hartley L. Assessment of functional communication. In: D. E. Tupper, K. D. Cicerone, editors. The neuropsychology of everyday life. Boston, MA: Kluwer Academic Publishers; 1990.
21. Eadie TL, Yorkston KM, Klasner ER, Dudgeon BJ, Deitz J, Baylor CR, Miller RM, Amtmann D. Measuring communicative participation: A review of self-report instruments in speech-language pathology. American Journal of Speech-Language Pathology 2006;15:307–320.

22. Baylor CR, Yorkston KM, Eadie T, Miller RM, Amtmann D. The levels of speech usage: A self-report scale for describing how people use speech. Journal of Medical Speech-Language Pathology 2008;16(4):191–198.
23. Yorkston KM, Baylor CR, Klasner ER, Deitz J, Dudgeon BJ, Eadie T, Miller RM, Amtmann D. Satisfaction with communicative participation as defined by adults with multiple sclerosis: A qualitative study. Journal of Communication Disorders 2007;40:433–451.
24. Baylor CR, Yorkston KM, Eadie T, Miller RM, Amtmann D. Developing the communication participation item bank: Rasch analysis results from a spasmodic dysphonia sample. Journal of Speech Language and Hearing Research 2009;52(5):1302–1320.
25. Yorkston KM, Baylor CR. Measurement of communicative participation. In: A. Lowitt, R. Kent, editors. Motor speech disorders. San Diego, CA: Plural Publishing; 2011.
26. Hartelius L, Wising C, Nord L. Speech modification in dysarthria associated with multiple sclerosis: An intervention based on vocal efficiency, contrastive stress, and verbal repair strategies. Journal of Medical Speech-Language Pathology 1997;5(2):113–140.
27. Yorkston KM, Beukelman DR. AAC intervention for progressive conditions: Multiple sclerosis, Parkinson's disease, and Huntington's disease. In: D. R. Beukelman, K. L. Garrett, K. M. Yorkston, editors. AAC intervention for adults in medical settings: Integrated assessment and treatment protocols. Baltimore, MD: Brookes Publishing; 2007.
28. Beukelman DR, Yorkston KM, Garrett KL. AAC decision-making teams: Achieving change and maintaining social support. In: D. R. Beukelman, K. L. Garrett, K. M. Yorkston, editors. AAC intervention for adults in medical settings: Integrated assessment and treatment protocols. Baltimore, MD: Brookes Publishing; 2007.
29. Hoffman JM, Yorkston KM, Shumway-Cook A, Ciol MA, Dudgeon BJ, Chan L. Effects of communication disability on satisfaction with health care: A survey of Medicare beneficiaries. American Journal of Speech-Language Pathology 2005;14(3):221–228.
30. Bartlett G, Blais T, Tamblyn R, Clermont RJ, MacGibbon B. Impact of patient communication problems on the risk of preventable adverse events in acute care settings. Canadian Medical Association Journal 2008;178(12):1555–1562.
31. Williams MV, Davis T, Parker RM, Weiss BD. The role of health literacy in patient-physician communication. Family Medicine 2002;34(5):383–389.
32. Kagan A, LeBlanc K. Motivating for infrastructure change: Toward a communicatively accessible, participation-based stroke care system for all those affected by aphasia. Journal of Communication Disorders 2002;35(2):153–169.
33. Iezzoni LI, Davis RB, Soukup J, O'Day B. Quality dimensions that most concern people with physical and sensory disabilities. Archives of Internal Medicine 2003;163:2085–2092.
34. Starks H, Morris M, Yorkston K, Gray R, Johnson K. Being in- or out-of-sync: A qualitative study of couples' adaptation to change in multiple sclerosis. Disability & Rehabilitation 2010;32(3):196–206.
35. Baylore C, Hula W, Donovan NJ, Doyle PJ, Kendall D, Yorkston K. An introduction to item response theory and Rasch models for speech–language pathologists. American Journal of Speech–Language Pathology 2011;20(3):243–259.
36. US Department of Health and Human Services. Aging America: Trends and projections. Washington DC: Department of Health and Human Services; 1991.

13

Mobility

Susan E. Bennett and Marcia Finlayson

CONTENTS

According to the *International Classification of Functioning, Disability and Health* (ICF), mobility refers to the ability to move around in one's environment by changing body position or location; transferring from one place to another; carrying, moving, or manipulating objects; walking, running, or climbing; and using various forms of transportation.[1] Most of what people do in their daily lives is dependent on one or more forms of mobility: getting up and dressing in the morning, getting to work or school, and engaging in the activities of the day (e.g., shopping, cleaning house, parenting, exercising). The wide range of MS-related impairments (e.g., spasticity, balance, fatigue, weakness) can contribute to mobility restrictions throughout the disease course.

Compared to other functions, mobility (particularly walking) is very important to people with MS.[2] In a mailed survey study, 162 people with MS (average age = 43, range 18–66) ranked gait as most important across 13 bodily functions. Among those with longer disease duration (>15 years), hand function was ranked the third most important function after gait and vision. Authors of two separate review articles also concluded that mobility plays an important role in the quality of life of people with MS.[3,4] This literature may explain why restrictions in mobility are a major predictor of the use of physical and occupational therapy services by people with MS.[5–7]

This chapter focuses on mobility restrictions that occur across the MS disease course and the ways in which rehabilitation professionals can assess and address these restrictions. After reading this chapter, you will be able to

1. Describe the nature and extent of mobility restrictions among people with MS,
2. Describe the impact of mobility restrictions on people with MS and their caregivers,
3. Describe the evaluation of mobility restrictions among people with MS,
4. Discuss various treatment options available to enhance mobility, and
5. Identify the gaps in knowledge in mobility assessment and treatment.

13.1 The Nature and Extent of Mobility Restrictions among People with MS

Although walking restrictions are most commonly associated with MS, epidemiological data indicate that people with this disease can experience high levels of restriction across all areas of mobility identified by the ICF.

13.1.1 Changing Body Position or Location and Transferring

Changing body position or location includes getting into and out of a body position (e.g., lying to sitting, sitting to standing) and moving from one location to another, such as getting out of a chair to lie down on a bed, and getting into and out of positions, such as kneeling or squatting.[1] Transferring involves moving from one surface to another, such as sliding along a bench or moving from a bed to a chair, without changing body position.[1] Although questions about these activities are often included in MS studies, findings are often reported only in aggregate form (e.g., total score across multiple activities). Consequently, knowledge about restriction for individual activities is limited to a few studies.[8-11] Across these studies, investigators used variable samples (e.g., level of disability, age, living arrangements) and different data collection methods, and addressed only a limited number of activities. Therefore, our collective knowledge about the extent of restriction for this aspect of mobility has large gaps.

In one study, 47 people with MS in Sweden (aged 22–69, EDSS from 1.0 to 8.5) were interviewed using the Canadian Occupational Performance Measure.[9] Difficulties in functional mobility accounted for 71 (20%) of the 366 performance problems identified by respondents. Across these 71 problems, transfers accounted for the largest proportion ($n = 22$, or 31%) and included six different types of actions: moving to and from the toilet, getting in and out of the bath tub, getting in and out of a car, moving from one chair to another, getting in and out of bed, and turning in bed. Three of these six actions were addressed in a study of 1282 middle-aged and older adults with MS in the Midwest of the United States.[10] Table 13.1 provides previously unpublished data from this study. Based on these data, transfers into and out of the shower are most likely to be restricted, followed by toilet transfers and bed transfers.

In a Canadian survey of 416 community-dwelling adults with MS, Finlayson et al.[8] found that 14% of respondents reported difficulty getting in and out of bed on most days over the previous month. Fourteen percent also reported difficulty using the toilet, although it is unclear if these difficulties were a function of the ability to transfer or to

TABLE 13.1

Rates of Restriction across Three Types of Transfers among Middle-Aged and Older Adults with MS

		Age in Two Groups			
		45–64 years		65+ years	
		Count	%	Count	%
Transfer to and from a bed or chair	Never need help	427	76.8	452	62.3
	Sometimes need help or piece of equipment	52	9.4	80	11.0
	Always need help/ assistance	77	13.8	193	26.6
	Total respondents	556	100.0	725	100.0
Get on and off the toilet	Never need help	409	75.0	399	58.2
	Sometimes need help or piece of equipment	43	7.9	56	8.2
	Always need help/ assistance	93	17.1	230	33.6
	Total respondents	545	100.0	685	100.0
Get into and out of the shower or tub	Never need help	329	60.5	294	43.5
	Sometimes need help or piece of equipment	56	10.3	45	6.7
	Always need help/ assistance	159	29.2	337	49.9
	Total respondents	544	100.0	676	100.0

Source: Based on Finlayson M, Peterson E, Cho C. *Archives of Physical Medicine and Rehabilitation* 2006;87:1274–1279.

complete other aspects of the toileting process (e.g., managing clothing, cleaning oneself). Both of these rates are similar to the "always need help" rates among the 45- to 64-year-old group shown in Table 13.1. Respondents with progressive MS ($N = 140$) were most likely to report difficulty with bed transfers and toileting (25% and 22%, respectively). The 25% restriction in bed transfers comes close to figures reported by Buchanan et al. in 2004[12] in which 32% of 1309 people with MS residing in nursing facilities were dependent for bed mobility. These relatively large proportions (25% and 32%) lie in stark contrast to data from Switzerland and Germany.[13] There health care records and individual interviews were used to collect information about the functioning of 205 people with MS (average age = 44.7, mean Expanded Disability Status Scale [EDSS] = 3.7). Findings indicated that only 5% of respondents had some degree of restriction in transferring (type not described).

From the data available, it appears that changing body position or location and transferring are challenging activities for many people with MS, particularly those who have more advanced disease. There is a need for more comprehensive, detailed investigations to fully understand the nature and extent of restrictions in the ability of people with MS to change body positions or locations and complete transfers between different surfaces.

13.1.2 Carrying, Moving, or Handling Objects

Mobility also includes carrying, moving, or handling objects. These activities primarily involve the upper extremity (e.g., lifting a cup, carrying a child, picking up and releasing objects, turning a door knob, throwing and catching), but they can also involve the lower extremities, for example, kicking a ball or pushing the pedals of a bicycle. While there is growing understanding about impairments in grip, pinch, and manual dexterity among people with MS,[14–16] little has been documented about specific activity restrictions that occur as a consequence.

Data available suggest that people with MS experience substantial restriction in mobility activities involving the upper extremity. Holper et al.[13] found that 71% of 205 people with MS (average age = 44.7, mean EDSS = 3.7) had some level of restriction in hand and arm use, 67% in fine hand use, 59% in lifting and carrying objects, and 56% in writing. Finlayson et al.[8] reported that among 416 community-dwelling adults with MS, 33% experienced difficulties cutting their toenails and 27% had difficulty writing or typing. While Chen et al.[17] did not report the sample proportions experiencing difficulties, a Rasch analysis of the Manual Ability Measure-36 indicated that for 44 participants with MS (mean age = 50 years; range = 34–68), the five most difficult manual tasks were cutting nails (hardest), using a hammer, peeling fruits or vegetables, buttoning clothes, and shuffling and dealing cards. The easiest tasks were eating a sandwich, washing hands, using a remote control, drinking from a glass, and brushing teeth.

Given the wide range of everyday activities that involve carrying, lifting, and manipulating objects, knowledge about the proportion of people with MS who experience restrictions in this area of mobility is severely lacking.

13.1.3 Walking, Running, or Climbing

In the ICF, walking, running, and climbing are defined as moving along a surface on foot, step-by-step, so that one foot is always on the ground, as when strolling, sauntering, or walking forward, backward, or sideways.[1] Restriction of walking ability is a hallmark of MS. In a study of 1844 persons with MS,[18] factors associated with progression to difficulties in walking without an aid included being male, older age at onset of first attack, diagnosis of progressive MS versus relapsing–remitting, an incomplete recovery from the first attack, and a second attack within a short time frame (less than a year). These variables predicted faster progression to an EDSS of 4. However, once this level of limitation was reached, none of the variables predicted subsequent progression to further walking restriction.

Studies have shown that within 15 years after the onset of MS, 50% of individuals will require the use of an aid for walking (EDSS 6.0 and 6.5) and 10% will require a wheelchair[19] (EDSS 7.0 and up). Twenty-five years after disease onset, approximately 90% will have significant functional limitation and disability,[20] with 50% of those individuals requiring a wheelchair for mobility. These rates clearly indicate that the majority of people with MS will experience walking restriction at some point in their disease course.

Item-level detail from two other studies provides specific data about walking and climbing restrictions among people with MS.[8,21] Raw data from a study by Finlayson et al.[21] indicate that respondents (average age = 56) experienced limitations in vigorous activities (such as running), climbing multiple flights of stairs, and walking long distances (see Table 13.2). The average score on the Patient Determined Disease Steps[22] for the sample was 4, which reflects the need for a cane, crutch, or other form of support for walking all the time or part of the time, especially when walking outside. The rate of restriction for stair climbing shown in Table 13.2 is somewhat lower than in an earlier study[8] in which 38% of 416 people reported limitations in stair climbing. Those with progressive MS (*n* = 140) reported restriction at a much higher rate compared to those with relapsing–remitting MS (*n* = 79) (63% versus 29%, respectively).

Walking restrictions often lead to the use of mobility devices. Iezzoni et al.[23] collected data about walking limitations and use of mobility devices from 703 community-dwelling, working-age adults with MS. Fifty-nine percent of respondents reported that over the previous two weeks, MS had required them to expend greater effort to walk, and 40% indicated that every day they need to hold onto furniture, walls, or someone's arm when walking indoors. Of the 434 respondents who used at least one mobility device,

TABLE 13.2

Restrictions in Running, Stair Climbing, and Walking among 177 People with MS who Completed the SF-36 Health-Related Quality of Life Survey at Baseline for a Study

Individual SF-36 Items	Yes, Limited a Lot		Yes, Limited a Little		No, Not Limited at All		Total Respondents	
	n	%	n	%	n	%	N	w%
Vigorous activities such as running, lifting heavy objects, strenuous sports	159	87.8	14	7.7	4	2.2	177	100.0
Climbing several flights of stairs	116	64.1	39	21.5	21	11.6	176	100.0
Climbing one flight of stairs	56	30.9	72	39.8	49	27.1	177	100.0
Walking more than a mile	133	73.5	31	17.1	13	7.2	177	100.0
Walking several blocks	97	53.6	45	24.9	35	19.3	177	100.0
Walking one block	55	30.4	54	29.8	68	37.6	177	100.0

Source: Based on Finlayson M et al. Multiple Sclerosis 2011;17(9):1130–1140.

137 reported having two devices and 192 reported having three or more. Among individuals with multiple devices, the most common combination was manual wheelchair, cane, and rollator walker. The researchers hypothesized that people with MS "mix and match" their mobility aids to accommodate variability in walking ability in different environments.

Studies using accelerometers to track step counts during natural living offer another perspective on walking restrictions. In a study by Cavanaugh et al.,[24] researchers reported average daily step counts of 21 people: the 10 with EDSS >4.5 took an average of 3177 steps per day and the 11 people with EDSS ≤ 4.5 took an average of 8860 steps. Another study[25] reported average daily step counts as 5902 ($N = 193$). Findings indicate that some people with MS may exceed definitions for being sedentary (i.e., <5000 steps/day), but the overall levels of activity are low relative to acceptable levels for healthy, community-dwelling adults.[26]

Despite available research, relatively few studies document walking restrictions in the general community. Because of the interaction between MS symptoms and environmental demands, walking in the community can be challenging for people with MS. Community walking requires adequate speed to cross the street, adequate strength and balance to ascend and descend a curb, and the ability to adjust to uneven surfaces without falling. Gait velocity can be a factor in determining if an individual can be a safe community ambulator. Studies have reported gait velocities required to cross a rural street (44.5 m/min), urban street (47.5 m/min),[27] and a busy city street (70–73 m/min).[28] Findings from an exploratory study examining alterations in gait parameters due to speed and use of a cane suggest that people with MS who have mild impairment (EDSS 4.0–5.5) may not meet gait-velocity requirements for community ambulation.[29] Of 11 people with MS (six of whom used canes), unassisted gait velocity at maximum speed was 58.6 m/min, enough to meet requirements for rural and urban street crossings. At preferred speed, gait velocity dropped to 44.0 m/min, which is inadequate for all crossings. The assisted velocities for the six cane users suggest that their community ambulation would be restricted at both maximal and preferred speeds (55.8 and 42.7 m/min, respectively). This study did not address issues of spasticity or balance deficits, or whether these additional factors might influence community ambulation safety. Based on the clinical experience of the first author, the faster the person attempts to walk the greater the difficulty he or she will have clearing the leg in swing phase (spasticity limiting flexor activation) and the less time he or she will have to use balance–maintenance strategies. Community mobility restriction needs further investigation.

13.1.4 Moving Around Using Various Forms of Transportation

Whether to continue driving can be a challenging decision for people with MS, although driving is only one form of moving around using transportation. According to the ICF, using transportation includes being a passenger in a car, bus, taxi, train, subway, aircraft, and so on. It also includes driving one's own car, bicycle, boat, or animal-powered vehicle.[1] While there is a growing number of studies examining driving-related issues among people with MS,[30–36] most of them have focused on examining correlations between driving difficulties and specific MS impairments (e.g., vision, fatigue, cognition, spasticity). Recently, two different research teams investigated the driving behaviors and routines of people with MS.[32,37]

Schultheis et al.[32] studied 66 people with MS who had an average age of 43 years (sd = 8), an average EDSS of 3.4 (range 1.5–6.5), and over 24 years of driving experience.

People with the highest EDSS (5.5–6.5, $n = 16$) reported the lowest average driving frequency at 5.19 days per week (sd = 1.56). Participants with lowest EDSS (1.5–3.0; $n = 39$) and moderate EDSS (3.5–5.0; $n = 11$) reported average driving frequencies of 6.28 days (sd = 1.23) and 5.86 days per week (sd = 1.38), respectively. Weekly averages for total miles driven were 102 (high EDSS), 151 (moderate EDSS), and 140 (low EDSS). The most common reasons for driving were doing errands, participating in leisure, and going to work. Bad weather and heavy traffic were most likely to lead to self-limiting driving behaviors.

In another study,[37] 18 (23%) of 78 people with MS had already stopped driving, with 44% of them making this decision on their own. Many of the findings specific to driving frequency and differences by level of disability were similar to those of Schultheis et al.[32] Seven percent of the drivers in the study by Ryan et al.[37] used vehicle modifications in order to drive (e.g., hand controls). Overall, participants who continued to drive had fewer physical and cognitive impairments, showed greater awareness of their deficits, were more likely to engage in compensatory behaviors, and reported fewer social barriers to driving than nondrivers with MS.[37]

These studies provide useful information about driving among people with MS but do not indicate utilization of other forms of transportation. In a study of 1282 people aging with MS in the Midwest, Finlayson[10] asked participants about their transportation methods. Table 13.3 provides previously unpublished data from this study. Data highlight that driving is the most common method of transportation, followed by being a passenger with a family member driving. They also show that transportation problems are experienced by about 20% of this sample.

TABLE 13.3

Transportation Methods Reported by 1282 People with MS in the Midwest (US)

		Age in Two Groups			
		45–64 years		65+ years	
		Number	%	Number	%
What is your usual means of transportation?	Drive myself	337	60.5	285	38.9
	Family member drives	163	29.3	322	44.0
	Friend drives	11	2.0	32	4.4
	Public transportation	5	0.9	3	0.4
	Taxi	0	0.0	4	0.5
	Transportation service for people with disabilities	25	4.5	58	7.9
	Other	16	2.9	28	3.8
	Total respondents	557	100.0	732	100.0
Do you experience any problems with this means of transportation?	No	431	77.9	586	81.3
	Yes	122	22.1	135	18.7
	Total respondents	553	100.0	721	100.0

Source: Based on Finlayson M, Peterson E, Cho C. Archives of Physical Medicine and Rehabilitation 2006;87:1274–1279.

13.2 Impact of Mobility Restrictions

Given the breadth of mobility activities, restrictions can have a major impact on the lives of people with MS. In addition to the obvious influences on basic and instrumental activities of daily living, mobility restrictions can also influence health and safety, housing and transportation options, financial status, employment and domestic life, and the experiences of family members and caregivers.

13.2.1 Impact on Health and Safety

Each type of mobility restriction has the potential to influence the health and safety of a person with MS. Difficulties with changing position or transferring or spending long periods of time sitting because of walking limitations can increase the risk of pressure ulcers.[38] Limitations in the ability to manipulate objects can negatively affect adherence to disease-modifying therapies that require injection.[39] Inability to drive or problems accessing other forms of transportation can influence access to physicians, rehabilitation services, and other health and social care resources.[40,41] Mobility restrictions can also contribute to the risk of secondary conditions such as reductions in bone mineral density, falls and fractures, and compromised respiratory function.

13.2.1.1 Bone Mineral Density

Ambulatory status may play a role in maintaining total body bone mineral content, total body fat mass, and total body fat-free mass.[42] In a study of 71 people with MS (32 nonambulatory), Formica et al.[42] found that the severity of bone-mass deficit was related to the degree of physical disuse by nonambulatory patients with MS. This finding is consistent with an earlier study of 30 people with relapsing–remitting MS.[43] Further supporting the association between mobility and bone mineral density, the study by Formica et al.[42] also found that ambulatory patients with MS had no difference in bone mass or body composition compared with age-comparable controls. Since bone-mass reductions have been linked to increase in fracture risk among people with MS[44] and healthy older adults,[45] the potential impact of mobility restrictions on the bone mineral density of people with MS is worthy of attention.

13.2.1.2 Falls and Fractures

Mobility requires an individual to maintain his or her balance. Balance requires a complex integration of information from the visual, somatosensory, and vestibular systems and a musculoskeletal system that provides adequate range of motion and strength (see Chapter 6). People with MS often present with symptoms in one or more of these systems. Moving from sit to stand, standing and reaching during bathroom or kitchen activities, and walking can be very challenging from a balance perspective, and this challenge may increase the risk of falls.

Fall rates in MS have been reported in four cross-sectional studies[10,46–48] and one longitudinal study.[49] The lowest rate reported was 45% of 31 people (average 42 years, EDSS range 3.0–5.0) over a six-month period.[48] The highest rate was at least 2 falls per year among 64% of 354 people (aged 55–94 years), 30% of whom reported falling once a month or more.[47] In addition to falls, near falls are common. In one nine-month prospective study,[49] 32 participants fell 2 times or more, and 11 fell 10 times or more; a total of 2352 near-fall incidents was reported.

Although a number of factors appear to increase the risk of falls among people with MS,[10,46,48,49] restrictions in ambulation[46,49] and use of one or more mobility devices are two of the more consistent findings.[10,23,46,49] While the findings may simply reflect that mobility device users are more disabled and, therefore, at greater risk, both Finlayson et al.[10] and Iezzoni et al.[23] also suggest that making decisions about mobility devices (e.g., to use any device, to use a more supportive device under specific circumstances) may also be playing a role. Research has not yet examined the role of other mobility activities (e.g., transfers, carrying objects) in MS fall risk.

Since falls can lead to fractures, particularly of the hip and wrist, mobility restrictions among people with MS may have serious health consequences. Two separate studies of injurious falls in this population[47,50] demonstrated that falls requiring medical attention are a serious and potentially costly problem. For people who are already experiencing mobility restrictions, an injurious fall may compromise their mobility even more.

13.2.1.3 Respiratory Function

Although no causal links have been established, greater mobility restrictions as measured by the EDSS have been associated with respiratory functions in MS, including expiratory and inspiratory muscle weakness,[51–53] reduced lung function,[54] lower cough efficacy,[55] and lower cardiorespiratory fitness.[56] These findings must be considered during rehabilitation assessment and treatment planning. Since coughing is an important airway defense mechanism[51] and expiratory muscle strength training has been shown to improve the strength of the expiratory muscles in a small study of people with MS,[57] further research on respiratory function, mobility restriction, and the potential of rehabilitation is warranted.

13.2.2 Impact on Housing and Transportation Options

When restrictions in walking require an individual with MS to use a wheelchair all or part of the time, access to suitable housing and transportation can be challenging.[58] Unfortunately, relatively little has been written specifically to the needs and concerns of people with MS regarding housing and transportation or the impact of mobility restriction on these issues. What is known is that people with MS are concerned about losing their mobility and worry that loss of mobility will mean having to live in a nursing home.[59] Mobility restriction tends to be the primary driver of home modifications (e.g., barrier-free entry, roll-in showers, external ramps, modified kitchens). Chapter 21 discusses physical environment assessment and interventions to facilitate participation.

In terms of transportation, restrictions in transferring, walking, or carrying may limit options available to a person with MS. For a person who is able to transfer into a vehicle and drive him- or herself, restrictions in the ability to fold and lift a walker or wheelchair and place it in the back seat or trunk may limit transportation options. Difficulties with pivot or sliding board transfers would further limit a person's transportation options. Yet, using a motorized wheelchair or scooter or having a vehicle with hand controls may not be an option for some people with MS, further limiting transportation options. Rehabilitation providers must consider how each component of mobility might influence transportation options for people with MS.

13.2.3 Impact on Finances

Several studies have investigated the costs of MS on individuals, families, and health care systems.[60–65] Within these studies, different direct and indirect costs are considered, including medications, use of rehabilitation and other support services, mobility equipment, and home modifications. Research has shown that people with higher levels of MS disability, usually measured by the EDSS, experience greater disease-related costs compared to people with lesser disability.[62,63,65,66] Based on these findings, it is safe to conclude that restrictions in mobility, particularly walking, can have a significant impact on the financial welfare of an individual with MS. This impact may be greater in countries where mobility equipment and home modifications are not provided through government programs or when the options provided are severely restricted and necessitate out-of-pocket expenses to meet needs.

13.2.4 Impact on Family Members and Caregivers

Mobility restrictions often mean that people with MS have difficulty performing daily activities. These difficulties, together with the variability of MS symptoms, can affect the family and other caregivers. In addition, the unpredictability of the disease affects the family and caregivers in a way unlike other chronic diseases.[67] The onset of a relapse, the progression of the disease, and daily fluctuations in functional ability can contribute to emotional stress[68] and additional responsibilities in terms of physical caregiving.[69,70] Chapter 22 discusses the full range of implications of MS caregiving on the caregiver and the assessments and interventions that can be employed by rehabilitation professionals to address these issues.

13.2.5 Summary of the Implications of Mobility Restriction

Given the central role of mobility in everyday life, it is important for rehabilitation providers to think broadly about the potential implications of mobility restrictions on people with MS. These implications may influence engagement in rehabilitative programs and the organization and delivery of comprehensive services, supports, and referrals.

13.3 Evaluating Mobility Restriction and Its Specific Components

The rehabilitation therapist must comprehensively evaluate mobility in order to identify the individual's abilities and how changes in ability may influence safety, independence, and participation in life activities. Underlying changes in body structure and function (symptoms) must also be identified to establish the best treatment program for the client. The symptoms of MS that most often limit safe and independent performance in mobility activities are weakness, spasticity, balance dysfunction, ataxia, pain, cognition, and fatigue. These impairments need to be evaluated to understand the contributors to mobility restriction. Since other chapters in this book address these topics in detail, they will not be addressed here. The focus in this section will be on assessment at the level of activity and participation.

13.3.1 Overview of Assessment of Mobility Functions

Although the EDSS[71] is commonly used to describe mobility of people with MS, it focuses almost exclusively on walking ability rather than mobility as a whole (see Chapter 2 for more detail). An EDSS score of 6.0 reflects a major change in walking ability and marks the point at which an individual requires a unilateral device and is limited in the distance he or she can walk. An EDSS score of 5.0 reflects that an individual is limited in performing a full day of activities, and walking distance is restricted to less than 200 m without an aid. However, because an assistive device is not used at an EDSS of 5.0, an individual may not be *perceived* as having a mobility restriction by some health care providers.

Despite the common use of the EDSS, the rehabilitation therapist needs to assess a full range of mobility tasks for all individuals. Table 13.4 provides a list of the various mobility tasks that should be evaluated as part of a comprehensive mobility assessment. The case at the end of this chapter presents the findings of an assessment incorporating many of these tasks. For individuals who are nonambulatory, the therapist should examine their ability to stand with hand support and assist as indicated. If an individual is limited in any of the tasks identified in Table 13.4, the therapist must then determine if an assistive device or personal assistance will be needed to enable performance. Table 13.5 contains standard definitions for the level of assistance required to perform a task based on the Functional

TABLE 13.4

List of Tasks that Should be Evaluated as Part of a Comprehensive Rehabilitation Assessment for Potential Mobility Restrictions

	Mobility Activities for Assessment
Bed mobility	Supine to sit; sit to supine; static sitting; dynamic sitting
Transfers	Dependent transfer—2 man lift or Hoyer lift; lateral with/without sliding board; stand pivot sit; car transfer
Home ambulation	Static stand; dynamic stand; ambulation with/without assistive device; with/without assist
Community ambulation	Speed; distance; head scanning; ascend/descend curb

TABLE 13.5

Summary of Definitions for Level of Assistance from the Functional Independence Measure

Levels of Assistance	Definition
Independent	Fully independent; patient is safe and timely; no helpers or device
Modified independence	Requiring the use of a device but no physical help, may need more than reasonable time; concern for safety
Supervision	Requiring only standby supervision; verbal prompting or help with set-up
Minimal contact assistance	Subject performs >75% of the effort
Moderate assistance	Subject performs 50–75% of the effort
Maximum assistance	Subject provides less than half the effort to complete task (25–49%)
Total assistance	Subject contributes <25% of the effort or is unable to do the task; or requires assist of two persons

Source: Adapted from Granger CV, Deutsch A, Linn R. Archives of Physical Medicine and Rehabilitation 1998;79:52–57.

Independence Measure™ (FIM). Based on the clinical experience of the first author, the likelihood of the individual achieving full independence (i.e., no assistance) during rehabilitation is not favorable when maximal assistance is required at baseline (therapist or caregiver providing 75% or more of the effort to complete the task). The use of adaptive equipment (e.g., sliding board to transfer rather than a pivot transfer) is more likely to enable performance after discharge in these situations.

13.3.2 Changing Body Position or Location and Transferring

Several instruments with standard procedures can be used to evaluate the ability of a person with MS to change body position or location and safely transfer between surfaces. These activities are highly dependent on the ability to maintain balance, and, as such, balance assessments are an important component of assessing these aspects of mobility. Both the Berg Balance Scale (BBS)[72] and the Equiscale Balance Test[73] include items specific to changing body positions, for example, sit to stand, bending over to retrieve an object from the floor, turning around, and stepping up and down from a foot stool. The BBS can be obtained from the Stroke Center website.[74]

Other more general functional assessments also provide opportunities to assess position changes and transfers. The Barthel Index is a performance-based instrument that was first published in 1965.[75] It addresses 10 areas of function, 4 of which can be directly connected to the ICF definition of mobility (i.e., toilet use, transfers [bed to chair and back], mobility [on level surfaces], and stairs). The Barthel Index is commonly used as an outcome measure in rehabilitation studies in MS, particularly outside of North America.[76–80] Scoring on the Barthel varies across items, usually from 0 = not able to perform to 10 = independent. For most items, the only intermediate score is 5 (needs some form of assistance). This scoring system limits the sensitivity of the instrument, particularly for small but clinically important changes. The FIM grew out of the Barthel[81] and contains a more defined scoring system (1 through 7 ordinal scale) for the degree of assistance required to complete an activity. Transfer and mobility (wheelchair and ambulation) tasks listed in Table 13.4 are covered on the Barthel and FIM. Both measures are valid and reliable for a wide variety of populations, including MS. The Barthel can be downloaded through the Stroke Center website.[74] The FIM is available through a subscription from the Uniform Data System for Medical Rehabilitation.[82]

13.3.3 Carrying, Moving, or Manipulating Objects

There are several instruments available to assess upper-extremity mobility activities, although none of them address all aspects included in the ICF definition. The Nine-Hole Peg Test (NHPT)[83] is one of the three components of the MS Functional Composite.[84] The test is pictured in Figure 13.1 and is administered by asking the client to take the pegs, one by one, from the shallow dish at one end, place them into the holes, and then remove and return the pegs to the dish as quickly as possible. The NHPT score is based on the average time of two trials (in seconds). Norms for the NHPT in healthy individuals were published in 1985 and 2003[85,86] and can be used to calculate a z-score.

For both MS research and clinical practice, the NHPT is the most commonly used test of upper extremity function even though it is primarily a measure of finger dexterity. Besides this limitation, the NHPT is also subject to practice effects, which reduces its ability to accurately capture change in response to intervention.[87] For these reasons, the NHPT is best used together with other instruments. Table 13.6 provides a summary of some of the

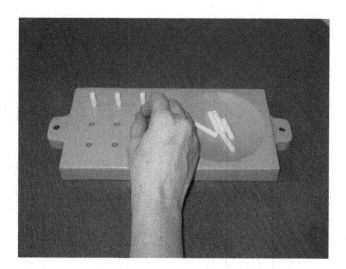

FIGURE 13.1
Administration of the Nine-Hole Peg Test.

key tools that have been described in the literature as having utility for MS rehabilitation research and practice and that can be used in combination with the NHPT.

13.3.4 Walking

There are several instruments available to objectively assess walking mobility and gait. Table 13.7 lists the characteristics of some of these instruments, their advantages, disadvantages, and psychometric properties. These tools are typically performed by occupational and physical therapists, but can be completed by other health care providers who follow the standard protocol. While most of these instruments are clinician administered, there are also self-reported tools as well.

From the instruments listed in Table 13.7, the most frequently used is the Timed 25-Foot Walk (TW25), which is a component of the MS Functional Composite [MSFC].[88–91] A limitation of the TW25 is that variations can occur as a result of the command given during testing (e.g., walk as fast as you can, or walk at your comfortable speed). Since time and distance are recorded, gait speed can be calculated. However, the influences of gait deviations or fatigue on ambulation might not be observed over this short distance. The TW25 does not assess the ability to modify gait during turns or identify fall risk.[92]

In 2007, the Consortium of MS Centers (CMSC) convened a consensus conference to examine measures for gait and fatigue. The recommendation was that the TW25, the Dynamic Gait Index (DGI), Timed Up and Go (TUG), the 12-Item Multiple Sclerosis Walking Scale (MSWS-12), and the 6-minute walk test should be used as gait measures in MS.

The validity and reliability of the TUG,[93] DGI[94] and the 2- and 6-minute walk were recently evaluated using the EDSS and TW25 as the gold standard.[95] In addition, the patient self-assessment report of walking as measured by the MSWS-12[96] was recorded. All the walking measures were highly correlated with the EDSS; the strongest were for the 2- and 6-minute walk tests at −0.808 and −0.811 (see Table 13.8). These values indicate that as the EDSS score increases, the distance that people with MS are able to walk in 2 or 6 minutes decreases. The walking measures with the best predictive validity for the EDSS (based on

TABLE 13.6

Assessment Instruments to Assess Mobility Activities of the Upper Extremity

Instrument	Purpose and Method of Administration	Time and Resources Needed	Scoring Method and Interpretation	Advantages	Disadvantages	Psychometric Properties
Manual Ability Measure-36 (MAM-36)[119]	36-item, self-report questionnaire to examine self-perceptions of manual ability.	Approximately 15–20 min to complete and score. Need questionnaire and pencil.	5 response options: 0 = almost never do, 1 = cannot do, 2 = very hard, 3 = a little hard, 4 = easy. Responses are summed. A higher score indicates greater perceived manual ability.	Easy and inexpensive to use. Addresses common everyday activities. Developed using Rasch measurement model.	None apparent	Correlates well with pinch and grip strength measures.[17] The 36 items measured a single construct with no misfitting items.[119] The items can reliably separate the people into five ability strata.[119]
Disabilities of the Arm, Shoulder and Hand (DASH)[120]	30-item, self-report questionnaire to examine self-perceptions of physical function and symptoms of the upper limb. Optional Sport/Music or Work sections are available.	Approximately 15-min to complete and score. Need questionnaire and pencil.	5 response options: 0 = no difficulty, 1 = mild difficulty, 2 = moderate difficulty, 3 = severe difficulty, 4 = unable. A higher score indicates worse functioning and symptoms.	Easy and inexpensive to use; addresses common everyday activities. Free online scoring available.[121]	Rasch analysis identified 13/30 misfitting items and disordered item-response-option thresholds for 9/30 items. This suggests that the tool needs additional development for MS.[122]	Overall, the traditional psychometric analysis supported the DASH as a reliable and valid measure of upper limb function in people with MS.[122]

Jebsen Hand Function Test[123]	Timed performance-based test to evaluate major aspects of hand functioning	10–20 min to administer, and minimal time to score. Need several common items for the subtests plus a stopwatch. Kits can be purchased from major rehab suppliers.	Activities are timed. Clients perform each activity with both the right and left hands, with the non-dominant hand tested first. Normative tables are available by age and sex. Longer times reflect greater disability.	Kits can be purchased or homemade. Addresses common everyday activities. Includes seven subtests: writing; turning over 3- by 5-inch cards (page turning); picking up small common objects; simulated feeding; stacking checkers; picking up large objects; and picking up large, heavy objects.	Assesses only time to complete subtests, not the quality of performance. The hands are tested separately, yet many tasks of daily living are bilateral.	Correlates well with the Nine-Hole Peg Test among people with MS.[124] No other psychometric data specific to people with MS could be identified.
Test d'Evaluation des Membres Supérieurs de Personnes Âgées (TEMPA)[125]	Timed performance-based test to assess upper extremity function in 9 everyday activities	Up to 30 min to administer, depending on level of restriction.	Each task is measured on three criteria: speed of execution, functional rating, and task analysis. The functional rating reflects autonomy in performance and is rated from 0 to –3. Task analysis rating quantifies difficulties in five areas: range of movement, strength, control of gross movement, prehension patterns, and fine movement.	Kits can be purchased from Physipro, a company in Quebec, Canada. Includes both bilateral and unilateral tasks relevant to daily life (e.g., open a jar and take a spoonful of coffee; unlock a lock; open a pill container; write on an envelope and stick a stamp on it; pick up a pitcher and pour water into a glass; handle coins).	Kits are ~$800 Cdn. Reliable scoring of the functional rating and task analysis criteria require training and practice. Length of time to administer is longer than some other hand-function tests.	Demonstrated concurrent validity with the NHPT and the Jebsen in an MS sample.[124] Moderate correlation with the Functional Independence Measure.[124] Reliability and internal consistency data specific to MS could not be located.

TABLE 13.7

Description of Gait Measures Recommended from the Consortium of MS Centers Consensus Conference "Towards a Consensus on Rehabilitation Outcomes in MS: Gait & Fatigue"

Gait Measure	Method of Administration	Purpose	Process of Administration	Time and Resources Needed	Scoring Method and Interpretation	Advantage	Disadvantages	Psychometric Properties
Timed 25-foot walk (T25FW)	Clinician administered	Measures gait velocity	Person is timed walking 25 ft, quickly but safely	Minimal time (1–2 min) Need tape measure, stop watch	Time to complete the task in seconds. Normative data: 8.5 s for median EDSS 2.5 (range 0–8)[126]	Easy to administer; inexpensive	Unable to assess gait deviations due to fatigue	High inter- and intra-rater reliability and validity[126]
6-min walk*	Clinician administered; training recommended	Measure of walking endurance	Distance walked is measured over 6-min period of time Walking is self-paced.	6 min of testing time; need to provide time for adequate recovery. Need stop watch; premeasured distance	Time to complete the task in minutes and seconds. Normative data: 499 m for healthy adults >60 years of age[127]	Easy to administer; inexpensive; provides valuable information on effects of fatigue on ambulation	Time to administer does not account for changes over the 6 min (i.e., first few minutes faster than last; qualitative changes not assessed)	High inter- and intra-rater reliability[127]

Dynamic Gait Index (DGI)	Clinician administered; training recommended	Measures patient's ability to respond to changing task demands during gait	Therapist instructs the patient in eight different walking tasks and scores performance.	Eight walking items with 4-point scale, 0 = unable, 4 = independent. Score range 0–24; < 19/24 correlates to high risk of falling among community-dwelling adults[128]	Minimal time (6–8 min; includes rest). Need stop watch, 40-foot walkway area	Easy to administer; inexpensive	Overlap of categories; does not assess fatigue due to short distances	Valid and reliable in people with MS[129]
Timed Up and Go (TUG)	Clinician administered	Measures time to get out of a chair, walk 10 ft, turn and return to chair; indicates fall risk	Individual is asked to stand from a seated position and walk 10 ft, turn around, and walk back to the chair as quickly and safely as possible.	Score of >30 s has been shown to indicate high risk of falls in community-dwelling older adults[130]	Minimal time (1–2 min). Need arm chair, tape measure, stop watch	Easy to administer; inexpensive; correlates with gait speed and balance[131,132]	Unable to assess gait deviations caused by fatigue	Data not currently available for MS
Multiple Sclerosis Walking Scale-12 (MSWS-12)	Self-report	Patient self-report of walking ability	Need paper and pencil, 12-item questionnaire completed by patient	1–5 score, 5 = limited a lot, 1 = not limited at all; total score range 12–60; higher score reflects greater limitations in walking.[96]	Time varies for individuals, but should be 10 min or less.	Easy to administer; inexpensive; provides patient's perspective	Self-report questionnaire	Valid, reliable, and responsive patient-based measure of the impact of MS on walking[96]

TABLE 13.8

Correlation between Selected Measures of Walking Ability and the EDSS among 50 People with MS

	Correlation with EDSS	p-value (two-tailed)
MS Walking Scale–12	0.748	<0.001
Timed 25-foot walk	0.748	<0.001
TUG	0.781	<0.001
DGI	−0.794	<0.001
2-min walk	−0.808	<0.001
6-min walk	−0.811	<0.001

Source: Adapted from Bennett S et al. Journal of MS Care 2010;12(Suppl 1):10.

the β estimate and adjusted r^2) were the DGI, 6-minute walk, 2-minute walk, and MSWS-12 (see Table 13.9).

Two useful self-report measures of walking are the MSWS-12 and the Patient Determined Disease Steps (PDDS).[97] The MSWS-12 is valid and reliable for both hospital- and community-based people with MS[96] and is suitable for measuring change over time.[56] It is also correlated with the oxygen cost of walking at a comfortable walking speed ($r = 0.641$, $p = 0.001$), a fast walking speed ($r = 0.616$, $p = 0.001$), and a slow walking speed ($r = 0.639$, $p = 0.001$).[98] The instrument includes items addressing stair climbing, running, indoor and outdoor walking, walking speed and smoothness, effort and distance. The PDDS is a surrogate measure of the EDSS used in the North American Research Committee on Multiple Sclerosis registry. The PDDS is scored from 0 = normal to 8 = bedridden, with scores between 3 and 7 specifically focused on patient-reported walking limitations. The PDDS correlates with the EDSS in cross-sectional and longitudinal studies.[97]

Lacking among the current walking measures is one that addresses community ambulation. Walking in the community is a complex process, involving the ability to negotiate uneven terrain, private venues, shopping centers, and other public venues.[99] Shumway-Cook and Patla[100] described eight environmental dimensions that influence an individual's community ambulation beyond the typical parameters of his or her motor capacity within a physical setting. They include ambient conditions (e.g., lighting and weather), terrain characteristics (e.g., uneven ground, compliance, friction), external physical load

TABLE 13.9

Predictive Validity of Selected Measures of Walking Ability on the EDSS among 50 People with MS

	Beta Estimate	p-value (two-tailed)	Adjusted r^2
MS Walking Scale–12	0.780	<0.001	0.601
Timed 25-foot walk	0.513	<0.001	0.248
TUG	0.545	<0.001	0.283
DGI	0.819	<0.001	0.663
2-min walk	−0.811	<0.001	0.651
6-min walk	−0.818	<0.001	0.663

Source: Adapted from Bennett S et al. Journal of MS Care 2010;12(Suppl 1):10.

(e.g., whether anything was being carried), attentional demands (e.g., talking while walking), postural transition (e.g., starting, stopping, changing direction), traffic level, time constraints, and walking distance. These dimensions emphasize that activity and participation should not be defined by the number of tasks a person can perform at home or within a defined community but rather by the range of environmental contexts or dimensions in which the tasks can be successfully completed. For the individual with an EDSS of 8.0, participation at a social event in an unrestrictive community environment (e.g., one with curb cutouts, automatic doors, accessible restrooms) may become impossible in winter months with snow and ice (i.e., terrain characteristics), or when navigating city streets with timed traffic lights (e.g., time constraints). This emphasizes that mobility restriction and assessment must be multifactorial and not focus solely on individual-level abilities.

13.3.5 Using Various Forms of Transportation

Although there are many forms of transportation that a person with MS might use, standard assessments tend to focus on the ability to operate a motor vehicle rather than taking the bus or using a taxi. Driving is a complex activity that requires abilities that go well beyond mobility (e.g., cognition, perception, vision). All rehabilitation therapists should screen for potential driving-related problems. The purpose of screening is to identify individuals who may require further assessment of their driving abilities and safety.[101]

When concerns are identified, the rehabilitation therapist has an ethical responsibility to recommend a referral for a full driving assessment, also known as a comprehensive driving evaluation (CDE). In some jurisdictions, the rehabilitation therapist may be able to make this referral directly. In others, the physician may have to make the referral based on the therapist's report and recommendation. Administration and interpretation of a CDE is an advanced skill and requires specialized training and certification in most jurisdictions (e.g., certified driver rehabilitation specialist). Therefore, this chapter will only describe the general process of CDE and will not provide detailed information about specific tools. It is important to note that "driving assessment" is a term that is used to describe both the medical evaluation of fitness-to-drive and a CDE. The medical evaluation is performed by a physician, often with input from an optometrist or ophthalmologist. In some jurisdictions, medical evaluation of fitness-to-drive is legislated after a certain age or for certain health conditions (e.g., after stroke).

A CDE involves two components: an office-based or pre-driving assessment and an on-the-road assessment. The pre-driving assessment generally includes the following: medical history (including medications); driving history and patterns; assessment of motor, sensory, and basic vision functions; visual–perceptual assessment; cognitive assessment; and behavioral assessment.[101,102] The on-road assessment typically involves a set evaluation route (up to 60 min in duration) that enables the driver assessor to evaluate a range of driving maneuvers and behaviors. Examples include turns, merges, reversing, responding to road signs, steering, using the pedals, controlling the vehicle at different speeds, using the signals, maintaining distance between vehicles, and using mirrors.[103] In addition, the driver assessor will observe entry and exit skills, the ability of the driver to load mobility aids safely into the vehicle, and complete pre-driving checks (e.g., adjust seat, mirrors, seat belt).

Some of the major issues with CDE are high cost, limited access, and high risk of the on-road test component. Increasingly, driving simulators are being used in both research and practice and offer a convenient and safe method for assessing driving behaviors. Their use appears to have great potential in MS.[104]

13.4 Interventions to Improve Mobility

The focus in neurologic rehabilitation is always to structure treatment programs to regain loss of functional mobility, improve activity performance, facilitate community participation, and maintain quality of life. An additional goal for physical therapists is to enhance overall cardiovascular fitness. To accomplish each of these goals, a comprehensive rehabilitation program focused on mobility functions should include task-specific repetitive training with the emphasis on motor learning and motor control, therapeutic exercise, aerobic exercise, and specific interventions to address underlying impairments. Treatment goals should be based on both the assessment findings and the client's priorities. Addressing underlying impairments should be followed by repetitive practice of the mobility tasks to be achieved.

13.4.1 Basic Principles of Task-Specific Repetitive Training

The basic principle of motor learning is that the best way to (re)learn a task is to practice it repeatedly.[105] Task-specific training is the repetitive practice of the task or activity that is the intended outcome of the rehabilitation process,[106] for example, walking, stair climbing, transferring into a car, cutting food, or handling coins. The research on task-specific training has its foundation in animal studies and brain plasticity.[105] Although clinical research has primarily been in the area of stroke rehabilitation, the basic principles are also being applied in the context of rehabilitation for MS-related mobility restrictions, particularly walking and upper-extremity mobility functions. According to Bayona et al.,[105] the basic principles of task-specific repetitive training include the following.

13.4.1.1 Use Tasks That Are Meaningful to the Client during Therapy

The meaningfulness of the task enhances therapy outcomes. This principle is consistent with the occupational therapy literature on the importance of client-centered and occupation-based practice.[107] Based on this principle, using real versus simulated tasks during therapy may be more effective for goal achievement.

13.4.1.2 Practice Repeatedly and Consistently

Repetition and consistent practice play a major role in inducing and maintaining changes obtained during therapy. Based on this principle, showing a client how to safely use a piece of adapted equipment to enhance mobility (e.g., transfer board, cane) a few times will be less effective than practicing with the client consistently over a period of time.

13.4.1.3 Grade the Task to Challenge Performance and Provide New, Meaningful Skill Learning

Repetitive motor activity alone is not sufficient to produce change. Significant changes are unlikely to occur when motor repetition occurs in the absence of learning new meaningful skills. Grading can occur by modifying different aspects of the task or the environment (e.g., walking faster, walking on uneven different surfaces).

The application of these principles can be seen throughout rehabilitation interventions for mobility restriction in MS care.

13.4.2 Reducing Restrictions for Position Changes, Transfers, and Upper-Extremity Mobility Functions

Treatment planning and goal setting for each of these aspects of mobility tend to focus on decreasing the need for assistance, increasing efficiency (reducing energy demands), decreasing the time needed to complete a task, or increasing safety during the task. Each of these goals can be pursued by designing treatment plans that include the following:

1. Apply the principles of task-specific repetitive training[108–110]

2. Manage underlying impairments contributing to the restriction, for example, balance (Chapter 6), weakness (Chapter 7), and fatigue (Chapter 5)

3. Prescribe and then train a client to use adaptive equipment to compensate for difficulties, reduce need for assistance, or improve overall safety (e.g., bath seat, transfer board, built-up utensils)

4. Recommend home modifications that reduce the need for assistance or improve overall safety (e.g., grab bar installation, see Chapter 21)

Both occupational and physical therapists are involved in designing treatment programs for position changes and transferring, while occupational therapists are more often responsible for developing plans related to upper extremity mobility functions. Through collaboration with the client, treatment planning needs to involve prioritization of the specific tasks to be accomplished based on restrictions identified during the assessment process. Intervention then involves integrating consistent and repeated practice into regular and meaningful routines, such as getting ready for work, making meals, or engaging in favorite pastimes. For clients on an inpatient unit, collaboration with and education of the nursing staff will be important for facilitating consistent practice. For clients receiving outpatient or home-based rehabilitation, education of family members and other caregivers will be necessary to achieve treatment goals.

13.4.3 Reducing Walking Restrictions

Treatment planning and goal setting for gait typically focus on decreasing the assist provided by the therapist or caregiver, increasing the distance walked, decreasing the time to walk a specific distance, and increasing overall safety. A guideline for writing goals is to decrease the level of assistance required by one level for the short-term goal (STG) and by two levels for the long-term goal (LTG). For example, if the individual walked 15 ft with a walker and moderate assist in the baseline assessment, then:

- STG—30 ft with a walker and minimal assist, and
- LTG—60 ft with a walker and contact guard.

A guideline for increasing the distance is doubling the number of feet walked at the initial examination for an STG and doubling that distance for an LTG. To achieve safe home ambulation, the goal is typically up to 100 ft/30 m and for community ambulation potentially up to 200 m or more (per EDSS scale). Increasing gait speed or velocity is most often driven by the goal to return the individual to community ambulation and, hence, the speed to cross a street (values identified earlier in this chapter). Consistent with the principles of task-specific repetitive training, interventions to achieve treatment goals need to be meaningful to the client. Physical therapists tend to be primarily responsible for providing rehabilitation specific to walking mobility. Several intervention methods are used to achieve walking-related goals.

13.4.3.1 Progressive Resistance Training

Progressive resistance training is an isotonic contraction of a muscle through the full range of motion with resistance being provided to the motion progressively over time. The individual may begin with a two-pound weight at baseline and, by the end of eight weeks, complete the active movement with 10 pounds of resistance. An eight-week progressive resistance-training program[111] enhanced walking in individuals moderately impaired by MS. In this specific study,[111] the main outcome measures were kinematic gait parameters, including knee range of motion; duration of stance, swing, and double-support phases in seconds; and percentage of time in each gait phase. Isometric strength, 3-minute stepping, fatigue, and self-reported disability were also measured. After two months of resistance training, there were significant increases ($p < 0.05$) in stride time and step length, and isometric leg strength improved ($p < 0.05$) in two of the four muscle groups tested. Subject self-report of fatigue indices also decreased ($p = 0.04$). The authors concluded that resistance training may be an effective intervention strategy for improving walking in moderately disabled persons with MS.

13.4.3.2 Treadmill Training

A four-week aerobic treadmill-training program was conducted with 16 people with MS who were randomly assigned to an immediate exercise group or a delayed exercise group. A significant difference in walking endurance was reported between the two groups.[112] In addition, a decrease in 10-m walk time was noted for both groups, with a significant change in the immediate exercise group ($p < 0.05$). The authors concluded that individuals with MS can benefit from aerobic treadmill training with potential for walking speed and endurance to improve following training. There was no increase in fatigue reported by the subjects. Another study[113] reported a similar finding with three subjects followed in a case study design completing a four-week aerobic treadmill-exercise program. The authors reported that aerobic treadmill-training is feasible, safe, and may improve early anomalies of posture and gait in mildly impaired individuals.

Bodyweight support treadmill training (BWSTT) has also been investigated in MS as it provides a means to unweigh a percentage of the body while the person is walking on the treadmill. A randomized control trial[76] compared robot-assisted BWSTT with a conventional walking program. Though the robot-assisted BWSTT did not demonstrate a significant change in gait, the treatment is feasible to perform in patients with MS who have severe walking disabilities. Lo et al.[114] used the BWSTT with and without robot assist. Again, no significant difference was found between groups; however, there was improvement noted in participants who completed the study. BWSTT is a form of repetitive, task-specific training described earlier in this chapter, which is beneficial to motor recovery but may be best utilized in conjunction with other treatment interventions.

13.4.3.3 Mobility Device Prescription and Training

Even after walking-related training, some people with MS will continue to need adaptive equipment to enable safe and independent mobility. Many different devices are available, ranging from Nordic walking poles to wireless functional electrical stimulators (FES). It is often the person with MS who first becomes aware of a new device to assist with mobility and brings it to the attention of the occupational or physical therapist. Once the device is identified, it is the responsibility of the therapist to determine if the device might be appropriate, to evaluate the use of the device specifically with the person, and then provide the training to ensure safe and efficient performance. Devices that are frequently prescribed by an occupational or physical therapist to assist with ambulation are described below. In the United States, the therapist, in consultation with the client and caregiver, recommends the device(s) to the physician, who then writes a prescription for insurance coverage. Determining if the patient's health insurance will cover the device should always be explored prior to initiating the purchase. This process is likely to be different in other countries.

13.4.3.3.1 Commonly Prescribed Devices for Ambulation and Walking Mobility

Straight canes and small- or wide-base quad canes (see Figure 13.2) are the most commonly prescribed unilateral assistive devices for ambulation. The cane is often utilized to compensate for balance deficits as it provides additional support and sensory feedback through the upper extremity to assist with balance processing. A cane may also be prescribed when weakness and spasticity are present in one lower extremity.

FIGURE 13.2
An example of a quad cane.

FIGURE 13.3
An example of Nordic walking poles.

Nordic walking poles can be used as a unilateral or bilateral device. They are beneficial for individuals with mild balance deficits, especially with head scanning, but without weakness or spasticity. Nordic walking poles are not recommended for people with ataxia or people whose balance deficits place them at risk for falling (see Figure 13.3).

Axillary crutches are rarely prescribed for people with MS; however, forearm crutches (i.e., Canadian or Loftstrand crutches) can assist individuals with balance dysfunction or weakness and spasticity in both lower extremities. The advantage of the forearm crutch is that the support provided around the wrist and elbow enhances the person's ability to bear weight through the crutch, which provides greater stability. Like the Nordic walking pole, the forearm crutch can be used as a unilateral or bilateral device (see Figure 13.4).

The standard walker, the two- or three-wheeled walker and the four-wheeled walker with a seat are prescribed for individuals who require bilateral upper-extremity support (see Figure 13.5). Use of a wheeled walker can enhance the community mobility of individuals with balance problems. A four-wheeled walker with a seat offers a person with fatigue the opportunity to sit down and rest when he or she needs to do so. A three-wheeled walker is easier to fold and transport and is good for some people with balance deficits, but it does not have a seat. The two-wheeled walker is used when more stability is required from the device or when the person is unable to control the handbrakes on three- or four-wheeled walkers. The standard walker provides the most stability; however, to advance the walker the person must pick it up and place it forward. A physical therapist should be involved in evaluating which type of walker is best suited for safe and independent mobility.

The ankle foot orthotic (AFO) is the device designed to assist in lifting the foot into dorsiflexion, which facilitates the swing phase of gait. It can be carbon or plastic. The

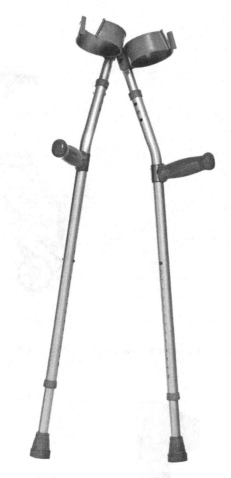

FIGURE 13.4
An example of forearm crutches.

plastic AFO may be a solid ankle or can be fitted with a joint to better control dorsiflexion and genu recurvatum. A carbon AFO is lighter than its plastic counterpart (see Figure 13.6). The weight of the AFO may influence the individual's ability to advance the extremity forward and must be considered by the physical therapist. The Dictus Band or the Foot-Up is a possible alternative to an AFO for some individuals (see Figure 13.7). These devices are very lightweight but only provide an assist to the anterior tibialis in dorsiflexing the foot for the swing phase. They do not provide any type of support to the ankle nor do they assist in controlling genu recurvatum. The Hip Flexion Assist Orthotic (HFAO) (see Figure 7.1 in Chapter 7) provides additional assistance in flexing the ankle, hip, and knee for the swing phase. Similar to the Dictus Band and Foot-Up, the HFAO does not support the ankle or control genu recurvatum. The patient, physical therapist, and orthotist must work closely to determine which orthotic provides the best gait assistance.

Wireless functional electric stimulation, such as the Walk-Aide and Bioness (Ness L300), predominantly function to activate the muscles for dorsiflexion (see Figure 7.3 in Chapter 7). Patients seeking these devices must be evaluated by a physical therapist and orthotist. Not

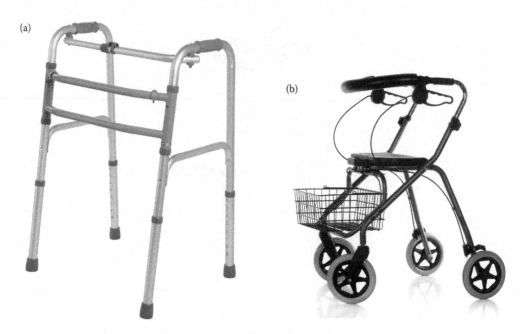

FIGURE 13.5
An example of a standard walker (a) and a four-wheeled walker with a seat (b).

FIGURE 13.6
An example of a carbon ankle foot orthotic (AFO). (Courtesy of Trulife.)

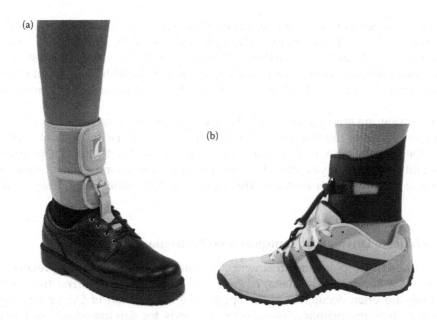

FIGURE 13.7
An example of the Foot-Up (a) and the Dictus Band (b). (Foot-Up image courtesy of Össur Americas, www.ossur. com. Dictus Band image courtesy of Erimed.)

everyone can tolerate the degree of electrical stimulation required to activate the muscle, and the effectiveness of the device must be evaluated with each individual patient. Ideally, each available device should be trialed with the person over four to five sessions in the clinic to determine which device is most effective.

13.4.3.3.2 *Commonly Prescribed Devices for Community Mobility (Scooters, Wheelchairs)*

For mobility in the community, there are many options. Examples include a transport chair, a manual wheelchair, a scooter, a motorized wheelchair with or without a tilt-in-space option for pressure relief, and wheelchair standers. A transport chair is most often prescribed for the individual who is limited in ambulation and may have a scooter or a manual or motorized wheelchair at home. The advantage of the transport chair is that it is lightweight and portable, so it can easily be placed in the trunk of a car. However, these chairs provide limited trunk support and, therefore, an individual should not be sitting in one for an extended period of time.

A manual wheelchair can be used in the home as well as in the community. These chairs come in different sizes that can accommodate people of different heights and weights. Typically, an individual propels him- or herself in the chair, but a caregiver can also push the chair. The manual wheelchair can be fitted with removal desk arms for lateral transfers, and leg rests may be elevated, swing-away, and removable. This type of chair should be considered for someone with adequate upper extremity range of motion and the strength needed to propel it.

The scooter is used most often for the individual who is safe and independent in walking at home but is limited in community ambulation due to fatigue and limited endurance. One disadvantage of the scooter is that the seat provides limited trunk support. Therefore, scooters should not be considered for people with poor trunk control.

Motorized or power wheelchairs are prescribed for individuals who are nonambulatory, have limited upper-extremity strength and function, and limited endurance. A motorized wheelchair can be customized to the person's needs for trunk support, risk of skin breakdown, and control methods (e.g., joy stick). As with the manual wheelchair, removal armrests can be added as well as elevating, swing-away, and removable leg rests.

When considering any type of wheeled device, the therapist must also take into consideration progression of the disease and the person's ability to use the mobility device over time. In the United States, health insurance companies will not authorize payment for a new mobility device each year; therefore, all recommendations must consider the patient's needs for at least two to three years. This process will be different in other parts of the world.

13.4.4 Reducing Driving and Transportation Restrictions

As with assessment, planning treatment and interventions for driving-related restrictions are most appropriately managed by a therapist with specialized training. These individuals are often occupational therapists. The Association of Driver Rehabilitation Specialists[115] lists the primary intervention methods for driving-related restrictions as follows:

1. Classroom driver education and training (e.g., driver improvement courses, simulator training)
2. Behind-the-wheel driver education and training
3. Reduction of impairments to optimize driving performance
4. License restrictions
5. Prescription and training for the use of specific vehicle modifications (e.g., hand controls) and adaptive equipment
6. Client fitting and training (the *CarFit* educational program is an example of this intervention strategy)[116]
7. Education of family members and caregivers for vehicle modifications and operation

To date, evaluations of driving-related interventions have not been documented in the MS research literature. Education on alternative transport options, such as driver-cessation programs,[117] also need to be developed and investigated for people with MS.

13.5 Rehabilitation for Mobility Restrictions: Case Description

Addressing mobility restrictions among people with MS is complex and multidimensional. Comprehensive management will involve multiple members of the rehabilitation team. The case of Megan illustrates and integrates the content that has been presented in this chapter.

CASE STUDY

Demographic/Social Overview/Medical History

Megan is a 45-year-old Caucasian female who was diagnosed with multiple sclerosis when she was 18 years of age. She resides in a downstairs apartment in a home owned by her sister (who resides on the second floor). There are two steps with a handrail at the side door to enter the home. Megan is single and lives alone, but her sister and brother-in-law see her at least twice day. She was independent in all activity and participation, was working part time, and was driving her own car to work. She used handicap parking as her ambulation is limited to 200 ft with a straight cane. Megan has Medicare and supplemental insurance through her part-time employment.

Recently Megan developed a systemic infection that exacerbated her disease process, requiring her to be hospitalized for eight days. She was transferred to a medical rehabilitation unit. Prior to the hospitalization, she was ambulatory with a straight cane and used a hip flexion assist orthotic (described in this chapter). Her EDSS prior to the exacerbation was 6.0, but community mobility was becoming limited.

Mobility Restrictions Leading to Home Care Rehabilitation

After the exacerbation, Megan received inpatient rehabilitation in a medical rehab unit for three weeks. She was discharged to home with nursing service, an aide two hours per day, occupational therapy, and physical therapy. Her ambulation is now limited to 50 ft with a two-wheeled walker, hip flexion assist orthotic, and an articulated ankle foot orthotic. She is able to ambulate independently in her home with the assistive/adaptive equipment. Minimal assistance/contact guard is needed to climb the two stairs to enter her apartment. All transfers, sit-to-stand, toilet, and bed mobility are safe and independent. She requires occasional assist with set-up for ADL and dressing activity and is independent in hygiene with set-up. She has stopped driving.

Providing Comprehensive Care: Concerns for the Rehabilitation Team

Megan has regressed considerably with this exacerbation. Prior to the systemic infection, she was stable though was showing very slow deterioration, consistent with secondary progressive MS. This recent illness and loss of independence in activity and participation has affected her physically, psychologically, and emotionally.

Family members have increased their surveillance of Megan from a functional and medical perspective and have intervened in her medical management. In this case, the family's involvement with Megan has caused more tension and stress. The family contacted Megan's neurologist to request a neuropsychology consult. It was determined that Megan was capable of managing her own affairs and medical management, and, though struggling with executive-function activity, she had strategies to manage her daily routine and activity.

Two home modifications were made prior to Megan's discharge from the medical rehab unit to home:

- Raised toilet seat with grab bars
- Bathtub bench with handheld showerhead

Megan is now looking for an apartment with a completely accessible entrance, wheelchair access in all rooms, if needed, and a fully accessible bathroom. Part of her motivation is to move out of her sister's home as the recent exacerbation has contributed to changes in family dynamics. The apartment Megan is pursuing has been specifically designed to accommodate adults living in the community who have physical limitations. Megan has also identified that she would like to be evaluated for driver retraining as she does not want to be a burden to her family with transportation needs to outpatient therapy and possible return to part-time employment.

Action Steps

1. Physical therapy evaluation and treatment with plan to progress to community mobility to enable referral to outpatient therapy
2. Occupational therapy evaluation and treatment with plan to progress to community participation, outpatient therapy, and possible return to work
3. Medical social work evaluation with Megan and her family to begin to address the family tension and Megan's concerns about burden of care
4. Occupational therapy referral to driver retraining program, if appropriate
5. Effective communication among all rehabilitation members, Megan, her homecare aide, and her family about the specific activities she is capable of doing independently and those for which she requires supervision or assist

It is important for the rehabilitation team to provide instruction concurrently to Megan and her caregiver regarding the consistent use of adaptive/assistive devices, how to get up from a fall, and how the caregiver can assist in all transfers if needed. In this situation, the therapist should physically demonstrate to Megan how the caregiver should assist her in getting up from the floor and assisting in basic transfers, such as bed to wheelchair and wheelchair to toilet, if the need arises. After the therapist has demonstrated the proper technique, Megan and the caregiver should practice with the therapist's supervision and guidance.

Physical Therapy Assessment Process, Findings, and Treatment Plan for Home Care Rehabilitation

Megan's sister was present during the examination, which was conducted in Megan's first-floor apartment.

1. Review of Systems: intact, blood pressure 136/92, heart rate 90
2. Neurologic impairments: Weakness is present in the left upper extremity with weakness and spasticity in the left lower extremity (patient is right-hand dominant).
3. Manual muscle test:
   Left upper extremity 3/5
   Left lower extremity iliopsoas, gluteus medius, and anterior tibialis 2/5
   Quadriceps, hamstrings, and gastroc-soleus 3/5
4. Modified Ashworth Scale:
   Left lower extremity adductors 2/4, quadriceps and hamstrings +1/4,
   Gastrocnemius and soleus 2/4 (4 the muscle is rigid).
5. Sensation and proprioception intact
6. Coordination intact, standing balance limited by weakness and spasticity in the left LE requiring one-hand support for static and dynamic standing

7. Equilibrium and righting reactions intact in sitting, trunk strength 4/5
8. The standard walking outcome measures were performed with the following scores:

Measures	Findings
25' Timed walk	19.4 s with two-wheeled walker (from living room down hallway)
TUG	32.3 s
DGI	12/24 *fall risk
2-min walk	NT
6-min walk	NT
MSWS-12	50/60 (12 = not limited, 60 = extremely limited)
PDDS	4 (0 = normal, 8 = bedridden)

9. Other findings: Cognitive processing is impaired primarily with inattention, and there is evidence of an inappropriate euphoria.

Treatment goals, intervention, outcome for home care PT: Megan was seen 3×/week × 4 weeks

Goals	Treatment
Ambulate safe and independently with an assistive device for 25 ft in 17.0 s *Achieved in 2 weeks*	Repeated step-up exercise to strengthen muscles of swing phase and endurance Repeated gait training in the home with verbal cues to vary speed Active range of motion of ankle anterior tibialis in sitting and iliopsoas (hip flexion), progress to standing with hand support at counter for balance
Perform Timed Up and Go test safe and independently with an assistive device in 29 s *Achieved at discharge*	Practice short-distance ambulation (5–10 ft) with repeated 180° turns in both directions (left and right) Repeated sit to stand off standard chair, progressing to therapy ball Gastrocnemius and soleus weight-bearing stretch on stair, incline board, or at wall (runner's stretch) Side-lying stretch and active range of motion of iliopsoas muscle
Move from a sitting to standing position 5× with two hands to push off the chair and no hand support in standing (This is an activity item on the BBS or Equiscale that would build her motor and balance skill to perform this specific task.) *Achieved in 2 weeks*	Bridging in supine progressing to unilateral bridge Step down one stair eccentrically working quadriceps muscles Sit to stand from elevated surface (sitting on a wedge on low mat table) Sit to stand off therapy ball
Ambulate outdoors 125 ft with assistive device safe and independently *Achieved at discharge*	Gait training outdoors with assistive device; verbal cues and assist as needed (repetitive practice on even and uneven surfaces)
Ascend and descend side stairs safe and independent with handrail *Achieved at discharge*	Instruct in stair climbing, repetitive practice with verbal cues and assist as needed

Epilogue: Megan did participate in outpatient therapy and driver retraining and has now returned to work part time driving her own car. She still lives in the first-floor apartment of her sister's home.

13.6 Gaps in Knowledge and Directions for the Future

As this chapter attests, rehabilitation for mobility restriction is a broad and complex area. While mobility for people with MS can be maintained or improved through aggressive rehabilitation and application of technology, much needs to be learned. Research is needed regarding the efficacy and effectiveness of task-specific repetitive training across the full range of mobility activities, including position changes, transfers, carrying and manipulating objects, and walking. Currently, the knowledge of this intervention approach comes primarily from stroke rehabilitation. Because of the progressive nature of MS, therapists need guidance about how to determine when motor retraining may be inappropriate and when compensatory strategies should be used.

Other directions for the future of MS-specific mobility research include driving screening, assessment, and interventions and determining the efficacy and effectiveness of ongoing "maintenance therapy" to slow down the disabling nature of this disease and maintain mobility functions as long as possible. Further work on interventions for preventing falls is also required, particularly because of the associations between restricted mobility and reductions in bone mineral density and fractures. Finally, there is a need for research into effective ways to promote and support the transition to using assistive devices (e.g., canes, grabbars, etc.) and progressing to more supportive devices over time. Individuals are often reluctant to begin using devices or to change to a more supportive device because the transition is often perceived as "giving in" to the disease. Education programs and strategies to assist with these transitions are needed to enable continued and safe mobility throughout the MS disease course.

References

1. World Health Organization (WHO). International classification of functioning, disability and health. Geneva: World Health Organization; 2001.
2. Heesen C, Böhm J, Reich C, Kasper J, Goebel M, Gold SM. Patient perception of bodily functions in multiple sclerosis: Gait and visual function are the most valuable. Multiple Sclerosis 2008;14:988–991.
3. Zwibel H. Contribution of impaired mobility and general symptoms to the burden of multiple sclerosis. Advances in Therapy 2009;26:1043–1057.
4. Sutliff M. Contribution of impaired mobility to patient burden in multiple sclerosis. Current Medical Research and Opinion 2010;26:109–119.
5. Mosley LJ, Lee GP, Hughes ML, Chatto C. Analysis of symptoms, functional impairments, and participation in occupational therapy for individuals with multiple sclerosis. Occupational Therapy in Health Care 2003;17(3/4):27–43.
6. Finlayson M, Dalmonte J. Predicting the use of occupational therapy services among people with multiple sclerosis in Atlantic Canada. Canadian Journal of Occupational Therapy 2002;69: 239–248.
7. Finlayson M, Plow M, Cho C. Use of physical therapy services among middle-aged and older adults with multiple sclerosis. Physical Therapy 2010;90:1607–1618.
8. Finlayson M, Winkler Impey M, Nicole C, Edwards J. Self-care, productivity and leisure limitations of people with multiple sclerosis in Manitoba. Canadian Journal of Occupational Therapy 1998;65(5):299–308.

9. Mansson Lexell E, Iwarsson S, Lexell J. The complexity of daily occupations in multiple sclerosis. Scandinavian Journal of Occupational Therapy 2006;13(4):241–248.

10. Finlayson M, Peterson E, Cho C. Risk factors for falling among people aged 45 to 90 years with multiple sclerosis. Archives of Physical Medicine and Rehabilitation 2006;87: 1274–1279.

11. Buchanan RJ, Martin RA, Wang S, Ju H. Analyses of nursing home residents with multiple sclerosis at admission and one year after admission. Multiple Sclerosis 2004;10:74–79.

12. Buchanan RJ, Martin RA, Zuniga M, Wang S, Kim M. Nursing home residents with multiple sclerosis: Comparisons of African American residents to white residents at admission. Multiple Sclerosis 2004;10:660–667.

13. Holper L, Coenen M, Weise A, Stucki G, Cieza A, Kesselring J. Characterization of functioning in multiple sclerosis using the ICF. Journal of Neurology 2010;257(1):103–113.

14. Iyengar V, Santos MJ, Ko M, Aruin AS. Grip force control in individuals with multiple sclerosis. Neurorehabilitation & Neural Repair 2009;23(8):855–861.

15. Iyengar V, Santos MJ, Ko M, Aruin AS. Effect of contralateral finger touch on grip force control in individuals with multiple sclerosis. Clinical Neurophysiology 2009;120(3):626–631.

16. Yozbatiran M, Baskurt F, Baskurt Z, Ozakbas S, Idiman E. Motor assessment of upper extremity function and its relation with fatigue, cognitive function, and quality of life in multiple sclerosis. Journal of Neurological Sciences 2006;246:117–122.

17. Chen CC, Kasven N, Karpatkin HI, Sylvester A. Hand strength and perceived manual ability among patients with multiple sclerosis. Archives of Physical Medicine & Rehabilitation 2007;88:794–797.

18. Confavreux C, Vukusic S, Adeleine P. Early clinical predictors and progression of irreversible disability in multiple sclerosis: An amnesic process. Brain 2003;126:770–782.

19. Courtney AM, Treadaway K, Remington G, Frohman E. Multiple sclerosis. Medical Clinics of North America 2009;93(2):451–476.

20. Simon J. MRI in multiple sclerosis. Physical Medicine and Rehabilitation Clinics of North America 2005;16(2):383–409.

21. Finlayson M, Preissner K, Cho CC, Plow MA. Randomized trial of a teleconference-delivered fatigue management program for people with multiple sclerosis. Multiple Sclerosis 2011;17(9):1130–1140.

22. Hohol MJ, Orav EJ, Weiner HL. Disease steps in multiple sclerosis: A simple approach to evaluate disease progression. Neurology 1995;45:251–255.

23. Iezzoni LI, Rao SR, Kinkel RP. Patterns of mobility aid use among working-age persons with multiple sclerosis living in the community in the United States. Disability and Health Journal 2009;2:67–76.

24. Cavanaugh JT, Gappmaier V, Dibble LE, Gappmaier E. Ambulatory activity in individuals with multiple sclerosis. Journal of Neurological Physical Therapy 2011;35:26–33.

25. Motl RW, Zhu W, Park Y, Mcauley E, Scott JA, Snook EM. Reliability of scores from physical activity monitors in adults with multiple sclerosis. Adapted Physical Activity Quarterly 2007;24:245–253.

26. Tudor- Locke C, Hatano Y, Pangrazi RP, Kang M. Revisiting "How many steps are enough?" Medicine & Science in Sports & Exercise 2008;40:S537–S543.

27. Robinett CS, Vondran MA. Functional ambulation velocity and distance requirements in rural and urban communities: A clinical report. Physical Therapy 1988;68(9):1371–1379.

28. Shumway-Cook A, Guralnik JM, Phillips CL. Age-associated declines in complex walking task performance: The walking InCHIANTI toolkit. Journal of the American Geriatrics Society 2007;55(1):58–65.

29. Gianfrancesco MA, Triche EW, Fawcett JA, Labas MP, Patterson TS, Lo AC. Speed- and cane-related alterations in gait parameters in individuals with multiple sclerosis. Gait & Posture 2011;33:140–142.

30. Chipchase SY, Lincoln NB, Radford KA. A survey of the effects of fatigue on driving in people with multiple sclerosis. Disability & Rehabilitation 2003;25:712–721.

31. Shawaryn MA, Schultheis MT, Garay E, DeLuca J. Assessing functional status: Exploring the relationship between the multiple sclerosis functional composite and driving. Archives of Physical Medicine & Rehabilitation 2002;83:1123–1129.
32. Schultheis MT, Weisser V, Manning K, Blasco A, Ang J. Driving behaviors among community-dwelling persons with multiple sclerosis. Archives of Physical Medicine and Rehabilitation 2009;90:975–981.
33. Schultheis MT, Manning K, Weisser V, Blasco A, Ang J, Wilkinson ME. Vision and driving in multiple sclerosis. Archives of Physical Medicine and Rehabilitation 2010;91:315–317.
34. Schultheis MT, Weisser V, Ang J, Elovic E, Nead R, Sestito N. Examining the relationship between cognition and driving performance in multiple sclerosis. Archives of Physical Medicine and Rehabilitation 2010;91:465–473.
35. Schultheis MT, Garay E, Millis SR, DeLuca J. Motor vehicle crashes and violations among drivers with multiple sclerosis. Archives of Physical Medicine and Rehabilitation 2002; 83:1175–1178.
36. Marcotte TD, Rosenthal TJ, Roberts E, Lampinen S, Scott JC, Allen RW. The contribution of cognition and spasticity to driving performance in multiple sclerosis. Archives of Physical Medicine and Rehabilitation 2008;89:1753–1758.
37. Ryan KA, Rapport LJ, Telmet Harper K, Fuerst D, Bieliauskas L, Khan O, Lisak R. Fitness to drive in multiple sclerosis: Awareness of deficit moderates risk. Journal of Clinical and Experimental Neuropsychology 2009;31(1):126–139.
38. Garber SL, Rintala DH, Holmes SA, Rodriguez GP, Friedman J. A structured educational model to improve pressure ulcer prevention knowledge in veterans with spinal cord dysfunction. Journal of Rehabilitation Research and Development 2002;39:575–588.
39. Deleu D, Alsharoqi I, Jumah MA, Tahan A, Bohlega S, Dahdaleh M, Inshasi J, Khalifa A, Szólics M, Yamout B. Will new injection devices for interferon beta-1a s.c. affect treatment adherence in patients with multiple sclerosis? An expert opinion in the Middle East. International Journal of Neuroscience 2011;121(4):171–175.
40. Minden SL, Frankel D, Hadden L, Hoaglin DC. Access to health care for people with multiple sclerosis. Multiple Sclerosis 2007;13:547–558.
41. Scheer J, Kroll T, Neri MT, Beatty P. Access barriers for persons with disabilities. Journal of Disability Policy Studies 2003;13:221–230.
42. Formica CA, Cosman F, Nieves J, Herbert J, Lindsay R. Reduced bone mass and fat-free mass in women with multiple sclerosis: Effects of ambulatory status and glucocorticoid use. Calcified Tissue International 1997;61:129–133.
43. Schwid SR, Goodman AD, Puzas JE, McDermott MP, Mattson DH. Sporadic corticosteroid pulses and osteoporosis in multiple sclerosis. Archives of Neurology 1996;53(8):753–757.
44. Cosman F, Nieves J, Komar L, Ferrer G, Herbert J, Formica C, Shen V, Lindsay R. Fracture history and bone loss in patients with MS. Neurology 1998;51:1161–1165.
45. Cumming RG, Nevitt MC, Cummings SR. Epidemiology of hip fractures. Epidemiologic Reviews 1997;19:244–257.
46. Cattaneo D, De Nuzzo C, Fascia T, Macalli M, Pisoni I, Cardini R. Risk of falls in subjects with multiple sclerosis. Archives of Physical Medicine and Rehabilitation 2002;83:864–867.
47. Peterson EW, Cho CC, von Koch L, Finlayson ML. Injurious falls among middle aged and older adults with multiple sclerosis. Archives of Physical Medicine and Rehabilitation 2008;89: 1031–1037.
48. Prosperini L, Kouleridou A, Petsas N, Leonardi L, Tona F, Pantano P. The relationship between infratentorial lesions, balance deficit and accidental falls in multiple sclerosis. Journal of the Neurological Sciences 2011;304:55–60.
49. Nilsagård Y, Lundholm C, Denison E, Gunnarsson LG. Predicting accidental falls in people with multiple sclerosis—A longitudinal study. Clinical Rehabilitation 2009;23:259–269.
50. Cameron MH, Poel AJ, Haselkorn JK, Linke A, Bourdette D. Falls requiring medical attention among veterans with multiple sclerosis: A cohort study. Journal of Rehabilitation Research and Development 2011;48:13–20.

51. Gosselink R, Kovacs L, Ketelaer P. Respiratory muscle weakness and respiratory muscle training in severely disabled multiple sclerosis patients. Archives of Physical Medicine & Rehabilitation 2000;81:747–751.

52. Howard R, Wiles C, Hirsh N. Respiratory involvement in multiple sclerosis. Brain 1992;115: 479–494.

53. Klefbeck B, Hamrah NJ. Effect of inspiratory muscle training in patients with multiple sclerosis. Archives of Physical Medicine & Rehabilitation 2003;84(7):994–999.

54. Buyse B, Demedts M, Meekers J, Vandegaer L, Rochette F, Kerkhofs L. Respiratory dysfunction in multiple sclerosis: A prospective analysis of 60 patients. European Respiratory Journal 1997;10:139–145.

55. Aiello M, Rampello A, Granella F, Maestrelli M, Tzani P, Immovilli P. Cough efficacy is related to disability status in patients with multiple sclerosis. Respiration 2008;76:311–316.

56. Motl RW, Goldman M. Physical inactivity, neurological disability, and cardiorespiratory fitness in multiple sclerosis. Acta Neurologica Scandinavica 2011;123(2):98–104.

57. Chiari T, Martin D, Davenport P. Expiratory muscle strength training in persons with multiple sclerosis having mild to moderate disability: Effect on maximal expiratory pressure, pulmonary function, and maximal voluntary cough. Archives of Physical Medicine and Rehabilitation 2006;87(4):468–473.

58. Green G, Todd J. 'Restricting choices and limiting independence': Social and economic impact of multiple sclerosis upon households by level of disability. Chronic Illness 2008;4(3):160–172.

59. Finlayson M, Van Denend T. Experiencing the loss of mobility: Perspectives of older adults with MS. Disability & Rehabilitation 2003;25:1168–1180.

60. Battaglia MA, Zagami P, Uccelli MM. A cost evaluation of multiple sclerosis. Journal of NeuroVirology 2000;6(Suppl 2):S191–S193.

61. Henriksson F, Fredrikson S, Masterman T, Jonsson B. Costs, quality of life and disease severity in multiple sclerosis: A cross-sectional study in Sweden. European Journal of Neurology 2001; 8:27–35.

62. Kobelt D, Berg J, Atherly D, Hadjimichael O. Costs and quality of life in multiple sclerosis: A cross-sectional study in the United States. Neurology 2006;66:1696–1702.

63. McCrone P, Heslin M, Knapp M, Bull P, Thompson A. Multiple sclerosis in the UK: Service use, costs, quality of life and disability. Pharmacoeconomics 2008;26:847–860.

64. Rotstein Z, Hazan R, Barak Y, Achiron A. Perspectives in multiple sclerosis health care: Special focus on the costs of multiple sclerosis. Autoimmunity Reviews 2006;5:511–516.

65. Whetten-Goldstein K, Sloan F, Goldstein L, Kulas ED. A comprehensive assessment of the cost of multiple sclerosis in the United States. Multiple Sclerosis 1998;4:419–425.

66. Naci H, Fleurence R, Birt J, Duhig A. Economic burden of multiple sclerosis. Pharmacoeconomics 2010;28:363–379.

67. Buhse M. Assessment of caregiver burden in families of persons with multiple sclerosis. Journal of Neuroscience Nursing 2008;40(1):25–31.

68. McKeown L, Porter-Armstrong A, Baxter G. Caregivers of people with multiple sclerosis: Experiences of support. Multiple Sclerosis 2004;10:219–230.

69. Finlayson M, Cho C. A descriptive profile of caregivers of older adults with MS and the assistance they provide. Disability and Rehabilitation 2008;30(24):1848–1857.

70. Buchanan RJ, Radin D, Chakravorty BJ, Tyry T. Informal caregiving to more disabled people with multiple sclerosis. Disability & Rehabilitation 2009;31:1244–1256.

71. Kurtzke JF. Rating neurological impairment in multiple sclerosis: An expanded disability status scale (EDSS). Neurology 1983;33:1444–1452.

72. Berg KO, Wood-Dauphinee SL, Williams JI, Maki B. Measuring balance in the elderly: Validation of an instrument. Canadian Journal of Public Health 1992;83:S7–S11.

73. Tesio L, Perucca L, Franchignoni FP, Battaglia MA. A short measure of balance in multiple sclerosis: Validation through Rasch analysis. Functional Neurology 1997;12(5):255–265.

74. Stroke Assessment Scales [Internet] [cited 2011 October 18]. Available from: http://www.strokecenter.org/trials/scales/index.htm.

75. Mahoney FI, Barthel DW. Functional evaluation: The Barthel Index. Maryland State Medical Journal 1965;14:61–65.
76. Beer S, Aschbacher B, Manoglou D, Gamper E, Kool J, Kesselring J. Robot-assisted gait training in multiple sclerosis: A pilot randomized trial. Multiple Sclerosis 2008;14:231–236.
77. Bovend'Eerdt TJ, Botell RE, Wade DT. Writing SMART rehabilitation goals and achieving goal attainment scaling: A practical guide. Clinical Rehabilitation 2009;23:352–361.
78. Khan F, Pallant JI, Brand C, Kilpatrick TJ. A randomised controlled trial: Outcomes of bladder rehabilitation in persons with multiple sclerosis. Journal of Neurology, Neurolsurgery, & Psychiatry 2010;81(9):1033–1038.
79. Bovend'Eerdt TJ, Dawes H, Sackley C, Izadi H, Wade DT. An integrated motor imagery program to improve functional task performance in neurorehabilitation: A single-blind randomized controlled trial. Archives of Physical Medicine and Rehabilitation 2010;91(6):939–946.
80. Grasso MG, Pace L, Troisi E, Tonini A, Paolucci S. Prognostic factors in multiple sclerosis rehabilitation. European Journal of Physical Rehabilitation Medicine 2009;45(1):47–51.
81. Granger C. Personal communication with the first author. 1990.
82. Uniform Data System for Medical Rehabilitation [Internet] [cited 2011]. Available from: http://www.udsmr.org/Default.aspx.
83. Kellor M, Frost J, Silberberg N, Iversen I, Cummings R. Hand strength and dexterity. American Journal of Occupational Therapy 1971;25:77–83.
84. Rudick RA, Cutter G, Reingold S. The multiple sclerosis functional composite: A new clinical outcome measure for multiple sclerosis trials. Multiple Sclerosis 2002;8:359–365.
85. Mathiowetz V, Weber K, Kashman N, Volland G. Adult norms for the Nine-Hole Peg Test of finger dexterity. Occupational Therapy Journal of Research 1985;5:24–38.
86. Oxford Grice K, Vogel KA, Le V, Mitchell A, Muniz S, Vollmer MA. Brief report—Adult norms for a commercially available nine hole peg test for finger dexterity. American Journal of Occupational Therapy 2003;57:570–573.
87. Rosti-Otajarvi E, Hamalainen P, Koivisto K, Hokkanen L. The reliability of the MSFC and its components. Acta Neurologica Scandinavica 2008;117:421–427.
88. Rudick R, Antel J, Confavreux C, Cutter G, Ellison G, Fischer J, Lublin F, Miller A, Petkau J, Rao S et al. Clinical outcomes assessment in multiple sclerosis. Annals of Neurology 1996;40:467–479.
89. Rudick R, Antel J, Confavreux C, Cutter G, Ellison G, Fischer J, Lublin F, Miller A, Petkau J, Rao S et al. Recommendations from the National Multiple Sclerosis Society Clinical Outcomes Assessment Task Force. Annals of Neurology 1997;42:379–382.
90. Cutter GR, Baier ML, Rudick RA, Cookfair DL, Fischer JS, Petkau J, Syndulko K, Weinshenker BG, Antel JP, Confavreux C et al. Development of a multiple sclerosis functional composite as a clinical trial outcome measure. Brain 1999;122:871–882.
91. Fischer JS, Rudick RA, Cutter GR, Reingold SC. The multiple sclerosis functional composite measure: An integrated approach to MS clinical outcome assessment. Multiple Sclerosis 1999;5:244–250.
92. Hutchinson B, Forwell SJ, Bennett SE, Brown T, Karpatkin H, Miller D. Towards a consensus on rehabilitation outcomes in MS: Gait and fatigue. International Journal of MS Care 2009;11(2):67–78.
93. Whitney SL, Marchetti GF, Schade A, Wrisley J. The sensitivity and specificity of the timed "up & go" and the dynamic gait index for self-reported falls in persons with vestibular disorders. Journal of Vestibular Research 2004;14(5):397–409.
94. McConvey JL, Bennett SE. Validity and reliability of the dynamic gait index in individuals with multiple sclerosis. Archives of Physical Medicine & Rehabilitation 2005;86(1):130–133.
95. Bennett S, Fisher N, Bromley LF, Decker C, Howley W, Rundell M. Validity reliability, and sensitivity of three gait measures in multiple sclerosis. International Journal of MS Care 2010;12(Suppl 1):10.
96. Hobart JC, Riazi A, Lamping DL, Fitzpatrick R, Thompson AJ. Measuring the impact of MS on walking ability: The 12 item MS walking scale (MSWS–12). Neurology 2003;60:31–36.

97. Hohol MJ, Orav EJ, Weiner HL. Disease steps in multiple sclerosis: A longitudinal study comparing disease steps and EDSS to evaluate disease progression. Multiple Sclerosis 1999;5(5):349–354.

98. Motl RW, Goldman MD, Benedict RH. Walking impairment in patients with multiple sclerosis: Exercise training as a treatment option. Journal of Neuropsychiatric Disease and Treatment 2010;16(6):767–774.

99. Lord S, Rochester L. Walking in the real world: Concepts related to functional gait. New Zealand Journal of Physiotherapy 2007;35(3):126–130.

100. Shumway-Cook A, Patla AE, Stewart A, Ferrucci L, Ciol MA, Guralnik JM. Environmental demands associated with community mobility in older adults with and without mobility disabilities. Physical Therapy 2002;82(7):670–681.

101. Korner-Bitensky N, Toal-Sullivan D, von Zweck C. Driving and older adults: Towards a national occupational therapy strategy for screening. Occupational Therapy Now 2007;9(4):3–5.

102. Vrkljan B, McGrath C, Letts L. Assessment tools for evaluating fitness to drive: A critical appraisal of evidence. Canadian Journal of Occupational Therapy 2011;78:80–96.

103. Patomella A, Tham K, Johansson K, Kottorp A. P-drive on-road: Internal scale validity and reliability of an assessment of on-road driving performance in people with neurological disorders. Scandinavian Journal of Occupational Therapy 2010;17:86–93.

104. Kotterba S, Orth M, Eren E, Fangerau T, Sindern E. Assessment of driving performance in patients with relapsing–remitting multiple sclerosis by a driving simulator. European Neurology 2003;50(3):160–164.

105. Bayona NA, Bitensky J, Salter K, Teasell R. The role of task-specific training in rehabilitation therapies. Top Stroke Rehabilitation 2005;12(3):58–65.

106. Sullivan KJ, Brown DA, Klassen T, Mulroy S, Tingting G, Azen SP. Effect of task-specific locomotor and strength training in adults who were ambulatory after stroke: Results of the STEPS randomized clinical trial. Physical Therapy 2007;8:1580–1602.

107. Kielhofner G. Model of human occupation. 4th ed. Philadelphia, PA: Wolters Kluwer/Lippincott Williams & Wilkins; 2008.

108. Gijbels D, Lamers I, Kerkhofs L, Alders G, Knippenberg E, Feys P. The armeo spring as training tool to improve upper limb functionality in multiple sclerosis. Journal of NeuroEngineering and Rehabilitation 2011;8(5), doi:10.1186/1743-0003-8-5.

109. Maitra K, Hall C, Kalish T, Anderson M, Dugan E, Rehak J, Rodriguez V, Tamas J, Zeitlin D. Five-year retrospective study of inpatient occupational therapy outcomes for patients with multiple sclerosis. American Journal of Occupational Therapy 2010;64(5):689–694.

110. Mark VW, Taub E, Bashir K, Uswatte G, Delgado A, Bowman MH, Bryson CC, McKay S, Cutter GR. Constraint-induced movement therapy can improve hemiparetic progressive multiple sclerosis. preliminary findings. Multiple Sclerosis 2008;14(7):992–994.

111. Gutierrez GM, Chow JW, Tillman MD, McCoy SC, Castellano V, White LJ. Resistance training improves gait kinematics in persons with multiple sclerosis. Archives of Physical Medicine & Rehabilitation 2005;86(9):1824–1829.

112. van den Berg M, Dawes H, Wade DT, Newman M, Burridge J, Izadi H, Sackley CM. Treadmill training for individuals with multiple sclerosis: A pilot randomised trial. Journal of Neurology, Neurosurgery & Psychiatry 2006;77(4):531–533.

113. Benedetti MG, Gasparroni V, Stecchi S, Zilioli R, Straudi S, Piperno R. Treadmill exercise in early multiple sclerosis: A case series study. European Journal of Physical Rehabilitation Medicine 2009;45(1):53–59.

114. Lo AC, Triche EW. Improving gait in multiple sclerosis using robot-assisted, body weight supported treatmill training. Neurorehabilitation and Neural Repair 2008;22(6):661–671.

115. CDRS Exam Handbook [Internet] [cited 2011]. Available from: http://www.driver-ed.org/files/Certification/2011_CDRS_Exam_handbook.pdf.

116. Carfit: helping mature drivers find their safest fit [Internet]; c2011 [cited 2011]. Available from: http://www.car-fit.org/carfit/FAQ.

117. Gustafsson L, Liddle J, Shunwei L, Hoyle M, Pachana N, Mitchell G. Participant feedback and satisfaction with the UQDRIVE groups for driving cessation. Canadian Journal of Occupational Therapy 2011;78:110–117.
118. Granger CV, Deutsch A, Linn R. Rasch analysis of the FIM mastery test. Archives of Physical Medicine and Rehabilitation 1998;79:52–57.
119. Chen CC, Bode RK. Psychometric validation of the manual ability measure-36 (MAM-36) in patients with neurologic and musculoskeletal disorders. Archives of Physical Medicine & Rehabilitation 2010;91(3):414–420.
120. Hudak P, Amadio PC, Bombardier C, The Upper Extremity Collaborative Group. Development of an upper extremity outcome measure: The DASH (disabilities of the arm, shoulder, and hand). American Journal of Industrial Medicine 1996;29:602–608.
121. The disabilities of the arm, shoulder, and hand (DASH) Score [Internet]; c2011 [cited 2011]. Available from: http://www.orthopaedicscore.com/scorepages/disabilities_of_arm_shoulder_hand_score_dash.html.
122. Cano SJ, Barrett LE, Zajicek JP, Hobart JC. Beyond the reach of traditional analyses: Using Rasch to evaluate the DASH in people with multiple sclerosis. Multiple Sclerosis 2011;17(2):214–222.
123. Jebsen RH, Taylor N, Trieschmann RB, Trotter MH, Howard LA. An objective and standardized test of hand function. Archives of Physical Medicine and Rehabilitation 1969;50:311–319.
124. Feys P, Duportail M, Kos D, Van Asch P, Ketelaer P. Validity of the TEMPA for the measurement of upper limb function in multiple sclerosis. Clinical Rehabilitation 2002;16(2):166–173.
125. Desrosiers J, Hebert R, Dutil E, Bravo G. Development and reliability of an upper extremity function test for the elderly: The TEMPA. Canadian Journal of Occupational Therapy 1993;60(1):9–16.
126. Drake AS, Weinstock-Guttman B, Morrow SA, Hojnacki D, Munschauer FE, Benedict RHB. Psychometrics and normative data for the multiple sclerosis functional composite: Replacing the PASAT with the symbol digit modalities test. Multiple Sclerosis 2010;16(2):228–237.
127. Bohannon RW. Six-minute walk test: A meta-analysis of data from apparently healthy elders. Topics in Geriatric Rehabilitation 2007;23(2):155–160.
128. Shumway Cook A, Baldwin M, Polissar N, Gruber W. Predicting the probability for falls in community-dwelling older adults. Physical Therapy 1997;77:812–819.
129. Hall CD, Schubery MC, Herdman SJ. Prediction of fall risk reduction as measured by dynamic gait index in individuals with unilateral vestibular hypofunction. Otology & Neurotology 2004;25:746–751.
130. Shumway-Cook A, Brauer S, Woollacott M. Predicting the probability for falls in community-dwelling older adults using the Timed Up &Go test. Physical Therapy 2000;80(9):896–899.
131. Podsiadlo D, Richardson S. The timed "UP and go": A test of basic functional mobility for frail elderly persons. Journal of the American Geriatrics Society 1991;39:142–148.
132. Nilsagård Y, Lundholm C, Denison E. Clinical relevance using timed walk tests and the 'Timed Up and Go' testing in persons with multiple sclerosis. Physiotherapy Research International 2007;12(2):105–114.

14

Self-Care

Susan Forwell and Amy Perrin Ross

CONTENTS

The ability to manage personal self-care is assumed from a very early age and is not a focus for healthy adults. Adults do not judge life successes, failures, fondest memories, or dire tragedies on their ability to participate in self-care occupations. When, however, an adult is unable to perform self-care or when he or she evaluates performance as unsatisfactory, exhausting, or arduous, the restriction to engage in self-care comes into sharp focus. One's dignity is challenged, identity as an adult is confronted, and ability to function in society is shaken. The immediacy of completing these tasks may supersede engaging in other meaningful occupations. When self-care participation restrictions emerge for persons with multiple sclerosis (MS), these become pressing concerns; meeting this challenge is where the weight of this chapter rests.

As defined by the World Health Organization (WHO) *International Classification of Functioning, Disability and Health* (ICF),[1] impairment of body functions (the physiological and psychological functions of body systems) and body structures (the anatomical parts of the body, such as organs and limbs) may contribute to restrictions in participating in self-care. In the lexicon of the ICF self-care participation restrictions are problems an individual may experience when involved in life activities and situations of self-care.

The term "self-care activities" represents a group of occupations including feeding, dressing, grooming, shaving, oral hygiene, showering, bathing, toileting (including bowel and bladder management), nail care, and medication management. Other terms that may be used to describe this group of occupations are activities of daily living (ADL), basic activities of daily living (B-ADL), personal care activities, daily life tasks, and self-maintenance occupations. This grouping of occupations is distinguished from instrumental activities of daily living (I-ADL) or activities related to home maintenance, such as meal preparation, laundry, vacuuming, and grocery shopping or caring for others, such as elder or child care. I-ADL is not included in this chapter but is addressed in Chapter 15.

For persons with MS, the range of self-care participation restrictions may be from occasional difficulty in one aspect of personal care to requiring full support to ensure that personal care needs are met. To gain a full understanding of both the scope of the restriction related to ability, importance, and temporal implications and the physical, social, and cultural context, it is essential to appropriately approach and intervene in self-care concerns. This chapter begins with a discussion of the epidemiology, symptoms, and context that affect self-care in MS, and then presents the evidence and best practices related to rehabilitation evaluation and intervention for self-care restrictions. In each of these sections, the various self-care occupations (feeding, dressing, grooming, etc.) will be highlighted. After reading this chapter, you will be able to:

1. Describe the self-care restrictions commonly experienced by people with MS, based on available evidence,

2. Discuss available assessment methods for identifying self-care restrictions, and

3. Discuss rehabilitation strategies that can be used to manage self-care restrictions, based on available evidence.

14.1 Epidemiology of Self-Care Participation Restrictions

Since the late 1990s, increased attention has been placed on enumerating abilities and restrictions of self-care and instrumental activities of daily living. Focusing on self-care activities, despite the sense of commonness of these occupations, has not been without complexities. Self-care participation research varies in tasks that might be included, the measures used to evaluate the level of self-care participation and their metrics, and the segments of the MS population included in studies. Several studies, however, have contributed to the scope of knowledge related to the self-care participation restrictions in MS. Table 14.1 summarizes a sample of these research studies. Collectively, these studies provide insight into the assistance required and restrictions experienced to engage in dressing, eating, toileting, grooming, bathing, and medication-management issues.

Based on these studies, cumulatively, the self-care occupations of bathing (38.6%) followed by (un)dressing (28.4%) and transfers (27.7%) present the most difficulty and need for assistance. It should also be noted, however, that across studies, 10–20% of the MS population experiences self-care restrictions in one or more self-care occupations. Given that the total MS population worldwide is approximately 2.5 million,[2] it can be conservatively estimated that 250,000–500,000 persons with MS experience self-care restrictions. The magnitude of the problem is underscored when there is a lack of services or assistance to provide support and much-needed assistance.

In a study of 47 adults with MS (mean age 49.4 years), a total of 366 occupations (median 8, range 3–15 per person) were reported as problematic.[3] Across all these occupations, 51% identified self-care difficulties, while 30% identified difficulties related to productivity and 19% to leisure. Of these, 21% were personal care occupations that were rated high for importance though scored low for both ability to perform the occupation(s) and satisfaction with performance. The personal care occupations that were most frequently reported as difficult in this study were dressing, showering, and grooming. It is interesting that men reported more problems with self-care occupations while no differences were found between men and women for productivity and leisure occupations.

A study by Rosenblum and Weiss[4] compared a group of healthy controls ($n = 26$) with a group of people with MS ($n = 50$) having a mean disease duration of 6.95 ± 5.3 years and Expanded Disability Status Scale (EDSS) of 3.18 ± 1.37 on performance in personal self-care activities, fatigue severity, and handwriting performance. No significant differences were found in performance for toileting, feeding, dressing, or grooming between groups. There was, however, a significant decreased ability in bathing for the MS group ($p = 0.02$). In another study, age and years since MS diagnosis were predictive of self-care restrictions,[5] although it is not clear that this finding was adjusted for age. The course of MS disease and the number of symptoms experienced were found to have a direct impact on participation in self-care activities. In terms of disease course, described for the study as relapsing–remitting, chronic progressive, and benign, persons in every group reported experiencing self-care restrictions, with the data showing that those participants with chronic progressive MS had significantly more self-care limitations.[5]

14.2 Symptoms That Influence Self-Care Participation

Research has shown that the number of limitations experienced correlated with the number of symptoms reported.[5] Several MS symptoms have been shown to affect performance in self-care activities or are influenced by participation in self-care. For example, self-care management was found to be related to the symptom of fatigue in work by Rosenblum and Weiss[4] while a study by Finlayson et al.[5] found that fatigue was not a factor in self-care

TABLE 14.1

Summary of Studies Investigating Frequency of Self-Care Participation Restrictions

	Studies								Total
	Finlayson et al., 1998[5]	Finlayson, 2002[74]	Månsson and Lexell, 2004[75]	Einarsson et al., 2006[76]		McDonnell and Hawkins, 2001[77]	Mosley et al., 2001[78]	Midgard et al., 1996[79]	
Study descriptors									
n	416	440	44	166		248	40	124	1478
Country	Canada	Canada and United States	Sweden	Sweden		Northern Ireland	United States	Norway	
MS population	Members of MS society with MS	Older than 55 years	EDSS[a] 6–8.5	Random sample of 2129 with MS		Selected counties, and definite MS	University MS center	Cohort of 139 with MS	
Measure	Study survey	Study survey	FIM[b]	Barthel[c]	Katz[d]	ISS[e]	Study survey	ISS[2]	
									Weighted mean %
Self-care activity	*Percentage of participants requiring assistance or dependent*								
Eating	7	15.5	16	22	8	5	—	9.4	12.4
Dressing/undressing	15	33.9	30	27	24	36	17.5	12	28.4
Bathing	24	40.3	32	26	26	68	10	16.2	38.6
Grooming	33 (cutting toenails)	—	2	9	—	16	—	8.6	20.4
Bladder management	—	—	18	23	20	—	—	—	21.9
Bowel management	—	—	7	14.5	—	—	—	—	12.9
Toilet use	14	—	11	20	18	—	—	—	15.4
Transfers	14 (bed)	—	25.7	16	16	48 (toilet, bed, chair)	—	—	27.7
Taking medications	15	24.2	—	—	—	—	—	—	19.8
Using telephone	7	12	—	—	—	—	—	—	9.6

a EDSS—Extended Disability Status Scale[80]; b FIM—Functional Independence Measure[91]; c Barthel—Barthel Index (BI)[87]; d Katz—Katz Extended ADL Index[76]; e ISS—Incapacity Status Scale from the Minimal Record of Disability.[81]

restrictions. In this latter study, however, the authors point out that this finding may be an artifact of the fatigue item used. Thorne and colleagues[6] found that self-care decisions or quality of daily life for persons with MS were markedly affected by available energy and reported that people with MS learned to "push through" fatigue to engage in valued occupations. Another symptom, intention tremor, has been shown to be extremely disabling and reported to interfere with the activities of daily life such as eating, drinking, grooming, dressing, and handwriting.[7,8] Impaired cognition is also viewed as a factor influencing self-reported participation[9] in self-care, particularly regarding medication management. In addition, the troublesome symptoms of spasticity, incoordination, speech, swallowing problems, and pain were found to adversely affect participation in self-care occupations.[5,10,11] There are several chapters in this book dedicated to the implications, assessment, and management of these and other symptoms experienced by persons with MS and are not repeated here. It should be emphasized, however, that accounting for these symptoms is essential in the overall assessment and management of self-care limitations.

14.3 Environments That Influence Self-Care Participation

Environmental factors related to health as defined by the ICF make up the physical, social, and attitudinal environment in which people live and conduct their lives.[1] With respect to self-care, environmental factors may result in the difference between ability to participate in self-care activities and being restricted from these tasks. When embarking on evaluation and intervention in the area of self-care, it is essential that the physical, social, attitudinal, and cultural contexts[12,13] as well as the health care environment surrounding the individual with MS are accounted for and built into rehabilitation strategies as their influence on the success and relevancy of intervention is considerable.[6,13]

14.3.1 Physical Environment and Self-Care Restrictions

The physical environment as related to self-care assessment and management with MS is central to gathering realistic information and ensuring appropriate strategies are integrated. The physical environment of one's home and that of public washrooms are examples of places that may present significant barriers and safety concerns in the performance of self-care activities. In the home, examples include attempting to bathe or shower safely, managing toileting tasks efficiently, and administering injectable medication with minimum difficulty. In public places, a good example is accessing and using public washroom facilities when the door weight, knob, or direction of the door swing create difficulties; the stall has inadequate space; and the location of toilet mechanisms are unreachable. Another physical environment issue for people with MS is related to temperature as there is evidence that an individual's heat sensitivity or intolerance may compromise performance, including self-care activities, and augment troubling symptoms of MS.[14,15]

Other aspects of the physical environment that should also be considered in self-care assessment and intervention include, but are not limited to, the amount of clutter, noise, and lighting in the presence of perceptual or visual–spatial disability; accessibility; and safety limitations. The scope of considerations for the effect of the physical environment on self-care participation can be enormous. In Chapter 21, an extensive discussion about the impact of environmental factors affecting people with MS is provided.

14.3.2 Social and Attitudinal Environment and Self-Care Restrictions

In terms of the social and attitudinal context of persons with MS, such as spouses, children, grandchildren, friends, neighbors, employers, and caregivers, it is imperative that their worries, goals, time availability, and perceived role and responsibility in self-care are considered.[16] As self-care occupations are engaged in daily, when assistance from the social network is required, their involvement in self-care assessment, decision making, goals, and subsequent strategies is essential.[17] Of critical importance is to understand the impact when those in the social network become the caregivers of their loved ones with MS. Typically, the caregiver is the spouse, who experiences a reduced quality of life and compromised physical and psychological well-being, and whose social life, financial situation, and career prospects are negatively affected as a result of the caregiver demands.[18] It is noteworthy that it has also been demonstrated that female caregivers tend to have more social support than male caregivers.[19] Given the implications for the social network, the rehabilitation self-care assessment and treatment must account for and routinely monitor the social network.[18]

14.3.3 Cultural Environment and Self-Care Restrictions

As a result of knowledge and sensitivity to the cultural context during self-care assessment, there may be the opportunity to overcome barriers. This is highlighted in the following: during a dysphagia assessment, a woman refused to cooperate when cold liquids or ice chips were introduced. To understand the reason(s) for refusal, the therapist considered the swallowing difficulty itself, food preferences, and language barriers. When talking with the woman's daughter, however, he realized that it was unrelated to any of these issues. Instead, it was based on the Chinese belief that warm food contributed to health and should be taken when ill while cold foods should not.[20] It is not expected that one understands diverse traditions and beliefs across cultures; rather, one needs an openness to consider culture as a relevant factor in self-care assessment and throughout the therapeutic process.[21] Chapter 23 focuses on cultural issues and provides a nuanced discussion on such considerations when working with people with MS and their families.

14.3.4 Health Care Environment and Self-Care Restrictions

In addition to the context of the individual and his or her world, the health care environment will affect the ability of the person with MS to have self-care limitations assessed and managed. There are diverse health care environments in which an individual with MS and his or her family may seek and receive services, including through a comprehensive team, a single or a limited number of professionals, as well as access to case management services, or self-management approaches.

For the comprehensive team approach, it is essential to have the right health care service in the right setting at the right time to maintain the highest level of function and quality of life. For example, the presence of weakness, spasticity, and pain may necessitate physical and occupational therapy services to enhance mobility and overall conditioning to decrease self-care restrictions. In the face of cognitive problems that interfere with self-care, speech therapists, occupational therapists, and psychologists may be consulted while social work services may assist when home care services are required to meet self-care needs.[22] A speech–language pathologist, occupational therapist, or dietician may be involved should swallowing problems be suspected. If self-care difficulties result from environmental restrictions, occupational therapy services may be used. A recent study that included a multidisciplinary, comprehensive rehabilitation program of both in- and outpatient interventions showed clinical and statistically significant improvement in several areas, including performance in self-care.[23]

Should the health care service be provided by a single physician or practitioner, it is necessary to establish access to a network of community rehabilitation services. An alternative is for the physician or practitioner to recommend local MS society resources that may assist in directing patients to rehabilitation services.

Case management is required when the coordination and management of multiple health and social services are required or when complex self-care and other needs demand the integration of numerous services. This comes into focus when there is a change in health, cognitive ability is compromised, the caregiver or social situation changes or is stressed, there is increased safety risk, or there is financial pressure.[22] The goal of case management is to improve functional and health outcomes and ensure efficient use of health care services and the system overall.

Self-management is the individual's ability to manage the symptoms, treatment, consequences, and lifestyle changes inherent with a long-term disorder.[24] Self-management of self-care issues is complementary and imperative to the success of comprehensive care, individual practitioners, and case-management health care environments. It has been shown that people with chronic conditions who have effective self-management skills have a higher quality of life and make better use of health care professionals' time.[25,26] On the other hand, those with MS who have poorly developed self-management skills feel less perceived control and higher uncertainty, which are associated with depression, hopelessness, and poor psychological adjustment.[27]

14.4 Self-Care Assessment Strategies

As a result of the magnitude of issues and factors that result from self-care restrictions, health care professionals, particularly occupational therapists, nurses, physical therapists, social workers, speech therapists, and psychologists, are best equipped to offer appropriate and specific assessments. Assessment of self-care restrictions are important to

- Ensure that issues are identified in the complex cognitive, social, affective, and physical MS context
- Determine the magnitude and impact of factors contributing to the disabling problem(s)
- Appropriately guide, establish, and target treatment priorities
- Have outcome measure(s) to attest to the efficacy of intervention
- Provide evidence of accountability and transparency as a licensed health professional

It is one thing to agree that a professional assessment of self-care ability is required, but it is a different matter to know the assessment or combination of assessments that should be employed. To adequately discuss the parameters of self-care assessment, it is important to understand the characteristics of self-care measures, available instruments, and the process for engaging in self-care assessment in MS. Each of the areas is discussed in this chapter.

14.4.1 Characteristics of Self-Care Measures

During a consensus conference, expert clinicians and researchers in MS developed criteria viewed to be essential for rehabilitation outcome assessments in MS.[28] These criteria

- Are time efficient in administration
- Have demonstrated metric integrity for MS
- Are applicable over the MS disease course
- Are self-report or clinician administered
- Have features of self-efficacy
- Identify factors contributing to issues
- Measure the impact on *activities* and *participation* as defined in the ICF[1]

Relating these criteria to self-care assessments, four characteristics should be considered when selecting and contemplating the use of self-care assessments. These characteristics include the level of detail of information collected (screen vs. in-depth); self-report versus performance based self-care assessments; standardization of self-care assessments; and MS-specific or across populations self-care assessments.

14.4.1.1 The "Screening" to "In-Depth" Assessment Continuum

This continuum refers to the level of detail captured in a self-care assessment. Screening assessments are used to identify the presence (or absence) of a difficulty and occasionally provide information on the severity of the problem. Screening tools do not provide detailed information about the problem; they merely confirm that one exists. These tools systematically cover the majority of topics across the target domain, can be carried out efficiently, and allow the practitioner to quickly focus on a problem and the need for a detailed assessment. The drawback to screening assessments is that they tend to be insufficient for directing intervention planning. Examples of self-care screening assessments include the Barthel Index (BI) and the MS ADL Scale (Table 14.2).

The other end of this continuum is the in-depth self-care assessment, the purpose of which is to identify contributing factors, details, safety issues, and quality of skills in engaging in self-care. The benefit of the in-depth assessment is that it isolates the factors that contribute to self-care restriction, is specifically relevant to the individual and context, and directs intervention goals and plans. A drawback to the in-depth assessment is that these tend to be time consuming. Examples of in-depth self-care assessments include the Assessment of Motor and Process Skills (AMPS) and the Klein Bell ADL Scale (Table 14.2).

14.4.1.2 Self-Report versus Performance-Based Self-Care Assessments

The assessment type can be characterized by the way the information is collected and the time frame of the assessment. In terms of methods of data collection, the first, the client report, is done in an interview format or a self-administered questionnaire and is based on the individual's subjective experience. This type of assessment uses recall of the past, such

TABLE 14.2

Examples of Self-Care Assessments

Assessment	Type of Assessment[a]	Type of Client	Time to Complete	Metrics in MS	Number of Items	Reference
MS ADL Scale	Screening, self-report	Adults with MS	5 min	Yes	15 items	Gulick[82,83]
Assessment of Motor and Process Skills (AMPS)	In-depth[b], performance based	For adults with motor or cognitive disability	30–45 min	Yes, for IADL but can be used for ADL, for example, dressing	2 tasks relevant to the person are chosen, demonstrated, and evaluated	Fisher,[84] Doble et al.,[85] Mansson & Lexell[75]
Barthel Index (BI)	Screening, self-report, although may be used as a performance-based assessment	Designed for adults with physical disabilities in acute, rehab, and community settings	5 min; with observation of performance more time is required	Yes	10 items in the original version; reduced to 5 items in large study with MS	Mahoney & Barthel,[86] Hobart & Thompson,[87] Salisbury et al.,[88] Van der Putten et al.[89]
Canadian Occupational Performance Measure (COPM)	Screening, self-report	Adults and children with disabilities	30–40 min (10–15 min for self-care items)	Yes	2 self-care items, each rated on three dimensions	Law et al.,[32] Dedding et al.,[90] Mansson Lexell et al.[3]
Dysphagia assessment	In-depth, performance based	Adults and children with swallowing problems	5–20 min (videofluroscopy is advised)	No	Varies according to the number of foods introduced	DePauw et al.[37]
Functional Independence Measure (FIM)	Screening, with self-report, medical record, or in-depth through performance observation	Adults and children with disabilities	15–40 min; with performance more time is required	Yes	18 items	Uniform Data System [online],[91] Keith et al.,[92] Sharrack et al.,[93] Van der Putten et al.[89]
Klein-Bell ADL Scale	In-depth, with observation of performance	Adults and children in the rehabilitation and community settings	Depending on person, 2 h if the majority of items are assessed	No; Tested with rehab population, spinal cord injury, and children	170 items	Klein & Bell,[94] Law & Usher[95]
Lawton and Brody's Physical Self-Maintenance Scale (PSMS)	Screening, self-report; in-depth with observation of performance	Elderly persons	20–30 min	No, although has been used with MS populations	6 items	Lawton & Brody,[96] Rosenblum & Weiss,[4] Edwards[97]
Self-Report Functional Measure (SRFM)	Screening, self-report	Adults with MS and spinal cord injury	10–15 min	Yes	20 items	Hoenig et al.[98,99]

a "Screening" refers to self-care assessments that identify the presence and potentially the severity of a self-care performance limitation.

b "In-depth" refers to self-care assessments that provide detail about the performance limitations and contributes to the intervention plan.

as in the last week or previous month. The benefit of this format is that it is convenient, efficient, and can be completed in most environments. In fact, when self-report assessments were compared to performance-based measures, the former demonstrated equivalent or better psychometric properties, ease of use and interpretation, and subject acceptability.[29] The limitation is accuracy in both the problems of recall and desire to give an optimistic rather than a realistic view of the situation. An example of client report self-care assessments is the Lawton and Brody's Physical Self-Maintenance Scale (PSMS) (Table 14.2).

The second type of self-care assessment is performance based; the individual demonstrates selected self-care occupations. The time frame for this assessment is here and now. The value of performance-based assessment is the accuracy of data and clarity for intervention while the drawback is the need for ensuring appropriate location, equipment, and sufficient time. Examples of performance-based self-care assessments are the AMPS and dysphagia assessment for determining the ability to swallow food and fluids.

14.4.1.3 Standardization of Self-Care Assessments

Assessments may range from highly tested and norm-referenced measures to nonstandardized tools. Self-care assessments that are standardized undergo various degrees of testing such that some tools may have established metric evaluation of some or all of the following features: reliability, validity, and responsivity. They may be tested with one population or several and in various settings, such as an acute hospital, rehabilitation center, outpatient clinic, or in the community. Ultimately, however, standardization refers to administrative practices in which a common protocol is adhered to across all administrators and a manual or training is provided as appropriate. The benefit, depending on the degree of standardization, is the ability to reproduce results and that they are widely understood.

Nonstandardized self-care assessments are typically "home grown," facility specific, or practitioner specific. While these target the precise issue, the marked limitation is the lack of confidence in reproducibility, difficulty in communicating findings, and difference between administrative practices. As a result of this limitation and the availability of several excellent tested measures, standardized assessments are strongly encouraged.

14.4.1.4 MS-Specific or Across-Populations Self-Care Assessments

A further characteristic to consider is the applicability of assessments to the MS population. Assessments may be (a) developed and tested for persons with MS (e.g., MS ADL Scale), (b) developed for another population and then tested with MS (e.g., Self-Reported Functional Measure), or (c) developed for another population and used with MS but not specifically tested (e.g., BI). Table 14.2 provides details about the applicability of a number of self-care assessments to the MS population.

14.4.2 Selection of Self-Care Assessment Used with People with MS

There is a broad array of self-care assessments available for use with the MS population. Table 14.2 provides an overview of assessments that have been specifically developed for, tested with, or used with the MS population. The references provided give the source for accessing the assessment as well as evidence for its metric testing.

There are two further groupings of assessments that should be noted and may contribute valuable information to understanding self-care restrictions. The first are assessments capturing overall functional ability to carry out roles and chosen activities and to have a perspective on quality of life. The second group includes assessments that evaluate specific aspects or components needed to carry out tasks. Assessments that collect information on overall functional ability and quality of life may have elements embedded in them that provide screening information about restrictions in self-care participation. Examples of such assessments include the Occupational Questionnaire,[30] the Occupational Self-Assessment,[31] the Canadian Occupational Performance Measure (COPM),[32] and the Multiple Sclerosis Impact Scale (MSIS-29).[33]

The second group of assessments are those that evaluate components that influence the ability to carry out tasks, like strength (manual muscle testing, dynamometry),[34] sensation,[35] fatigue (Fatigue Impact Scale),[36] swallowing (dysphagia assessment),[37,38] balance (Berg Balance Scale),[39] spasticity (Modified Ashworth Scale),[40] fine motor control (9-hole peg test),[41] and cognitive ability (Multiple Sclerosis Neuropsychological Screening Questionnaire [MSNQ], Minimal Assessment of Cognitive Function in Multiple Sclerosis [MACFIMS]).[42,43]

Both these types of assessments, those that capture overall functional ability and those related to the components needed to carry out tasks, are addressed in other chapters. It is critical, nonetheless, not to overlook these because they contribute to the assessment of self-care participation restrictions and determining appropriate interventions.

14.4.3 Process for Engaging in Self-Care Assessment

Evaluating self-care restriction(s) should adhere to a systematic process to confidently gather information required for treatment planning. Figure 14.1 provides an overview for the rehabilitation evaluation process as it relates specifically to self-care evaluation. There are three important stages shown in the figure—identify key concerns, screen for self-care restrictions, and assess self-care ability.

The first, identify key self-care concerns, is focused on getting to the primary problem(s) efficiently, honing in on the client's issues, understanding his or her perspective, and dealing with the most pressing issues. Ask the client or caregiver to describe the problem with questions such as: What is your primary concern about taking care of yourself? On what issues would you like to focus? Where would you like to begin? Once the key self-care concerns have been identified, the second stage is a screening to determine the presence of self-care restrictions and understand their severity, frequency, and impact. The skilled practitioner may complete this stage iteratively with the first stage. The third stage is completing thorough assessment(s) of self-care to obtain a comprehensive understanding and detailed information about factors contributing to restriction(s), to facilitate clinical reasoning, and to guide treatment decisions.

14.5 Rehabilitation Goals That May Arise from Self-Care Assessment Results

To illustrate the type of rehabilitation goals that may emerge as a result of self-care assessment, two vignettes are presented. From the outset, the setting of goals must be a collaboration between the individual with MS and his or her social network, as appropriate,

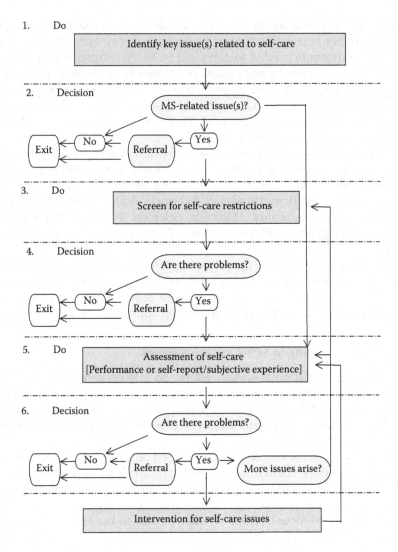

FIGURE 14.1
Rehabilitation evaluation process for self-care.

and with the assistance of the health professional (most commonly the occupational therapist or nurse, depending on the nature of the problem).

14.5.1 Vignette 1: Ellen

Ellen is a 36-year-old married woman who has two children, ages five and three; she works part time at the municipal library. Ellen continues to have residual problems related to pins and needles in her right, and dominant, hand (she calls it her "clumsy hand"), trouble walking because "balance is off," and marked fatigue. Ellen is worried that she is losing her abilities and wants to continue to work and make the "little things" in her daily life easier.

Based on a self-care screening using the Functional Independence Measure (FIM), augmented by the performance of elements in the FIM where difficulties were reported as well as administering component-based assessments (the Berg Balance Scale and a comprehensive fatigue assessment), the following self-care goals were developed:

- Establish a grooming routine that minimizes energy expenditure and accommodates her children's activities
- Perform nail care safely and efficiently
- Establish a safe, confident, energy-efficient method for showering

14.5.2 Vignette 2: Ricardo

Ricardo is 52-year-old recently divorced man who lives alone and has 2 h of home care services two times per week, primarily to assist with bathing and I-ADL activities. He works from home for a not-for-profit agency that provides him with a little extra income. He uses a walker or furniture to get around his apartment as he experiences increased lower extremity weakness. He has difficulty dressing himself and taking his medications due to upper extremity intention tremor. He also has difficulties because of disorganization and a lack of routine. Ricardo is adamant about living in his apartment and wants strategies so that he can continue to do so.

Based on a self-care screening using the MS ADL Scale, the performance-based AMPS, and several component-based assessments (one each for intention tremor, executive function, manual muscle testing, and 25-foot walking test), the following self-care goals were established:

- Develop strategies to facilitate independence, minimize embarrassment, and ensure success in eating and drinking
- Establish a routine to simplify (un)dressing with the assistance of the home care services
- Establish a system to ease medication management

To address the rehabilitation goals established for Ellen and Ricardo, the health care practitioners turned their attention to the intervention strategies for these self-care restrictions.

14.6 Interventions and Management of Self-Care Restrictions

To accomplish the goal(s) that resulted from the self-care screening and assessment, a diverse repertoire of interventions are used to enhance management of self-care performance. Self-managing self-care restrictions to maintain or improve participation is a daily challenge that requires decision making, planning, selecting from options related to symptom management, goal attainment, and seeking help from others.[6]

In a study designed to "examine the nature of everyday self-care decision making within its natural complexity,"[6] the goal was to better understand and support self-care management strategies. This qualitative study involved 21 participants, 7 with type-2 diabetes, 7 with HIV/AIDS, and 7 with MS. The results of this work suggested that the primary focus of decisions for those with MS was to gain control of the management of their disease using a process of fine-tuning and ongoing evaluation. Self-care decisions involved a need to maintain the aspects of life that were meaningful, balancing the health care providers' expectations against the practicality in their lives. There was a "commitment to controlling the disease rather than being controlled by it."[6] To assume control, active self-management strategies and routines were undertaken to assist others in becoming partners in helping to manage the disease. When self-care routines were successful, attention could shift away from managing their disease. When a disruption to these routines ensued, a focus on self-care needs would dominate.[6]

In the provision of self-care interventions, health professionals must use their expertise to assist the physical, social, and cultural context of individuals with MS to self-manage their disease, resulting symptoms, and participation restrictions. This section discusses self-care intervention strategies, addresses the important issues and changing landscape of medication management, and demonstrates the application of these for the vignettes of Ellen and Ricardo.

14.6.1 Self-Care Intervention Strategies

The focus of rehabilitation interventions with self-care strategies for persons with MS can be grouped into four types: *remediate and restore* ability to participate, *maintain* ability to perform self-care activities, *modify* approach to self-care to accommodate limitations, and *prevent* self-care restrictions. The strategies incorporated may involve not only the person with MS, but those with whom they live and others in their social network.

14.6.1.1 Remediate and Restore Participation Ability

This approach focuses on restoring and remediating function such that participation in self-care approximates the levels of independence similar to that prior to the onset of difficulties or decline. A retrospective chart review showed that persons with MS improved their FIM scores at discharge with increased intensity of occupational therapy using, in part, this remedial approach that included cognitive skills training, community reintegration, practice of specific self-care tasks, and neuromuscular education.[44] Results showed a positive effect for upper extremity dressing, bathing, and toileting transfers particularly related to self-care specific task practice. This study also showed that the neuromuscular education was not related to improved self-care performance.[44]

The following are examples of the remedial approach that might be used for persons with MS to restore self-care function:

- Establish a morning routine that is energy efficient and minimizes unnecessary activity
- Participate in a hand-function program that has dexterity, sensory, and fine motor coordination task training to facilitate independence in fine motor activities involved in dressing and grooming
- Engage in balance and gross motor exercises to enhance strength and range of motion to restore showering and bathing independence[45,46]

Another study that focused on attention deficits of persons with MS found that using the remedial approach of computer-assisted retraining produced significant beneficial effects that remained at the 9-week follow-up, reducing fatigue and physical slowness and enhancing mental processing.[47] The generalization of this technique to self-care, however, was not established, although it may be a consideration for future studies.

14.6.1.2 Maintain Ability to Participate

Using this approach, rehabilitation intervention provides supports that will allow clients with MS to preserve capabilities and prevent functional deterioration. A study examining the intensity of inpatient rehabilitation (occupational therapy and physical therapy) found a significant reduction in length of stay by 14 days for the group receiving intensive therapy.[48] A study that examined outpatient rehabilitation provided weekly for 1 year for persons with MS showed that it resulted in a reduction in fatigue, decrease in MS symptoms, and trend toward maintaining functional ability.[49] These studies suggest that with access to therapy those with MS are more able to maintain function and overcome limitations. From a rehabilitation perspective, strategies used are the provision of equipment/assistive devices and support.

14.6.1.2.1 Equipment and Assistive Devices

To maintain performance in self-care, a wide range of devices are used, the number and complexity of which increases with advanced stages of disease.[50] Prescriptions for mobility devices typically used by persons with MS to enhance self-care function include transfer boards, walkers, standing poles, and commodes. Self-care aids that may be recommended are grab bars, bath benches, shower chairs, toilet-safety arm supports, extended reachers, and weighted utensils. Weighted wrist cuffs to facilitate self-feeding in the presence of intention tremor have also demonstrated promise.[51] Further equipment that may reduce energy requirements include environmental controls such as on/off light sensors and door-locking/unlocking units.

Mathiowetz and Matuska[52] showed that 85% of the equipment/items recommended by occupational therapy to facilitate bathing and showering, toileting, feeding, grooming, and dressing were in continued use by persons with MS, and 65% used these items daily. This high use of equipment was explained by the strong mean satisfaction ratings of 4.6/5 with equipment. It has also been shown that the use of assistive devices, the most common being mobility aids and grab bars, was more likely if occupational therapy services had been provided, greater disability was present, and individuals were unemployed.[50] A meta-analysis showed that occupational therapy is efficacious in providing treatment to improve performance in the self-care activities of dressing and bathing.[45]

It should be noted that the success of equipment and assistive devices to maintain self-care performance lies in the individual's willingness and integration of these pieces into self-care routines. There may be resistance to incorporating these devices if there are feelings that these represent increasing disability, giving up, or simply that there is no need, despite the precarious methods used, to engage in self-care tasks. It has been shown, however, that the value of education is important for trying new strategies and for providing information on what works and does not work.[53] When faced with the person with MS who is opposed to using devices, the rehabilitation professional must reinforce that the individual has control over decisions and optimal methods to manage disability-related issues[22] while reinforcing that it is essential that the health care professional support good decision making by providing options for, education about, and practice experiences with recommended equipment and assistive devices.

14.6.1.2.2 Support

It is an imperative responsibility of the rehabilitation professional to listen and understand the gravity of concerns and the context of the individual with MS and his or her family.

There are various supports needed at any stage of the MS disease process related to self-care, including the following[21]:

- Providing the right amount of information,
- Clarifying misconceptions and misinformation,
- Coping with and managing the degenerative changes and potential changes, and
- Assisting with changing role changes.

Another form of support to maintain engagement in self-care is the use of service animals, 90% of which are professionally trained dogs. Research suggests both psychological and functional benefits, including increased self-esteem, independence in self-care and household activities, and improved community integration, such as attendance at school and part-time employment.[54] A further advantage was that the average number of paid assistance hours decreased by 68%.[54]

In addition to the individual with MS, caregivers also require support. When the burden of disability increases and family caregivers are involved in maintaining self-care, it is essential that their concerns are understood, that timely information is provided, and that there is an opportunity for respite.[17–19] Resources for local services and respite are valuable and should be understood and available in a timely manner.

14.6.1.3 Modify Approach to Self-Care

Using this approach, the rehabilitation intervention is primarily compensatory, modifying or adapting circumstances to facilitate participation. This is achieved by adapting the task, the environment, and the individual's approach to the task.[55] The compensatory approach does not attempt to restore function to previous levels but seeks to facilitate optimal engagement at the current level of function.

Benefits of a short-term (1 h/day for 2 weeks) occupational therapy and physiotherapy outpatient program for persons with MS that included compensatory strategies demonstrated improvement in activities of daily living through the provision of equipment, advice, and education on techniques; promoting normal posture and equilibrium reactions; and "damping" involuntary movement of the upper extremity and trunk.[56] In general, persons with MS may be able to compensate for self-care limitations through education on alternate ways of performing an occupation, modifying the environment, using adapted equipment, and splinting.

14.6.1.3.1 Education

In this context, education relates to applying general behaviors or principles across several self-care occupations. For example, one can address a series of energy-management principles like sitting rather than standing to engage in self-care[57]; selecting clothing that is easy to put on and take off such that there are no fasteners, or only those that are simple-to-use; employing proximal stabilization and hand-over-hand technique to self-feed with upper limb intention tremor[7]; and embedding routine and encouraging the development of habits that enhance efficacy and sense of control over daily life.[6]

Education can be completed using one-to-one, family, or peer sessions, and these can be done in person or using teleconference formats.[58,59] In particular, it was reported that providing two group sessions of 1–2 h over 1 month had a positive impact on mental health and fatigue levels of persons with MS as well as maintenance of self-care function. These groups included discussions on self-care strategies supported by an information booklet to be used outside the group.[60] In addition, print materials and online resources may also be helpful provided the sources are respected and reputable providers of MS information.

14.6.1.3.2 Physical Environment Modifications

The intervention plan may include modifications to relevant environments to allow persons with MS to continue to perform their self-care occupations. These modifications may be necessary to allow for wheelchair or scooter access; to augment safety; to reduce confusion or distraction in the presence of perceptual and cognitive difficulties; to minimize distances; and to support the work of a caregiver.[21] Most often modifications do not require structural changes but rather education related to altering furniture arrangement, changing the function of a room, rearranging the location of items, altering the height of frequently used items, minimizing the use of unsafe flooring coverings, reducing clutter, modifying lighting, and using devices for cueing.[21] Modifications to the physical environment have a substantial impact on function, as once the changes are made, the change is sustained if beneficial. There is no need to remember to do or wear something as the modification becomes part of the background in which performance occurs. Chapter 21 provides a detailed explanation of physical environmental modifications useful for people with MS.

14.6.1.3.3 Social Environment Modifications

The intervention plan may consider alterations to the routines and activities of the social network. For example, the routine of a spouse may change to accommodate to the assistance required of his or her loved one with MS, or a child may be asked to retrieve items upon request so as to reduce the amount of energy required to access needed articles.

14.6.1.3.4 Adaptive Equipment

The use of adaptive equipment to support continued engagement in self-care and compensate for limitations is advantageous as it minimizes demands in the presence of self-care restrictions related to MS symptoms. The following examples illustrate how adaptive equipment is integrated into a compensatory approach to optimize self-care function.

- Provide weighted utensils, a straw in the glass, dycem under the plate; elevate eating surface or precut portions to facilitate independence at mealtime[21]
- Support safety while showering by recommending use of soap on a rope and a small plastic chair in the tub to eliminate the need to stand and reach in the shower
- Recommend cooling garments including vests, wraps, and caps to cool or maintain body temperature, to be donned prior to or following engagement in self-care tasks[14,61,62]

14.6.1.3.5 Splinting

Both commercial and custom-made splints may be recommended or fabricated to facilitate self-care ability for persons with MS. These splints primarily involve the distal joints although there may be a need for neck support using a soft or semirigid collar. Splints for the wrist and fingers that may be used include volar wrist support in the presence of intention tremor or weakness; universal cuff or dorsal wrist support to eliminate the need to hold small items (i.e., fork or toothbrush) for extended time when weakness, sensory problems, or lack of coordination are present; and finger opposition splint to enable holding a pen.

14.6.1.4 Prevent Self-Care Restrictions

While preventing disease progression is outside the domain of rehabilitation, preventing or slowing functional deterioration is a unique and valued contribution. This is achieved through provision of equipment, support, and education. Empowering the client and family to use and build on strategies already in place will assist in preserving self-care function and preventing restrictions.[21]

A prevention focus in the management of self-care restrictions thwarts health problems and enhances safety for individuals with MS and those around them. A number of factors contribute to issues of health and safety, particularly related to preventing falls, supporting safe choices, and ensuring optimal urinary function, that may be related to cognitive and perceptual ability, vision, fear of falling, advanced physical disability, balance, gait or mobility impairment, and urinary incontinence.[63,64] For example, injurious falls have been reported to be over 50% in middle-aged and older adults with MS.[65] There are several programs that have been shown to have efficacy for fall prevention[66,67] that, while not tested for MS, may have potential application for the MS population. Another example of preventive management is learning to use a catheter when bladder problems are present to reduce the probability of bladder infections.

14.6.2 Medication Management

As a result of advances and developments of approved pharmaceuticals as well as widespread use of complementary and alternative methods, medication management has become an important and complex self-care focus. To account for and to ensure appropriate medication management, the issues related to its management are summarized.

14.6.2.1 Prescription Pharmaceuticals

For MS, these medications fall into three broad categories: acute/relapse management, symptomatic management, and disease-modifying therapies (DMTs). The cornerstone for managing acute MS exacerbations has long been corticosteroids to reduce inflammation in the brain and spinal cord.[68] The benefits of this course of therapy are that symptoms are typically relieved, severity and duration of a relapse is reduced, and they are generally taken for only 3–5 days, as tolerated. These do not usually result in self-management concerns due to their short duration.

For some symptoms of MS (e.g., spasticity, fatigue, urinary frequency, depression), there are pharmaceutical treatment options. These are typically in pill or tablet form and taken by mouth. Because these medications may be taken over months or years, adherence may be problematic due to the long duration, tolerance, side effects, organization of medications, and perceptions of efficacy. To ensure continued self-management of this group of medications, the multidisciplinary team approach includes education, easing organization (e.g., use of weekly or monthly dispensers), and maintaining communication to address concerns in a timely manner. The health care team must continually evaluate the need for, and response to, these medications. Adjustments in dose and types of medication are often needed as symptoms fluctuate.

The third group of prescription pharmaceuticals, DMTs, is relatively new, and while not curative, they are used to slow disease progression so that even with significant MS disease, there can be improved outlook and quality of life. Since 1993, six injectable therapies have been approved by the U.S. Food and Drug Administration. See Table 14.3 for a

TABLE 14.3

2010 Approved Disease-Modifying Therapies

Generic Name of Therapy (Trade Name)	FDA Approval (year)	Type(s) of MS[a]	Dosage	Schedule of Administration	Route[b] of Administration
Interferon beta–1b (Betaseron®)	1993	RR, CIS	0.25 mg, single vial	Every other day	Injected subcutaneously
Interferon beta–1b (Extavia®)	2009	RR, CIS	0.25 mg, single vial	Every other day	Injected subcutaneously
Interferon beta–1a (Avonex®)	1996	RR, CIS	6.6 miu or 33 mcg, single prefilled syringe	1× per week	Injected intermuscularly
Interferon beta–1a (Rebif®)	2001	RR	22 or 44 mcg, single prefilled syringe	3× per week	Injected subcutaneously
Glatiremer acetate (Copaxone®)	1997	RR, CIS	20 mg, single prefilled syringe	Daily	Injected subcutaneously
Natalizumab (Tysabri®)	2004/2006[c]	RR	330 mg	1× every 28 days	Intravenously
Fingolimod (Gilenya™)	2010	RR	0.5 mg tablet	Daily	Oral

[a] RR, relapsing–remitting form of MS; CIS, clinically isolated syndrome.
[b] Each of the subcutaneously injected therapies may be self-injected using an autoinjector or manually.
[c] In November 2004, the FDA approved Tysabri for use in the treatment of relapsing forms of MS. In February 2005, however, the drug was voluntarily recalled due to three cases of progressive multifocal leukoencephalopathy (PML). In July 2006, Tysabri was reintroduced under a restricted access classification and administered through the TOUCH program. This program requires patients, prescribers, pharmacists, nurses, as well as infusion centers to be registered to administer Tysabri.

summary of these medications and type of MS treated, dosage, and schedule and route of administration.

Collectively, the proven DMTs present MS patients and their families with new challenges, including self-injection and self-management of side effects. Side effects can influence adherence to regimens and include pain, skin breakdown, swelling, nausea, and interstitial necrosis. In addition to adverse effects, perceived lack of efficacy is a barrier to adherence such that it might be assumed by persons with MS that the medications are not working when symptoms do not resolve or new symptoms are experienced. The health care provider plays a pivotal role and must understand the priorities of the patient.

The decision to begin a DMT should include a discussion of approved treatment options, schedules of administration, and potential side effects. A proactive approach to address factors affecting adherence to medication regimens is essential and includes a multidisciplinary approach combining pharmacologic and nonpharmacologic treatment modalities. Proactive measures include a discussion about the evidence for DMTs, that treatment will be continuous and long term, that there are strategies to minimize adverse effects, and that involvement in support groups may be beneficial. These strategies must be accompanied with maintaining open and timely communication and managing treatment expectations. Comprehensive education of both the patient and their family members is imperative to set realistic expectations and discuss effective management strategies.

To assist health care professionals, a guideline was developed[69] that describes the challenging skin conditions that may arise (e.g., pain, erythema, lipoatrophy, infections, swelling, necrosis) with the use of these injectable agents and techniques to manage these conditions. This guideline also provides recommendations for injection techniques to minimize injection-site reaction. These recommendations include, for example, wash hands and clean skin thoroughly prior to injection, avoid massage before and after injection, avoid injection-site sun exposure, rotate site of injection, remove cap from needle immediately before injection to avoid contamination, apply ice or cold compress to site after injection, and do not use an area on the body for the injection where clothing may irritate the site.[69]

Recently, another DMT, fingolimod (Gilenya), has become available that, unlike the injectable DMTs, is taken orally (refer to Table 14.3). Yet to be approved are numerous other oral DMTs that are in various stages of research clinical trials. These include teriflunomide, daclizumab, and alemtuzumab (Campath).

14.6.2.2 Complementary and Alternative Methods/Therapies

Generally, people with MS hold the view that health professionals are resistant to the use of complementary and alternative (CAM) therapies.[6] Despite this, individuals with MS regularly make self-care decisions that include participating in trials or using numerous CAM therapies[6] such that 65–90% of persons with MS use at least one type of CAM, typically in combination with prescribed MS medications.[70–72]

Complementary therapies are used *in addition to* approved prescribed pharmaceutics while alternative therapies are used *in place of* prescribed pharmaceutics.[70] Reasons for choosing CAM are that conventional treatment did not seem effective; hearing anecdotal reports of CAM's efficacy; wanting to integrate an approach that involves the mind, body, and spirit; and doctor referral.[6,70] The most common types of CAM used by people with MS include exercise, vitamins, herbal and mineral dietary supplements, relaxation techniques, acupuncture, cannabis, and massage.[71] While health care professionals want to encourage patients to maintain a sense of hope, identifying realistic expectations is vital when discussing complementary and alternative therapies. A way to approach this is by educating the individual with MS and his or her family about evaluating the evidence for the efficacy in MS. A discussion about types of evidence to evaluate should include anecdotal evidence, laboratory evidence, animal evidence, observational evidence, and experimental clinical evidence. CAM therapy should provide a convincing level of evidence for its use while weighing the risks, costs, and effort involved. CAM therapy should not be used instead of proven medication for treatment of MS. The health care professional's respecting, however, the client's choice to use CAM therapy and providing information and resources about CAM may offer marked benefits to persons with MS as well as supporting ongoing treatment decisions and the therapeutic process.[70,73]

14.6.3 Applying Self-Care Intervention Strategies

To demonstrate the application of self-care intervention strategies, we return to the vignettes that illustrated rehabilitation goals related to self-care. These interventions are one course of action that the health care professional and the client may undertake to address these goals. As each individual's context, his or her expectations, and the resources available will vary, the strategies described must be taken as an example only.

14.6.3.1 *Vignette 1: Ellen*

To address Ellen's self-care goals, remedial, maintenance, modification, and prevention intervention approaches are used.

Goal 1. Establish a grooming routine that minimizes energy expenditure and accommodates to her children's activities

   Self-care intervention strategies

- *Support and education*: Determine the levels of grooming standards that are important for her (e.g., for days she is going to work, days she is not working, and special events)
- *Education and modification of approach*: Address the time and energy required and develop management strategy (e.g., plan time each day to do grooming, simplify grooming tasks doing more with less, sit, rest arm on raised surface to apply make-up)
- *Physical environment modification*: Mount mirror so that position during grooming is ergonomically optimal and least physically demanding
- *Social environment modification*: Involve the children in their own grooming or another activity near her while she is engaged in her grooming activities

Goal 2. Safely and efficiently perform nail care
Self-care intervention strategies

- *Adapt equipment*: Mount a nail clipper or recommend one that is stabilized on a base to reduce the need for fine motor manipulation when cutting nails
- *Modify approach*: Lay emery board on table to reduce need for fine manipulation when filing nails

Goal 3. Establish a safe, confident, energy-efficient method for showering
Self-care intervention strategies

- *Physical environment modification and adaptation of equipment*: Recommend a hand-held showerhead that is mounted on a pole to allow for height adjustment; to avert a loss of balance or a fall in the shower, advise on the use and location of grab bars, nonskid mat for the shower floor, and chair or bench to be placed in the shower to allow showering in a seated position

14.6.3.2 *Vignette 2: Ricardo*

To address the self-care goals of Ricardo, remedial, maintenance, and modification intervention approaches are used.

Goal 1. Develop strategies to facilitate independence, minimize embarrassment, and ensure success in eating and drinking

   Self-care intervention strategies

- *Support, education, and modification approach*: using the Step-Wise Approach for Treatment of Intention Tremor (SWAT-IT),[7] determine optimal behavioral techniques (e.g., proximal positioning, proximal stabilization, hand-over-hand) and assistive devices (e.g., weighed cuff, cooling wrap, splint) to minimize the effects of intention tremor while eating and drinking

Goal 2. Establish a routine to simplify (un)dressing with the assistance of the home care services
   Self-care intervention strategies

- *Education and modification approach*: Select clothes for daily use that are easiest to put on/take off (e.g., garments without fasteners, clothes that are easily pulled on, and fabrics with little resistance). Dress and undress in a seated position to avoid falls and save energy
- *Social environment modification*: Coaching the support workers to make clothing easily accessible for the next 2–3 days at the end of each of their visits

Goal 3. Establish a system to ease medication management
   Self-care intervention strategies

- *Support, education, and modification of approach*: Open communication to understand the barriers to taking medication and take action accordingly; an example of recommendations might include weekly or monthly pill dispensers/organizers, coach support workers to remind Ricardo to take medications, establish routine and/or cues that can be maintained. If there are difficulties related to DMT injection, it is imperative that the problem is identified to ensure appropriate recommendations (i.e., practice routine used for injection and make recommendations related to optimal techniques to be used for injections, or, if necessary, arrange to have someone else perform the injection, such as family member, home care nurse, or health care provider at an MS clinic)

14.7 Directions for the Future

In the last 15 years, there has been an exponential growth in research in the area of MS rehabilitation, a part of which has been dedicated to understanding and advancing issues related to self-care. Along with the advent of DMTs and the easy access to knowledge and networking through technology, the ability and resources to self-manage self-care restrictions has improved.

Notwithstanding these advancements, there remains much work and advocacy for the next 15 years. Beginning with the clinical assessment, there exists a tension between accessing the detailed information required to make informed clinical decisions and the limited time to collect this information. While some of the following notions are not new, they are necessary to consider in MS rehabilitation:

1. Patient self-report measures and assessments (including overall function, quality of life, specific components of function, and functional assessments related to self-care) are completed prior to or after meeting with the therapist or health care practitioner.
2. Patient-report measures are available in multiple locations and formats, including
   - In a computer application—an "app"

- On facility, public domain, or password-protected websites
- In paper format

3. Face-to-face sessions are reserved for
   - Initial interview and rapport building
   - Administration of performance-based measures
   - Interpretation of patient-report measures
   - Provision of intervention

4. Interactive computer technology can be used to observe patients' performance in their homes or work environments

5. Practitioners create an assessment toolbox to include a broad repertoire of assessments that are easily retrieved

6. Practitioners continue to "grow" their menu of assessments and outcome measures based on sound research in MS rehabilitation to support efficacy of treatment and clinical reasoning

In terms of MS rehabilitation interventions, there is a dire need to demonstrate the efficacy of treatments. Research must focus on providing the evidence for current intervention, when they should be used and understanding the reasons for success or failure. Topics requiring investigation related to self-care include, but are not limited to, the development of optimal self-management strategies and systems, fall prevention, generalization of remedial strategies (e.g., computer-assisted cognitive training, hand-function program, and balance exercises and vestibular training) for the benefit of self-care performance, and behavioral strategies to manage involuntary movement to facilitate self-care performance.

Gaps in the research related to self-care restrictions in MS include demonstrating the efficacy of task-specific practice strategies; showing the merits, drawbacks, and mechanisms for incorporating technology into easing daily occupations; integrating risk analysis; and providing information on cost-benefit aspects of treatment. To do this requires the use of study methods that are appropriate to rehabilitation intervention, including multiple single-subject designs, case studies, and controlled trials.

References

1. World Health Organization (WHO). International classification of functioning disability and health. Geneva, Switzerland: WHO; 2001.
2. Multiple Sclerosis [Internet]; c2010 [cited 2010 August 10]. Available from: http://www.cleveland-clinicmeded.com/medicalpubs/diseasemanagement/neurology/multiple_sclerosis/#cesec4.
3. Mansson Lexell E, Iwarsson S, Lexell J. The complexity of daily occupations in multiple sclerosis. Scandinavian Journal of Occupational Therapy 2006;13(4):241–248.
4. Rosenblum S, Weiss PL. Evaluating functional decline in patients with multiple sclerosis. Research in Developmental Disabilities 2010;31:577–586.
5. Finlayson M, Winkler Impey M, Nicole C, Edwards J. Self-care, productivity and leisure limitations of people with multiple sclerosis in Manitoba. Canadian Journal of Occupational Therapy 1998;65(5):299–308.
6. Thorne S, Paterson B, Russell C. The structure of everyday self-care decision making in chronic illness. Qualitative Health Research 2003;13(10):1337–1352.

7. Hawes F, Billups C, Forwell SJ. Interventions for upper limb intention tremor in multiple sclerosis: A feasibility study. International Journal of Multiple Sclerosis Care 2010;12(3):122–131.

8. Feys P, Romberg A, Ruutiainen J, Ketelaer P. Interference of upper limb tremor on daily life activities in people with multiple sclerosis. Occupational Therapy in Health Care 2003;17(3/4):81–95.

9. Goverover Y, Kalmar J, Gaudino-Goering E, Shawaryn M, Moore NB, Halper J, DeLuca J. The relationship between subjective and objective measures of everyday life activities in persons with multiple sclerosis. Archives of Physical Medicine and Rehabilitation 2005;86:2303–2308.

10. Gilmore R, Strong J. Pain and multiple sclerosis. British Journal of Occupational Therapy 1998;61(4):169–172.

11. Barnes MP, Kent RM, Semlyen JK, McMullen KM. Spasticity in multiple sclerosis. Neurorehabilitation and Neural Repair 2003;17:66–70.

12. Cooper B, Rigby P, Letts L. Evaluation of access to home, community, and workplace. In: C. A. Trombly, editor. Occupational therapy for physical dysfunction. 4th ed. Baltimore, MD: Williams & Wilkins; 1995.

13. Radomski MV. Assessing context: Personal, social and cultural. In: M. V. Radomski, C. Trombly Latham, editors. Occupational therapy for physical dysfunction. Baltimore, MD: Lippincott Williams & Wilkins; 2007.

14. Flensner G, Lindercrona C. The cooling-suite: Case studies of its influence on fatigue among eight individuals with multiple sclerosis. Journal of Advanced Nursing 2002;37(6):541–550.

15. Ku YT, Montgomery LD, Lee HC, Luna B, Webbon BW. Enhancing of neuroperformance by cooling in multiples sclerosis patients. Scandinavian Journal of Rehabilitation Medicine 2000;32(20):4–16.

16. Mansson Lexell E, Lund ML, Iwarsson S. Constantly changing lives: Experiences of people with multiple sclerosis. American Journal of Occupational Therapy 2009;63:772–781.

17. Finlayson M, Dahl Garcia J, Preissner K. Development of an educational programme for caregivers of people aging with multiple sclerosis. Occupational Therapy International 2008;15(1):4–17.

18. McKeown LP, Porter-Armstrong AP, Baxter GD. The needs and experiences of caregivers of individuals with multiple sclerosis: A systematic review. Clinical Rehabilitation 2003;17(3):234–248.

19. Good DM, Bower DA, Einsporn RL. Social support: Gender differences in multiple sclerosis spousal caregivers. Journal of Neuroscience Nursing 1995;27(5):305–311.

20. Forwell SJ, Whiteford G, Dyck I. Cultural competence in New Zealand and Canada: OT students' reflections on class and fieldwork curriculum. Canadian Journal of Occupational Therapy 2001;68(2):90–103.

21. Forwell SJ. Occupational therapy practice guidelines for adults with neurodegenerative diseases: Multiple sclerosis, Parkinson's disease, amyotrophic lateral sclerosis & transverse myelitis. 2nd ed. American Occupational Therapy Association Inc; 2006.

22. Northrop DE. Addressing the specialized needs of younger adults with multiple sclerosis. The Case Manager 2005;16(1):66–69.

23. Khan F, Pallant JF, Brand C, Kilpatrick TJ. Effectiveness of rehabilitation intervention in persons with multiple sclerosis: A randomized controlled trial. Journal of Neurology, Neurosurgery, & Psychiatry 2008;79:1230–1235.

24. Foster G, Taylor SJ, Eldridge SE, Ramsay J, Griffiths CJ. Self management education programmes by lay leasers for people with chronic conditions (review). Cochrane Database of Systematic Reviews 2007;(4). Art. No.: CD005108. DOI: 10.1002/14651858.CD005108.pub2.

25. Barlow J, Turner AP, Wright CC. A randomized controlled study of the arthritis self-management programme in the UK. Health Education Research 2000;15(6):665–680.

26. Lorig KR, Sobel DS, Stewart AL, Brown BW Jr, Bandura A, Ritter P. Evidence suggesting that a chronic disease self-management program can improve health status while reducing hospitalization: A randomized trial. Medical Care 1999;37(1):5–14.

27. Bishop M, Frain M, Trschopp M. Self-management, perceived control, subjective quality of life in multiple sclerosis: An exploratory study. Rehabilitation Counseling Bulletin 2008; 54(1):45–56.

28. Hutchinson B, Forwell SJ, Bennett SE, Brown T, Karpatkin H, Miller D. Towards a consensus on rehabilitation outcomes in MS: Gait and fatigue. International Journal of MS Care 2009;11(2):67–78.

29. Myers A, Holliday P, Harvey K, Hutchinson. Functional performance measures: Are they superior to self assessments? Journal of Gerontology 1993;48(5):M196–M206.

30. Kielhofner G. A model of human occupation: Therapy and application. 3rd ed. Philadelphia: Lippincott Williams & Wilkins; 2002.

31. Kielhofner G, Forsyth K. Measurement properties of a client self-report for treatment planning and documenting therapy outcomes. Scandinavian Journal of Occupational Therapy 2001;8:131–139.

32. Law M, Baptiste S, Carswell A, McColl MA, Polatajko H, Pollock N. The Canadian-occupational performance measure. 4th ed. Ottawa, Canada: CAOT Publications; 2005.

33. Riazil A, Hobart JC, Lamping DL, Fitzpatrick R, Thompson AJ. Multiple sclerosis impact scale (MSIS-29): Reliability and validity in hospital based samples. Journal of Neurology, Neurosurgery, & Psychiatry 2002;73:701–704.

34. Flinn N, Trombly Latham C, Robinson Podolski C. Assessing abilities and capacities: Range of motion, strength, and endurance. In: M. V. Radomski, C. Trombly Latham, editors. Occupational therapy for physical dysfunction. Baltimore, MD: Lippincott Williams and Wilkins; 2008.

35. Bentzel K. Assessing abilities and capacities: Sensation. In: M. V. Radomski, C. Trombly Latham, editors. Occupational therapy for physical dysfunction. Baltimore, MD: Lippincott Williams and Wilkins; 2008.

36. Fisk JD, Ritvo PG, Ross L, Haase DA, Marrie TJ, Schlech WF. Measuring the functional impact of fatigue: Initial validation of the fatigue impact scale. Clinical Infectious Diseases 1994;18: S79–S83.

37. De Pauw. Dysphagia in multiple sclerosis. Clinical Neurology and Neurosurgery 2002; 104(4):345–351.

38. Calcagno P, Ruoppolo G, Grasso MG, De Vincentis M, Paolucci S. Dysphasia in multiple sclerosis: Prevalence and prognostic factors. Acta Neurologica Scandinavica 2002;105(1):40–43.

39. Berg KO, Wood-Dauphinee SL, Williams JI, Maki B. Measuring balance in the elderly: Validation of an instrument. Canadian Journal of Public Health 1992;2:S7–S11.

40. Nuyens G, De Weerdt W, Ketalaer P, Feys H, De Wolf L, Hantson L, Nieuwboer A, Spaepen A, Carton H. Inter-rater reliability of the Ashworth Scale in multiple sclerosis. Clinical Rehabilitation 1994;8(4):286–292.

41. Fischer S, Rudick RA, Cutter GR, Reingold SC, National MS Society Clinical Outcomes Assessment Task Force. The multiple sclerosis functional composite measure (MSFC): An integrated approach to MS clinical outcome assessment. Multiple Sclerosis 1999;5(4):244–250.

42. Benedict RHB, Munschauer F, Linn R, Miller C, Murphy E, Foley F, Jacobs L. Screening for multiple sclerosis cognitive impairment using a self-administered 15-item questionnaire. Multiple Sclerosis 2003;9(1):95–101.

43. Benedict RHB, Coofair D, Gavett R, Gunther M, Munschauer F, Garg N, Weinstock-Guttman B. Validity of the minimal assessment of cognitive function in multiple sclerosis. Journal of the International Neuropsychological Society 2006;12(4):549–558.

44. Maitra K, Hall C, Kalish T, Anderson M, Dugan E, Rehak J, Rodriguez V, Tamas J, Zeitlin D. Five-year retrospective study of inpatient occupational therapy outcomes for patients with multiple sclerosis. American Journal of Occupational Therapy 2010;64(5):689–694.

45. Baker NA, Tickle-Degnan L. The effectiveness of physical, psychological and function interventions in treating clients with multiple sclerosis: A meta-analysis. American Journal of Occupational Therapy 2001;55:385–392.

46. Gauthier L, Dalziel S, Gauthier S. The benefits of group occupational therapy for patients with Parkinson's disease. American Journal of Occupational Therapy 1987;41:360–365.

47. Plohmann AM, Kappos L, Ammann W, Thordai A, Wittwer A, Huber S, Bellaiche Y, Lechner-Scott J. Computer assisted retraining of attentional impairments in patients with multiple sclerosis. Journal of Neurology, Neurosurgery, & Psychiatry 1998;64:455–462.

48. Slade A, Tennant A, Chamberlain MA. A randomized controlled trial to determine the effect of intensity of therapy upon length of stay in a neurological rehabilitation setting. Journal of Rehabilitation 2002;34:260–266.
49. DiFabio RP, Choi T, Soderburg J, Hansen CR, Schapiro RT. Extended outpatient rehabilitation: Its influence on symptom frequency, fatigue and functional status for person with progressive multiple sclerosis. Archives of Physical Medicine & Rehabilitation 1998;79:141–146.
50. Finlayson M, Guglielmello L, Liefer K. Describing and predicting the possession of assistive devices among persons with multiple sclerosis. American Journal of Occupational Therapy 2001;55:545–551.
51. McGruder J, Cors D, Tiernan AM, Tomlin G. Weighted wrist cuffs for tremor reduction during eating in adults with static brain lesions. American Journal of Occupational Therapy 2003;57:507–516.
52. Mathiowetz V, Matuska KM. Effectiveness of inpatient rehabilitation on self-care abilities of individuals with multiple sclerosis. NeuroRehabilitation 1998;11:141–151.
53. Holberg C, Finlayson M. Factors influencing the use of energy conservation strategies by persons with multiple sclerosis. American Journal of Occupational Therapy 2007;61(1): 96–107.
54. Allen K, Blascovich J. The value of service dogs for people with severe ambulatory disabilities. JAMA 1996;275:1001–1006.
55. Holm MB, Rogers JC, James AB. Treatment of activities of daily living. In: M. E. Neistadt, E. B. Crepeau, editors. Willard and Spackman's occupational therapy. 9th ed. Philadelphia: Lippincott; 1998.
56. Jones L, Lewis Y, Harrison J, Wiles CM. The effectiveness of occupational therapy and physiotherapy in multiple sclerosis patients with ataxia of the upper limb and trunk. Clinical Rehabilitation 1996;10:277–282.
57. Mathiowetz V, Matuska KM, Murphy ME. The effects of an energy conservation course for persons with multiple sclerosis. Archives of Physical Medicine and Rehabilitation 2001;82:449–456.
58. Finlayson M, Holberg C. Evaluation of a teleconference-delivered energy conservation education program for people with multiple sclerosis. Canadian Journal of Occupational Therapy 2007;74(4):337–347.
59. Mathiowetz VG, Matuska KM, Finlayson ML, Luo P, Chen HY. One-year follow-up to a randomized controlled trial of an energy conservation course for persons with multiple sclerosis. International Journal of Rehabilitation Research 2007;30(4):305–313.
60. O'Hara L, Cadbury H, De Souza L, Ide L. Evaluation of the effectiveness of professionally guided self-care for people with multiple sclerosis living in the community: A randomized controlled trial. Clinical Rehabilitation 2002;16(2):119–128.
61. Coyle PK, Krupp LB, Doscher C, Deng Z, Milazzo A. Clinical and immunological effects of cooling in multiple sclerosis. Journal of Neurologic Rehabilitation 1996;10:9–15.
62. Kinnman J, Andersson U, Wetterqvist L, Kinnman Y, Andersson U. Cooling suit for multiple sclerosis: Functional improvement in daily living?. Scandinavian Journal of Occupational Therapy 2000;32(1):20–24.
63. Finlayson M, Peterson E, Cho C. Risk factors for falling among people aged 45 to 90 years with multiple sclerosis. Archives of Physical Medicine and Rehabilitation 2006;87:1274–1279.
64. Cattaneo D, De Nuzzo C, Fascia T, Macalli M, Pisoni I, Cardini R. Risk of falls in subjects with multiple sclerosis. Archives of Physical Medicine and Rehabilitation 2002;83:864–867.
65. Peterson EW, Cho CC, von Koch L, Finlayson ML. Injurious falls among middle aged and older adults with multiple sclerosis. Archives of Physical Medicine and Rehabilitation 2008;89:1031–1037.
66. Yates SM, Dunnagan TA. Evaluating the effectiveness of a home-based fall risk reduction program for rural community-dwelling older adults. Journal of Gerontology 2001;56(4):M226–M230.
67. Pang MYC, Eng JJ, Dawson AS, McKay HA, Harris JE. A community-based fitness and mobility exercise program for older adults with chronic stroke: Randomized controlled trial. Journal of the American Geriatrics Society 2005;53:1667–1674.

68. Murray TJ. Multiple sclerosis: The history of a disease. New York: Demos Medical Publishing; 2005.

69. McEwan Lea. Best practices in skin care for the multiple sclerosis patient receiving injectable therapies. International Journal of MS Care 2010;12:177–1789.

70. Bowling AC. Complementary and alternative medicine in multiple sclerosis: Dispelling common myths about CAM. International Journal of Multiple Sclerosis Care 2005;7(2):42–44.

71. Olsen SA. A review of complementary and alternative medicine (CAM) by people with multiple sclerosis. Occupational Therapy International 2009;16(1):57–70.

72. Complementary and alternative medicine [Internet]; c2010 [cited 2010 April 28]. Available from: http://www.nationalmssociety.org/about-multiple-sclerosis/what-we-know-about-ms/treatments/complementary–alternative-medicine/index.aspx.

73. Bowling AC. Complementary and alternative medicine and multiple sclerosis. 2nd ed. New York: Demos Medical Publishing; 2007.

74. Finlayson M. Health and social profile of older adults with multiple sclerosis: Findings from three studies. International Journal of MS Care 2002;4(3):139–143, continued 148–151.

75. Månsson E, Lexell J. Performance of activities of daily living in multiple sclerosis. Disability and Rehabilitation 2004;26(10):576–585.

76. Einarsson U, Gottberg K, Fredrikson S, von Koch L, Holmqvist LW. Activities of daily living and social activities in people with multiple sclerosis in Stockholm County. Clinical Rehabilitation 2006;20(6):543–551.

77. McDonnell GV, Hawkins SA. An assessment of the spectrum of disability and handicap in multiple sclerosis: A population-based study. Multiple Sclerosis 2001;7:111–117.

78. Mosley LJ, Lee GP, Hughes ML, Chatto C. Analysis of symptoms, functional impairments, and participation in occupational therapy for individuals with multiple sclerosis. Occupational Therapy in Health Care 2003;17(3/4):27–43.

79. Midgard R, Gronning M, Riise T, Kvale G, Nyland H. Multiple sclerosis and chronic inflammatory diseases. A case-control study. Acta Neurologica Scandinavica 1996;93:322–328.

80. Kurtzke JF. Rating neurologic impairment in multiple sclerosis: An expanded disability status scale (EDSS). Neurology 1983;33:1444–1452.

81. LaRocca NG. Field testing of a minimal record of disability in multiple sclerosis: The United States and Canada. Acta Neurologica Scandinavia 1984;70(101):126–138.

82. Gulick EE. Parsimony and model of confirmation of the ADL self-care scale for multiple sclerosis persons. Nursing Research 1987;36(5):278–283.

83. Gulick EE. The self-administered ADL self-care scale for persons with multiple sclerosis In: The measurement of clinical and educational outcomes in nursing. New York, NY: Springer Publishers; 1988.

84. Fisher AG. Assessment of motor and process skills, vol 1. Development, standardization, and administration manual. Ft. Collins, CO: Three Star Press; 2003.

85. Doble SE, Fisk JD, Fisher AG, Murray TJ. Functional competence of community-dwelling persons with multiple sclerosis using the assessment of motor and process skills. Archives of Physical Medicine and Rehabilitation 1994;75(8):843–851.

86. Mahoney FI, Barthel DW. Functional evaluation: The Barthel Index. Maryland State Medical Journal 1965;14(61):65.

87. Hobart JC, Thompson AJ. The five item Barthel Index. Journal of Neurology 2001;71:225–230.

88. Salisbury A, Seebass G, Bansal A, Young JB. Reliability of the Barthel Index when used with older people. Age and Ageing 2005;34:228–232.

89. Van der Putten JJMF, Hobart JC, Freeman JA, Thompson AJ. Measuring change in disability after inpatient rehabilitation: Comparison of the responsiveness of the Barthel Index and the Functional Independence Measure. Journal of Neurology, Neurosurgery, & Psychiatry 1999;66(4):480–484.

90. Dedding C, Cardol M, Eyssen CJM, Beelen E, Beelen A. Validity of the Canadian Occupational Performance Measure: A client-centred outcome measurement. Clinical Rehabilitation 2004;2004(18):660–667.

91. Uniform Data System for Medical Rehabilitation [Internet]. Available from: http://udsmr.org/WebModules/FIM/Fim_About.aspx.
92. Keith R, Hamilton BB, Sherwin FS. The functional independence measure: A new tool for rehabilitation In: M. G. Eisenberg, R. D. Grzesiak, editors. Advances in clinical rehabilitation. New York, NY: Springer-Verlag; 1987.
93. Sharrack B, Hughes RA, Soudin S, Dunn G. The psychometric properties of clinical rating scales used in multiple sclerosis. Brain 1999;122(2):141–159.
94. Klein RM, Bell B. Self-care skills: Behavourial measurement with the Klein-Bell ADL scale. Archives of Physical Medicine and Rehabilitation 1982;63:335–338.
95. Law M, Usher P. Validation of the Klien-Bell Activities of Daily Living Scale for children. Canadian Journal of Occupational Therapy 1988;55(2):63–68.
96. Lawton MP, Brody EM. Assessment of older people: Self-maintaining and instrumental activities of daily living. Gerontologist 1969;9:179–186.
97. Edwards M. The reliability and validity of self-report activities of daily living scales. Canadian Journal of Occupational Therapy 1990;57(5):273–278.
98. Hoenig H, McIntyre L, Sloane R, Branch LG, Truncali A, Horner RD. The reliability of a self-reported measure of disease, impairment, and function in persons with spinal cord dysfunction. Archives of Physical Medicine and Rehabilitation 1998;79(4):378–387.
99. Hoenig H, Hoff J, McIntyre L, Branch LG. The self-reported functional measure predictive validity for health care utilization in multiple sclerosis and spinal cord injury. Archives of Physical Medicine and Rehabilitation 2001;82(5):613–618.

15

Domestic Life

Marcia Finlayson, Eva Månsson Lexell, and Susan Forwell

CONTENTS

Participation in domestic life activities is fundamental to survival and lifestyle maintenance, both at an individual level and within family units. Engaging in domestic life activities contributes to social order and integration and is infused with social expectations and standards. In general, adults are expected to be able to perform or oversee the completion of tasks that are determined by one's assigned roles within the home environment. Each domestic life activity requires a different set of knowledge and abilities to meet the specific physical, cognitive, social, and emotional demands of that activity. For example, paying the rent or mortgage on time requires a person to remember the due date, to know how to make the payment, and then to have the ability and resources to act on that knowledge. Depending on the method of payment, different knowledge and skills will be required. Online payments require knowledge and skills to use a computer and access the Internet, whereas paying by postal mail requires the ability to write a check, address and seal the envelope, and get the envelope to the post office. Using an automatic payment system requires that the individual budget and remember to have adequate funds in the account from which the transfer is authorized.

The task of paying the rent or mortgage is only one example in a wide range of activities that fall under the rubric of "domestic life" in the *International Classification of Functioning, Disability and Health* (ICF).[1] Carrying out domestic life activities is embedded in contextual and personal resources, cultural norms, familial role modeling, personal preferences, and values. Restrictions, such as having multiple sclerosis (MS), that encumber engagement in domestic activities may threaten fundamental routines, ways of doing, and standards of performance, and may disrupt household organization, strain relationships, and challenge a person's competency. Therefore, this chapter focuses on domestic life participation and the rehabilitation services that may be brought to bear to mitigate or manage these restrictions. After reading this chapter, you will be able to:

1. Describe the nature, scope, and impact of domestic life restrictions experienced by people with MS,

2. Review key assessment tools that can be used to evaluate domestic life restrictions,

3. Discuss approaches to intervention that are used to reduce restrictions to participating in domestic life, and

4. Identify areas for future clinical and research development specific to supporting domestic life participation among people with MS.

15.1 Nature, Scope, and Impact of Domestic Life Restrictions

The ICF identifies three broad areas within the domain of domestic life participation: (1) acquiring necessities such as housing, goods, and services; (2) completing household tasks such as making meals, doing housework and laundry, doing yard work, managing money, and repairing clothing; and (3) assisting other people and caring for animals and household objects. Beyond the ICF, other common rubrics for these types of tasks include instrumental activities of daily living (IADL), or domestic productivity. Regardless of the terminology used, most individuals have some responsibilities across one or more of these areas, irrespective of age, gender, marital status, disability level, or social situation. The extent of these responsibilities will be greatly influenced by the social and cultural expectations associated with these various personal characteristics (see Chapters 18 and 23).

Current knowledge about domestic life restrictions among people with MS is primarily based on survey studies with samples that are comprised mainly of middle-aged women with relapsing–remitting disease (see Table 15.1). Across these studies, a fairly narrow range of domestic life activities has been addressed relative to the range that is included in the ICF. During the preparation of this chapter, very limited epidemiological information could be located about the extent of restriction experienced by people with MS for a number of specific tasks (see Table 15.2). For those areas about which some restriction estimates are available, the instruments used to evaluate the level of restriction are highly variable. Estimates are often based on dichotomous (limited/not limited) scales, ordinal (e.g., not limited, somewhat limited, very limited) scales, or summative scales that provide an overall restriction score but no information about restrictions in specific activities. With these limitations in mind, Table 15.2 identifies the range of domestic life activities identified in the ICF, the number of MS-specific studies that provide information about rates of restrictions for these activities, and the lowest and highest restriction rates published. Studies were located in MEDLINE and CINAHL using keyword searches for "multiple sclerosis" in combination with terms such as "domestic life," "instrumental activities of daily living,"

TABLE 15.1

Characteristics of Samples in Studies Reporting Task-Specific Domestic Life Restrictions among People with MS

	Holper et al.[2]	Svestkova et al.[3]	Paltamaa et al.[4,5]	Khan et al.[6,7]	Minden et al.[8]	Finlayson et al.[9]	Aronson et al.[10]
Sample size	N = 205	N = 100	N = 120 (year 1) N = 109 (year 2)	N = 101	N = 2156	N = 430	N = 697
Average age	44.7 ± 12.4 (SD)	41.7 ± 11.3 (SD)	45 ± 10.8 (SD)	49.5 ± 9.9 (SD)	Young (20–64): 48.6 Old (65+): 70.8	Women: 49 ± 13 (SD) Men: 52 ± 12 (SD)	48.3
% Female	72.2	70	75	71.3	77.0	75.8	70
Type of MS	54.6% (RRMS)	73% (RRMS)	88% (RRMS)	Not reported	57.6% (RRMS)	Women: 20% (RRMS) Men: 16% (RRMS)	21% (RRMwS) 39% (S or PPMS)

Note: SD = standard deviation; RRMS = relapsing remitting MS; S or PPMS = secondary or primary progressive MS.

TABLE 15.2

Extent of Task-Specific Domestic Life Restrictions Identified in the MS Literature

Domestic Life Activities Identified by the ICF	Number of Studies Reporting Level of Restriction among People with MS	Low Estimate	High Estimate
Acquiring a place to live	None identified	—	—
Shopping for household items	5	40.2%[8]	91%[6]
Shopping for groceries			
Preparing simple meals	4	28%[9]	88.1%[6]
Preparing complex meals			
Doing housework	6	41.3%[8]	93%[6]
Doing laundry	None identified	—	—
Washing dishes	None identified	—	—
Vacuuming	None identified	—	—
Using appliances	None identified	—	—
Storing daily necessities	None identified	—	—
Disposing of garbage	None identified	—	—
Caring for household objects	3	60%[3]	83.1%[6]
Making and repairing clothes	None identified	—	—
Maintaining dwelling and furnishings	None identified	—	—
Maintaining appliances	None identified	—	—
Maintaining vehicles	None identified	—	—
Maintaining assistive devices	None identified	—	—
Taking care of plants, indoors and outdoors	1		75%[a9]

continued

TABLE 15.2   (continued)

Extent of Task-Specific Domestic Life Restrictions Identified in the MS Literature

Domestic Life Activities Identified by the ICF	Number of Studies Reporting Level of Restriction among People with MS	Low Estimate	High Estimate
Taking care of animals	None identified	—	—
Assisting others with self-care	2	63%[2]	86.1%[6]
Assisting others in movement			
Assisting others in communication			
Assisting others in interpersonal relations			
Assisting others in nutrition			
Assisting others in health maintenance			

[a] Estimate is based on restrictions doing snow removal and yard work.

or "household activities." Studies that did not provide estimates for individual activities (i.e., had summative scores only) were excluded.

Based on the data presented in Table 15.2, it is apparent that domestic life restriction among people with MS is high. For example, for all but one task, the lowest restriction estimate is 40%. Given that few studies include participants with progressive forms of MS, these estimates likely underestimate the level of restriction that is present for the MS population as a whole. In addition, the studies do not include children and adolescents with MS. While young people tend to have fewer domestic life responsibilities, it is unclear to what extent they are engaging in household chores, whether they are meeting expectations for involvement, or what implications any restriction might have for their future lives as adults.

Given the limited epidemiological data available, qualitative data from various projects conducted by the authors of this chapter were compiled (see Table 15.3). These quotations provide additional insights into the domestic life restrictions that people with MS experience and the ways in which symptoms and contextual factors interact to generate these restrictions. In particular, these quotations illustrate the complexity of domestic life and the ways in which restrictions influence the ways in which people lead their everyday lives.

15.2 Assessment of Domestic Life Restrictions

Understanding and addressing the scope of domestic life requires rehabilitation providers to account for a magnitude of issues that may contribute to these restrictions (e.g., specific symptoms, nature of the physical and social environment, cultural expectations). Therefore, a client-centered approach to the assessment process is critical to enable the development of relevant treatment goals and plans. According to the multiple sclerosis clinical practice guidelines produced by the National Institute of Clinical Excellence (NICE),[11] a comprehensive assessment of activities of daily living, which includes domestic life, should

- Address the goals, aspirations, priorities, interests, and potential of the person with MS
- Occur across time and in different environments

TABLE 15.3

Quotations on Domestic Life Restrictions Experienced by People with MS

Participant Description	Area of Domestic Life as per the ICF	Quotations
Oscar, 41 years old, EDSS = 6.5, divorced, lives alone, has shared custody of two young daughters who live 200 km away	Caring for others	"My physician didn't think it was a good idea for me to drive such a long distance ... you don't want to fall asleep when you have a valuable load such as your children in the car. I took the train half the way and the ex-wife drove the children the other half, and then I took the train back to my place with my children. By the time we got there it was late ... and then the same procedure going back on Sunday ... taking the train ... of course, one can socialize on the train, but it was not the same, it was a bit disturbed I think."
Sara, 43 years old, EDSS = 8.0, married with young son, has personal care attendant during the day	Caring for others	"The municipality said, "No ... when one has a travel pass [including taxi rides partly paid for by the municipality] to get to work, [the journey] should be from the home to the place of work" ... You're not allowed to go past daycare [to pick up a child], even if the personnel at the daycare center dressed him [the son], and brought him out to the car so I didn't have to leave the car."
Birgitta, 67 years old, EDSS = 8.0, lives alone in an apartment with a small garden/outdoor space	Gardening	"I have great plans for my terrace ... I want to have it nice with lots of flowers and such, and I know exactly how I want it to be, but I can't manage to do it by myself, and I have asked for help but there is no one that has the time ... and then I think, darn it, that I can't manage it myself."
Nina, 52 years old, EDSS = 6.0, lives alone in an apartment, but spends a lot of time in her boyfriend's summerhouse with a garden	Gardening	"The last time I tried to manage some gardening there [in the summer house], I worked on the front side of the house ... and I noticed that I couldn't move, I was stuck and couldn't walk, so I had to record a message on his [answering] machine that he had to come out and help me. If he hadn't heard the message, I would have had to stay there for hours."
Margaret, 73 years old, walks with a cane, lives with husband in a bungalow	Shopping	"When I get into some of the stores, it would be nice if they had carts that people that were handicapped could use ... So that's too bad because I have to hold on to something when I go into a store. I just can't walk freely into a store and to just walk with a cane, it gets too tiring. I prefer to hold on to something and let the wheels roll and go through the store."
Anne, 54 years old, walks with a cane, lives with husband in a bungalow	Shopping	"I don't go shopping anymore by myself ... I don't walk the malls anymore. 'Cause even if I have the [wheel] chair with the basket on the back or a tote that I put things in, still it's harder to carry the stuff around."
Jane, 59 years old, walks with a cane or rollator, lives with husband in a two-story home	Laundry	"I can do stairs, uh, but I can't do them too frequently. And getting things up and down, like laundry, carrying up and down, I can't do that. So I usually just wait and have my husband carry it up for me or whatever. We've tried doing a pulley system where we have like a box to pull up and down, and it just doesn't work very well."

continued

TABLE 15.3   (continued)

Quotations on Domestic Life Restrictions Experienced by People with MS

Participant Description	Area of Domestic Life as per the ICF	Quotations
Beth, 50 years old, EDSS = 6.5, lives with husband in a house on the countryside, uses manual wheelchair indoors	Cooking	"I have to trust my husband to do everything for me … I baked and cooked, and I used to love to cook …, one has to accept what one's partner thinks about food, because I was a vegetarian before but now I have started to eat meat again. He [the partner] tries to satisfy some of my wishes, but it is tough at times."
	Cleaning	"I used to take on the whole responsibility for chores inside the house, such as cleaning, washing windows, and laundry. I had full control over when things had to be done … but now I can only tell him [the husband]; I can't do anything by myself."

- Consider environmental factors, including social supports
- Address current and future needs

In order to pursue and meet these guidelines, rehabilitation providers must recognize the characteristics of available domestic life measures from which they can select, understand the scope and selection of available instruments, and then engage in a systematic assessment process.

15.2.1   Characteristics of Domestic Life Measures

Like the previous chapter on self-care, there are many different assessments that can be used to evaluate the extent of domestic life restrictions that are not MS specific. Together, these instruments fulfill a variety of purposes, including the following:

- Identifying priorities for goal setting and treatment planning
- Measuring outcomes of treatment (efficacy and effectiveness)
- Determining readiness for discharge from rehabilitation services
- Determining level of postdischarge care required to inform the provision of appropriate supports and resources
- Supporting the preparation of funding applications through third parties (e.g., for home modifications or durable medical equipment)

Across each of these purposes, the assessment process must address the physical, cognitive, social, and contextual complexity of domestic life restrictions and generate data that can guide interventions that facilitate optimal domestic life participation after rehabilitation.[12–17]

Like the self-care assessments described in Chapter 14, instruments to assess domestic life can be characterized in several ways. First, instruments vary in their depth; for example, some instruments screen for potential restrictions while others provide detailed and in-depth information to guide goal setting and treatment planning. Second, instruments vary in their approach to data collection; for example, some depend on client self-report while others involve objective evaluation of actual performance. Some instruments combine both approaches. Third, domestic life instruments vary in the degree to which they rely on recall of past or typical performance (common in self-reported tools) while others may target present abilities (focus of performance-based tools). Finally, instruments vary in the extent to which they provide standard protocols, have been examined for their psychometric properties, or are specific to the MS population (vs. a generic tool). To illustrate these characteristics, Table 15.4 provides a listing of a selection of domestic life assessments and their characteristics.

Table 15.5 provides additional details about the instruments listed in Table 15.4 and identifies the specific domestic life activities that are included in each tool (a check mark on the table indicates the item is addressed in the particular tool). In Table 15.5, assessments are grouped into three categories: (1) client-centered assessments, (2) assessments that are specific to domestic life issues, and (3) broader assessments that include items or sections addressing domestic life. The contents of this table indicate that not all domestic life assessments address the full range of activities within this domain of functioning. Consequently, rehabilitation providers may need to use more than one assessment to capture this area of functioning comprehensively, or, alternatively, ensure that they ask additional questions of their clients in order to cover all content domains.

15.2.2 Assessment Process and Tool Selection

As Tables 15.4 and 15.5 indicate, there is a broad array of domestic life assessments available for use with the MS population and each tool has strengths and limitations in terms of characteristics and content. Because of the range available, it can be challenging to determine which instrument(s) to use in a given setting and for a particular client. This is particularly the case because participation in domestic life is influenced by the person's physical and cognitive abilities (i.e., impairments) and the various individual actions (e.g., standing, reaching, carrying) required by a specific activity. In addition, the context (e.g., done alone or with others) and demands (e.g., physical, social, cultural) of the activity and environment interact to influence the level of restriction that the person with MS experiences. Consequently, the process of selecting the assessment(s) to use must be informed by the client's situation and the intended purpose of the data generated (e.g., setting goals vs. planning postdischarge services). For example, one client's circumstances may require that the assessment capture the impact of cognitive function on domestic life and, therefore, selecting a measure like the Independent Living Scale (ILS) or the Timed Instrumental Activities of Daily Living (TIADL) would be appropriate. For another client, concern about safety at home due to physical or cognitive limitations may lead the therapist to select the Safety Assessment of Function and the Environment for Rehabilitation (SAFER-HOME) instead.

Regardless of the assessments selected, the evaluation of domestic life restriction(s) should adhere to a systematic process so that the rehabilitation process is fully and confidently informed. Figure 14.1 in Chapter 14 provides an overview for the rehabilitation evaluation process focused on self-care. This same process can be applied to domestic life evaluation. The initial step in the process is to identify potential concerns the client may have related to domestic life restriction. This information can be obtained through an initial interview to understand the client's life roles, responsibilities, and daily routines. Tools such as the Canadian Occupational Performance Measure (COPM) can be useful in this regard.[22]

Once specific concerns have been identified, the next step is to conduct an in-depth evaluation in those areas identified as problematic to determine the nature and extent of

TABLE 15.4

Examples of Domestic Life Assessments

Assessments	Assessment Characteristics					
	Type of Assessment[a]	Type of Client	Time to Complete	Metrics in MS	Number and Type of Items	References
Assessment of Motor and Process Skills (AMPS)[b]	Performance based; in-depth	For adults and children, including people with MS	30–45 min	Yes, for specificity and convergent and divergent validity	At least two tasks that are culturally relevant to the person are chosen, demonstrated, and evaluated	Fisher[18], Doble et al.[19], Månsson and Lexell[12]; Law et al.[13]
Assessment of Living Skills and Resources (ALSAR)	Client based (self-report); may be supplemented with observing performance	For elderly individuals and may be useful for MS	30 min	None specifically reported for MS	11 tasks rating skills and resources; focus on accomplishment of task (meal preparation, finances, telephone use, leisure, reading, shopping, laundry, housekeeping, home maintenance, medication management, transportation)	Williams et al.[20], Clemson et al.[21]
Canadian Occupational Performance Measure (COPM)[b]	Client based (self-report) through semistructured interview	For adults and children with disabilities, including MS	30–40 min (10–15 min for domestic items)	Yes, for convergent and divergent validity	Client identifies performance concerns across self-care, domestic life, and leisure. Client rates each concern for importance (1–10), and then client rates top five concerns on performance (1–10) and satisfaction (1–10)	Law et al.[22]; Dedding et al.[23]; Månsson Lexell et al.[24]; Law et al.[13]
Disability Rating Index (DRI)	Client based (self-report): questionnaire	Developed for persons with pain and muscular or neurological deficits and applied to MS	2–4 min	Yes, for test–retest and face and construct validity	12 physical function items, of which 5 are related to domestic life	Salén et al.[25]
Executive Function Performance Test (EFPT)[b]	Performance-based; in-depth	For persons with cognitive impairment associated with functional disability and then tested with MS	1–2 h	Yes, interrater reliability and validity	Five tasks are assessed (cooking, hand-washing, telephone use, bill payment, medication management). Some applications delete hand washing. Tasks evaluated on initiation, sequencing, organization, task completion, judgment, and safety	Baum, et al.[26]; Baum et al.[27]; Goverover et al.[28,29]; Voelbel et al.[30]

Extended ADL Index	Client based (self-report): questionnaire	Developed and tested for post-CVA patients and may be useful in MS	5–8 min	None specifically reported for MS	22 items grouped in four categories (mobility, leisure, kitchen, domestic; the last two also termed household)	Lincoln and Gladman[31]
Frenchay Activities Index (FAI)	Client based (self-report): questionnaire	Developed and tested for post-CVA patients and applied in MS	5–10 min	None specifically reported for MS	15 items of domestic life assessing frequency of participation	Holbrook and Skilbeck[32]; Einarsson et al.[16]
Functional Autonomy Measuring System (SMAF)[b]	Performance based; in-depth	For adults with impairments, disabilities, or handicaps; may be useful for MS	42 min (average reported)	None specifically reported for MS	29 items are evaluated for (in)dependence and are grouped into five categories: ADL, mobility, communication, mental functioning, IADL (eight items)	Hébert et al.[33,34]
Goal Attainment Scale (GAS)	Client based (self-report)	Used with children, adults, and older adult population, including persons with MS	20 min–1 h	Yes, for sensitivity, responsivity, and construct and divergent validity	As per the number of goals set by the individual (5–10), assessment is individualized according to goals	Becker et al.[35]; Rockwood et al.[36]; Khan et al.[7]
Independent Living Scale (ILS)	Performance based; in-depth	Developed and tested for persons with cognitive impairment (mental retardation, TBI, dementia, psychiatric population); may be useful for MS	45 min	None specifically reported for MS	70 items within five subscales (memory/orientation, managing money, managing home and transportation, health and safety, and social adjustment). Items are contextually designed from an American perspective	Loeb[15]; Ashley et al.[37]
Instrumental Activities of Daily Living (IADL)[b]	Client based (self-report), proxy, observation, medical records	For a range of populations, including elderly individuals; has been applied to MS	10–15 min	None specifically reported for MS	Eight tasks essential to living independently in community (meal preparation, finances, telephone use, travel, shopping, laundry, housekeeping, medication management)	Law et al.[13]; Lawton and Brody[38]; Salter et al.[39]

continued

TABLE 15.4   (continued)

Examples of Domestic Life Assessments

Assessments		Assessment Characteristics				
	Type of Assessment[a]	Type of Client	Time to Complete	Metrics in MS	Number and Type of Items	References
Katz Extended ADL Index[b]	Client based (self-report) that may be supplemented with observing performance	Developed and tested for post-CVA patients and applied in MS	Not reported	None specifically reported for MS	Six ADL items and four domestic life items (shopping, cooking, cleaning indoors, outdoor transportation)	Asberg and Sonn[40]; Einarsson et al.[16]
Kitchen Task Assessment (KTA)	Performance based; in-depth	For persons with Alzheimer's disease and applied to MS	30 min	None specifically reported for MS	One cooking or meal preparation task assessing level of independence or support required	Baum and Edwards[41]; Law et al.[13]
Kohlman Evaluation of Living Skills (KELS)[b]	Performance based in combination with interview; in-depth	For persons with TBI and cognitive impairment as well as the elderly; may be useful for MS	30–45 min	None specifically reported for MS	17 skills are assessed in five areas (self-care, safety and health, money management, transportation and telephone, work and leisure)	Kohlman Thomson[42]; Zimnavoda et al.[43]
Multidimensional Functional Assessment Questionnaire— Instrumental Activities of Daily Living (MFAQ-IADL) developed by the Older Americans Resources and Services (OARS)	Client based (self-report): questionnaire	Developed and tested with elderly persons and may be useful in MS	5–10 min	None specifically reported for MS	Seven items of domestic life assessing level of (in)dependence (telephone use, shopping, meal preparation, housekeeping, taking medication, finances, and transportation)	Whittle and Goldenberg[44]; Fillenbaum[45]; Fillenbaum and Smyer[46]; Ottenbacher et al.[47]

Assessment	Type	Population	Time	Validity for MS	Description	Reference
Safety Assessment of Function and the Environment for Rehabilitation (SAFER-HOME)[b]	Performance based (observation) and interview; in-depth	For individuals with multiple diagnoses, complex functional and environmental problems; may be useful for MS	45–60 min	None specifically reported for MS	74 items grouped under 12 categories that cover safety and function in the home. Two categories: household and communication collective have 12 performance items in domestic life	Chiu et al.[17]
Timed Instrumental Activities of Daily Living (TIADL)[b]	Performance based	For populations with slower visual-processing speed or cognitive problems and has been applied to MS	20 min	Yes, predictive and criterion validity for MS	Five tasks (finding a name and number in the phone book, making change, reading ingredients on cans, findings items on a shelf, reading directions on a medicine bottle)	Owsley et al.[48]; Goverover et al.[49]

[a] "Screening" refers to assessments that identify the presence and potentially the severity of a limitation to performing domestic life tasks. "In-depth" refers to assessments that provide details about the performance limitation and contributes to the intervention plan.

[b] Assessments that contain ADL items.

TABLE 15.5

Items Included in Selected Domestic Life Assessments

Assessment	Meal Preparation	Laundry	Cleaning/ Housekeeping	Home Maintenance	Shopping	Finances	Telephone	Reading	Outdoor Transportation	Other
Client-Centered Assessments										
Assessment of Motor and Process Skills (AMPS)	✓	✓	✓	✓	✓	✓	✓	✓	✓	Can include any or all of these components based on the priorities and goals of each individual
Canadian Occupational Performance Measure (COPM)	✓	✓	✓	✓	✓	✓	✓	✓	✓	
Goal Attainment Scale (GAS)	✓	✓	✓	✓	✓	✓	✓	✓	✓	
Domestic Life-Specific Assessments										
Assessment of Living Skills and Resources (ALSAR)	✓	✓	✓	✓	✓	✓	✓	✓	✓	Leisure Medications
Executive Function Performance Test (EFPT)	✓					✓	✓			Hand washing Medications
Extended ADL Index	✓	✓	✓	✓	✓	✓	✓	✓	Walking outside In/Out/Drive car Public transport	Socialize Write letters
Instrumental Activities of Daily Living (IADL)	✓	✓	✓		✓	✓	✓	✓	✓ Travel	Medications

Specific Items

Assessment					Items
Kitchen Task Assessment (KTA)	✓				
Multidimensional Functional Assessment Questionnaire—Instrumental Activities of Daily Living (MFAQ-IADL)	✓	✓	✓	✓	Medications
Timed Instrumental Activities of Daily Living (TIADL)		✓	✓	✓	Finding items on a shelf
Assessments with Items or Sections Related to Domestic Life					
Disability Rating Index (DRI)	✓	Making a bed / Light work	✓	Walking outside	Dressing / Running / Heavy work
Frenchay Activities Index (FAI)	✓	✓	✓	Walking outside / In/Out/Drive car / Public transport	Socialize / Hobby / Employment
Functional Autonomy Measuring System (SMAF)	✓	✓	✓	Moving around outside / Transportation	Seven ADL tasks / Five mobility tasks / Three communication tasks / Five mental functions

continued

TABLE 15.5 (continued)

Items Included in Selected Domestic Life Assessments

Assessment	Meal Preparation	Laundry	Cleaning/ Housekeeping	Home Maintenance	Shopping	Finances	Telephone	Reading	Outdoor Transportation	Other
Independent Living Scale (ILS)				✓	✓	✓	✓	✓	✓ Taking bus/ cab Transportation	Memory Orientation Social adjustment
Katz Extended ADL Index	✓		✓		✓				✓	Six ADL tasks
Kohlman Evaluation of Living Skills (KELS)						✓	✓		Transit system Community mobility	Leisure Work Safety Self-care
Safety Assessment of Function and the Environment for Rehabilitation (SAFER-HOME)	✓	✓	✓	✓	✓	✓	✓		✓ Transit system Community mobility Driving	Mobility Hazards Kitchen safety Self-care Medications Leisure Wandering

Specific Items

restriction and possible limiting factors (e.g., specific symptoms and activity restrictions, environmental barriers). This part of the assessment process typically involves a combination of tools, including those that provide information about specific impairments that may be hindering the performance of domestic life activities (e.g., balance, strength). Finally, data from all assessments are examined and interpreted to set goals and determine priorities for intervention, in partnership with the client and his or her family, as appropriate.

15.3 Interventions for Domestic Life Restrictions

Despite the extent and impact of domestic life restrictions experienced by people with MS and the number of instruments available to evaluate them, there is a remarkable lack of documentation of interventions—evidence based or otherwise—specific to domestic life participation in the MS rehabilitation literature. This situation has remained virtually unchanged since the publication of the NICE MS clinical practice guidelines[11] and the MS rehabilitation recommendations from the European Multiple Sclerosis Platform[50] nearly 10 years ago. Both these documents provide only general recommendations regarding rehabilitation for domestic life. For example, the NICE[11] guidelines recommend that interventions should be goal directed, be consistent with the priorities of the person with MS, and focus on increasing and maintaining independence. The guidelines also recommend regular follow-up as well as communication with social services (e.g., home care), if the person with MS is agreeable. No further guidance is provided about how to implement these recommendations in the course of everyday clinical practice.

Even recent literature searches combining the keywords "multiple sclerosis" with each of the instrument titles from Table 15.4 generated only a single intervention study that examined whether the use of an Odstock drop foot stimulator (ODFS) led to improvements in activities of daily living as measured by the COPM.[51] Although a few domestic life activities were identified as concerns by the participants through the COPM process before intervention (e.g., carrying shopping bags, preparing meals), the concerns that were scored and reevaluated after the intervention focused on walking. Consequently, it is unknown as to whether the intervention contributed to changes in domestic life participation.

Given the lack of documented evidence specific to MS, the following paragraphs summarize common rehabilitation practices assumed to positively influence participation in domestic life among people with MS. The person–environment–occupation (PEO) model[52] has been used to structure this information and link it to the concepts of the ICF.

15.3.1 Person-Level Approaches

In the PEO model, the person is viewed as having attributes and experiences that influence participation in everyday activities. Attributes include motor, sensory, and cognitive capabilities; self-concept; confidence; motivation; and general health.[52] These attributes overlap with the ICF domains of body structure and function (impairment) and personal context.

MS-related impairments (e.g., weakness, fatigue, cognition) have the potential to negatively affect engagement in domestic life activities. Interventions to remediate these impairments are assumed to lead to increased capacity to perform everyday domestic activities. For impairment-specific rehabilitation strategies, see Chapters 5 through 11.

Remediation of impairments may not always be possible, depending on the level of disability of the person with MS or the personal context of the individual. Therefore, rehabilitation providers should also consider educationally based compensatory approaches, for example, building the person's knowledge, skills, and confidence to self-manage new domestic life challenges as they arise. Additional compensatory approaches are discussed in the subsequent sections on environment and occupation/activity approaches.

Educationally based compensatory approaches build the client's ability to identify problems, communicate his or her needs to others, and select and evaluate potential solutions(s) to reduce or eliminate domestic life restrictions based on personal values, priorities, and context (e.g., requesting assistance, delegating, pacing, using equipment). By developing personal skills and confidence for problem solving, decision making, and goal setting, these interventions enable people with MS to continue to engage in domestic life activities despite ongoing impairments over time.

15.3.2 Environment-Level Approaches

As part of a comprehensive plan to reduce or eliminate domestic life restrictions for people with MS, consideration should be given to possibilities for reducing the demands of the physical environment (e.g., removing restrictions and barriers). Environment-level approaches focus on minimizing or compensating for the impact of impairments on activity and participation. Using low-tech assistive devices, such as built-up cooking utensils; a long-handled reacher; a small tea cart with wheels; and lightweight appliances (e.g., iron, vacuum, lawn mower), may be beneficial for some people with MS. The appropriateness and feasibility of these options will depend on available funding, acceptability to the client and the client's family, and availability of storage within the person's home. More elaborate options such as environmental control units or more extensive home modifications (see Chapter 21) can also be considered.

In addition to considering the physical environment, the social environment can also be a target of intervention when addressing domestic life restrictions. Educating caregivers, enhancing family communication, and helping the individual and family build and maintain social networks may support domestic life participation for the person with MS. These interventions may involve enabling the person with MS to negotiate roles and responsibilities, shift to a cooperative rather than independent model for engaging in domestic life activities, and finding ways to effectively and efficiently use the assistance of others. Chapters 12 and 22 provide additional information about addressing these issues in the context of rehabilitation practice.

15.3.3 Occupation- or Activity-Level Approaches

A final option to consider when addressing the domestic life restrictions experienced by a person with MS is to modify the activity or adapt how the activity fits within the person's habits and routines. Task or activity analysis[53] can be used to break down activity patterns and the steps, sequences, movements, and demands of a given task (e.g., doing the laundry). Performance analysis[54] involves analyzing these components as a task is completed by a specific individual. Information from these analyses can then be used to identify potential modifications (e.g., changing the way a task is done, changing how an activity fits within a person's routine, altering activity steps). Through collaboration and partnership, rehabilitation providers and clients can work together to identify alternative ways to get necessary activities accomplished. Throughout the modification process, careful consideration needs to be given to enhancing the efficiency of movements and energy expenditures, safety, and independence.

15.3.4 Summary

Interventions for domestic life restrictions often involve interventions that cross two or three of these intervention approaches. Often, interventions involve remediation of impairments, modification of the environment to compensate for residual impairments, and modification of the activity to maximize domestic life participation. Whichever interventions are recommended and implemented, each one must be acceptable and relevant to the person's context and consistent with his or her values, priorities, and preferences. To illustrate the use of rehabilitation assessments and interventions that address participation restrictions in domestic life activities for those with MS, a case study is provided in the following section.

15.4 Case Study

Carl has been referred to outpatient rehabilitation. Carl has MS, is married, and has two children (4 and 8 years old). They live in a terraced house with a 10-step staircase indoors. Carl is employed in a grocery store, but has been on full-time sick leave for about 4 months. He is 43 years old and was diagnosed with MS 8 years ago. At the time of admission, Carl was experiencing lower-extremity spasticity and fatigue, and had difficulty climbing stairs and walking indoors and outdoors. He was not leaving the house by himself. He expressed a desire to be more involved in household activities, but his wife, who works evenings and nights, was worried that he would not be able to manage activities for himself and the children, such as preparing light meals. Currently, Carl is independent in self-care tasks, although he has bladder symptoms that result in frequent urinary infections. This problem, in turn, prevents him from going shopping and taking his children to soccer practice.

15.4.1 Assessment Process

On admission to rehabilitation, each member of the team engaged Carl in a discipline-specific assessment in order to inform interdisciplinary goal setting and the development of a treatment plan.[55] Throughout their work, the team used ICF terminology[1] to guide the selection of assessment tools, interpretation of results, and identification of activity and participation goals.

The occupational therapist administered the COPM.[22,56] The COPM is a semistructured interview that enables a client to identify and prioritize areas of performance that he or she wants to address during rehabilitation. Through the COPM process, Carl and his wife were able to identify several concerns with domestic life activities:

- Cooking (e.g., preparing snacks and sandwiches for the children; cooking a hot meal for the whole family)
- Shopping (e.g., going to the grocery store to purchase small items)
- Caring for others (e.g., taking the children to soccer practice)

- Caring for the house (e.g., tidying up; making the bed; doing basic home maintenance such as changing light bulbs or the gasket in the kitchen tap)

Across these concerns, Carl prioritized five specific tasks through the COPM process
(see Table 15.6). For each of these five tasks, Carl evaluated his performance on a scale of
1 to 10, with 1 being completely unable to perform and 10 being able to perform extremely
well. He also evaluated his satisfaction with his performance, with 1 being not satisfied at
all and 10 being extremely satisfied. For both performance and satisfaction, Carl rated
himself low; his average performance rating was 2.0 and his average satisfaction score
was 2.6.

While the COPM helped the occupational therapist understand *which* tasks Carl and his
wife found problematic to perform, she needed additional performance information
to understand *why* these activities might be difficult. This information was critical for
tailoring an intervention specific to Carl's needs. Therefore, the occupational therapist conducted a performance analysis using the Assessment of Motor and Process Skills (AMPS),
which is an instrument that provides a systematic method of observing task performance
(see Figure 15.1). During the AMPS, Carl chose to prepare a sandwich with his sons'
favorite spread and to make the bed, both of which were tasks he identified as problematic
during the COPM. The occupational therapist observed Carl and rated the quality of his
performance for each of the 16 motor skills and 20 process skills addressed by the AMPS
instrument. She then determined which actions and skills had the greatest impact on
Carl's performance.

The AMPS performance analysis indicated that in the area of motor skills, Carl had a
moderate increase in *physical effort* during the performance of both tasks. For example,
during the sandwich-making task, he was unstable when he was carrying objects, needed
to continuously hold on to the counter, fumbled with utensils, and had problems opening
wrappings. When he was making the bed, he had to move between the work spaces constantly. His increased physical effort became more pronounced as the bed-making task
proceeded, and he had to sit down and rest during the activity.

TABLE 15.6

Admission and Discharge Findings from Carl's COPM

		On Admission			At Discharge	
	Problems in Performance	Rating of Importance	Rating of Performance	Rating of Satisfaction	Rating of Performance	Rating of Satisfaction
1	Very difficult to prepare a snack for the children	10	2	1	9	10
2	Cannot take children to soccer practice	10	1	1	10	10
3	Cannot cook lighter meals for the family	9	1	3	7	8
4	Very difficult to make the bed	8	4	5	8	8
5	Would like to be able to go to the grocery to shop and buy a few things	7	1	1	8	10
	Mean scores		2.0	2.6	8.4	9.2

Note: Rating scale—1 (lowest) to 10 (highest).

The AMPS[54] is an observational assessment that measures the activity component in the ICF,[1] focusing on the quality of a person's task performance in activities related to self-care and domestic life. Occupational therapists can complete an intense 1-week training to administer the AMPS (see the AMPS Project International website for more information[58]). The administration of the AMPS involves four steps:

1. Interviewing the client and deciding which tasks to perform
2. Setting up the test environment
3. Observing task performance
4. Scoring performance and interpreting the results[59]

When the tasks are chosen, it is crucial that they are relevant and familiar to the client and his or her life situation. There are over 100 standardized tasks included in the AMPS that can be selected during the assessment process.

During task performance, the occupational therapist observes the smallest components of performance, that is, the motor and process skills. Motor skills are actions a person carries out to move him- or herself or objects during the performance of the task. Process skills are logical actions a person carries out to perform tasks over time, to select and use appropriate tools and materials, and to adapt his or her performance when problems occur. Motor and process skills are observable goal-directed actions that people always need to complete or fulfill the performance of a task.

The occupational therapist evaluates the quality of the task performance in terms of effort, efficiency, safety, and independence for each of the 16 motor skills and the 20 process skills by scoring 1–4. A score of 1 represents markedly deficient performance that impedes the action; 2 = ineffective performance; 3 = questionable performance; and 4 = competent performance. The raw ordinal scores determined during the observation are converted into interval data by a many-faceted Rasch analysis,[60] and are presented graphically as motor and process ability measures (called logits) on two linear scales for each of the tasks performed. High-positive motor and process ability measures indicate that clients are more able, whereas low- or negative-ability measures indicate that clients are less able.[61] There are cut-off criteria for both the motor (2.0 logits) and process ability (1.0 logits) scales. Clients whose motor performance falls below this cut-off are likely to use more effort when they perform tasks, and clients whose process skills fall below this cut-off are more likely to be at risk of performing tasks unsafely, inefficiently, and less independently. Clients below the cut-off scores are likely to have difficulty managing on their own in the community.

There has been extensive research on the psychometric properties of the AMPS.[58] Two studies used the AMPS to assess performance in household tasks in MS populations.[12,19] Doble et al.[19] found that more than half of the 22 participants had problems performing household tasks, even though most of them had only mild disease. Månsson and Lexell[12] studied participants with moderate to severe disease and found that all of them had motor ability measures below the cut-off criteria of 2.0 (range −0.7 to 1.5). For the process ability measures, approximately two thirds of participants had so much difficulty performing household tasks that their independence in the community was at risk. Both studies concluded that it is important to go beyond self-care assessment for people with MS and also assess their abilities to perform complex household activities.

FIGURE 15.1
The Assessment of Motor and Process Skills (AMPS).

In the area of process skills, the performance analysis also indicated that during both tasks, Carl was moderately *inefficient*, especially when organizing space and objects. Inefficiency was the major reason behind Carl's increased physical effort in the bed-making task. However, throughout both tasks, Carl was *independent* and fulfilled the tasks in a *safe* manner. His overall performance also showed several strengths. For example, Carl had no problems searching for objects for either task and was able to replace all the equipment upon conclusion of each task. Together, the information on both motor and process skills provided the occupational therapist with detailed and multidimensional performance-based information about the reasons behind Carl's domestic life restrictions.

The observations that the occupational therapist made regarding Carl's difficulties and strengths were confirmed through the AMPS computer-generated report, which converts the therapist's ordinal observation ratings into interval-level measures (called logits) using Rasch analysis (see Figure 15.1). Carl's report indicated that he had a motor-ability logit of −0.1 and a process-ability logit of 0.9. Since both these scores are below the established cut-off criteria of 2.0 (motor skills) and 1.0 (process skills), it indicated that Carl's motor and process limitations made it difficult for him to manage domestic life activities on his own.

The assessments by the physiatrist and nurse focused primarily on Carl's impairments, specifically his problems with his bladder, fatigue, and spasticity. They concluded that he had a urinary tract infection, likely caused by difficulty emptying his bladder, and that recurrent urinary tract infections were likely contributing to his spasticity problems (i.e., body's response to infection). Carl's bladder problems were also affecting his ability to go into the community as he was afraid he would not find a toilet when he needed one. The physiatrist and nurse also noted that Carl was not taking medication for fatigue.

The physical therapist assessed Carl's motor functions and concluded that he was generally weak, particularly on his left side, and that his balance was limited, contributing to difficulties managing meal preparation in the kitchen. The physical therapist also determined that the spasticity Carl experienced in his legs was negatively affecting walking and was more pronounced when he was tired. The social worker's assessment focused primarily on Carl's wife since she was very worried about Carl's difficulties and the impact his restrictions would have on his ability to share in the daily domestic life responsibilities for the family.

15.4.2 Goal Setting and Treatment Planning

Using the results of their individual assessments together with Carl's priorities, the rehabilitation team collaborated with Carl to set goals that were important to and realistic for him. These goals were entered in the rehabilitation plan as interdisciplinary team goals (see Figure 15.2). The impairments, activity limitations, and participation restrictions that were described in the rehabilitation plan then guided treatment planning and implementation.

15.4.3 Interventions

When an interdisciplinary approach (see Chapter 4) is used during rehabilitation, every profession contributes its particular expertise during intervention delivery. In Carl's case, the physiatrist and the nurse treated his urinary infection and taught him how to independently manage clean intermittent self-catheterization. When his bladder problem was managed, his spasticity problems decreased. This then allowed the physical therapist to focus on an exercise program for Carl that would enhance his balance and strength. Carl and the physical therapist also worked to improve his stair climbing and walking indoors without falling. During the rehabilitation, the social worker supported Carl's wife and made sure that she felt confident leaving some of the household responsibilities to her husband.

Most of the rehabilitation team contributed to addressing Carl's problems with his fatigue: the physiatrist prescribed medication, the nurse provided lifestyle education, and the physical therapist provided education and encouragement to support his engagement in regular exercise. The occupational therapist provided education and support regarding the use of energy-management strategies, such as planning, pacing, and activity modification. The occupational therapist also worked with Carl on the specific activities he wanted to achieve (as identified on the COPM). Together they found different strategies for performing his valued domestic life activities that compensated for impairments influencing his performance. Building on the

Name: Carl	Initial: January 17
Planned length of stay: 4 weeks	Reassessment: February 11

Client's expectations

After rehab: Better manage in my home *For the future*: Return to work

Signature: Carl	Coach: Karin, OT

Resources	Hindrances/barriers
Supportive family, motivated	Two-story housing

Impairments	Activity limitations/participation restrictions
Urinary infection, leakage	Difficult to walk indoors and outdoors
Generally tired, fatigue	Difficulties when preparing snack
Spasticity in legs	Difficulties when making the bed
Weak in left side, reduced balance	Cannot prepare family dinner
Reduced sensibility in left leg/foot	Difficult to climb stairs inside house
Mood swings	Cannot go outdoors by himself
	Cannot take children to their leisure activities
	Cannot shop for smaller amounts of groceries

Team goals	Goal time frame	Reevaluation findings
Walk a 10-step staircase without rest or risk of falling	Week 2	Done
Walk safely inside the house	Week 2/3	Needs more practice
Prepare a snack for the children	Week 3	Yes, with assistive devices
Manage to make his and his wife's bed	Week 3	Yes, with assistive devices
Prepare a warm meal for the family	Week 4	Yes, with assistive devices
Be able to take his children to soccer practice	Week 4	Yes, with scooter
Be able to go to the store and buy five items	Week 4	Yes, with scooter

Interventions	Responsible
Practice clean intermittent self-catheterization	Physiatrist, nurse
Mobility training, exercises for balance and strength	Physical therapist
Fatigue management	Physiatrist, occupational therapist, nurse, physical therapist
Advice regarding: activity modification and compensatory techniques	Occupational therapist
Information and provision of assistive devices	Occupational therapist, physical therapist
Support to family	Social worker

FIGURE 15.2
Overview of Carl's rehabilitation plan.

results from the AMPS assessment, the occupational therapist showed Carl different ways to organize objects, to plan his actions to minimize the amount of effort needed during task performance, and to sit during meal preparation to minimize his fatigue. Carl was given the opportunity to try different assistive devices (e.g., scooter) that made it possible for him to accompany his sons, who were biking, to their soccer practice.

15.4.4 Evaluation of Rehabilitation Goals

All the goals that were included in Carl's rehabilitation plan were achieved, and when Carl reevaluated the scores in the COPM, his performance and satisfaction ratings had improved, exceeding a 2-point change (necessary for a clinically significant change with this instrument)[56,57] (see Table 15.6). A reevaluation with the AMPS showed that Carl had improved his motor ability from −0.1 to 0.6, even though he was still below the cut-off of 2.0 logits. His process ability had also improved from 0.9 to 1.2, and he was no longer below the cut-off of 1.0 logits. His *physical effort* no longer increased during the performance of tasks because he had learned to organize space and objects that he needed. He was still *independent* and fulfilled the tasks in a *safe* manner. Carl was discharged from rehabilitation with an improved ability to engage in the domestic life activities that were meaningful and important to him and his family.

15.5 Gaps in Knowledge and Directions for the Future

Since the publication of the ICF in 2001, there has been a growing interest in rehabilitation research and practice on participation and participation outcomes. Despite the number of assessments that include domestic life items, there is very little documented in the literature about the nature and impact of restriction experienced by the general MS population in this area of functioning. Even less is known about differences in the level of restriction across subgroups of people with MS (e.g., early vs. late disability, adolescents, single adults, parents, older adults, men vs. women, specific cultural groups) or what differences might mean for intervention delivery and evaluation. These knowledge gaps limit the collective ability of rehabilitation researchers and providers to determine whether current assessment instruments are adequate for capturing the challenges facing people with MS as they carry out their daily domestic responsibilities. They also make it difficult to set priorities for the development of targeted interventions that could be subjected to rigorous evaluation. These areas will be crucial to pursue in future research if rehabilitation providers are to convince funders that rehabilitation for domestic life participation is warranted and effective for people with MS.

The gaps in the literature that were uncovered through the preparation of this chapter also point to the need for rehabilitation researchers and practitioners to better articulate the assumptions implicit in their work. Currently, there is an implicit assumption that the reduction in, or compensation for, impairments and improvement in activity capacity will enable clients to engage in complex activities, such as those falling under the rubric of domestic life. Both practitioners and future clients would benefit greatly if researchers moved forward to rigorously test these assumptions and determine the most promising directions for intervention development. For example:

1. If balance is improved and fatigue managed better, are people with MS more able to fulfill necessary and important domestic life responsibilities in ways that are consistent with cultural and personal expectations?

2. Do modifications to activities and environments enable people with MS to fulfill necessary and important domestic life responsibilities in ways that are consistent with cultural and personal expectations?

3. What factors influence participation in domestic life activities for persons with MS?

4. What are the optimal ways of incorporating the social network to enable continued participation in domestic life activities for persons with MS?

To date, studies examining functional changes among people with MS after rehabilitation have focused on basic or personal activities of daily living.[6] To ensure that MS rehabilitation addresses all areas of functioning, it is necessary to answer questions such as those identified above. There is much work to be done related to rehabilitation for domestic life restrictions for people with MS.

References

1. World Health Organization (WHO). International Classification of Functioning, Disability and Health. Geneva, Switzerland: World Health Organization; 2001.

2. Holper L, Coenen M, Weise A, Stucki G, Cieza A, Kesselring J. Characterization of functioning in multiple sclerosis using the ICF. Journal of Neurology 2010;257(1):103–113.

3. Svestkova O, Angerova Y, Sladkova P, Keclikova B, Bickenbach JE, Raggi A. Functioning and disability in multiple sclerosis. Disability and Rehabilitation 2010;32(1):S59–S67.

4. Paltamaa J, Sarasoja T, Leskinen E. Measuring deterioration in international classification of functioning domains of people with multiple sclerosis who are ambulatory. Physical Therapy 2008;88:176–190.

5. Paltamaa J, Sarasoja T, Leskinen E, Wikstrom J, Malkia E. Measures of physical functioning predict self-reported performance in self-care, mobility, and domestic life in ambulatory persons with multiple sclerosis. Archives of Physical Medicine and Rehabilitation 2007;88:1649–1657.

6. Khan F, Pallant JF. Use of international classification of functioning, disability and health (ICF) to describe patient-reported disability in multiple sclerosis and identification of relevant environmental factors. Journal of Rehabilitation Medicine 2007;39(1):63–70.

7. Khan F, Pallant JF, Turner-Stokes L. Use of goal attainment scaling in inpatient rehabilitation for persons with multiple sclerosis. Archives of Physical Medicine and Rehabilitation 2008;89(4):652–659.

8. Minden SL, Frankel D, Hadden LS, Srinath KP, Perloff JN. Disability in elderly people with multiple sclerosis: An analysis of baseline data from the Sonya Slifka longitudinal multiple sclerosis study. NeuroRehabilitation 2004;19(1):55–67.

9. Finlayson M, Winkler Impey M, Nicole C, Edwards J. Self-care, productivity and leisure limitations of people with multiple sclerosis in Manitoba. Canadian Journal of Occupational Therapy 1998;65(5):299–308.

10. Aronson KJ, Cleghorn G, Goldenberg E. Assistance arrangements and use of services among persons with multiple sclerosis and their caregivers. Disability and Rehabilitation 1996;18(7):354–361.

11. National Institute of Clinical Excellence. Multiple sclerosis: National clinical guideline for diagnosis and management in primary and secondary care. London, UK: Royal College of Physicians of London; 2003.

12. Mansson E, Lexell J. Performance of activities of daily living in multiple sclerosis. Disability and Rehabilitation 2004;26(10):576–585.

13. Law M, Baum C, Dunn W. Measuring occupational performance: Supporting best practice in occupational therapy. Danvers, MA: Slack Incorporated; 2001.

14. Goverover Y, Chiaravalloti N, DeLuca J. The relationship between self-awareness of neurobehavioral symptoms, cognitive functioning, and emotional symptoms in multiple sclerosis. Multiple Sclerosis 2005;11(2):203–212.

15. Loeb PA. Independent living scales San Antonio, TX: Psychological Corporation; 1996.

16. Einarsson U, Gottberg K, Fredrikson S, von Koch L, Holmqvist LW. Activities of daily living and social activities in people with multiple sclerosis in Stockholm County. Clinical Rehabilitation 2006;20(6):543–551.

17. Chiu T, Oliver R, Ascott P, Choo LC, Davis T, Gaya A, Goldsilver P, McWhirter M, Letts L. SAFER Home–Version 3 manual. Toronto, ON: Comprehensive Rehabilitation and Mental Health Services; 2006.

18. Fisher AG. Assessment of motor and process skills–Volume 1: Development standardization, and administration manual. Ft. Collins, CO: Three Star Press; 2003.

19. Doble SE, Fisk JD, Fisher AG, Murray TJ. Functional competence of community-dwelling persons with multiple sclerosis using the assessment of motor and process skills. 1994;75(8):843–851.

20. Williams JH, Drinka TJK, Greenberg JR, Farrell-Holtan J, Euhardy R, Schram M. Development and testing of the assessment of living skills and resources (ALSAR) in elderly-community-dwelling veterans. Gerontologist 1991;31(1):83–91.

21. Clemson L, Bundy A, Unsworth C, Fiatarone Singh M. Validation of the modified assessment of living skills and resources: An IADL measure for old people. Disability and Rehabilitation 2009;31(5):359–369.

22. Law M, Baptiste S, Carswell A, McColl MA, Polatajko H, Pollock N. The Canadian Occupational Performance Measure. 4th ed. Ottawa, ON, Canada: CAOT Publications; 2005.

23. Dedding C, Cardol M, Eyssen CJM, Beelen E, Beelen A. Validity of the Canadian Occupational Performance Measure: A client-centered outcome measurement. Clinical Rehabilitation 2004;18(6):660–667.

24. Månsson Lexell E, Iwarsson S, Lexell J. The complexity of daily occupations in multiple sclerosis. Scandinavian Journal of Occupational Therapy 2006;13(4):241–248.

25. Salén BA, Spangfort EV, Nygren AL, Nordemar R. The disability rating index: An instrument for the assessment of disability in clinical settings. Journal of Clinical Epidemiology 1994;47(12): 1423–1434.

26. Baum C, Edwards DF, Morrow-Howell N. Identification and measurement of productive behaviors in senile dementia of the Alzheimer type. Gerontologist 1993;33: 403–408.

27. Baum CM, Morrison T, Hahn M, Edwards DF. Test manual for the Executive Function Performance test. St. Louis, MO: Washington University; 2003.

28. Goverover Y, Kalmar J, Gaudino-Goering E, Shawaryn M, Moore NB, Halper J, DeLuca J. The relationship between subjective and objective measures of everyday life activities in persons with multiple sclerosis. Archives of Physical Medicine and Rehabilitation 2005;86:2303–2308.

29. Goverover Y, Genova HM, Hillary FG, DeLuca J. The relationship between neuropsychological measures and the timed instrumental activities of daily living task in multiple sclerosis. Archives of Physical Medicine and Rehabilitation 2009;86:2303–2308.

30. Voelbel GT, Goverover Y, Gaudino EA, Moore NB, Chiaravalloti N, DeLuca J. The relationship between neurocognitive behavior of executive functions and the EFPT in individuals with multiple sclerosis. Occupational Therapy Journal of Research 2011;31(1):S30–S42.

31. Lincoln NG, Gladman JR. The extended ADL scale: Further validation. Disability and Rehabilitation 1992;14:41–43.

32. Holbrook M, Skilbeck CE. An activities index for use with stroke patients. Age and Ageing 1983;12:166–170.

33. Hébert R, Carrier R, Bilodeau A. The functional autonomy measurement system (SMAF): Description and validation of an instrument for the measurement of handicaps. Age and Ageing 1988;17(5):293–302.

34. Hébert R, Guilbeault J, Pinsonnault E. Functional autonomy measuring system: User guide. Sheerbrook, QC: Expertise Centre of Sheerbrooke Geriatric University Institute; 2002.

35. Becker H, Stuifbergen A, Rogers S, Timmerman G. Goal attainment scaling to measure individual change in intervention studies. Nursing Research 2000;49(3):176–180.

36. Rockwood K, Howlett S, Stadnyk K, Carver D, Powell C, Stolee P. Responsiveness of goal attainment scaling in a randomized controlled trial of comprehensive geriatric assessment. Journal of Clinical Epidemiology 2003;56(8):736–743.

37. Ashley MG, Persel CS, Clark MC. Validation of an independent living scale for post acute rehabilitation applications. Brain Injury 2001;15(5):435–442.

38. Lawton MP, Brody EM. Assessment of older people: Self maintaining and instrumental activities of daily living. Gerontologist 1969;9:179–186.

39. Salter AR, Cutter GR, Tyry T, Marrie RA, Vollmer T. Impact of loss of mobility on instrumental activities of daily living and socioeconomic status in patients with MS. Current Medical Research and Opinion 2010;26(2):493–500.

40. Asberg KH, Sonn U. The cumulative structure of personal and instrumental ADL. A study of elderly people in a health service district. Scandinavian Journal of Rehabilitation Medicine 1989;21:171–177.

41. Baum C, Edwards DF. Cognitive performance in senile dementia of the Alzheimer's type: The kitchen task assessment. American Journal of Occupational Therapy 1993;47(5):431–436.

42. Kohlman Thomson L. The Kohlman Evaluation of Living Skills. 3rd ed. Bethesda, MD: American Occupational Therapy Association; 1992.

43. Zimnavoda T, Weinblatt N, Katz N. Validity of the Kohlman Evaluation of Living Skills (KELS) with Israeli elderly individuals living in the community. Occupational Therapy International 2002;9(4):312–325.

44. Whittle H, Goldenberg D. Functional health status and instrumental activities of daily living performance in noninstitutionalized elderly people. Journal of Advanced Nursing 1996;23: 220–227.

45. Fillenbaum G. Screening the elderly—A brief instrumental activities of daily living measure. Journal of the American Geriatrics Society 1985;33(10):698–706.

46. Fillenbaum G, Smyer M. The development, validity and reliability of the OARS Multidimensional Functional Assessment Questionnaire. Journal of Gerontology 1981;36(4):428–434.

47. Ottenbacher KJ, Mann WC, Granger CV, Tomita M, Hurren D, Charvat B. Inter-rater agreement and stability of functional assessment in the community-based elderly. Archives of Physical Medicine and Rehabilitation 1994;75(12):1297–1301.

48. Owsley C, McGwin Jr G, Sloane ME, Stalysy BT, Wells J. Timed instrumental activities of daily living tasks: Relationship to cognitive function and everyday performance assessments in older adults. Gerontology 2002;48:254–265.

49. Goverover Y, Genova HM, Hillary FG, DeLuca J. The relationship between neuropsychological measures and the timed instrumental activities of daily living task in multiple sclerosis. Archives of Physical Medicine and Rehabilitation 2007;13:636–644.

50. European Multiple Sclerosis Platform. Recommendations for rehabilitation services for persons with multiple sclerosis in Europe. Genoa, Italy: Associazione Italiana Sclerosi Multipla; 2004.

51. Esnouf JE, Taylor PN, Mann GE, Barrett CL. Impact on activities of daily living using a functional electrical stimulation device to improve dropped foot in people with multiple sclerosis measured by the Canadian Occupational Performance Measure. Multiple Sclerosis 2010;16(9): 1141–1147.

52. Law M, Cooper BA, Strong S, Stewart D, Rigby P, Letts L. The person-environment-occupational model: A transactive approach to occupational performance. Canadian Journal of Occupational Therapy 1996;63(1):9–23.

53. Watson D, Wilson SA. Task analysis: An individual and population approach. 2nd ed. Bethesda, MD: AOTA Press; 2003.

54. Fisher AG, Bray Jones K. Assessment of motor and process skills: Volume 1–Development, standardization, and administration manual. 7th ed. Fort Collins, CO: Three Star Press; 2010.

55. Korner M. Interprofessional teamwork in medical rehabilitation: A comparison of multidisciplinary and interdisciplinary team approach. Clinical Rehabilitation 2010;24(8):745–755.

56. Law M, Polatajko H, Pollock N, McColl MA, Carswell A, Baptiste S. Pilot testing of the Canadian Occupational Performance Measure: Clinical and measurement issues. Canadian Journal of Occupational Therapy 1994;61(4):191–197.

57. Phipps S, Richardson P. Occupational therapy outcomes for clients with traumatic brain injury and stroke using the Canadian Occupational Performance Measure. American Journal of Occupational Therapy 2007;61(3):328–334.

58. AMPS Project International [Internet]; c2011 [cited 2011 November 2]. Available from: http://www.ampsintl.com.

59. Fisher AG, Bray Jones K. Assessment of motor and process skills—Volume II: User manual. 7th ed. Fort Collins, CO: Three Star Press; 2010.

60. Linacre JM. Many-faceted Rasch measurement. 2nd ed. Chicago, IL: MESA; 1994.

61. Fisher AG. The assessment of IADL motor skills: An application of many-faceted Rasch analysis. American Journal of Occupational Therapy 1993;47:319–329.

16

*Employment**

Kurt L. Johnson and Robert T. Fraser

CONTENTS

Employment is an important component of community participation for people in general, and people with multiple sclerosis (MS) are no exception. People with MS value employment for a number of reasons. Probably first and foremost is that it provides an income. Many individuals also rely on employment-related health care benefits. They see continuation of employer health benefits as critical since they have high health care costs, and immunomodulatory medications can be very expensive. But employment is important for other reasons as well. It provides an opportunity to demonstrate and experience competence, is a valuable social outlet, and has been associated with a higher overall perceived

---

quality of life. It is important for people with MS to note that it is easier to ignore symptoms, such as pain, when they are working.[1] Not only is employment important for the person with MS, but it has been found that caregiver distress also rises significantly with unemployment.[2,3] After reading this chapter, you will be able to:

1. Identify barriers to full participation in education and employment for people with MS,
2. Describe the vocational rehabilitation process,
3. Implement vocational rehabilitation interventions, and
4. Understand the process of accommodations to increase participation.

16.1 Epidemiology of Education and Employment Participation Restrictions in MS

The proportion of people with disabilities in general who were employed in 2009 was only 19.2% as compared to the population in general, which is 64%, according to the U.S. Department of Labor.[4] There are a number of reasons for this discrepancy. For example, according to census data, people with disabilities tend to be older, more likely to be female, and more likely to be from underrepresented backgrounds than those in the general population—these are all factors associated with lower employment rates. But in many surveys, people with disabilities consistently say they would prefer to work. Many of the barriers to employment that people with disabilities face are related to subsidy and health care benefit systems, and one of the significant factors associated with disability is a high rate of poverty.[5]

It is difficult to get a fully representative picture of the status of people with MS with respect to employment, education, or many other variables because no surveys representative of the general population in the United States have been conducted, and all published surveys have significant sample bias. Johnson et al.[6] reported that virtually all respondents to their survey ($N = 1125$) had graduated from high school and most had at least some college education, with 30.6% holding baccalaureate degrees and 17.7% reporting that they had graduate or professional degrees. In another study, the median education level for clients seeking vocational rehabilitation services was 15 years.[7] Education can be critical for people with MS since higher levels of education are associated with more flexibility on the job; more options about how one performs the job, such as flexible schedules; and a higher salary.

In terms of employment, historically it has been reported in the literature that 90% of people with MS have a history of employment.[8] However, the number of people who were working 5–15 years from diagnosis of MS has been reported to be between 20% and 30%.[9,10] More recently, Johnson et al.[6] reported that, in their survey, 40% of the respondents with MS were working at least 20 h per week, 13.4% were working fewer than 20 h per week, and of those most worked only 2–3 h per week. Employment status, although still poor for this disability group, may be improving slightly due to improvements in medication, increased allowable earnings under Social Security Disability Income (SSDI) (currently up to $1000 monthly), the need to earn funds given a recession, and other more subtle factors.

MS can have a significant impact on individuals' careers, leading them to retire prematurely, or they may find that they must leave more challenging and demanding occupations for lower-stress jobs.[11,12] As has been found in the literature on disability in general, nearly half of the unemployed people with MS would prefer to work if they could.[8]

Volunteer work is often seen as a way to engage in meaningful activity whether one is working or not. Volunteer work is often sought by people who retire from employment due to age. However, a review of the data from the survey reported by Johnson et al.[6] revealed that few people with MS in that sample were engaged in volunteer activity. The majority of respondents, 67%, reported that they were not engaged in any volunteer activity, and another 21% reported that they engaged in an average of 2.5 h of volunteer activity a week. Volunteers were about evenly split between employed and not employed.

16.2 Barriers to Employment

A number of factors contribute to an individual's employability, ranging from educational level to temperament and from ability to labor market forces. In addition to these, people with MS may face barriers to employment that are related to personal factors, such as the progression of their MS and systemic factors related to health insurance and disability subsidies.

16.2.1 Individual Factors

In recent investigations, the most significant predictors of employment of people with MS in the United States have been self-reported cognitive changes and limitations in mobility.[6] It appears that the higher the "disease burden" that people with MS experience, the less likely they are to work. In other studies, it has been found that other variables, such as disease progression and affective status, may contribute.[13] These studies are difficult to compare because of a lack of uniformity in the definition of both independent and dependent variables, including employment. Prediction studies, however, do not shed much light on the factors that make employment difficult or accommodations that might be made for people still working or seeking to return to work.

As noted in survey studies described above and work done by Rao et al.,[14] people with cognitive changes are much less likely to be employed. Certainly, the whole range of symptoms experienced by people with MS can serve as barriers to maintaining employment. In addition to cognitive changes, these symptoms include fatigue, depression, anxiety, pain, heat intolerance, changes in bowel and bladder function, reduced mobility, difficulty with fine motor coordination, and changes in vision. A "metasymptom" that can be very difficult to manage is the unpredictability of MS. Many people report that they cannot predict their function from one day to the next.[15] In a qualitative study by Johnson et al.,[1] interviewees reported that it was really the convergence of cognitive changes, fatigue, concern about negative evaluation by others, and pain that proved to be difficult to handle. For example, interviewees reported that "thinking" made them tired, and when they were tired, "thinking" became more difficult. For purposes of vocational rehabilitation intervention, it can be difficult to parse out these concerns because they are interwoven.

16.2.2 Social and Programmatic Factors Related to Employment

Individuals with MS may find that as their symptoms worsen, people in their support network and health care professionals recommend that they leave employment.[16,17] Although these recommendations may be well intended, the consequences of leaving employment may be significant for the people with MS. Individuals report that even when work is difficult, it is often preferable to reduced income, lack of health care insurance, and loss of vocational identity. In addition, work offered them a distraction from symptoms such as pain. Individuals living with MS in the workplace also report that the attitudes of others may have an impact on them. Coworkers and supervisors may not understand the functional limitations with which they live or may overestimate the impact of MS on their ability to work. In reference 1, respondents reported that they either had faced disability discrimination in the workplace or were concerned that they could.

Issues involving financial and health care benefits can serve as significant barriers to full employment as well, especially in the United States, where health care benefits are tied either to employment or to disability benefits. For many people who have long-term disability insurance, the option to work part time while receiving benefits is not available. Even with the U.S. federal long-term disability insurance, "Social Security Disability Insurance," a program that has some built in incentives (e.g., capacity for some earnings, such as allowances for Individual Work-Related Expenses [IWRE]) to allow people to work while still keeping their benefits, the options are somewhat limited. Often, continuation of health care benefits is tied to the wage subsidy. Furthermore, it may be difficult economically and psychologically for people to move into less demanding jobs if they cannot continue their current jobs. The result is that people fall into the "disability trap," in which they perceive that they have no options other than to work in their current jobs or retire.[2,15]

As one might imagine, when people struggle to participate at work, they have fewer resources for home, family, and community life. Shopping after work or doing routine household chores or even socializing with family members may be impossible because of exhaustion. People living alone report that they are isolated because of their fatigue after work.[1] Stress and family role changes are often associated with changes in employment status as well.[3]

When individuals must leave employment, it can cause financial stress for the whole family although often people move from wages associated with employment to either private or federal disability subsidies that may mitigate the impact. There is no question, however, that there is a marked loss of income, and even more so if federal disability subsidies, such as SSDI in the United States, is the only option. Evidence also exists to indicate that workers with MS seem to move to long-term disability status at both a dramatically higher percentage and a faster rate than a general-disability or other neurological-disability group.[18] In some cases, this may be a direct function of disease symptoms and complications. Unfortunately, this could also be a function of faulty advice from medical advisers, human resource departments, and similar sources of advice.

16.3 Vocational Rehabilitation

In the United States, vocational rehabilitation (VR) programs are offered at the state level but are funded with $4 dollars of federal funds for each $1 of state funds. State VR programs are staffed with (vocational) rehabilitation counselors who are mostly trained at the master's degree level. Individuals qualify for VR services if they have functional limitations associated with disability that interfere with participation in employment and if changing their employment status is a primary goal. Rehabilitation counselors also work in private, nonprofit hospital, and other settings. In other countries such as Canada, Australia, and members of the European Union, similar programs are in place at a national or provincial level, and there are vocational specialists or occupational therapists who provide services, although their level of training (i.e., bachelor's or master's degree) varies by country. The history of vocational rehabilitation is relatively limited in MS to demonstration projects chiefly funded by the U.S. Rehabilitation Services Administration (RSA). These projects have included MS Back to Work or Operation Job Match (1980), the Job Raising Program (1983), the Return to Work, the Career Possibilities Program (1994), and Project Alliance (1997). The projects were often funded in an affiliation between the National MS Society and its chapters and university research groups. The early projects involved more direct placement assistance coupled with job-seeking skills while later projects emphasized more early intervention and accommodation at the workplace before the job is lost.

The last multisite Project Alliance effort is of particular interest. It involved a very structured intervention, including early intervention and open discussion of issues between employer and employee, jobsite analysis, report formulation with accommodation recommendations, and follow-up by vocational rehabilitation staff relative to retention. The project was successful, with 85% of clients who fully engaged in the intervention still working at follow-up. Only one third of the participants, however, fully engaged in the process to the point of allowing project staff to even complete their job analyses.[17] One significant limitation of all these projects was that they were "time-limited" demonstration projects without ongoing funding.[17] A later review by Fraser[19] indicated that these projects, nevertheless, underscored a number of needs relative to clients with MS and vocational rehabilitation.[19]

1. Clarification of financial/disability subsidy issues (success was not realized for clients on SSDI)
2. Importance of individualized vocational assessment, including neuropsychological status
3. Importance of effective strategizing with respect to disclosure and accommodation
4. The need for a range of effective job placement models, including home-based models

In a recent Cochrane review[20] on the effectiveness of vocational rehabilitation interventions on employment and returning to work for people with MS, only two trials met the review criteria (total $n = 80$), both scoring poorly on methodological quality. There was insufficient evidence to support VR intervention due to the lack of relevant randomized and properly controlled clinical trials. There were also inadequate data to support recommendations relative to intervention type, setting, duration, or other program components. The lack of evidence is a concern, but it may reflect the difficulties of conducting randomized, controlled trials of vocational rehabilitation and raises the question about the appropriateness of the current standards of evidence for these types of complex interventions.

16.3.1 Employability/Vocational Assessment

The major questions for the adult with MS are whether one can work and, if so, full time or part time, and with or without accommodations. This assessment process is always collaborative with the individual with MS and not "diagnostic." To begin with, it is helpful for the person with MS to "take inventory" of his or her current situation and resources with respect to employment. For example, they can respond to the Work Experience Survey, which reviews the essential functions of the job, accessibility issues, concerns about job mastery, and so forth, and asks the respondent to identify barriers to employment and solutions[21] (see Table 16.1). The Personal Capacities Questionnaire (PCQ)[22] can be helpful in clarifying functional strengths and limitations that might have an impact on work. Response to these self-report inventories can be reviewed with family or significant others to inform and provide context to the evaluations conducted by a rehabilitation counselor and neuropsychologist. Obviously, financial subsidy, in addition to cognitive and physical capacities, feed into the decision to return to work.

16.3.2 Baseline Vocational Evaluation

A baseline vocational evaluation is important as a starting point to set new vocational goals and is often funded by state vocational rehabilitation when a client needs a new job goal. It involves a thorough review of physical-, cognitive-, and emotional-functioning variables as well as the client's financial status. Psychometric testing can include vocational interests (e.g., the Career Assessment Inventory), work values or reinforcers (e.g., the Work Preference Match), emotional/personality functioning, and academic achievement and abilities testing.

Synthesis of cognitive concerns in the development of vocational options is generally provided through review of neuropsychological assessment data. Briefer neuropsychological batteries have been developed[14,23] that focus on executive functioning, processing speed, learning and memory, language skills, and attentional capacities. Chapter 8 discusses these assessments in greater detail. These types of abilities are often very germane to jobs requiring higher levels of education and that are more demanding cognitively. As an example, for people working at a higher level in one study, neuropsychological measures related to speed of information processing, cognitive flexibility, and word fluency were most significantly related to employment retention across all other medical and psychosocial variables.[24] Consideration of these variables can be critical in arriving at viable job goal options.

When possible, members of a formal or ad hoc interdisciplinary team should participate in the vocational evaluation. In some settings, the evaluation will be led by the rehabilitation or vocational counselor, while in other settings, it may be led by a psychologist or

TABLE 16.1

Examples of Key Elements of Roessler and Gottcent's Work Experience Survey

Environmental and built-environment accessibility
Requirements to perform essential functions of the job
• Physical abilities
• Cognitive abilities
• Task-related abilities
• Social abilities
• Working conditions

Source: Adapted from Roessler R, Gottcent J. Journal of Applied Rehabilitation Counseling 1994;25:16–21.

TABLE 16.2

Examples of Roles in the Vocational Rehabilitation Assessment Process[a]

Physician/advanced registered nurse practitioner (ARNP)	Provides the verification to the employer of disability when required; assists in maximizing symptom management
Neuropsychologist/rehabilitation psychologist	Documents the cognitive strengths and limitations and recommends accommodations and compensatory strategies
Physical therapist	Assists with improving mobility and endurance and recommends fitness and exercise programs
Occupational therapist	Assists with maximizing motor control and upper extremity function and designing accommodation strategies, including computer access, which may involve modifications to seating and positioning
Speech pathologist	Assists in designing and implementing compensatory strategies for cognitive deficits
Rehabilitation counselor	Coordinates and implements the vocational rehabilitation process

[a] Recognizing that there is considerable disciplinary overlap.

occupational therapist. In any case, the attending physician, advanced registered nurse practitioner (ARNP) or other health care provider will provide important verification of disability and information about medical management, prognosis, and course of disease. Other team members, such as occupational therapists, physical therapists, and speech pathologists, will contribute to the evaluation as described in Table 16.2, depending on the setting and organization of the team.

Based upon clinical interview and test findings, some job goals can be established by considering transferability of skills. Often, an outcome of the baseline vocational evaluation is a recommendation for a community-based job tryout or situational assessment to gain further information (see Table 16.3). The vocational evaluation might also recommend short-term training, given the impact of the disability and the midcareer background of many of these individuals (e.g., medical billing and coding). Finally, the evaluation might recommend a placement approach and valuable accommodation considerations.

16.3.3 Community-Based Tryouts

For a number of potential workers with MS, it is difficult to assess employability and accommodation needs without a job tryout or situational assessment. This type of performance-based tryout is of value in assessing an individual's capacity to return to some form of work within a prior company or perform a new type of work in a novel setting. Situational assessment is often a critical part of the vocational evaluation process. An individual can either volunteer for a specific position within a nonprofit or take advantage of the U.S. Department of Labor (1993) waiver for unpaid work. The waiver allows individuals with disabilities to work in the private sector on an unpaid basis for purposes of vocational exploration, assessment, and job training. The total number of hours allowed is 215 h, 5 h for vocational exploration, 90 h for vocational assessment, and 120 h for actual vocational training. There are, of course, several caveats (e.g., not displacing other workers), but this is a tremendous benefit to the worker with MS in establishing whether returning to a former job is possible, what type of modifications are important, or viability in a new position. The hours worked can be varied as necessary, and if an individual starts a new job tryout (even within the same company), the 215-h allowance begins anew. This arrangement is usually

TABLE 16.3

Vocational Evaluation Process: Guiding Questions for People with MS When Making Decisions about Employment

---

What do I want to do?
- Keep my current job?
- Find another job?
- Leave employment?

What factors should I consider in making an employment decision?
- What are my interests and preferences?
- My personal style or temperament?
- My financial needs?
- My physical and cognitive status?
- My potential changes as MS progresses?

What do I need to know about how to be successful on the job?
- What overall strengths to I bring to a job?
- What kinds of barriers can I anticipate?
- Related to my MS, such as fatigue, cognitive changes, difficulty with mobility? This is important for planning accommodations.
- Continuing my employer-based health insurance?
- Impact of employment on my subsidy benefits, such as SSDI in the United States?

How can I gather additional information?
- Through self-reflection
- Through the counseling process
- By responding to inventories asking me questions about my interests, skills, and values to help me think of new ways to consider my options
- Evaluating the labor market
- By working for short periods of time in different environments doing different tasks to evaluate what would be a good fit for me

---

engineered through a vendor for a state vocational rehabilitation agency, and the vendor covers any worker's compensation costs. In the United States and other countries, work tryouts are often set in a hospital, university, or other community setting.

16.3.4 Finding or Keeping Employment

Individuals with MS who are already working but having difficulty may, through the assessment process, decide that it is feasible to keep that job if they can make some changes or add some support features (see the following section on accommodations). Some individuals can manage this on their own, but many will profit from guidance and technical assistance from a rehabilitation professional. If the decision is to seek a new job, then in vocational rehabilitation, the focus will be on the job placement process.

There is no good evidence base to support choosing a placement approach for a specific client with MS. Through projects funded by the U.S. RSA, it has been found that it is very difficult to move clients on Social Security subsidies to competitive employment. Fraser et al. found that neuropsychological variables were the most predictive of job retention; therefore, financial and cognitive variables appear to be important.[24] Level of physical impairment, emotional status, and work-related variables (time on the job, level of complexity, availability of supervision) were also significant. As noted earlier, disease burden and cognitive status are associated with unemployment.[6]

Simplistically, the more functional limitations an individual experiences with respect to life functioning, the less support will be available at the worksite (especially with increasing job complexity), and the more significant the subsidy barriers, the more intensive the vocational rehabilitation intervention will need to be. The following would be the sequence of options available through a state vocational rehabilitation agency at increasing levels of support.

16.3.4.1 Vocational Counseling and Advisement Only

Clients with MS who are receiving this service from the state vocational rehabilitation agency or other resource generally lack significant cognitive and physical impairment, are able to organize a job search, profit from informational interviews, and so forth. They benefit from the vocational rehabilitation staff's advisement, job-seeking skills, and other training while not requiring actual brokering of their skills to the employment community.

16.3.4.2 Selective Placement

This approach can involve persons with significant physical or cognitive impairment. They require brokering because certain accommodations may be necessary, financial incentives are being offered to the employer (e.g., tax credits, preliminary tryout), or the job is part time and is being "carved out" for the individual. The majority of individuals with MS coming to a vocational rehabilitation program are likely to fall into this category. Developing a home-based work option is, at times, an option. Diverse types of work can all be done from home using the Internet, including customer service, ordering and billing, information and referral, and similar jobs. For people with fatigue, working from home may allow a flexible schedule. State vocational rehabilitation can also assist some individuals set up self-employment or independent contract work. Because of the potential for fraud on the part of some offering home-based work opportunities as well as other difficulties, home-based employment has never "taken off" to the degree initially envisioned.[25]

The process of preserving employment or seeking new employment, or even retiring, should be based on empirical data derived from assessment. The stakes are high in terms of self-esteem, financial status, and health status. It is critical to determine what kinds of employment supports will be necessary, including workplace accommodations. The process of getting or keeping a job is highly individualized, depending on the specific needs of the person with MS. The more personal and systems barriers confronting an individual, the higher the level of support required to secure and maintain employment.

16.3.4.3 Natural Supports in the Workplace

In this case, a vocational rehabilitation counselor or specialist works with coworkers or jobsite supervisors so that a client with MS is mentored or, in some cases, has a physical assistant in order to be proficient on the job. In some cases, the company is reimbursed and receives On-the-Job Training (OJT) funds for this effort.

Curl and colleagues have formalized a procedure in directly using paid coworkers as trainers.[26] The paid coworker model could be implemented by a vocational rehabilitation agency. In consideration of the higher educational level completed by many clients with MS, this paid coworker–as trainer approach appears highly applicable. Often, a young job coach hired by a rehabilitation agency simply will not have the skill base to coach individuals on the higher end of semiskilled to skilled work, such as an architect, electronic technologist, mortgage processor, and other work requiring more education or training. The paid coworker can meet the on-the-spot mentoring need, even for professionals, at very reasonable costs.[27]

16.3.4.4 Agency-Based Jobsite Support

Although there are a number of forms of supported employment, the one-on-one job-coach model as described by Wehman and colleagues[28] would be most applicable in the MS field, specifically for individuals with more involved cognitive impairment.[28] Due to the more complex level of jobs held by many midcareer professionals with MS, the traditional job-coach model will often be inappropriate and a coworker/trainer optimal.

Some types of jobsite support provided by an agency are directly physical in nature (e.g., transportation to work, assistance in toileting). A physical assistant may initially be provided by an agency but often will eventually become a paid employee of the worker with MS. The cost of this type of physical-assistive support may be paid under the IWRE mechanism, which reduces net income and results in Social Security financial subsidy levels not being affected.

16.3.5 Workplace Accommodations

Accommodations can be either self-taught or formalized, and may be managed by the individual with MS him- or herself or collaboratively with the employer. The prohibition of discrimination in employment for people with disabilities and the legal basis for their right to reasonable accommodations is governed by the Americans with Disabilities Act (ADA) of 1990 and the Americans with Disabilities Restoration Act of 2007 in the United States. The ADA protects "qualified" employees (e.g., those with a disability that limits ability to perform major life activities, including work) from discrimination in employment. Specifically, if a qualified individual with MS is able to do the essential functions of the job (e.g., core functions) with or without accommodation, the employer may not discriminate against the employee in hiring, retaining, or similar aspects of employment. Employers with 15 or more employees are covered under ADA although many states have laws that obligate smaller employers as well. What is "reasonable" depends on a variety of factors, but from a practical perspective, "reasonable" depends on the size of the employer and the resources available.[29] Additional information on the ADA can be obtained from the regional network of ADA technical assistance centers.[30] Similar obligations exist under disability rights legislation in the European Union and Canada as well.

Accommodations in the workplace may include procedural modifications (schedule, tasks, supervisory style, workplace tools), physical environment modifications, or the person's technical equipment (e.g., voice-activated software). Examples of accommodations are provided as follows by functional limitation, but the reader should recognize that combinations of accommodations are often required. A good place to find out more about workplace accommodations, including assistive technology is at the website and database maintained by ABLEDATA. Before considering accommodations, one must decide whether to disclose to the employer since under U.S. law, this determines who bears financial responsibility.

16.3.6 Disclosure

Requesting an accommodation requires disclosure to the employer, which may seem risky to some people with MS who fear subtle discrimination if they disclose. But under the law, if the employer is to provide the accommodation, the worker must disclose the functional limitations that require accommodation. This is best disclosed to the human resources department rather than an immediate supervisor, if possible. The request for accommodation is intended to be collaborative and interactive between the employee and the employer.

DISCLOSURE TIPS

- Disclosure mantra: Too much information is TMI!
- Decide whether disclosure is advisable. Does the individual need services such as accommodation or family leave from the employer?
- Careful planning and role-play are recommended prior to disclosure.

Disclosure involves the worker's describing the functional limitations that he or she has due to a disability and the specific accommodations (procedural, workstation modifications, or assistive technology/equipment) needed to perform effectively. The disability, per se, does not need to be disclosed. As discussed earlier, the Work Experience Survey[21] can be helpful in facilitating the identification of barriers that need to be accommodated. If the worker with MS cannot identify appropriate accommodations, brainstorming with a rehabilitation counselor, a representative from the regional ADA Centers,[30] or West Virginia's Job Accommodation Network[31] can be extremely helpful.

It is also important to not only develop a disclosure/accommodations request script, but to practice, practice, practice! Rumrill described a Progressive Request Model to include role-playing the person with MS making the request and the employer's response.[32] Fraser[33] also emphasized a Disclosure Script Worksheet.[33] If the individual has cognitive issues related to the MS, this practice piece is even more important.

16.3.7 Fatigue

Fatigue is consistently cited by people with MS as a major barrier to performance of many preferred activities, including work, and, as noted above, fatigue may interact with cognitive changes and stress, so managing fatigue in the workplace may help improve cognitive function as well. People with MS report different patterns of fatigue. Some find that their fatigue progresses during the day while others find their fatigue to be unpredictable. For those with predictable fatigue, it may be possible to schedule more cognitively demanding tasks in the morning and less demanding tasks later in the day. For all, scheduling rest breaks may be critical. For some this involves "power naps," or periods of relaxation. For others, it may mean taking an hour off during lunch to nap by requesting (as an accommodation) to split the day with a two-hour break in the middle. Also, for some people with heat sensitivity, managing the local temperature climate may reduce fatigue. Chapter 5 describes rehabilitation strategies to manage fatigue in detail.

16.3.8 Cognitive Changes

As noted above, managing fatigue can have a significant impact on perceived cognitive status. People with cognitive changes may benefit from increasing the structure and predictability of their work environment. For some, restricting distractions will be important, so a private office, acoustical and/or visual screening, or segmenting time away from answering telephones may help. Low-tech options such as memory books and "yellow stickies" (e.g., for returning to task after being interrupted) are used by many people with MS. Developing routines helps to focus cognitive energy on specific tasks.

Organizing the physical environment to reduce visual clutter may help with concentration. Similarly, using strategic organization of folders for e-mail and file structure on servers may help avoid "inbox chaos" and lost files. Integrated e-mail/calendar/task-list systems such as Microsoft Outlook Exchange and Google Calendar may be enlisted as cognitive prosthetics. Alarms may be enabled in calendar entries that will alert the user to upcoming events or tasks. These alarms may be audio from a computer, but users can also set these up so that e-mail alerts are sent or SMS (text messages) text pages to their mobile phone. For some people, learning new material is more difficult, and they will require additional time to review material, outline key elements, and reread to acquire mastery. They may require ongoing use of "cheat sheets" so that they do not depend on new learning for key points. In some cases, a supervisor or coworker might volunteer or be paid to highlight the critical sections of a training manual, sequence written operative steps, and so forth. Chapter 8 describes rehabilitation strategies to manage cognitive limitations in detail.

16.3.9 Heat Sensitivity

Many people with MS have heat sensitivity to some extent and find that they do not function as well with higher temperatures. In hotter climates, this may require that people use air-conditioned vehicles and avoid lengthy walks in the heat to and from parking or other buildings. Within the workplace, they may be able to manage by positioning their workstation in a cooler area away from direct sunlight or using portable personal air-conditioning systems (e.g., a moveable canister) or personal heat-extraction systems, such as cooling jackets. Additional strategies for managing heat sensitivity are addressed in Chapter 5.

16.3.10 Mobility

Mobility limitations may be accommodated in a number of ways. First, positioning the workstation to require the least amount of walking may help. This means that the workstation is located nearer to a building's entrance, the restroom, key meeting spaces, or closer to a parking spot, as desired. For others, with significant mobility demands in the workplace, a mobility aid such as a scooter may be used in the workplace. For some, this will be their personal mobility device, but others may only require a device in the workplace. Chapter 13 describes rehabilitation strategies to manage mobility restrictions in detail.

16.3.11 Seating and Positioning

Ensuring that the workspace is ergonomically well designed can significantly reduce positional fatigue and pain. This includes not only the chair, but also the position of the keyboard, monitor, and other tools in the workplace.

16.3.12 Vision

Vision changes may be accommodated by modifying the overhead or task lighting to improve contrast of reading materials or reduce glare on the computer screen. High-quality computer monitors that can be positioned to maximize access may be important. Screen enlargement software may be useful as well. For people with difficulty reading print materials, a closed-circuit TV (CCTV) to enlarge print materials may help. It may also be helpful to the person with MS to receive electronic copies of materials so that text-to-speech features on the computer may be used to "read" the text to the employee. Chapter 11 describes rehabilitation strategies to manage visual symptoms in detail.

16.3.13 Assistive Technology

As noted above, assistive technology (AT) often plays a key role in accommodations. AT can be "personal technology." For example, if an individual uses a power wheelchair for mobility across a variety of environments, including at work, this is his or her personal AT. On the other hand, the AT may be specific to the work environment. For example, an individual might use speech-to-text software to make it possible to dictate material into the computer; in the workplace, this technology might be considered an accommodation. It is important that any AT be considered in the context of the overall function of the individual and the context of any other accommodations.

16.4 When Continuing (or Returning to) Employment Is Not an Option

For perhaps the majority of people living with MS, there comes a time when continuing employment is no longer the best option. One "tipping point" is when the accumulation of functional limitations limits the efficiency of job performance enough so that the individual is no longer competitive. Another tipping point is when, because of disease progression, working takes so much energy from an individual that he or she is too depleted to perform other valued life roles, such as participating in important family relationships. Whatever the reason, people anticipating leaving employment will have significant concerns about their identity, financial security, status of their health care coverage, loss of social outlet, and other important factors. When possible, encouraging individuals to anticipate disability retirement before their hands are forced by the employer or by their health may give them more control over the process. Health care providers may assist in this by routinely asking about employment status during provider visits and referring individuals who express some concern about maintaining their current status for evaluation and counseling as above.

16.5 Conclusions for Health Care Providers

Employment is highly valued. In the United States, where health insurance is often tied to employment status, careful consideration about maintaining or leaving employment is critical. Even in countries where health care is universal, leaving or entering employment can have significant consequences for individuals and their families. We recommend that when serving people with MS health care providers, including physicians, nurses, therapists, and others, routinely ask those people about employment. Early identification of problems or risks may allow for early intervention while there is still time to salvage a valued job or to retire with dignity.

16.6 Knowledge Gaps and Future Directions

In closing, there are a number of persistent concerns relative to people with MS and employment. These issues include the following:

- Vocational rehabilitation has the potential to provide highly cost-effective approaches to finding employment for adults with MS, but additional research is needed to establish the critical components of intervention that are most likely to benefit people with MS.[20] Advocacy is required to establish federal and private support for this research in the future.

- The National Multiple Sclerosis Society (NMSS) website has considerable material on maintaining employment, including advice on disclosure, employment rights, accommodation strategies with accompanying DVDs, workbooks, and other relevant materials. In some local chapters of the NMSS, there may be individualized employment services available as well. There may be a logical convergence between advocacy, self-management, and vocational rehabilitation for people with MS that can be brokered through NMSS chapters and other organizations.

There needs to be more seamless coordination across medical, psychosocial, and vocational rehabilitation providers relative to employment needs and services.[34] For people living with MS and their families, employment can be a very significant issue.

References

1. Johnson K, Yorkston K, Klasner E, Kuehn C, Johnson E, Amtmann D. The costs and benefits of employment: A qualitative study of experiences of individuals living with multiple sclerosis. Archives of Physical Medicine and Rehabilitation 2004;85(2):201–209.
2. Aronson KJ. Quality of life among persons with multiple sclerosis and their caregivers. Neurology 1997;48(1):74–80.
3. Starks H, Morris M, Yorkston K, Gray R, Johnson K. Being in- or out-of sync: A qualitative study of couples' adaptation to change in multiple sclerosis. Disability and Rehabilitation 2010;32(3):196–206.
4. Persons with a Disability: Labor Force Characteristics Summary. [Internet]; c2010 [cited 2010 July 28]. Available from: http://www.bls.gov/news.release/disabl.nr0.htm.
5. National Council on Disability [Internet]; c2008 [cited 2010 July 28]. Available from: http://www.ncd.gov/newsroom/publications/2008/FinancialIncentives.html.

6. Johnson K, Bamer A, Fraser R. Disease and demographic characteristics associated with unemployment among working age adults with multiple sclerosis. International Journal of MS Care 2009;11(3):137–143.

7. Fraser RT, Johnson EK, Clemmons DC, Getter A, Johnson KL, Gibbons L. Vocational rehabilitation in multiple sclerosis (MS): A profile of clients seeking services. Work 2003;19:1–9.

8. LaRocca N, Kalb R, Scheinberg L, Kendall P. Factors associated with unemployment of patients with multiple sclerosis. Journal of Chronic Diseases 1985;38(2):203–210.

9. Busche KD, Fisk JD, Murray TJ, Metz LM. Short term predictors of unemployment in multiple sclerosis patients. International Journal of Rehabilitation Research 1986;9(2):155–165.

10. Kornblith AB, LaRocca NG, Baum HM. Employment in individuals with multiple sclerosis. International Journal of Rehabilitation Research 1986;9(2):155–165.

11. Jackson MF, Quaal C, Reeves MA. Effects of multiple sclerosis on occupational and career patterns. Axone 1991;13(1):16–27.

12. Hassink G, Manegold U, Poser S. Early retirement and occupational rehabilitation of patients with multiple sclerosis. Rehabilitation (Stuttg) 1993;32(2):139–145.

13. Roessler RT, Fitzgerald SM, Rumrill PD, Koch LC. Determinants of employment status among people with multiple sclerosis. Rehabilitation Counseling Bulletin 2001;45:31–39.

14. Rao SM, Leo GJ, Ellington L, Nauertz T, Bernardin L, Unverzagt F. Cognitive dysfunction in multiple sclerosis, II: Impact on employment and social functioning. Neurology 1991;41(5):692–696.

15. Johnson K, Amtmann D, Yorkston K, Klasner E, Kuehn C. Medical, psychological, social, and programmatic barriers to employment for people with multiple sclerosis. Journal of Rehabilitation 2004;70:38–50.

16. O'Day B. Barriers for people with multiple sclerosis who want to work: A qualitative study. Journal of Neurological Rehabilitation 1998;12(3):139–146.

17. Rumrill P. Employment and multiple sclerosis: Policy, programming, and research recommendations. Work 1996;6:205–209.

18. Fraser RT, McMahon B, Danczyk-Hawley CE. Progression of disability benefits: A perspective on multiple sclerosis (MS). Journal of Vocational Rehabilitation 2003;19:173–179.

19. Fraser R. Vocational rehabilitation intervention. New York: Demos Vermonde Press; 2002.

20. Khan F, Ng L, Turner-Stokes L. Effectiveness of vocational rehabilitation intervention on the return to work and employment of persons with MS. Cochrane Library 2009;(1). Art. No.: CD007256. DOI: 10.1002/14651858.CD007256.pub2.

21. Roessler R, Gottcent J. The work experience survey: A reasonable accommodation/career development strategy. Journal of Applied Rehabilitation Counseling 1994;25:16–21.

22. Crewe N, Athelston G. Personal capacities questionnaire. Menomonee, WI: University of Wisconsin-Stout, Materials Development Center; 1984.

23. Clemmons D, Fraser R, Rosenbaum G, Getter A, Johnson E. An abbreviated neuropsychological battery in multiple sclerosis vocational rehabilitation: Findings and implications. Rehabilitation Psychology 2004;49(2):100–105.

24. Fraser RT, Koepnick D, Getter A, Johnson E, Gibbons L. Predictors of employment stability in MS vocational rehabilitation. Journal of Vocational Rehabilitation 2009;31:1–7.

25. Rumrill P, Fraser R, Anderson J. New directions in home-based employment. Journal of Vocational Rehabilitation 2000;14:3–4.

26. Curl R, Fraser RT, Cook R, Clemmons D. Traumatic brain injury vocational rehabilitation: Preliminary findings from the co-worker as trainer project. Journal of Head Trauma Rehabilitation 1996;11:75–85.

27. Fraser R, Cook R, Clemmons D, Curl R. Work access in traumatic brain injury rehabilitation: A perspective for the physiatrist and allied health team. Physical Medicine and Rehabilitation Clinics of North America 1997;8:371–387.

28. Wehman P, Bricout J, Targett P. Supported employment for persons with traumatic brain injury: A guide for implementation. In: R. T. Fraser, D. Clemmons, editors. Traumatic brain injury

rehabilitation: Practical vocational, neuropsychological, and psychotherapy. Boca Raton, FL: CRC Press; 2000.

29. Johnson K, Fraser R. Mitigating the impact of multiple sclerosis on employment. Physical Medicine and Rehabilitation Clinics of North America 2005;16(2):571–582.
30. National Network: Information, Guidance, and Training on the Americans with Disabilities Act [Internet]; c2010 [cited 2010 August 24]. Available from: http://adata.org.
31. Job Accommodation Network [Internet]: Office of Disability Employment Policy, U.S. Department of Labor; c2010 [cited 2010 August 24]. Available from: http://askjan.org.
32. Rumrill P. Employment issues and multiple sclerosis. New York: Demos Vermonde Press; 1996.
33. Fraser R. Employment strategies and community resources. In: R. Fraser, editor. The MS workbook: Living fully with MS. Oakland, CA: New Harbinger; 2006.
34. Rumrill P. Introduction to employment and MS: Key concerns. MS in Focus 2010;16:4–5.

17

Physical Activity and Leisure

Matthew Plow and Robert W. Motl

CONTENTS

Physical activity can be described as any bodily movement produced by voluntary skeletal muscle contraction that results in a substantially greater expenditure of energy compared to resting levels.[1] Physical activity is now recognized as an important lifestyle behavior with meaningful outcomes among individuals with multiple sclerosis (MS). Routine engagement in physical activity might help preserve physical function and prevent the development of secondary conditions in adults with MS. However, those with MS are often sedentary as they face many barriers to engaging in physical activity. Rehabilitation professionals are in an ideal position to provide support in overcoming barriers to engaging in physical activity. There are five learning objectives of this chapter on physical activity and leisure participation among individuals with MS. After reading this chapter, you will be able to:

1. Describe the benefits of physical activity and exercise,
2. Identify average levels of physical activity participation,
3. Give examples of facilitators and barriers to participating in physical activity,
4. Understand methods to measure physical activity, and
5. Implement strategies to help increase physical activity.

17.1 Types of Physical Activity

Physical activity can be described in terms of intensity, frequency, duration, and type. Although there are many different types of physical activity, for the purposes of this chapter, we will describe four types: leisure physical activity, occupational physical activity, exercise, and therapeutic exercise. Leisure physical activity can be defined as activities that an individual performs during free time, whereas occupational physical activity can be defined as activities that an individual participates in while doing his or her job.[2] Examples of leisure physical activity include hiking, running, swimming, dancing, bicycling, walking, gardening, yoga, household chores, and playing sports. Examples of occupations that involve moderate-to-high amounts of physical activity include construction, fire fighting, mail delivery, and waiting tables. Exercise can be considered a subtype of leisure physical activity, but it is structured with a primary purpose of maintaining or improving health-related fitness (e.g., muscle strength or aerobic capacity).[1] Therapeutic exercise is the prescription of a physical activity program for the purpose of improving, alleviating, or treating health problems.[3] There have been recent efforts to evaluate prehabilitation programs, which are home-based therapeutic exercise programs for the prevention of functional decline.[4,5] However, traditionally the focus of therapeutic exercise has been to alleviate or treat health-related problems. This chapter focuses on leisure physical activity and exercise among adults with MS; therapeutic exercise and activities related to occupation are discussed in other chapters.

17.2 Epidemiology of Physical Activity and Leisure Participation

Despite the benefits of physical activity (described below), persons with MS are largely inactive and sedentary.[6] In a seminal study, Ng and Kent-Braun[7] reported that, in comparison with a group of sedentary healthy controls, individuals with MS were less physically active when measured using an accelerometer. However, the differences between physical activity levels of the MS and sedentary control groups became nonsignificant when physical activity was measured using a seven-day recall interview. In a large survey across different chronic conditions, Nortvedt et al.[8] found that individuals with MS reported engaging in significantly less strenuous and light physical activity as compared to people with other types of diseases and the general population. Furthermore, 63% of the research sample with MS reported engaging in no strenuous leisure physical activity.

That overall picture of inactivity in MS has been confirmed in a narrative review and a meta-analysis. Motl et al.[6] confirmed, in a literature review, that individuals with MS generally engage in low amounts of physical activity and that this amount of physical activity was less than that of nondiseased participants. The most common types of physical activity among those with MS include walking and swimming, and the least common types of physical activity involve jogging and aerobics. Individuals with MS typically engage in less than 1 hour of aerobic endurance exercise or strength training per week and between 1 and 2.9 hours of walking per week, and the intensity of the physical activity commonly results in some shortness of breath and perspiration. Therefore, the overall level and pattern of physical activity would support a relatively inactive and sedentary lifestyle among individuals with MS.

The meta-analysis quantified the difference in physical activity behavior between individuals with MS and other populations.[6] Overall, 53 effects were retrieved from 13 studies with 2360 MS participants that yielded a weighted mean effect size of −0.60 (95% CI = −0.44, −0.77), which indicated that individuals with MS were significantly and moderately less active than the overall comparison group. This effect was largest (a) when comparing those with MS to individuals without MS or any other apparent disease ($d = -0.96$); (b) when using objective measures of physical activity ($d = -1.27$); and (c) in individuals with primary-progressive MS ($d = -0.87$). Indeed, the degree of inactivity is alarming given the rate of sedentary behavior in the general population of adults. These findings underscore the importance of developing and testing behavioral interventions that are delivered through widely generalizable and implementable approaches for increasing lifestyle or free-living physical activity in persons with MS. Such an approach would be more ecologically valid than the promotion of physical activity under supervised conditions that typically coincides with most of the literature in individuals with MS.

17.3 Physical Activity and Leisure Participation Assessment

Much of our knowledge regarding physical activity in persons with MS relies upon valid and reliable measures of this behavior. Moreover, there have been criticisms regarding the reliability and validity of physical activity measures, even for objective devices such as accelerometers and pedometers, in those with MS. As an example, cognitive deficits and mobility impairments, which are prevalent problems in persons with MS, have raised concerns over the psychometric properties of self-reported and objectives measures of physical activity, respectively. However, there is now emerging evidence for the reliability and validity of objective and self-report measures of physical activity among ambulatory individuals with MS. Based on research from Motl,[9] along with others, such as Kayes et al.,[10] clinicians can feel confident that they can accurately measure physical activity levels—when using the appropriate devices and tools. The two validated objective devices are ActiGraph accelerometers (measures acceleration of the body while engaged in physical activity) and Yamax pedometer (measures number of steps while walking); the three validated questionnaires are the Godin Leisure-Time Exercise Questionnaire [GLTEQ],[11] the 7-day Physical Activity Recall [7dPAR],[12] and the Physical Activity and Disability Survey [PADS][13] (see Table 17.1). The GLTEQ includes three items that measure the frequency of strenuous, moderate, and mild exercise for periods of more than 15 minutes during one's free time in a typical week. The 7dPAR and PADS include more questions than the GLTEQ and specifically ask about duration, type, and intensity of activities. Subscale scores can be calculated for the different types of activity (e.g., exercise, leisure-time physical activity, and general activity).

Motl et al.[14] have demonstrated that both the Yamax SW-200 and SW-401 pedometers exhibit good accuracy at walking speeds of 67, 80, and 94 m min⁻¹, but poor accuracy at walking speeds of 41 and 54 m min⁻¹. Thus, pedometers can provide an accurate quantification of physical activity behavior among individuals with MS who can walk at normal or fast speeds. Motl et al.[15] have also confirmed that a pedometer (Yamax SW-200), an ActiGraph accelerometer, and the GLTEQ and 7dPAR are moderately to strongly correlated with each other. In other words, different types of methods for measuring physical activity behavior are in agreement despite unique strengths and limitations of each, which further

TABLE 17.1

Physical Activity Measures with Support for Validity and Reliability among Individuals with MS

Objective measures
- ActiGraph accelerometers
  - Measures acceleration forces while engaged in physical activity
- Yamax pedometers
  - Measures number of steps while walking

Self-report questionnaires
- Godin Leisure-Time Exercise Questionnaire
  - Measures frequency of strenuous, moderate, and mild exercise
- Seven-Day Physical Activity Recall
  - Measures duration, frequency, and type of physical activity
- Physical Activity and Disability Survey
  - Measures duration, frequency, and type of physical activity with specific questions on rehabilitation programs and wheelchair use

---

provides support for the validity of all four methods of measuring physical activity. An additional study examined the reliability of scores from a pedometer and an accelerometer among individuals with MS.[16] It was found that a minimum of 3 days of recording is necessary to achieve adequate reliability for the pedometer and accelerometer. Taken together, evidence supports the accurate quantification of physical activity among ambulatory individuals with MS using validated objective devices and using the GLTEQ and 7dPAR. We should note that accelerometers are often considered the "gold standard" for measuring physical activity in all populations, including those with MS.[9] Nonetheless, recent evidence suggests that accelerometers may measure both physical activity behavior and mobility impairments, whereas self-report questionnaires may only be measuring physical activity behavior.[17]

In addition to the GLTEQ and 7dPAR, the PADS may also be a valid measure of physical activity behavior in adults with MS. The PADS was originally developed by Rimmer et al.[13] This measure is comprised of six subscales, which include planned exercise, leisure-time physical activity, general physical activity, employment, and rehabilitation programs. A total score can be calculated from the six subscales. Kayes et al.[10,18] have conducted two studies on the PADS. In both studies, the scale was easily understood by people with MS and had adequate test–retest reliability. However, the PADS did not accurately predict accelerometer counts. The PADS was revised in the second study to facilitate understanding and to improve the method of scoring the scale. Although the PADS needs to be further tested, it is the most comprehensive scale to date that has been developed specifically for persons with disabilities, and there is at least some evidence supporting its psychometric properties among individuals with MS.

17.4 Benefits of Physical Activity and Exercise in MS

Individuals with MS were once told that they should not engage in physical activity because it could make their symptoms worse and exacerbate the disease process itself.[19]

This initial recommendation was, perhaps, based on the observation that some persons with MS experience a temporary worsening of symptoms (e.g., fatigue) after a single bout of physical activity. Individuals with MS may further have abnormal physiological responses to exercise. In comparison to the general population, individuals with MS differ in metabolic,[20] muscular,[21] cardiovascular,[22] and sweat responses to exercise.[23] Smith et al.[24] found that over 40% (*n* = 34) of individuals with MS experienced an increase in the number and frequency of sensory symptoms immediately after exercising. However, increases in symptoms usually subside after rest, and there is no published evidence that the abnormal physiological responses make it unsafe for individuals with MS to engage in physical activity. In fact, over the last 20 years, many studies have documented the benefits of exercise within laboratory/outpatient settings for individuals with MS. Review articles and meta-analyses have indicated that individuals with mild-to-moderate symptoms of MS experience benefits from routine participation in exercise programs.

There is now cumulative evidence from meta-analyses documenting the effect of exercise training on mobility and quality of life (QOL) among individuals with MS. Snook and Motl[25] reported a clinically meaningful effect of exercise training on walking mobility, defined as an index of neurological disability, in persons with MS. Forty-three published papers were identified and reviewed, of which 22 provided enough data to compute effect sizes expressed as Cohen's *d*. From the 22 publications that included 609 MS participants, 66 effect sizes were retrieved. There was an overall weighted mean effect size of *g* = 0.19 (95% CI = 0.09, 0.28), indicating that exercise training was associated with an improvement in ambulation. There were larger effects associated with supervised exercise training, exercise programs that were less than three months in duration and mixed samples of relapsing–remitting and progressive MS. In another meta-analysis, evidence from 13 studies with 109 effect sizes from 484 individuals with MS indicated that exercise training was associated with approximately 1/4 standard deviation improvement in QOL.[26] An effect size of this magnitude is potentially clinically meaningful when compared, for example, to the overall effectiveness (*d* = 0.30) of disease-modifying medications for reducing exacerbations in individuals with MS.[27]

Systematic reviews of the literature have further documented the benefits of exercise. Rietberg et al.[28] reviewed nine high-quality, randomized, controlled trials that included 260 participants with MS and reported that there was strong evidence that exercise improved muscle power function, exercise tolerance, and mobility-related activities compared with no exercise at all. There was also moderate evidence for exercise's improving mood. In another review by Dalgas et al.,[29] there was good evidence for the benefits of prescribing endurance training, but there was little evidence to support the benefits of resistance training and even less evidence to support the combination of endurance and resistance training. However, Garrett and Coote[30] noted in their review of the literature that a combination of aerobic exercise and progressive resistance exercise could offer substantial benefits to reducing MS-related symptoms, and further noted that yoga and aqua exercise could be of benefit. For balance in reporting we note that Asano et al.[31] concluded that there were very few methodologically sound studies to guide the prescription of exercise programs in individuals with MS. This group further noted that it was difficult to determine which types of exercise programs were most beneficial because of the wide range of exercise programs that had been evaluated using different outcomes. Table 17.2 provides a summary of potential benefits for engaging in physical activity among persons with MS, while Table 17.3 summarizes studies that highlight the diversity in the types of exercise programs that have been evaluated in people with MS.

TABLE 17.2

Potential Benefits of Physical Activity/Exercise among Individuals with MS Arranged by ICF Framework

Body function and structure	Improved mood
	Improved muscle strength
	Decreased fatigue
	Increased cardiovascular fitness
	Decreased body fat
	Decreased triglycerides
	Decreased very-low-density lipoprotein
	Impeded development of secondary conditions
Activity	Improved mobility
	Increased health-related QOL
	Slowed functional decline
Participation	Improved social functioning
	Improved role functioning
	Improved community integration

TABLE 17.3

Research Design, Outcomes, and Description of Exercise Interventions

First author (year)	Design	# with MS	Type of Exercise Program	Intensity	Frequency	Duration	Positive Outcomes[a]
Petajan (1996)[70]	RCT	46	Combined arm and leg stationary bicycle	60% of VO_{2max}	3×40 min per week	15 weeks	VO_{2max}, lipid profile, strength, % body fat, mood, fatigue
Mostert (2002)[71]	RCT	26	Stationary bicycling	Training heart rate at aerobic threshold	5×30 min per week	3–4 weeks	Aerobic threshold, QOL, fatigue
Romberg (2004)[41]	RCT	95	Resistance training with elastic bands	Not indicated	3–4 times per week	26 weeks	MS-functional composite
Oken (2004)[72]	RCT	57	Yoga	Modified to patients' needs	1 time per week for 90 min	6 months	Fatigue

Note: RCT = Randomized controlled trial.

[a] Improvements are across time or between groups.

17.5 Impact of Physical Activity and Leisure Participation on Everyday Life

In keeping with the theme of this book, the focus will now be on the relationship between physical activity and activity restrictions/functional limitations, participation in life roles, and QOL. One of the first large quantitative studies to explore the relationship between health-promoting behaviors, functional limitations, and QOL in individuals with MS was a study by Stuifbergen et al.[32] They found that health-promoting behaviors, including engagement in physical activity, mediated the relationship between functional limitations and QOL. Stuifbergen et al.[33] also found that 66% of variance in QOL could be explained through a path model that included health-promoting behaviors, severity of illness, barriers, self-efficacy, and resources. The relationship between QOL and severity of illness was mediated by barriers, self-efficacy, and resources. The authors concluded that interventions to enhance social support, decrease barriers, and increase self-efficacy could result in improved health-promoting behaviors and QOL. In a subsequent five-year longitudinal study, Stuifbergen et al.[34] helped confirm this conclusion. It was found that decreases in physical activity level were associated with the trajectory of functional decline. Thus, people with MS who routinely engage in physical activity may be able to delay functional decline.

These results have been confirmed and expanded upon by Motl and colleagues. In two separate studies they found that physical activity was indirectly associated with QOL. One study found that self-efficacy for function and control mediated the relationship between physical activity and QOL,[35] whereas the other study found that self-efficacy and functional limitations mediated the relationship between physical activity and QOL, independent of general perceptions of social support.[36] In a more recent study, Motl and colleagues tested a path model that is consistent with Nagi's disablement model for examining the relationship between physical activity and disability. The path model indicated that individuals with MS who are more physically active have better function and fewer symptoms; those who have fewer symptoms have better function, and those with better function have less disability.[37] Similarly, in a longitudinal study, it was found that changes in physical activity were associated with changes in function, changes in symptoms were associated with changes in function and disability, and that changes in function were associated with changes in disability.[38] These studies indicate that increasing physical activity behavior may be a distinct theoretical entry point to interrupt the worsening of QOL and accumulation of disability in individuals with MS.

In addition to the work of Motl and colleagues, three other noteworthy studies on physical activity and participation were conducted by Turner et al.,[39] Crawford et al.,[40] and Romberg et al.[41] Turner et al.[39] found in a sample of 2995 veterans with MS that those who exercised regularly were more likely to report better social functioning as measured with the SF-36. Crawford et al.[40] found that people with mobility impairments, including a large subsample of individuals with MS who engaged in physical activity, were more likely to report greater participation, better health, and a higher level of reintegration into normal community living compared to inactive participants. Romberg et al.[41] found that peak oxygen consumption was negatively associated with disability level as measured by the Expanded Disability Status Scale (EDSS), and this finding was recently confirmed and extended by Motl and Goldman,[42] who reported that disability level and physical activity were independently associated with peak oxygen consumption in persons with MS. Such findings indicate that there is a relationship between exercise capacity and MS-related disability and suggest that cardiovascular training in a physical activity intervention may be an important component to reduce disability among individuals with MS.

Finally, we note an important, yet underexplored area of MS physical activity research, the relationship between participation restrictions, deconditioning, and the development of secondary conditions. Indeed, Motl and colleagues provide evidence of a linear relationship between physical activity, functional limitations, and disability. However, this relationship may be much more pronounced if we were to follow an individual with MS over an extended period of time to capture the development of secondary conditions. Specifically, there may be a disabling cyclical relationship between inactivity, deconditioning, functional limitations, development of secondary conditions, and participation restrictions.

MS-related symptoms may make it difficult to engage in physical activity. Inactivity increases deconditioning, exacerbates common MS symptoms (e.g., fatigue and mobility impairments), and increases the risk of developing secondary conditions, which all further augment inactivity and participation restrictions. Such a model has been proposed by Durstine et al.[43] for understanding physical activity for people with chronic illness and disability and has recently been developed for persons with MS.[44] Promoting physical activity may help interrupt this disabling cyclical relationship by improving fitness and decreasing the likelihood of secondary conditions, which in return will prevent or reduce participation restrictions.

17.6 Correlates of Physical Activity and Leisure Participation

Given the rates of physical activity among individuals with MS, the question becomes why are rates of participation so low? Do MS-related symptoms make it difficult to engage in physical activity? Do symptoms of the disease influence or compromise the skills and cognitions necessary to engage in physical activity? Are levels of physical activity influenced by whether one considers his or her disease as hopeless or manageable? These questions are fundamental to understanding physical activity behavior in individuals with MS, and only in the past five to eight years have there been concerted efforts to begin to answer such questions (with a few exceptions from Stuifbergen[45,46]). We have summarized this literature by dividing it into two parts: (1) facilitators or variables that have positively correlated with physical activity behavior and (2) barriers that have negatively correlated with physical activity behavior.

Although we are dividing this section into facilitators and barriers, the relationship between facilitators and barriers can be conceptualized in terms of decision-making theory. This theory states that people are more likely to engage in a behavior when pros/facilitators outweigh cons/barriers. Because of the daily variability in MS-related symptoms, this theory may be particularly useful in understanding physical activity behavior in individuals with MS.[47] For example, an individual with MS may have every intention of engaging in physical activity, but there may be a threshold or tipping point when barriers, such as fatigue, overwhelm and compromise motivation and ability to engage in physical activity.

Efforts to identify facilitators and barriers to physical activity behavior in individuals with MS have typically originated in cross-sectional studies that identify correlates of physical activity behavior. Although we will provide suggestions on how to change facilitators and barriers that are associated with physical activity behavior, readers should be reminded of the fact that correlations do not imply causation, that is, modifying correlates of physical activity behavior does not necessarily result in physical activity behavior change. Instead, cross-sectional studies should be interpreted as a starting point or as a blue print for designing and evaluating interventions that can truly test a cause-and-effect relationship.

17.6.1 Facilitators

Cross-sectional studies generally support a social cognitive approach to promoting physical activity behavior in individuals with MS. Similar to the general population, social support, self-efficacy, stages of change, outcome expectations, enjoyment, history of physical activity behavior, and self-identity have been positively associated with physical activity behavior in individuals with MS.[35,48–50] Which of these correlates or combination of correlates are most important to target when promoting physical activity behavior is difficult to determine with the limitations of the cross-sectional studies and differences in study methods. Furthermore, it is difficult to determine which behavior-change theory is most relevant to promoting physical activity behavior in individuals with MS because studies rarely evaluate an entire theory or compare constructs between theories.

Nonetheless, self-efficacy has been found to be a strong and robust correlate of physical activity behavior in individuals with MS. Self-efficacy for physical activity can be defined as one's confidence in his or her ability to engage in physical activity on a regular basis. Motl and colleagues have found that self-efficacy for physical activity is not only impor-tant in directly influencing physical activity behavior but also mediates several key rela-tionships between symptoms, function, QOL, and physical activity behavior.[35,51] Bandura[52] has suggested that self-efficacy is influenced through performance accomplishment, vicar-ious experience, verbal persuasion, and physiological or emotional arousal. Thus, strate-gies to enhance self-efficacy could include helping an individual differentiate MS-related fatigue from fatigue felt after engaging in physical activity, persuading individuals that they can successfully engage in physical activity on a regular basis, encouraging individu-als to reward themselves when they accomplish their physical activity goals, and inform-ing individuals of seminars or Internet resources where they are likely to meet other individuals with MS who are physically active.

Another correlate of physical activity in individuals with MS is social support. Motl et al.[48] have established that global social support is indirectly related to physical activity through self-efficacy. Plow et al.[49] have found that social support specifically for exercise is directly correlated with physical activity behavior even after accounting for such factors as self-efficacy, self-identity, and intentions for engaging in physical activity. Social support can be divided into three subcategories: informational, emotional, and tangible. Encouraging individuals to attend seminars on physical activity, enlisting friends and family members to reinforce engagement in physical activity, and encouraging individuals to exercise with a buddy may help create a supportive social environment for engaging in physical activity. Indeed, the social environment may also interfere with engaging in phys-ical activity, which is why creating a positive social environment through the enlistment of social support is so important to promoting physical activity in individuals that may already encounter numerous barriers to physical activity.

Although not as well tested as self-efficacy or social support, two other noteworthy facilitators of physical activity may be outcome expectancy and goal setting. Outcome expectancy is a person's belief concerning the likely consequences (positive or negative) of a behavior. Ferrier et al.[53] found that outcome expectations directly influenced physical activity behavior and that self-efficacy directly influenced outcome expectations. Plow et al.[47] found in a qualitative study that most participants expected exercise to be benefi-cial but had mixed-to-negative feelings regarding whether exercise would help manage or prevent the exacerbation of MS-related symptoms; hence, the importance of emphasiz-ing the benefit of exercise specific for individuals with MS. McAuley et al.[54] recently demonstrated the validity of the Multidimensional Outcome Expectations for Exercise Scale in persons with MS, which should help facilitate future studies that explore the rela-tionship between outcome expectancy and physical activity behavior. A construct that has been explored even less than outcome expectancy is goal setting, which shows extreme promise in influencing physical activity behavior. Goal setting can be conceptualized and measured as the frequency of setting, monitoring, and committing to meet goals. Suh et al.[55] found that goal setting was a stronger cross-sectional correlate of physical activity in persons with MS than self-efficacy, outcome expectations, or impediments. Indeed, the simple strategy of encouraging people with MS to set realistic and measurable goals may be a cost-effective approach to promote physical activity.

17.6.2 Barriers

MS-related symptoms can be major barriers to physical activity participation (see Table 17.4). Motl and colleagues have published many studies on the relationship between MS-related symptoms and physical activity behavior. Their work has demonstrated that MS-related symptoms are associated with inactivity and that self-efficacy mediates the relationship between symptoms and physical activity behavior. They have demonstrated that overall symptoms, fatigue, and problems with walking are associated with lower levels of physical activity and that problems with walking may partially explain the relationship between overall symptoms and physical activity.[56,57] They have also demonstrated that a symptom cluster, that is, three or more concurrent symptoms, is strongly associated with physical activity behavior.[58] Fatigue, depression, and pain represented a symptom cluster, and functional limitation, but not self-efficacy, helped explain the relationship between the symptom cluster and physical activity. Not surprisingly, they have also found that worsening of symptoms is also associated with lower levels of physical activity.[59] These studies by Motl and colleagues emphasize the importance of monitoring symptoms during a physical activity intervention and that it may be necessary to provide additional strategies to help individuals overcome MS-related health barriers to physical activity. These additional strategies may include education on self-managing symptoms, such as fatigue management, as well as increasing self-efficacy to self-manage symptoms.

The nature of the physical environment may interfere with participating in physical activity. However, few studies have explored this premise. Morris et al.[60] found that the Neighborhood Environment Walkability Scale, a comprehensive account of the perceptions of the physical environment, was not associated with physical activity behavior in individuals with MS. However, clearly, factors such as climate and accessibility of a gym may influence physical activity behavior in persons with MS. Due to the fact that many

TABLE 17.4

Potential Facilitators and Barriers to Physical Activity in Persons with MS Arranged by ICF Framework

	Facilitators	Barriers
Body function and structure		Fatigue
		Pain
		Depression
		Declining health
Activity		Functional limitations
Participation	Social support	Social obligations
		Mobility impairments
		Domestic obligations
		Money/costs
Personal and environmental factors	Self-efficacy	Climate
	Stages of change	Inaccessibility of health clubs
	Outcome expectations	Social stigma
	Enjoyment	Stress
	Self-identity	
	Goal setting	

individuals with MS have thermoregulation difficulties, hot weather can make MS-related symptoms worse; thereby making it difficult to engage in physical activity. Plow et al.[47] had noted in a qualitative study that individuals with MS reported difficulty exercising indoors, even in air-conditioned rooms, when it was hot outside. In addition to climate, individuals with MS often experienced many physical and social environmental barriers when trying to exercise at gymnasiums. They reported feeling uncomfortable exercising at a gym with people who were "healthy" and being fatigued even before starting to exercise because of having to park and walk long distances or having to maneuver through a crowded and cluttered gym environment. In an evaluation of 35 health clubs, Rimmer et al.[61] has confirmed the poor accessibility of workout gyms for individuals with functional limitations. Potential strategies for overcoming physical environmental barriers can include wearing a cooling vest, taking a cool shower before engaging in physical activity, letting individuals know that finding a suitable recreational gym may be a trial-and-error process, or, if they fail to find a suitable gym, they could engage in a home exercise program as a reasonable alternative.

Few physical activity studies, even within the general population, measure the full spectrum of the social environment. Although social support is often measured in physical activity studies, factors related to an unsupportive social environment are rarely measured. Indeed, we probably have all experienced situations when we have intended to engage in physical activity, but, because we are "too busy," we fail to do so. Social obligations, stress, employment, and activities associated with domestic life roles often interfere with intentions of persons with MS to engage in physical activity. Vanner et al.[62] found that lack of energy, high costs, lack of motivation, and lack of support were commonly cited as barriers to physical activity in individuals with MS. In addition to encouraging individuals to enlist social support for engaging in physical activity, it may be beneficial to teach skills related to time and stress management. People who are better able to manage their time and are able to successfully cope with stress may be more likely to overcome social environmental barriers to engage in physical activity.

17.7 Treatment Planning

Although there is some evidence regarding the benefits of exercise in individuals with MS, there are very few studies that have explored strategies to promote physical activity or exercise adherence. This is because most studies to date have been conducted in laboratory or outpatient settings, which helps control for extraneous factors to establish the benefits of exercise but does little to help evaluate ecologically valid intervention strategies to promote physical activity. The recommendation for individuals with MS to increase or maintain cardiovascular function is to exercise 30 minutes per day at 60–85% of maximum heart rate three days per week.[63] Recommendations for weight training are one to two sets of 10–12 reps, two to three times a week, alternating the days for weight training with days for cardiovascular training. However, there is little evidence to show that this is the ideal exercise prescription for individuals with MS. Thus, besides the need to develop and evaluate physical activity intervention strategies that promote adherence, there is still a need to conduct laboratory or outpatient research that compares exercise programs differing in type, intensity, and duration to determine which exercise program is most effective in promoting health and function.

TABLE 17.5

Research Design, Outcomes, and Description of Interventions that Implemented Behavior Change Strategies to Promote Physical Activity

First Author (year)	Purpose	Design	# with MS	Intervention Topics	Delivery Formats	Length of Intervention	Positive Outcomes[a]
Bombardier (2008)[64]	Evaluated a motivational interviewing intervention	RCT	130	Exercise, fatigue, anxiety, social support, stress, or alcohol/drug abuse	Face-to-face and phone	12 weeks	Increased health-promoting behaviors
McAuley (2007)[65]	Evaluated an efficacy-enhancement intervention	RCT	26	Physical activity	Face-to-face	12 weeks	Increased self-efficacy and physical activity
Motl (2011)[42]	Evaluated Internet physical activity intervention	RCT	54	Physical activity	Internet	12 weeks	Increased self-efficacy and physical activity
Stuifbergen (2003)[68]	Examined the effects of a wellness intervention program	RCT	113	Lifestyle adjustment, physical activity, healthy eating, stress, intimacy and sexuality, women's health issues	Face-to-face and telephone calls	8 weeks of classes and telephone follow-up	Increased health-promoting behaviors and QOL
Plow (2009)[47]	Compared the effects of a wellness intervention to physical therapy	RT	42	Lifestyle adjustment, physical activity, healthy eating, stress, fatigue management	Face-to-face and telephone calls	4 PT sessions or 8 weeks of classes and telephone	Improved QOL, fitness, mental health, physical activity

Note: RCT = Randomized controlled trial, RT = Randomized trial.
[a] Improvements are across time or between groups, QOL = quality of life.

Table 17.5 summarizes studies that have been effective in promoting physical activity using behavior change strategies. Preliminary evidence generally supports the efficacy of goal setting, motivational interviewing, encouraging social support, and increasing self-efficacy in promoting physical activity in individuals with MS.[64-68] Research suggests that broad holistic approaches that include intervention topics besides physical activity promotion (e.g., self-management tasks and skills) and approaches that only focus on physical activity promotion (i.e., efficacy enhancement, self-monitoring, and goal setting for physical activity) across a range of delivery formats (e.g., face-to-face, in a group, in a one-to-one session, or telecommunication) may be effective in promoting physical activity behavior in those with MS. Specifically, encouraging exercise with a buddy, providing information on community resources, teaching individuals how to set and monitor realistic physical activity goals, using verbal persuasion to foster confidence to engage in physical activity, and teaching problem-solving skills to overcome such barriers as fatigue and heat sensitivity may be effective in promoting physical activity behavior in those with MS.

Rehabilitation professionals can also utilize their expertise to help promote physical activity. Promoting physical activity includes (1) prescribing a physical activity program that is tailored to the health needs of the patient, (2) teaching symptom management strategies to help overcome MS symptoms as barriers to physical activity, and (3) educating patients how to safely engage in physical activity to receive health benefits.

1. Clinicians can tailor physical activity programs to a patient's health problems, which is important in minimizing potential risks and reducing frustration caused by a physical activity program that is too difficult. For example, clinicians can help patients with significant balance and mobility problems find alternative types of physical activity that are not primarily related to walking (e.g., water aerobics, yoga, stationary cycling, etc.). If it is necessary to prescribe a challenging exercise, ensuring appropriate safety precautions are in place and encouraging/enlisting social support from family and friends may be helpful in reducing frustrations (e.g., teaching the family member or friend how to help with the stretching of a spastic muscle). See Table 17.6.

TABLE 17.6

Potential Strategies to Promote Physical Activity in Persons with MS

---

1. Highlighting MS-specific benefits of physical activity.
2. Differentiating MS-related fatigue from fatigue felt after engaging in physical activity.
3. Persuading individuals that they can successfully engage in physical activity.
4. Teaching individuals how to set measureable and realistic physical activity goals.
5. Teaching individuals how to self-monitor their goals.
6. Encouraging individuals to reward themselves when they accomplish their goals.
7. Informing individuals of seminars or Internet resources where they are likely to meet other individuals with MS who are physically active.
8. Encouraging exercise with someone and enlisting social support from family and friends.
9. Teaching self-management tasks and skills to overcome MS-related barriers such as fatigue and spasticity (e.g., fatigue-management strategies and unloaded cycling, respectively).
10. For heat sensitivity, recommend cooling vest or taking cool shower before physical activity.
11. Motivational interviewing.
12. Prevent excessive fatigue by recommending reducing intensity, frequency, or duration of physical activity if patient reports increased in symptoms two hours after a bout of physical activity.
13. Prescribe a tailored physical activity program that meets the health needs of the patient.

---

2. Teaching patients how to self-manage symptoms that act as barriers to physical activity may also help promote long-term physical activity maintenance. For example, encouraging patients to take frequent rest breaks throughout the day, re-organize their home or work environment, and use proper ergonomics may help the patient conserve energy so they have enough energy to expend on planned physical activity. Stretching, icing, or engaging in unloaded stationary cycling[69] may help temporarily to reduce spasticity so patients can engage in other types of physical activity. Encouraging the use of a cooling vest may help patients with heat sensitivity. Other self-management strategies that might help patients with MS engage in physical activity can be found throughout this book.

3. Clinicians can also help educate patients on how to properly engage in physical activity. Because of variability in heart rate response to physical activity in MS, using a rate of perceived exertion scale (e.g., Borg) to prescribe a physical activity program may be useful. Encouraging patients to keep track of their perceived exertion during activity might help provide guidance on when it is appropriate to reduce the intensity of physical activity. Although not an established guideline, we tell our research participants that if they cannot engage in daily activities two hours after a bout of physical activity because of an increase in their symptoms, they should reduce the intensity of activity for the next time. During physical activity, one should recommend that the patient take frequent rest breaks, particularly if his or her gait becomes unsteady. It is also important to remind the patient to drink plenty of cool water during physical activity. When patients are first trying to increase their physical activity levels, they should be told to have a post-physical activity plan, whether it is a planned rest break or having someone nearby to provide tangible support. Clinicians should emphasize that finding an activity that is fun and at an appropriate intensity may be a trial-and-error process, and the important thing is not to give up and to keep in mind that some activity is better than no activity.

17.8 Conclusions and Future Directions

There has been considerable progress over the last 20 years in demonstrating the benefits of physical activity and understanding physical activity behavior in those with MS. However, as evident from this chapter, there is still much work to be done. How best to prescribe a physical activity program (i.e., intensity, duration, and type), and which strategies should be used to promote routine physical activity engagement, is still unclear. Thus, there is a need to conduct both laboratory-based exercise studies and ecologically valid physical activity studies. A fundamental problem with the existing literature is that adults with an EDSS score ≥6 have typically not been included in physical activity studies; physical activity research is urgently needed in those with MS who need a walker for mobility or who have restricted mobility requiring a wheelchair. What is clear from the research literature is that physical activity should be encouraged for patients with mild-to-moderate MS symptoms and that symptoms need to be addressed and mitigated to help promote routine engagement in physical activity.

References

1. Bouchard CJ, Shephard RJ. Physical activity, fitness, and health. International proceedings and consensus statement. Champaign, IL: Human Kinetics; 1994.
2. Caspersen CJ, Powell KE, Christenson GM. Physical activity, exercise, and physical fitness: Definitions and distinctions for health-related research. Public Health Reports 1985;100:126–131.
3. Taylor NF, Dodd KJ, Shields N, Bruder A. Therapeutic exercise in physiotherapy practice is beneficial: A summary of systematic reviews 2002–2005. Australian Journal of Physiotherapy 2007;53:7–16.
4. Gill TM, Baker DI, Gottschalk M, Peduzzi PN, Allore H, Byers A. A program to prevent functional decline in physically frail, elderly persons who live at home. The New England Journal of Medicine 2002;347:1068–1074.
5. Gill TM, Baker DI, Gottschalk M, Peduzzi PN, Allore H, Van Ness PH. A prehabilitation program for the prevention of functional decline: Effect on higher-level physical function. Archives of Physical Medicine and Rehabilitation 2004;85:1043–1049.
6. Motl RW, McAuley E, Snook EM. Physical activity and multiple sclerosis: A meta-analysis. Multiple Sclerosis 2005;11:459–463.
7. Ng AV, Kent-Braun JA. Quantitation of lower physical activity in persons with multiple sclerosis. Medicine & Science in Sports & Exercise 1997;29:517–523.
8. Nortvedt MW, Riise T, Maeland JG. Multiple sclerosis and lifestyle factors: The Hordaland health study. Neurological Sciences 2005;26:334–349.
9. Motl RW. Physical activity and its measurement and determinants in multiple sclerosis. Minerva Medica 2008;99:157–165.
10. Kayes NM, Schluter PJ, Mcpherson KM, Taylor D, Kolt GS. The physical activity and disability survey-revised (PADS-R): An evaluation of a measure of physical activity in people with chronic neurological conditions. Clinical Rehabilitation 2009;23:534–543.
11. Godin G, Shephard RJ. A simple method to assess exercise behavior in the community. Canadian Journal of Applied Sport Sciences 1985;10:141–146.
12. Sallis JF, Haskell WL, Wood PD, Fortmann SP, Rogers T, Blair SN, Paffenbarger RS, Jr. Physical activity assessment methodology in the five-city project. American Journal of Epidemiology 1985;121:91–106.
13. Rimmer JH, Riley BB, Rubin SS. A new measure for assessing the physical activity behaviors of persons with disabilities and chronic health conditions: The physical activity and disability survey. American Journal of Health Promotion 2001;16:34–42.
14. Motl RW, McAuley E, Snook EM, Scott JA. Accuracy of two electronic pedometers for measuring steps taken under controlled conditions among ambulatory individuals with multiple sclerosis. Multiple Sclerosis 2005;11:343–345.
15. Motl RW, McAuley E, Snook EM, Scott JA. Validity of physical activity measures in ambulatory individuals with multiple sclerosis. Disability and Rehabilitation 2006;28:1151–1156.
16. Motl RW, Zhu W, Park Y, McAuley E, Scott JA, Snook EM. Reliability of scores from physical activity monitors in adults with multiple sclerosis. Adapted Physical Activity Quarterly 2007;24:245–253.
17. Snook EM, Motl RW, Gliottoni RC. The effect of walking mobility on the measurement of physical activity using accelerometry in multiple sclerosis. Clinical Rehabilitation 2009;23:248–258.
18. Kayes NM, Mcpherson KM, Taylor D, Schluter PJ, Wilson BJ, Kolt GS. The physical activity and disability survey (PADS): Reliability, validity and acceptability in people with multiple sclerosis. Clinical Rehabilitation 2007;21:628–639.
19. Ponichtera-Mulcare JA. Exercise and multiple sclerosis. Medicine & Science in Sports & Exercise 1993;25:451–465.
20. Kent-Braun JA, Sharma KR, Weiner MW, Miller RG. Effects of exercise on muscle activation and metabolism in multiple sclerosis. Muscle & Nerve 1994;17:1162–1169.

21. Ng AV, Miller RG, Gelinas D, Kent-Braun JA. Functional relationships of central and peripheral muscle alterations in multiple sclerosis. Muscle & Nerve 2004;29:843–852.

22. Acevedo AR, Nava C, Arriada N, Violante A, Corona T. Cardiovascular dysfunction in multiple sclerosis. Acta Neurologica Scandinavica 2000;101:85–88.

23. Davis SL, Wilson TE, Vener JM, Crandall CG, Petajan JH, White AT. Pilocarpine-induced sweat gland function in individuals with multiple sclerosis. Journal of Applied Physiology 2005;98:1740–1744.

24. Smith RM, Adeney-Steel M, Fulcher G, Longley WA. Symptom change with exercise is a temporary phenomenon for people with multiple sclerosis. Archives of Physical Medicine and Rehabilitation 2006;87:723–727.

25. Snook EM, Motl RW. Effect of exercise training on walking mobility in multiple sclerosis: A meta-analysis. Neurorehabilitation and Neural Repair 2009;23:108–116.

26. Motl RW, Gosney JL. Effect of exercise training on quality of life in multiple sclerosis: A meta-analysis. Multiple Sclerosis 2008;14:129–135.

27. Filippini G, Munari L, Incorvaia B, Ebers GC, Polman C, D'amico R, Rice GP. Interferons in relapsing remitting multiple sclerosis: A systematic review. Lancet 2003;361:545–552.

28. Rietberg MB, Brooks D, Uitdehaag BMJ, Kwakkel G. Exercise therapy for multiple sclerosis. Cochrane Database of Systematic Reviews. 2004;(3). Art. No.: CD003980. Doi: 10.1002/14651858. CD003980.pub2.

29. Dalgas U, Stenager E, Ingemann-Hansen T. Multiple sclerosis and physical exercise: Recommendations for the application of resistance-, endurance- and combined training. Multiple Sclerosis 2008;14:35–53.

30. Garrett M, Coote S. Multiple sclerosis and exercise in people with minimal gait impairment: A review. Physical Therapy Reviews 2009;14:169–180.

31. Asano M, Dawes DJ, Arafah A, Moriello C, Mayo NE. What does a structured review of the effectiveness of exercise interventions for persons with multiple sclerosis tell us about the challenges of designing trials? Multiple Sclerosis 2009;15:412–421.

32. Stuifbergen AK, Roberts GJ. Health promotion practices of women with multiple sclerosis. Archives of Physical Medicine and Rehabilitation 1997;78:S3–S9.

33. Stuifbergen AK, Seraphine A, Roberts G. An explanatory model of health promotion and quality of life in chronic disabling conditions. Nursing Research 2000;49:122–129.

34. Stuifbergen AK, Blozis SA, Harrison TC, Becker HA. Exercise, functional limitations and quality of life: A longitudinal study of persons with multiple sclerosis. Archives of Physical Medicine & Rehabilitation 2006;87(7):935–943.

35. Motl RW, Snook EM. Physical activity, self-efficacy, and quality of life in multiple sclerosis. Annals of Behavioral Medicine 2008;35:111–115.

36. Motl RW, McAuley E, Snook EM. Physical activity and quality of life in multiple sclerosis: Possible roles of social support, self-efficacy, and functional limitations. Rehabilitation Psychology 2007;52:143–151.

37. Motl RW, Snook EM, McAuley E, Scott J, Gliottoni RC. Are physical activity and symptoms correlates of functional limitations and disability in multiple sclerosis? Rehabilitation Psychology 2007;52:463–469.

38. Motl RW, McAuley E. Longitudinal analysis of physical activity and symptoms as predictors of change in functional limitations and disability in multiple sclerosis. Rehabilitation Psychology 2009;54:204–210.

39. Turner AP, Kivlahan DR, Haselkorn JK. Exercise and quality of life among people with multiple sclerosis: Looking beyond physical functioning to mental health and participation in life. Archives of Physical Medicine and Rehabilitation 2009;90:420–428.

40. Crawford A, Hollingsworth H, Morgan K, Gray D. People with mobility impairments: Physical activity and quality of participation. Disability and Health Journal 2008;1:7–13.

41. Romberg A, Virtanen A, Ruutiainen J, Aunola S, Karppi SL, Vaara M, Surakka J, Pohjolainen T, Seppanen A. Effects of a 6-month exercise program on patients with multiple sclerosis: A randomized study. Neurology 2004;63:2034–2038.

42. Motl RW, Goldman M. Physical inactivity, neurological disability, and cardiorespiratory fitness in multiple sclerosis. Acta Neurologica Scandinavica 2011;123:98–104.
43. Durstine JL, Painter P, Franklin BA, Morgan D, Pitetti KH, Roberts SO. Physical activity for the chronically ill and disabled. Sports Medicine 2000;30:207–219.
44. Motl RW. Physical activity and irreversible disability in multiple sclerosis. Exercise and Sport Sciences Reviews 2010;38:186–191.
45. Stuifbergen AK. Physical activity and perceived health status in persons with multiple sclerosis. Journal of Neuroscience Nursing 1997;29:238–243.
46. Stuifbergen AK. Barriers and helath behaviors in rural and urban persons with MS. American Journal of Health Behavior 1999;23:415–425.
47. Plow MA, Resnik L, Allen SM. Exploring physical activity behavior of persons with multiple sclerosis: A qualitative pilot study. Disability Rehabilitation 2009;31:1652–1665.
48. Motl RW, Snook EM, McAuley E, Scott JA, Douglass ML. Correlates of physical activity among individuals with multiple sclerosis. Annals of Behavioral Medicine 2006;32:154–161.
49. Plow M, Mathiowetz V, Resnik L. Multiple sclerosis: Impact of physical activity on psychosocial constructs. American Journal of Health Behavior 2008;32:614–626.
50. Kasser S, Stuart M. Psychological well-being and exercise behavior in persons with and without multiple sclerosis. Clinical Kinesiology 2001;55:81–86.
51. Motl RW, Snook EM, McAuley E, Gliottoni RC. Symptoms, self-efficacy, and physical activity among individuals with multiple sclerosis. Research in Nursing & Health 2006;29:597–606.
52. Bandura A. Self-efficacy: The exercise of control. New York, NY: W.H. Freeman; 1997.
53. Ferrier S, Dunlop N, Blanchard C. The role of outcome expectations and self-efficacy in explaining physical activity behaviors of individuals with multiple sclerosis. Journal of Behavioral Medicine 2010;36:7–11.
54. McAuley E, Motl RW, White SM, Wojcicki TR. Validation of the multidimensional outcome expectations for exercise scale in ambulatory, symptom-free persons with multiple sclerosis. Archives of Physical Medicine and Rehabilitation 2010;91:100–105.
55. Suh Y, Weikert M, Dlugonski D, Sandroff B, Motl RW. Social cognitive correlates of physical activity: Findings from a cross-sectional study of adults with relapsing–remitting multiple sclerosis. Journal of Physical Activity and Health 2011;8(5):626–635.
56. Snook EM, Motl RW. Physical activity behaviors in individuals with multiple sclerosis: Roles of overall and specific symptoms, and self-efficacy. Journal of Pain and Symptom Management 2008;36:46–53.
57. Motl RW, Snook EM, Schapiro RT. Symptoms and physical activity behavior in individuals with multiple sclerosis. Research in Nursing & Health 2008;31:466–475.
58. Motl RW, McAuley E. Symptom cluster as a predictor of physical activity in multiple sclerosis: Preliminary evidence. Journal of Pain and Symptom Management 2009;38:270–280.
59. Motl RW, Arnett PA, Smith MM, Barwick FH, Ahlstrom B, Stover EJ. Worsening of symptoms is associated with lower physical activity levels in individuals with multiple sclerosis. Multiple Sclerosis 2008;14:140–142.
60. Morris KS, McAuley E, Motl RW. Self-efficacy and environmental correlates of physical activity among older women with multiple sclerosis. Health Education Research 2008;23:744–752.
61. Rimmer JH, Riley B, Wang E, Rauworth A. Accessibility of health clubs for people with mobility disabilities and visual impairments. American Journal of Public Health 2005;95: 2022–2028.
62. Vanner E, Block P, Christodoulou C, Horowitz B, Krupp L. Pilot study exploring quality of life and barriers to leisure-time physical activity in persons with moderate to severe multiple sclerosis. Disability and Health Journal 2008;1:58–65.
63. American College of Sports Medicine. ACSM's exercise management for persons with chronic diseases and disabilities. Champaign, IL: Human Kinetics; 2003.
64. Bombardier CH, Cunniffe M, Wadhwani R, Gibbons LE, Blake KD, Kraft GH. The efficacy of telephone counseling for health promotion in people with multiple sclerosis: A randomized controlled trial. Archives of Physical Medicine and Rehabilitation 2008;89:1849–1856.

65. McAuley E, Motl RW, Morris KS, Hu L, Doerksen SE, Elavsky S, Konopack JF. Enhancing physical activity adherence and well-being in multiple sclerosis: A randomised controlled trial. Multiple Sclerosis 2007;13:652–659.

66. Motl RW, Dlugonski D, Wojcicki TR, McAuley E, Mohr DC. Internet intervention for increasing physical activity in persons with multiple sclerosis. Multiple Sclerosis 2011;17:116–128.

67. Plow MA, Mathiowetz V, Lowe DA. Comparing individualized rehabilitation to a group wellness intervention for persons with multiple sclerosis. American Journal of Health Promotion. 2009;24(1):23–26.

68. Stuifbergen AK, Becker H, Blozis S, Timmerman G, Kullberg V. A randomized clinical trial of a wellness intervention for women with multiple sclerosis. Archives of Physical Medicine and Rehabilitation 2003;84:467–476.

69. Motl RW, Snook EM, Hinkle ML. Effect of acute unloaded leg cycling on spasticity in individuals with multiple sclerosis using anti-spastic medications. International Journal of Neuroscience 2007;117(7):895–901.

70. Petajan JH, Gappmaier E, White AT, Spencer MK, Mino L, Hicks RW. Impact of aerobic training on fitness and quality of life in multiple sclerosis. Annals of Neurology 1996;39:432–441.

71. Mostert S, Kesselring J. Effects of a short-term exercise training program on aerobic fitness, fatigue, health perception and activity level of subjects with multiple sclerosis. Multiple Sclerosis 2002;8:161–168.

72. Oken BS, Kishiyama S, Zajdel D, Bourdette D, Carlsen J, Haas M, Hugos C, Kraemer DF, Lawrence J, Mass M. Randomized controlled trial of yoga and exercise in multiple sclerosis. Neurology 2004;62:2058–2064.

Section IV

Influence of Personal and Environmental Contexts on Multiple Sclerosis and Multiple Sclerosis Rehabilitation

18

Influences of Age and Sex

Marcia Finlayson, Tanuja Chitnis, and Elizabeth A. Hartman

CONTENTS

According to the *International Classification of Functioning, Disability and Health* (ICF),[1] personal factors are internal features of an individual that are not part of the health condition itself and can be used to distinguish individuals who are similar in their health condition, functioning, and environment. Personal factors can interact with, and influence, the functional consequences of a person's health condition.[1] The focus of this chapter is on age and sex and how these two factors can influence a client's functioning, experience with multiple sclerosis (MS), and his or her rehabilitation process. Sex refers to the anatomical and associated hormonal differences that distinguish a person as being male or female. Since functioning is heavily influenced by social and cultural expectations about what constitutes appropriate roles, responsibilities, and performance for men and women, this chapter also indirectly addresses the role of gender. Gender is defined as the behavioral, cultural, or psychological traits typically associated with one's sex. To illustrate the role of personal factors on the health and functioning of people with MS, four cases will be introduced in the next section and then woven throughout this chapter in order to meet the following objectives. After reading this chapter, you will be able to:

1. Highlight key developmental stages and associated roles across the life course and the potential implications for people with MS,

2. Describe the unique features of pediatric MS and its implications for rehabilitation,

3. Describe the unique features of late-life MS, the interactions between aging and MS, and the implications of both late-life diagnosis and aging on MS rehabilitation, and

4. Describe sex-specific issues that influence course and progression of MS and how the gender expectations may influence MS rehabilitation.

18.1 Cases

Imagine four people—David (age 67), Melissa (age 48), Jack (age 28), and Amy (age 16)—each of whom was recently diagnosed with MS. Medically they exhibit many similarities in the presentation of their MS. For example, all of them scored 3.5 on the Expanded Disability Status Scale (EDSS) (see Chapter 2) because they were fully ambulatory and exhibited moderate ataxia of the lower extremities (see Chapter 7), mild decreases in sensation (particularly touch and position sense), and mild decreases in cognition (particularly attention and new learning (see Chapter 8). Their assessments also indicated severe fatigue (average Fatigue Severity Scale [FSS] score of 5 [see Chapter 5]) and some perceived difficulties with walking (Multiple Sclerosis Walking Scale 12 Score of 23/60 [see Chapter 13]). None of them reported a recent MS-related fall, although all of them reported several stumbles and near falls over the past 6 months. Concerns about performance in the areas of self-care activities were nonspecific, primarily because of daily variations in symptom severity. All four of them reported emerging difficulties with domestic life activities (keeping up with their responsibilities in their respective households [see Chapter 15]) and with social and leisure–activity participation (see Chapter 17).

Despite the similarities in their profiles, their levels of disability and participation and their personal experiences of having MS are very different. David, Melissa, Jack, and Amy have different roles, responsibilities, and activity preferences that influence their daily routines and choices about engagement in specific activities. Roles, responsibilities, and preferences influence the rehabilitation goal-setting and treatment-planning processes because these aspects of the person help to define what activities are important and meaningful (see Table 18.1). While many other personal factors may also play a role in the rehabilitation process (e.g., social background, education, experience), the age and sex of the client must be considered to ensure an individually tailored approach in MS rehabilitation.

18.2 Age Influences on MS

Regardless of whether a discussion is focused on children with MS or older adults with MS, it must take into account the current age of the person, the age at which the person was

TABLE 18.1

How Age and Sex Differences Can Influence Functioning

	Examples of Possible Roles, Responsibilities, and Social and Leisure Activities			
	David	Melissa	Jack	Amy
Roles	Husband Grandfather Volunteer	Mother Wife Part-time social worker	Manual laborer Baseball player Boyfriend	Student Sister Friend
Domestic life responsibilities	Mowing grass and shoveling snow Doing home repairs Vacuuming	Preparing meals Taking children to school Gardening	Shopping for groceries Managing finances Taking out the garbage	Studying Caring for dog Learning to drive Babysitting siblings
Social or leisure activities	Playing cards Playing golf Fixing the car Playing with grandchildren	Going to book club Having lunch with friends Going out with husband	Playing baseball Dating Hanging out at the sports bar with friends	Hanging out at the mall Texting friends Playing soccer Playing the flute

diagnosed with MS, and the historical time period during which diagnosis occurred. These considerations are roughly equivalent to what epidemiologists and gerontologists refer to as age, cohort, and period effects.[2] Current age will influence an individual's specific life concerns and the social and cultural expectations about "appropriate" roles and responsibilities given his or her age (e.g., being a parent vs. a grandparent; going to grade school vs. going to college). The cases of David, Melissa, Jack, and Amy illustrate this point.

A person's current age and experience will also influence his or her perceptions about ability, disability, and independence. As an older adult, David may know people his age who have difficulty walking, perhaps because of arthritis, Parkinson's disease, or complications from diabetes. Consequently, it is possible that he will interpret and cope with MS-related walking difficulty differently than Jack, who is unlikely to have peers with similar physical challenges. In addition to current age, age of diagnosis influences how long a person's disease experience is likely to be (based on life expectancy) and what life experiences will occur at different points of the disease course. Melissa has already had her children and parented them without MS symptoms. If either Amy or Jack chooses to become a parent, they will have to cope with a variable disease while simultaneously parenting young children.

Finally, discussions on age, aging, and MS must also consider the historical time period during which a person was diagnosed. This period defines the nature and extent of treatments available based on the scientific discoveries of the time, not only for MS but also for general age-related health conditions.

18.2.1 Children and Teens with MS

The earliest article about MS in children indexed to the PubMed database was published in 1946 in the *American Journal of Diseases of Children*.[3] Nevertheless, attention to pediatric MS is a relatively recent phenomenon in MS care and research. Most of this attention has focused on describing pediatric MS, its clinical features, and options for medical treatment. As of fall 2011, there were no studies reporting the results of rehabilitation interventions (e.g., physical or occupational therapy, neuropsychology) specifically targeting children or teens with MS that were indexed to Medline or CINAHL, although calls for such studies were being made.[4-6] Before such studies can be developed and implemented, rehabilitation providers must understand the features of pediatric MS and then consider this information together with their knowledge of normal childhood development.

18.2.1.1 Epidemiology of Pediatric MS

It is increasingly recognized that MS can present in childhood or adolescence. The youngest onset of MS in the medical literature is 2 years of age; however, the majority of children are diagnosed in their early teens.[7] Studies have estimated the prevalence of pediatric-onset MS (POMS) to range from 2.7% to 10.5% of the total MS populations (2.7%[8]; 3.6%[9]; 4.4%[10]; 5%[11]; 10.5%[12]). Recently, Chitnis and colleagues found that 3.06% of 4399 MS patients with an electronic record of their MS history experienced a first attack when they were younger than 18 years old,[13] suggesting that these patients may represent a pediatric-onset population. Another recent study suggested that there are significant differences in ethnic and racial characteristics of children with MS as compared to an adult MS population.[14] The Chitnis et al.[13] study further supported this finding since these researchers found a higher proportion of non-Caucasian patients in the pediatric-onset group (11.7%) when compared to an adult-onset group (6.18%; $p = 0.014$). Moreover, of patients from one pediatric MS Center, 33.3% were non-Caucasian compared to 5.5% in the comparative adult MS cohort.[15]

18.2.1.2 Clinical Presentation and Medical Treatment

Children with MS overwhelmingly experience a relapsing–remitting course at onset.[8,10] Primary progressive MS is extremely rare. One study recently showed that children with MS experience two to three times as many relapses as adults with MS.[15] Moreover, recovery from relapses appears to be more rapid in children than in adult MS patients (mean time of relapse-related symptoms: 4.3 weeks in pediatric MS vs. 6–8 weeks in adult MS[16]). Common presenting symptoms of MS in children include sensory deficits (26.0%), optic neuritis (21.6%), cranial nerve or brainstem symptoms (12.9%), and gait disorders (8.6%).[9] Children with MS typically have a polysymptomatic presentation, although monosymptomatic presentations are not uncommon. Other features that are more common in children compared to adults with MS are encephalopathy and seizures, likely representing the overlap of acute disseminated encephalomyelitis (ADEM) and MS, which is almost exclusive to children. In addition to physical disability, children with MS are particularly vulnerable to fatigue, as shown in a comparison of the FSS scores from a pediatric MS group compared to age-matched healthy controls. When the fifth percentile of the healthy controls was used as the cutoff for fatigue, a total of 73% of the pediatric MS group met the criterion for severe fatigue.[17]

Several studies have demonstrated that initial disease progression is slower as measured by the EDSS scale in individuals with POMS compared with geographically region-matched individuals with adult-onset MS (AOMS).[9,12,18,19] This finding likely represents a difference in accrual of locomotor disability; however, these studies also demonstrated that at any given age a child with POMS will have a higher disability score than a person with AOMS.

Cognitive dysfunction can present early in the course of pediatric multiple sclerosis.[4,17,20,21] One study of 37 children with MS found that 60% experienced cognitive difficulties in one major domain, and 35% had difficulties in two domains.[21] The areas that were most commonly affected were complex attention, naming, delayed recall, and visual memory. In contrast, verbal fluency and immediate recall were relatively intact. Another study of 61 children found that 31% of participants exhibited significant impairment in three assessed domains of cognitive functioning, and 53% failed at least two tests.[17] Longitudinal data over a 2-year period have shown that over 60% of children continue to accrue cognitive deficits.[22] Neuropsychological testing should be considered for all children with MS and related demyelinating diseases, since symptoms may occasionally be subtle.[22] In a subset of children with MS who underwent psychiatric evaluation, almost half were diagnosed with depression or an anxiety disorder.[4] These findings underline the need for routine neuropsychological testing and psychiatric evaluation as a part of the management of childhood MS.

Although there are no FDA-approved therapies for pediatric MS, standard injectable MS therapies, including beta-interferon (Avonex®, Betaseron®, and Rebif®) and glatiramer acetate (Copaxone®), are commonly used for children and teens with MS. Three- to five-day courses of daily intravenous corticosteroids are usually used to treat relapses. As described above, physical disability is generally limited to periods of relapses, and recovery tends to be good in children and teens with MS. Notwithstanding these characteristics of pediatric MS, subtle changes may be present over the long term that are not measured on standard neurological scales such as the EDSS. Consequently, referral to rehabilitation services should be considered to maximize function during these episodes and to return the child back to his or her usual and developmentally appropriate level of activity. Occasionally, improvement may take several weeks to months, and reassurance that symptoms will likely improve is crucial for both the client and the parent. Further research is required to assess the long-term outcomes of pediatric MS, including physical, cognitive, and psychosocial outcomes, as well as more sensitive and pediatric-specific outcome measures.

18.2.1.3 Activity Participation and Implications for Rehabilitation

At this point in time, there are no empirical studies describing the impact of pediatric MS on participation in roles, routines, or daily activities of children or teens with the disease. However, knowledge of the clinical presentation of pediatric MS, normal development, and typical activities of childhood and adolescence can inform the clinical reasoning of rehabilitation providers with respect to assessment, setting goals, and planning treatment for this subgroup of clients with MS.

Since children with MS tend to have more frequent relapses, rehabilitation providers should assess the consequences of relapses on the child's school attendance, ability to keep up with school work, ability to engage in organized activities that require consistent participation or practice (e.g., music lessons, team activities), and opportunities to engage in the spontaneous play of childhood. In addition to relapses, fatigue and cognitive symptoms can add further challenges that will vary based on the age of the child and his or her developmental stage and preferred activities. Disclosure of MS to peers, teachers, and others may be challenging and require support from the rehabilitation team, particularly the social worker or psychologist. Teachers may require education, information, and support to help them adapt classroom strategies and participation expectations to accommodate the variable symptoms experienced by students with MS. Occupational therapists can play an important educational and collaborative role in this regard.

For a 16-year-old like Amy, attendance and success in high school may have long-term implications for her postsecondary educational opportunities and job options. Consequently, rehabilitation assessment should explore whether Amy, her parents, or her school teachers have specific concerns about the impact of her symptoms on academic progress and on the performance of specific school-related tasks (e.g., computer keyboarding, studying, concentrating). If concerns exist, rehabilitation providers need to work with Amy, her parents, and her school teachers to develop goals and treatment strategies that minimize the impact of her symptoms on her school performance and ability to progress. Intervention may include cognitive rehabilitation (remedial and compensatory), education on energy management (e.g., planning, pacing) and study skills (e.g., alternative methods of note taking), exercise, and environmental modification (e.g., ergonomic desk arrangement). This range of interventions calls for involvement of occupational and physical therapists and neuropsychologists. As she nears graduation, involvement of a rehabilitation counselor may also be warranted (see Chapter 16).

If Amy's MS leads to frequent absences from school and missed social and leisure activities, she may experience difficulties developing new friendships or maintaining existing ones. Given the importance of social relationships during adolescence, the rehabilitation team needs to inquire about this issue during assessment and offer intervention if indicated. Counseling from a psychologist or a social worker may be warranted if Amy is struggling with her feelings about having MS and being different from her peers. Training from an occupational therapist or speech-language pathologist may be considered if Amy's cognitive or sensory symptoms are making it difficult for her to use technology to interact with her friends (e.g., texting or using Facebook on a tablet or cell phone). A recreation therapist may become involved to help Amy explore new leisure opportunities that can support her desire for social participation with her peers.

As Amy grows older and her MS impairments accumulate, she will need information and support to make important life decisions. Skills in chronic disease self-management (e.g., problem solving, decision making, action planning)[23] and maintaining a healthy lifestyle will be important to ensure ongoing and satisfying participation in daily life. Rehabilitation providers can play an important role in supporting Amy's transition into adulthood.[24]

Rehabilitation providers have a significant part to play in enabling children and teens to engage in the activities that are important and meaningful to them, and that allow them to continue to develop their physical, social, cognitive, and emotional abilities. Nevertheless, practice with this subgroup is currently based on clinical reasoning, experience, and evidence from adult populations. Further research is required to develop rehabilitation assessments and intervention protocols specific to children and teens with MS and to evaluate their efficacy and effectiveness. This research will need to address physical, cognitive, and psychosocial issues at multiple levels: the child, family, school system, and society.

18.2.2 Older Adults with MS

MS is usually diagnosed among people in their late 20s or early 30s,[25,26] but incident cases have been reported in people over the age of 80.[27] Therefore, any discussion on older adults with MS must distinguish between people who are diagnosed with MS at an early age and move into middle age and older adulthood with the disease (*aging with* MS) and those individuals who acquire MS at an older age (*late-onset* MS). People from both of these groups contribute to the total population of older adults with MS. In 2004, researchers estimated that 9.4% of MS cases in the United States were individuals over 65 years of age.[28]

18.2.2.1 Life Expectancy

While the proportion of older adults with MS is relatively small, the general and disability-specific trends toward increasing life expectancy suggest that this proportion may increase over time. In the past, it was commonly believed that MS reduced life expectancy. Over the past 40 years, longer life expectancies in the general population[29] plus improvements in MS-specific care have contributed to longer average survival times for people with MS.[30] In addition, it has been recognized that reductions in life expectancy of people with MS tend to be associated with advanced disability, progressive disease course, and related complications.[31] Overall, life expectancy reductions reported in the literature range from 6 to 10 years,[32,33] but estimates tend not to account for other medical or age-related comorbidities or secondary conditions.

18.2.2.2 Late-Onset MS

The case of David, introduced earlier in this chapter, is an example of someone who would be considered as late-onset MS. Late-onset MS is defined as clinical presentation of the disease after age 50.[34] Late-onset MS tends to be monosymptomatic[34,35] and to present initially with motor or cerebellar symptoms.[35–37] Individuals with late-onset MS rarely experience visual disturbances.[34,35,37] They tend to have a progressive disease course[34–37] and more prominent degenerative pathology in the spinal cord.[27,34,38] Individuals with late-onset MS also tend to have a poor prognosis,[34] perhaps because of their older age and greater potential for age-related comorbidities (e.g., cardiovascular disease, respiratory disease, diabetes). Multiple comorbidities can make the diagnosis of late-onset MS challenging.[34,35] In one study, investigators found that MS diagnosis was delayed in 40% of individuals with late-onset disease.[36] This same research team also found that misdiagnosis occurred in 33% of cases of late-onset MS (i.e., another diagnosis was given initially), which compared to only 15% misdiagnosis among younger adults.[36] To date, there are no guidelines for the pharmaceutical treatment of late-onset MS.[34]

18.2.2.3 Age-Related Changes and MS

In a general medical review on aging with MS, Stern et al.[39] described a series of normal age-related changes (e.g., muscle atrophy, reduced cardiopulmonary reserve, impaired temperature regulation, depression) and their potential implications for people with MS. Many of the symptoms experienced by people with MS (e.g., pain, fatigue, depression, cognitive changes, visual disturbances, and mobility losses) are also commonly experienced by people aging without MS. Normal age-related changes, together with MS symptoms, can contribute to the functional challenges of an older adult with MS. It is often difficult to distinguish between MS-related symptoms and disabilities and those that are a consequence of normal age-related changes.[40,41]

The complex interaction between normal age-related changes and MS-related progression may partially explain the growing epidemiological evidence that accumulation of disability in MS is age accelerated and age dependent.[42–45] For example, Trojano and colleagues found that EDSS scores increased significantly with increasing age, and that older patients showed accelerated rates of disability progression compared to younger patients.[44] In multivariate analysis examining disease severity, current age explained 59% of variance in the model ($\beta = 0.1325$, $p < 0.01$). When duration of disease was controlled for, older age was associated with a greater probability of reaching an irreversible EDSS rating of 4 or 6, and a faster rate of progression of disability.

Functionally, comparisons between younger (<65 years) and older (>65 years) people with MS have indicated that older adults with MS report greater activity limitations than younger adults with MS in the areas of bathing, dressing, toileting, getting out of bed, getting around a room, taking medication, shopping, and getting around the community.[28] Consistent with these differences, another study found that older adults with MS identified transportation, accessible housing, professional home care, social well-being programs, pharmaceutical services, nutrition programs, and physical wellness/exercise programs as the most important services they needed to maintain their health and well-being.[46] Other research has indicated that age appears to play a role in receipt of formal support services among people with MS. For example, Putnam and Tang[47] found that older age increased the likelihood that a person with MS would receive formal support services, which included disability-specific transportation, in-home chore assistance, home health care, and care in a skilled nursing facility. Similarly, Minden and colleagues[28] found that older adults with MS were more likely to have received home care in the past year compared to younger adults with MS (40% vs. 30%, $p = 0.01$).

David and Melissa will be faced with these challenges as they continue to age. Comprehensive, multidisciplinary teamwork can help both of them evaluate whether emerging issues are age related, MS related, or due to other health conditions. Rehabilitation providers can play a critical role in promoting lifestyle habits that support physical and mental health, social interaction, and community participation. While these efforts may occur in the context of outpatient or home-based rehabilitation, partnerships with local community centers, seniors' programs, or MS societies provide additional opportunities to promote health and well-being of people with MS as they age.

18.2.2.4 Implications for Rehabilitation

In recent years, several qualitative research studies have been published exploring the experience of aging with MS.[40,41,46,48,49] Findings suggest that people aging with MS are particularly concerned about the continued loss of their mobility,[40,46] becoming a burden on family members and other caregivers,[40,49] and the possibility of having to move into a nursing home.[40] Factors that enable older adults with MS to maintain and promote healthy aging include engagement with others, good health care, financial flexibility, resilience, and healthy lifestyle.[50]

All these issues point to the need for multidisciplinary perspectives when working with older adults with MS. As with all MS clients, assessment must focus on their issues and concerns and actively involve the client in the rehabilitation process (see Chapter 4). Physical and occupational therapists will play an important role in enabling older adults with MS to maintain their self-care capacity, home and community mobility, and to ensure that caregivers have the necessary skills to provide physical assistance safely and effectively (see Chapters 13 and 22). Nurses may also play an important role in educating caregivers, particularly if bladder, bowel, or medication adherence issues are a concern. Social workers may need to provide information about available community supports and help families find ways to pay for those that are needed. Involvement of several disciplines may be required as home modifications become necessary to enable a person with MS to age-in-place (see Chapter 21).

Because of comorbidities such as cardiovascular or respiratory disease, or advanced MS-related disability, older adults with MS are often passively excluded from rehabilitation studies. Consequently, there are few rehabilitation-research intervention studies that specifically target older adults with MS. One area that has received preliminary attention is falls prevention using a self-management approach.[51] To date, exercise intervention research has not focused on people with long-standing MS who have significant disabilities. Nevertheless, research has shown beneficial effects of multiple exercise regimens on fatigue, muscle strength, and functional capacity in adults with MS.[52,53] Rehabilitation providers must, therefore, take existing evidence about the efficacy and effectiveness of rehabilitation interventions for adults with MS, consider it in light of their knowledge of geriatrics and gerontology, and combine this knowledge judiciously to inform their clinical reasoning.

For an older client like David, rehabilitation providers must include questions about other health conditions in the baseline assessment because the presence of comorbidities may influence goal setting or selection of treatment approaches. Goals may need to be more conservative if a client has multiple comorbidities, and some interventions may be contraindicated or need modification to ensure client safety. While questions about available social supports should be asked of all rehabilitation clients, asking more specific questions about the health and disability status of caregivers of older adults with MS may be warranted. Older MS clients often have older caregivers who have their own health and disability issues or they may have caregivers who do not live with them.[54] Caregiver situations may influence the frequency and intensity of rehabilitation (due to transportation issues or other issues), recommendations for home modifications, or the types of postdischarge supports that are recommended after rehabilitation is complete.

For clients like Melissa, who is currently middle aged, rehabilitation providers need to be knowledgeable about both the age-related service system (e.g., Area Agencies on Aging in the United States) and the disability-related system (e.g., Independent Living Resource Centers). Middle-aged clients may need to have supportive services parceled together from more than one service system because of variations in eligibility requirements (e.g., minimum age). Because of the complexity of working across systems and payment sources, social workers often play a critical role on the rehabilitation team serving people with MS who are middle aged and older. In addition, education on long-term care planning is an important part of the rehabilitation process for this subgroup of people with MS to ensure that individuals are able to stay in the least restrictive environment for as long as possible.[47] Multiple team members may be involved in this process, depending on specific client issues.

18.3 Sex Influences on MS

Sex is determined by inherited chromosomes (XY = male, XX = female) and is characterized by sex-specific phenotypic development (male development of testes, female development of ovaries) and sex hormone production (male increased levels of testosterone and lower levels of estrogen and progesterone, female vice versa). The relative contribution of sex chromosomes (and associated gene expression) versus the contribution of sex hormones to a disease state is difficult to ascertain in a complex autoimmune disease such as MS. Female sex confers an increased risk of autoimmune disease that exceeds the risk contribution of any known genetic or environmental factor.[55] A female predominance is noted in multiple autoimmune diseases, including MS, rheumatoid arthritis, lupus, and thyroid disease. Sex hormones including estrogens, progesterone, and testosterone influence immune activity in autoimmune diseases such as MS.[56] The effects of sex hormones in MS are complex and incompletely understood. Human studies have shown that sex and sex hormones (along with genetic and environmental factors) affect individual MS risk and may exert effects on disease course and MS symptoms, as discussed in more detail below.

Gender may also affect how men and women experience and adapt to a chronic and often disabling disease such as MS. For example, gender has been shown to have a role in health care utilization,[57] with women more likely than men to utilize medical services. In a survey of over 2000 MS patients, women reported better awareness of MS disease symptoms and more positive perceptions of the use of disease-modifying therapy than men.[58] Gender-specific roles and expectations may affect coping, help-seeking behaviors, adaptation, and learning.[59] It is important for rehabilitation professionals to be familiar with some of the key ways that the sex and gender of an individual with MS may affect his or her functioning and rehabilitation.

18.3.1 Epidemiology

MS predominantly affects women, with a 2–3:1 female-to-male ratio.[60] In younger MS patients, the ratio is higher at 3.2:1.[61] Eighty-five percent of individuals have relapsing–remitting MS, and over 10–15 years at least 50% develop secondary progressive MS. Approximately 10% of individuals have primary progressive MS. Male sex is associated with an increased risk of primary progressive MS, with an overall equal (1:1) or higher likelihood of developing primary progressive MS compared to females.[62,63] Women comprise 70–75% of individuals with MS, yielding approximately 700,000 men affected by MS worldwide.[64] These numbers reflect the higher female incidence in relapsing–remitting MS patients, as well as the relatively shorter overall life expectancy of men compared to women. Prior to puberty, the incidence of pediatric MS is equal between the sexes; after puberty, more females develop MS.[65] In individuals presenting with an initial symptom of concern for MS (referred to as a clinically isolated syndrome), men are at lower risk than women to have a second attack and develop clinically definite MS.[66] The majority of women are diagnosed in their 20s or 30s, which also corresponds to ages when most women have children. Men tend to be older at diagnosis,[67] and age is associated with disability progression in MS. In addition to the difference in incidence and prevalence of MS between the sexes, MS may also manifest differently over the lifetime of men and women.

18.3.2 Unique Features of MS in Men

Males with multiple sclerosis are more likely than females to be older (30–40) at disease onset.[55] Some researchers have found that males have more rapid progression of the disease,[56] but others have not found a clear sex effect on disease progression.[68–70] Age effects may explain the difference in findings, as older age at MS onset is associated with a shorter time to progression.[67] This is further supported by the finding that both males and females reach an EDSS of 6 (unilateral assistance for gait) at similar ages.[37]

Males with MS appear to be more susceptible to MS-related cognitive impairment, especially in areas of memory and visuospatial ability.[71] Poor health-related quality of life, especially mental well-being, is more significantly associated with increasing disability in men than in women.[72] Some small studies have noted differences in MRI lesions between sexes, such as females having more active MRI lesions[73] and males having more destructive brain MRI lesions.[56] However, larger studies have not found significant sex differences in MRI findings.[74] No significant sex differences in responsiveness or tolerability to MS disease-modifying therapy have been noted to date.[75] At an individual level, there is a large degree of variability in clinical manifestations of MS in men, reflecting the overall heterogeneity of the disease.

The role of testosterone in men with MS is incompletely understood. Testosterone levels naturally decline with age (around age 30), which is also when men are more likely to have onset of MS.[76] Males with MS have been noted to have lower testosterone levels as well as sperm quality compared to healthy controls.[77] Symptoms of low testosterone, including depression, loss of lean muscle mass, and lower bone density,[78] may be inappropriately attributed to MS. Testosterone treatment in a study of 11 male relapsing–remitting MS patients was associated with slowed cognitive decline and brain atrophy,[79] consistent with a protective role. The potential beneficial effect and safety of testosterone therapy in males with MS is yet to be confirmed in larger studies.

18.3.3 Unique Features of MS in Women

As previously noted, MS is more prevalent in women than in men. Across a lifetime, women experience sex-specific and hormonally driven physiological changes that may pose unique challenges for women with multiple sclerosis.

18.3.3.1 Menses

Onset of menarche is associated with an increased risk of MS for females compared to males, as noted above (see Section 18.3.1). Some women note that preexisting MS symptoms temporarily worsen a few days before and during menses.[80] Such worsening is temporary, does not affect overall disease course, and in general does not require additional intervention. No significant difference in MS risk has been noted with use of oral contraceptive pills in women.[81]

18.3.3.2 Pregnancy

Fertility is not affected by multiple sclerosis,[82] although hormonal therapies for infertility may increase risk of MS exacerbation. Preconception, parents should be counseled regarding MS risk to their offspring. With a single parent affected by MS, the risk is approximately 3–4% for female offspring and 1.5% for male offspring.[83] Risk increases to 30% if both parents have MS. No increased risk of birth defects[84] or pregnancy-related complications has been noted in women with MS.

MS disease-modifying medications have varying safety levels during pregnancy, and a woman should discuss treatment options with her physician prior to attempting to conceive. In most cases, disease-modifying therapy is interrupted prior to attempting to conceive due to possible fetal effects. Pregnancy may be associated with worsening of preexisting MS symptoms such as fatigue, bladder dysfunction, or gait instability.[82] Many symptomatic medications are contraindicated or should be used with caution in pregnant women. For example, commonly used medications for fatigue and bladder dysfunction are pregnancy category C (animal studies show adverse fetal effects but no controlled human studies) and some neuropathic pain medications and antidepressants are category D (positive evidence of human fetal risk). Women with MS may be more likely to require assisted deliveries,[85] which may relate to fatigue, weakness, and spasticity, especially in patients with higher levels of disability. There is no clear contraindication or adverse effect on MS relapses associated with any form of anesthesia during labor.[82]

During pregnancy, especially the second and third trimesters, women are less likely to experience MS exacerbations.[86] The reduction in MS disease activity during pregnancy is believed to be moderated by multiple hormonal changes that create an overall less inflammatory immune state. Postpartum, women are at increased risk for relapses. Nearly one in three women has a relapse in the 6 months postpartum.[86] Patients with higher levels of disability and frequent relapses prior to or during pregnancy are at increased risk for postpartum relapses. Women with MS who have had at least one child appear to have an overall lower risk of long-term disability compared to women with MS who were never pregnant.[87]

After delivery, women with MS face important decisions and challenges. In women with active MS activity prior to or during pregnancy, resumption of disease-modifying therapy and consideration of prophylactic use of steroids or other therapies to prevent relapses should be considered. Per FDA prescribing information, current MS disease-modifying treatments are not recommended in breast-feeding women due to possible medication transmission and unknown effects in breast milk. Recent studies suggest that exclusive breast feeding may provide some protection against postpartum MS relapses,[88] but interpretation of this benefit is limited by the bias of nonrandomized trials and reflects the reality that women with less disability and less active disease may be more likely to elect to breastfeed than patients with more disability and active MS.[89] Poor sleep and the demands of caring for an infant also pose special challenges to women with MS, especially those with fatigue and impairments in sensation, strength, and coordination.

Estriol, an estrogen that increases in pregnant women, is under investigation as a potential MS therapy. A small trial of estriol in 10 women with MS showed a decrease in new MRI brain lesions,[90] although safety and reproducibility needs to be confirmed in larger studies.

18.3.3.3 Menopause

There is little research on menopause and MS. Some women notice worsening of MS symptoms while others note no change or improvement.[82] Hot flashes may temporarily worsen MS symptoms due to Uhthoff's phenomenon, which is a temperature-dependent slowing down of nerve conduction. Cognitive and mood changes associated with menopause may sometimes be mistaken for worsening MS. The hormone changes associated with menopause are associated with increased bone loss, and women with disability or prior steroid exposure due to MS are at increased risk for osteopenia and osteoporosis.[91] Individuals with MS are more likely to develop progressive disease with age. MS is less likely to develop in women over age 50, and thus MS risk is lower in postmenopausal women.

18.4 Implications for Rehabilitation

Men and women have different roles, routines, and responsibilities in society and have different expectations placed on them as a result. The age of the person also influences these differences. Baseline assessments need to explore what clients need and want to do, how they are evaluating their own performance, and what their goals are for the future (see Chapter 4). While there may be similarities between men and women (e.g., need to work, desire to be a parent), there are likely to be important differences associated with gender roles and expectations. Cultural background will also be an influencing factor (see Chapter 23). Appreciating the complex interaction among sex, gender, age, and culture on a client's activities and goals will allow the members of the rehabilitation team to engage in comprehensive assessments and develop relevant goals together with individual clients.

Women with MS who are pregnant or want to become pregnant may be unable to take symptomatic medications for MS. Rehabilitation providers can play an important role in providing alternative treatments for these clients. Rehabilitation strategies for self-managing symptoms such as fatigue, weakness and spasticity, cognitive limitations, depression, and pain (see Chapters 5 through 10) should be considered before problems become disabling. In addition, activity and environmental modifications (see Chapters 12 through 17, 21) can be used to reduce some of the demands of parenting, particularly the physical demands. Postpartum relapses pose special challenges to women because they may be breastfeeding or sleep deprived and will have the demands of caring for an infant. Therefore, members of the rehabilitation team should consider preventive intervention and education with the woman and the members of her support network. Regular follow-up for ongoing problem solving will also be important to minimize physical and emotional consequences on the woman with MS, her child, and the family.

For MS rehabilitation in general, studies examining the influence of sex differences on the process and outcomes of MS rehabilitation interventions are limited. One recent study examined whether age, sex, employment status, or level of MS impairment had a moderating influence on the outcomes of a teleconference fatigue-management intervention.[92] Findings indicated that younger participants experienced greater reductions in fatigue impact and greater improvements in self-efficacy over time compared to older participants, but no age differences were found in physical or mental health. Individuals with less impairment experienced greater mental health gains and were more likely to retain these gains over time compared to individuals with more impairment. Although women experienced greater benefits in terms of fatigue impact than men, men experienced greater mental health benefits. Work status did not moderate outcomes. These findings suggest that this particular intervention does not have uniform impacts across clients, and this is likely the case for other interventions as well. Since MS risk, age of onset, disease type, and disease severity are moderated by sex hormones and other genetic and environmental factors, rehabilitation providers must consider differences between men and women in their rehabilitation plans. Men have a higher risk of primary progressive disease, and, due to later age at onset of relapsing–remitting disease, may also develop secondary progressive disease after shorter disease duration. Studies also suggest that men with MS may be more likely to develop MS-related cognitive impairment, which may require intervention to identify, improve, or manage deficits. Women with MS face special challenges related to pregnancy, many of which may be addressed by a multidisciplinary care team. Comprehensive, client-centered assessment and collaborative goal setting can lead to individually tailored rehabilitation interventions that address the issues and concerns of the client, regardless of sex and gender roles.

18.5 Gaps in Knowledge and Directions for the Future

Rehabilitation providers receive education on normal development and appreciate the impact of societal expectations and demands on a client's roles, activities, and routines across the life span and differences that may occur between men and women. Nevertheless, there is very little rehabilitation research that targets specific subgroups of people with MS (children, older adults) or examines whether age or sex moderates or mediates the efficacy or effectiveness of interventions. Future research needs to document the functional abilities and needs of these subgroups using instruments common to rehabilitation clinical practice and test whether interventions vary in effectiveness across groups. Interdisciplinary, collaborative, and mixed methods studies are necessary to address these issues and obtain comprehensive understandings that can inform clinical practice. Collaborative work with researchers and clinicians outside of MS—for example, experts in child development, school systems and school therapy, gerontology, gender studies, and the like—would add important depth to these investigations. As this text illustrates, there is already a great deal of evidence to support rehabilitation for people with MS—future work can determine if and how intervention approaches and strategies can or should be developed or modified to best serve subgroups such as children, older adults, men, and women.

References

1. World Health Organization (WHO). International Classification of Functioning, Disability and Health. Geneva, Switzerland: WHO; 2001.
2. Young TK. Population health: Concepts and methods. New York, NY: Oxford University Press; 1998.
3. Carter HR. Multiple sclerosis in childhood. American Journal of Diseases of Children 1946;71:138–149.
4. MacAllister WS, Christodoulou C, Milazzo M, Krupp LB. Longitudinal neuropsychological assessment in pediatric multiple sclerosis. Developmental Neuropsychology 2007;32(2):625–644.
5. Portaccio E, Goretti B, Zipoli V, Hakiki B, Giannini M, Pasto L. Cognitive rehabilitation in children and adolescents with multiple sclerosis. Neurological Sciences 2010;31(2):S275–S278.
6. Squillace M. Pediatric multiple sclerosis and occupational therapy intervention. OT Practice 2011;16(14):19–22.
7. Chitnis T. Pediatric multiple sclerosis. Neurologist 2006;12(6):299–310.
8. Duquette P, Murray TJ, Pleines J, Ebers GC, Sadovnick D, Weldon P. Multiple sclerosis in childhood: Clinical profile in 125 patients. Journal of Pediatrics 1987;111(3):359–363.
9. Boiko A, Vorobeychik G, Paty D, Devonshire V, Sadovnick D. Early onset multiple sclerosis: A longitudinal study. Neurology 2002;59(7):1006–1010.
10. Ghezzi A, Deplano V, Faroni J, Grasso MG, Liguori M, Marrosu G. Multiple sclerosis in childhood: Clinical features of 149 cases. Multiple Sclerosis 1997;3(1):43–46.
11. Sindern E, Haas J, Stark E, Wurster U. Early onset MS under the age of 16: Clinical and paraclinical features. Acta Neurologica Scandinavia 1992;86(3):280–284.
12. Simone IL, Carrara D, Tortorella C, Liguori M, Lepore V, Pellegrini F. Course and prognosis in early-onset MS: Comparison with adult-onset forms. Neurology 2002;59(12):1922–1928.
13. Chitnis T, Glanz B, Jaffin S, Healy B. Demographics of pediatric-onset multiple sclerosis in an MS center population from the northeastern United States. Multiple Sclerosis 2009;12(6):299–310.
14. Kennedy J, O'Connor P, Sadovnick AD, Perara M, Yee I, Banwell B. Age at onset of multiple sclerosis may be influenced by place of residence during childhood rather than ancestry. Neuroepidemiology 2006;26(3):162–167.
15. Gorman MP, Healy BC, Polgar-Turcsanyi M, Chitnis T. Increased relapse rate in pediatric-onset compared with adult-onset multiple sclerosis. Archives of Neurology 2009;66(1):54–59.

16. Ruggieri M, Iannetti P, Polizzi A, Pavone L, Grimaldi LM. Multiple sclerosis in children under 10 years of age. Neurology Sciences 2004;4:S326–S335.

17. Amato MP, Goretti B, Ghezzi A, Lori S, Zipoli V, Portaccio E. Cognitive and psychosocial features of childhood and juvenile MS. Neurology 2008;70(20):1891–1897.

18. Renoux C, Vukusic S, Mikaeloff Y, Edan G, Clanet M, Dubois B. Natural history of multiple sclerosis with childhood onset. The New England Journal of Medicine 2007;356(25):2603–2613.

19. Trojano M, Paolicelli D, Bellacosa A, Fuiani A, Cataldi S, Di Monte E. Atypical forms of multiple sclerosis of different phases of a same disease? Neurology Sciences 2004;25(4):S323–S325.

20. Banwell BL, Anderson PE. The cognitive burden of multiple sclerosis in children. Neurology 2005;64(5):891–894.

21. MacAllister WS, Belman AL, Milazzo M, Weisbrot DM, Christodoulou C, Scherl WF. Cognitive functioning in children and adolescents with multiple sclerosis. Neurology 2005;64:1422–1425.

22. Amato MP, Goretti B, Ghezzi A, Lori S, Zipoli V, Moiola L, Falautano M, De Caro MF, Viterbo R, Patti F et al. Cognitive and psychosocial features of childhood and juvenile MS: Two- year follow-up. Neurology 2010;75(13):1134–1140.

23. Lorig KR, Holman H. Self-management education: History, definition, outcomes, and mechanisms. Annals of Behavioral Medicine 2003;26:1–7.

24. Cronin A, Mandich M. Human development and performance throughout the lifespan. Clifton Park, NY: Thomson/Delmar Learning; 2005.

25. Vukusic S, Confavreux C. Natural history of multiple sclerosis: Risk factors and prognostic indicators. Current Opinion in Neurology 2007;20:269–274.

26. Burks JS, Johnson KP. Multiple sclerosis: Diagnosis, medical management, and rehabilitation. New York, NY: Demos Medical Publishing; 2000.

27. Abe M, Tsuchiya K, Kurosa Y, Nakai O, Shinomiya K. Multiple sclerosis with very late onset: A report of a case with onset at age 82 years and review of the literature. Journal of Spinal Disorders 2000;13(6):545–549.

28. Minden SL, Frankel D, Hadden LS, Srinath KP, Perloff JN. Disability in elderly people with multiple sclerosis: An analysis of baseline data from the Sonya Slifka longitudinal multiple sclerosis study. NeuroRehabilitation 2004;19(1):55–67.

29. Kemp BJ. What the rehabilitation professional and the consumer need to know. Physical Medicine and Rehabilitation Clinics of North America 2005;16:1–18.

30. Ragonese P, Aridon P, Salemi G, D'Amelio M, Savettieri G. Mortality in multiple sclerosis: A review. European Journal of Neurology 2008;15(2):123–127.

31. Kantarci OH, Weinshenker BG. Natural history of multiple sclerosis. Neurologic Clinics 2005;23(1):17–38.

32. Sadovnick AD, Ebers GC, Wilson RW, Paty DW. Life expectancy in patients attending multiple sclerosis clinics. Neurology 1992;42(5):991.

33. Bronnum-Hansen H, Stenager E, Nylev Stenager E, Koch-Henriksen N. Suicide among Danes with multiple sclerosis. Journal of Neurology, Neurosurgery and Psychiatry 2006;76(10):1457–1459.

34. Martinelli V, Rodegher M, Moiola L, Comi G. Late onset multiple sclerosis: Clinical characteristics, prognostic factors and differential diagnosis. Neurological Sciences 2004;25(4):s350–s355.

35. Polliack ML, Barak Y, Achiron A. Late-onset multiple sclerosis. Journal of the American Geriatrics Society 2001;49(2):168–171.

36. Kis B, Rumberg B, Berlit P. Clinical characteristics of patients with late-onset multiple sclerosis. Journal of Neurology 2008;255:697–702.

37. Tremlett H, Paty D, Devonshire V. Disability progression in multiple sclerosis is slower than previously reported. Neurology 2006;66:172–177.

38. Schultheiss T, Reichmann H, Ziemssen T. Rapidly progressive course of very late onset multiple sclerosis presenting with parkinsonism: Case report. Multiple Sclerosis 2011;17(2):245–249.

39. Stern M, Sorkin L, Milton K, Sperber K. Aging with multiple sclerosis. Physical Medicine and Rehabilitation Clinics of North America 2010;21(2):403–417.

40. Finlayson M. Concerns among people aging with multiple sclerosis. American Journal of Occupational Therapy 2004;58:54–63.
41. Dilorenzo TA, Becker-Feigeles J, Halper J, Picone MA. A qualitative investigation of adaptation in older individuals with multiple sclerosis. Disability & Rehabilitation 2008;30:1088–1097.
42. Confavreux C, Vukusic S, Moreau T, Adeleine P. Relapses and progression of disability in multiple sclerosis. New England Journal of Medicine 2000;343(20):1430–1438.
43. Confavreux C, Vukusic S, Adeleine P. Early clinical predictors and progression of irreversible disability in multiple sclerosis: An amnesic process. Brain 2003;126:770–782.
44. Trojano M, Liguori M, Zinatore GB, Bugarini R, Avolio C, Paolicelli D. Age-related disability in multiple sclerosis. Annals of Neurology 2002;51:475–480.
45. Koch M, Mostert J, Heersema D, De KJ. Progression in multiple sclerosis: Further evidence of an age dependent process. Journal of Neurological Sciences 2007;255:35–41.
46. Finlayson M, Van Denend T. Experiencing the loss of mobility: Perspectives of older adults with MS. Disability & Rehabilitation 2003;25:1168–1180.
47. Putnam M, Tang F. Long-term care planning and preparation among persons with multiple sclerosis. Home Health Care Services Quarterly 2008;27:143–165.
48. Fong T, Finlayson M, Peacock N. The social experience of aging with a chronic illness: Perspectives of older adults with multiple sclerosis. Disability and Rehabilitation 2006;28:695–705.
49. DalMonte J, Finlayson M, Helfrich C. In their own words: Coping processes among women aging with multiple sclerosis. Occupational Therapy in Health Care 2004;17(3–4):115–137.
50. Ploughman M, Austin MW, Murdoch M, Kearney A, Fisk JD, Godwin M, Stefanelli M. Factors influencing healthy aging with multiple sclerosis: A qualitative study. Disability and Rehabilitation 2012;34(1):26–33.
51. Finlayson M, Peterson EW, Cho C. Pilot study of a fall risk management program for middle aged and older adults with MS. NeuroRehabilitation 2009;25(2):107–115.
52. Dalgas U, Ingemann-Hansen T, Stenager E. Physical exercise and MS recommendations. International Multiple Sclerosis 2009;16(1):5–11.
53. Oken BS, Kishiyama S, Zajdel D, Bourdette D, Carlsen J, Haas M, Hugos C, Kraemer DF, Lawrence J, Mass M. Randomized controlled trial of yoga and exercise in multiple sclerosis. Neurology 2004;62(11):2058–2064.
54. Finlayson M, Cho C. A descriptive profile of caregivers of older adults with MS and the assistance they provide. Disability and Rehabilitation 2008;30(24):1848–1857.
55. Voskuhl R. Sex differences in autoimmune disease. Biology of Sex Differences 2011;2:1–21.
56. Tomassini V, Pozzilli C. Sex hormones: A role in the control of multiple sclerosis? Expert Opinion Pharmacotherapy 2006;7(7):857–868.
57. Green C, Pope C. Gender, psychosocial factors and the use of medical services: A longitudinal analysis. Social Sciences Medicine 1999;48(10):1363–1372.
58. Vlahiotis A, Sedjo R, Cox E, Burroughs T, Rauchway A, Lich R. Gender differences in self-reported symptom awareness and perceived ability to manage therapy with disease-modifying medications among commercially insured multiple sclerosis patients. Journal of Managed Care Pharmacy 2010;16:206–210.
59. Addis ME, Mahalik JR. Men, masculinity, and the contexts of help-seeking. American Psychologist 2003;58(1):5–14.
60. Noonan C, Kathan S, White M. Prevalence estimates for MS in the United States and evidence of an increasing trend for women. Neurology 2002;58:136–138.
61. Duquette P, Pleines J, Girard M, Charest L, Senecal Quevillon M, Masse C. The increased susceptibility of women to multiple sclerosis. Canadian Journal of Neurological Sciences 1992;19(4):66–71.
62. Miller D, Leary S. Primary-progressive multiple sclerosis. Lancet Neurology 2007;6(10):903–912.
63. Runmarker B, Anderson O. Prognostic factors in a multiple sclerosis incidence cohort with twenty-five years of follow-up. Brain 1993;116:117–134.
64. Orton SM, Herrera BM, Yee IM, Valdar W, Ramagopalan SV, Sadovnick AD. Sex ratio of multiple sclerosis in Canada: A longitudinal study. Lancet Neurology 2006;5(11):932–936.

65. Tintore M, Arrambide G. Early onset multiple sclerosis: The role of gender. Journal of Neurological Sciences 2009;286:31–34.

66. Thrower B. Clinically isolated syndromes: Predicting and delaying multiple sclerosis. Neurology 2007;68(24 Suppl 4):S12–S15.

67. Greer JM, McCombe PA. Role of gender in multiple sclerosis: Clinical effects and potential molecular mechanism. Journal of Neuroimmunology 2011;234(1–2):7–18.

68. Eriksson M, Anderson O, Runmaker B. Long term follow-up of patients with clinically isolated syndromes, relapsing remitting and secondary progressive multiple sclerosis. Multiple Sclerosis 2003;9(3):260–274.

69. Koch M, Uyttenboogaart M, van Harten A, De Keyser J. Factors associated with the risk of secondary progression in multiple sclerosis. Multiple Sclerosis 2008;14(6):799–803.

70. Koch M, Kingwell E, Rieckman P, Tremlett H. The natural history of primary progressive multiple sclerosis. Neurology 2009;73(23):1996–2002.

71. Beatty WW, Auperle RL. Sex differences in cognitive impairment in multiple sclerosis. Clinical Neuropsychology 2001;16(4):472–480.

72. Casetta I, Riise T, Wamme Nortvedt M, Economou NT, De Gennaro R, Fazio P, Cesnik E. Gender differences in health-related quality of life in multiple sclerosis. Multiple Sclerosis Journal 2009;15(11):1339–1346.

73. Weatherby S, Mann C, Davies M, Fryer A, Haq N, Strange R. A pilot study of the relationship between gadolinium-enhancing lesions, gender effects and polymorphisms of antioxidant enzymes in multiple sclerosis. Journal of Neurology 2000;247(6):476–480.

74. Fazekas F, Enzinger C, Wallner-Blazek M, Ropele S, Pluta-Fuerst A, Fuchs S. Gender differences in MRI studies on multiple sclerosis. Journal of Neurological Sciences 2009;286:28–30.

75. Rudick RA, Kappos L, Kinkel R, Clanet M, Phillips JT, Herndon RM, Sandrock AW, Muschauer FE. Gender effects on intramuscular interferon beta-1a in relapsing-remitting multiple sclerosis: Analysis of 1406 patients. Multiple Sclerosis 2011;17(3):353–360.

76. Gray A, Berlin JA, McKinlay JB, Longcope C. An examination of research design effects on the association of testosterone and male aging: Results of a meta-analysis. Journal of Clinical Epidemiology 1991;44(7):671–684.

77. Safarinejad MR. Evaluation of endocrine profile, hypothalamic-pituitary-testis axis and semen quality in multiple sclerosis. Journal of Neuroendocrinology 2008;20(12):1368–1375.

78. Corona G, Rastrelli G, Forti G, Maggi M. Update in testosterone therapy for men. Journal of Sex Medicine 2011;8(3):639–654.

79. Sicotte NL, Giesser BS, Tandon V, Klutch R, Steiner B, Drain AE, Shattuck DW, Hull L, Wang H, Elashoff RM et al. Testosterone treatment in multiple sclerosis: A pilot study. Archives of Neurology 2007;64(5):683–688.

80. Zorgdrager A, De Keyser J. Premenstrual exacerbations in multiple sclerosis. European Journal of Neurology 2002;48:204–206.

81. Hernan MA, Hohol MJ, Olek MJ, Spiegelman D, Ascherio A. Oral contraceptives and the incidence of multiple sclerosis. Neurology 2000;55(6):848–854.

82. Ghezzi A, Zaffaroni M. Female-specific issues in multiple sclerosis. Expert Review of Neurotherapeutics 2008;8(6):969–977.

83. Sadovnick AD, Dircks A, Ebers GC. Genetic counseling in multiple sclerosis: Risks to sibs and children of affected individuals. Clinical Genetics 1999;56:118–122.

84. Van der Kop ML, Pearce MS, Dahlgren L, Synnes A, Sadovnick D, Sayao AL, Tremlett H. Neonatal and delivery outcomes in women with multiple sclerosis. Annals of Neurology 2011;70(1):41–50.

85. Kelly VM, Nelson LM, Chakravarty EF. Obstetric outcomes in women with multiple sclerosis and epilepsy. Neurology 2009;73(22):1831–1836.

86. Confavreux C, Hutchinson M, Hours MM, Cortinovis-Tourniare P, Moreau T. Rate of pregnancy-related relapse in multiple sclerosis: Pregnancy in multiple sclerosis group. New England Journal of Medicine 1998;339:285–291.

87. D'Hooghe M, Nagels G, Uitdehaag B. Long-term effects of childbirth in MS. Journal of Neurology, Neurosurgery, and Psychiatry 2010;81:38–41.

88. Langer-Gould A, Huang S, Gupta R, Leimpeter A, Greenwood E, Albers K, Van Den Eeden S, Nelson L. Exclusive breastfeeding and the risk of postpartum relapses in women with multiple sclerosis. Archives of Neurology 2009;66(8):958–953.

89. Portaccio E, Ghezzi A, Hakiki B, Martinelli V, Moiola L, Patti F, La Mantia L, Mancardi GL, Solaro C, Tola MR et al. Breastfeeding is not related to postpartum relapses in multiple sclerosis. Neurology 2011;77(2):145–150.

90. Sicotte NL, Liva SM, Klutch R, Pfeiffer P, Bouvier S, Odesa S, Wu TC, Voskuhl RR. Treatment of multiple sclerosis with the pregnancy hormone estriol. Annals of Neurology 2002;52(4):421–428.

91. Hearn AP, Silber E. Osteoporosis in multiple sclerosis. Multiple Sclerosis 2010;16(9):1031–1043.

92. Finlayson M, Preissner K, Cho C. Outcome moderators of a fatigue management program for people with multiple sclerosis. American Journal of Occupational Therapy 2012;66(2):187–197.

19

Other Health Concerns

Ruth Ann Marrie

CONTENTS

Multiple sclerosis (MS) is a disabling disease with variable manifestations; these have significant adverse effects on the affected individuals, their families, and society. Many affected individuals also experience comorbid diseases and complications that potentially influence health outcomes. In order to optimize rehabilitation programs and outcomes, a comprehensive approach is needed; one that accounts for these comorbid conditions. This requires knowledge of which conditions are most frequent in MS and their influences on outcomes; the purpose of this chapter is to review available data on this subject. After reading this chapter, you will be able to:

1. Identify the most common physical and mental comorbidities experienced by persons with MS,

2. Recognize the impact of comorbidity on disability and quality of life, and

3. Understand how the presence of one or more comorbidities may influence rehabilitation strategies in MS.

19.1 Definitions

Comorbidity typically refers to the total burden of illness other than the specific disease of interest; for example, hypertension would be a comorbidity when MS is the disease of interest, while MS could be a comorbidity if hypertension were the specific disease of interest.[1] Complications and secondary conditions are physical and mental disorders that arise directly or indirectly from a primary disease, for example, osteoporosis due to reduced mobility or urinary tract infections secondary to neurogenic bladder.[2]

TABLE 19.1

Definitions of Comorbidity and Complications and their Implications for Rehabilitation

Term	Definition	Possible Implication for Rehabilitation
Comorbidity	The total burden of illness other than the specific disease of interest, in this case MS	Same treatment could be indicated for index disease and comorbidity, for example, physical exercise to reduce fatigue in MS is also helpful for hypertension control.
		Treatment for index disease could have antagonistic effect on comorbidity, for example, corticosteroids for MS relapse have a negative impact on diabetes control.
Complication	A condition that ensues as a result of the primary disease of interest.	Focus on treating the complication and the primary disease/underlying factors, for example, pressure ulcers—although the pressure ulcer should be treated directly, the spasticity and urinary incontinence that increase the risk for pressure ulcers also need attention.

Complications of disease need to be distinguished from comorbidities; both could independently influence outcomes (see case study 1). Recurrent urinary tract infections secondary to neurogenic bladder may precipitate relapses, augment disability progression, or precipitate death[3-5]; comorbidities could have similar effects. From the perspective of the person with MS, the distinction between complications and comorbidities may seem irrelevant if the ultimate outcome is the same. From a clinical standpoint the distinction is valuable because the interventions may differ. If the comorbidity worsens the outcome, the focus should be on treating the comorbidity, but if the complication worsens outcome, the focus should be on treatment of the primary disease in addition to treating the complication (see Table 19.1).

BOX 1   CASE STUDY 1

A 50-year-old woman with multiple sclerosis presents with concerns regarding declining mobility, increasing spasticity, and fatigue. Her disease is complicated by neurogenic bladder with recurrent urinary tract infections (UTIs) occurring four to five times per year. She also has comorbid arthritis. Her rehabilitation program included daily stretching, beneficial for her spasticity and arthritis. Given the potential adverse impact of UTIs on spasticity, she was also advised to follow-up with her neurologist for further evaluation and treatment of her neurogenic bladder.

19.2 Comorbidity

Comorbidity is common in the general population and increases with age. Based on the 1987 National Medical Expenditure Survey, 88.5 million Americans had one or more chronic health conditions associated with the use of medical care or disability days (days in which illness or injury caused the person to stay in bed more than half a day or otherwise cut down on activities for as much as a day).[6] By 2005, 133 million Americans had one or more chronic health conditions.[7] An analysis of seven high-impact chronic conditions reported in the 2005 Canadian Community Health Survey suggested that 33% of Canadians had at least one chronic condition.[8] Similar findings have been reported in other regions of the world.[9] Comorbidity is associated with a broad range of adverse health outcomes, including increased disability, reduced functional status, higher mortality, increased health care utilization and costs, and reduced quality of life.[10–13]

Like the general population, people with MS are at increasing risk of comorbidity with advancing age; this occurs at a point in the disease course when MS-related disability is also increasing. Thus, comorbidity may assume greater importance as MS progresses and management of MS becomes more complex.

19.2.1 Comorbidity in MS

Much of the early work on comorbidity in MS focused on whether a specific type of condition occurred with greater or lesser frequency in MS than in the general population,[14–6] but more recent studies have had a broader scope.[17] Studies conflict as to whether autoimmune diseases occur more frequently with MS than expected.[18,19] At least two population-based studies suggest that inflammatory bowel disease is more frequent in people with MS than in the general population.[20,21] Rheumatoid arthritis may affect up to 4.4% of people with MS. Although it is uncertain as to whether thyroid disease is more common among people with MS than in the general population, it is reported to affect up to 10% of the MS population in some studies,[15,16,18,22–26] and it is clinically important because of its potential contributions to fatigue.

Examination of a broader range of conditions suggests that physical comorbidities are quite common in people with MS (see Table 19.2). In 2006, 8983 participants in the North American Research Committee on Multiple Sclerosis Survey (NARCOMS) reported their comorbidities.[17] The most common conditions reported were hypercholesterolemia (37%), hypertension (30%), arthritis (16%), irritable bowel syndrome (13%), and lung disease (13%). Such findings deserve attention; in the general population, the leading causes of disability are hypertension, arthritis, back or spine problems, lung disease, and heart disease.[27] The NARCOMS study did not include a control group, but age-standardized comparisons with self-reported data from National Health Interview Survey (NHIS), National Health and Nutrition Examination Survey (NHANES), and other population-based surveys of the general population suggest that people with MS suffer from many of these conditions at least as often as the general population.[28]

Sleep disorders may also complicate chronic neurological disease.[29] Although the literature is not entirely consistent, several studies suggest that persons with MS have a higher frequency of sleep disorders than the general population, and these often go unrecognized.[30,31] One study of US Veterans found that those with MS had a two-fold higher frequency of sleep disturbance (5.9%) as compared to those without MS (2.7%).[32] In that study, sleep disturbance was defined as the presence of at least one diagnostic polysomnogram or sleep disorder during the study period. Restless legs syndrome and narcolepsy were more common among Veterans with MS as compared to those without MS. A review of 116 published symptomatic cases of narcolepsy found that 8.6% were associated with MS.[33] Finally, the frequency of restless legs syndrome ranges from 13.3% to 37.5% among those with MS.[34,35] All these studies suggest that sleep disorders are common among people with MS. Given the potential contribution of sleep orders to fatigue, they deserve particular attention (see Chapter 5 for further discussion on fatigue in MS).

It is likely, therefore, that the individual with MS referred for rehabilitation or other therapies will have one or more comorbid physical conditions. Along with physical comorbidities, mental comorbidities are also very common among persons with MS (see Table 19.2). While the lifetime prevalence of depression in the general population is

TABLE 19.2

Common Comorbidities with MS and Sample Plans of Care

Comorbidities	Possible Referrals/Other Team Members Involved	Plan of Care Elements[a]
Physical		
Hypertension	Family physician Dietitian	Education Promotion of physical activity Diet modification Medication
Hypercholesterolemia	Family physician Dietitian	Education Promotion of physical activity Diet modification Medication
Arthritis	Family physician Occupational therapist Physical therapist Dietitian Orthotist	Education Weight reduction Activity modification Promotion of physical activity Analgesic medication Assistive devices Braces Surgery in some cases
Chronic lung disease	Family physician Respirologist Respiratory therapist Physical therapist Occupational therapist	Education Smoking cessation Medication Oxygen supplementation Activity modification Promotion of physical activity (pulmonary rehabilitation)
Irritable bowel syndrome	Dietitian	Education Treatment for depression/anxiety, if present Symptom diary Medication
Sleep disorders	Sleep specialist Nurse	Education Sleep hygiene Medication or medical device, if appropriate
Thyroid disease	Family physician	Education Medication
Mental		
Depression	Mental health professional (e.g., psychiatrist, psychologist, social worker)	Education Psychotherapy Medication Exercise
Anxiety	Mental health professional (e.g., psychiatrist, psychologist, social worker)	Education Psychotherapy Medication Exercise

[a] Example for illustrative purposes.

estimated to be 15–17%,[36] it is 50% among persons with MS.[37] The 12-month prevalence of depression in the general population is 5.9–7.3%,[36] while it is as high as 14% in MS population.[37] Psychosis, bipolar disorder, and anxiety are also more common among people with MS compared to the general population.[38,39] The reported frequency of psychosis is 0.8%, while bipolar disorder affects 0.30–13%.[14,38,40,41] Anxiety disorders and social phobias may affect up to 35% of persons with MS over their lifetime.[42,43] In the NARCOMS population, the presence of other physical comorbidities was associated with an increased risk of depression (hazard ratio 2.20; 95% confidence interval [CI]: 2.04–2.38). The management of depression in MS is discussed in Chapter 10.

19.2.2 Impact of Comorbidity on MS

Comorbidity potentially influences several aspects of MS relevant to rehabilitation, including fatigue, physical disability, functional status, and health-related quality of life (HRQoL). As discussed in Chapter 5, fatigue affects at least 70% of persons with MS and is a common cause of unemployment.[44] Undiagnosed sleep disorders, abnormal sleep cycles, and other sources of sleep disruption adversely influence fatigue and quality of life.[45,46] There is a lack of data on the impact of non-sleep-related comorbidities on fatigue and on the impact of comorbidities on other symptoms, such as spasticity, pain, and cognitive impairment.

A study of NARCOMS Registry participants in 2006 found that those affected by vascular, musculoskeletal, or mental comorbidities, or with obesity reported more severe ambulatory disability at diagnosis than unaffected participants.[47] After adjustment, the odds of moderate as compared to mild disability at diagnosis were 1.5-fold higher among participants with vascular comorbidity (odds ratio [OR] 1.51; 95% CI: 1.12–2.05) and approximately 1.4-fold higher in those with obesity (OR 1.38; 95% CI: 1.02–1.87). Further, the odds of severe as compared to mild disability were more than 1.5-fold higher in participants with musculoskeletal (OR 1.81; 95% CI: 1.25–2.63) or mental (OR 1.62; 95% CI: 1.23–2.14) comorbidity.

Following diagnosis, comorbidity may also adversely influence the progression of disability. In a three-year observational study of 146 people with recently diagnosed MS, those with musculoskeletal comorbidities experienced a 5-point decline on the motor scale of the Functional Independence Measure while those without such comorbidities experienced only a 2-point decline.[48] Participants in the NARCOMS Registry who reported having vascular comorbidities, including hypertension, hypercholesterolemia, diabetes, heart disease, and peripheral vascular disease, experienced more rapid progression of ambulatory disability than those without comorbidities. The median time from MS onset to needing a unilateral assistive device to walk was 18.8 (95% CI: 18.4–19.3) years for participants without comorbidities but 6.0 years earlier for participants with comorbidities.[49] The mechanism underlying these findings is unknown. Comorbidities could influence the pathophysiology of MS, or the findings could reflect the additive impact of multiple diseases and their influence on a common outcome, such as mobility.

Comorbidity is known to be associated with reduced health-related quality of life (HRQoL) in other chronic diseases[1,50] regardless of the specific condition.[13] Persons with MS report lower HRQoL as compared with general and other chronic disease populations,[51–54] with factors such as age, sex, socioeconomic status, and disability status influencing HRQoL.[55] The limited data available suggest that comorbidity adversely affects HRQoL in MS too. A study of 262 participants with relapsing–remitting MS found poorer physical HRQoL, as measured by the SF-36, in those who had comorbid musculoskeletal and respiratory conditions.[56] In the Canadian Community Health Survey, responders with MS and one or more comorbidities had poorer HRQoL than those without comorbidities[57]; arthritis, hypertension, chronic fatigue syndrome, depression, and urinary incontinence were associated with worse HRQoL.

19.3 Potential Impact of Comorbidity on Rehabilitation and Symptom Management

Although relatively few studies have examined the impact of comorbidity on MS, the existing data suggest a broad, negative impact consistent with the findings in other chronic diseases. Also, comorbidity may influence management of MS. Although data regarding this issue are limited for MS, findings in other chronic diseases support this contention.

Comorbidities may influence access to care and quality of care. Patients with multiple chronic health conditions report numerous barriers to self-care.[58] In a qualitative study of adults with multiple chronic conditions, barriers to self-care included the compound effects of medications, the difficulties of coordinating multiple medications at different times, the total burden of medications, and financial constraints.[58]

In the presence of comorbidity, treatment of some chronic conditions may differ in frequency or intensity, but the direction of this effect is not uniform.[59-61] For example, general practitioners are less likely to initiate antidepressant treatment for people with cardiac disease newly diagnosed with depression.[62] About 30–50% of people with illness adhere poorly to treatment, irrespective of disease state or clinical setting[63]; comorbidity could potentially limit adherence even further by acting as a barrier to adherence.[58,64,65] Poor adherence, then, has the effect of reducing the benefits of therapies that already are only partially effective.

Although many pharmacological and nonpharmacological interventions exist for managing the chronic symptoms of MS, such as fatigue, spasticity, and impaired mobility, studies of symptomatic interventions often exclude people with major comorbid conditions.[66-68] Oken et al.[66] randomized people with clinically definite MS to a weekly yoga class, weekly exercise class using a stationary bicycle, or a wait-list control group. They excluded participants with insulin-dependent diabetes; symptomatic lung disease; uncontrolled hypertension; liver or kidney failure; alcoholism/drug abuse; symptoms or signs of congestive heart failure, ischemic heart disease, or symptomatic valvular disease; or corrected visual acuity worse than 20/50 binocularly. Wiles et al.[68] conducted a controlled trial to determine whether physiotherapy could improve mobility in MS; participants were excluded if they had other major general medical or surgical disorders. While these exclusions were made to ensure participant safety, the consequence is that we lack knowledge about the safety, tolerability, and efficacy of the studied regimens in people with common comorbidities.

To date, no studies have examined the influence of comorbidity on the outcomes of specific rehabilitation programs in MS or how rehabilitation programs should be modified to optimize participation and good outcomes. Nonetheless, we can examine findings in other diseases in which the presence of comorbidities is associated with smaller improvements in functional status after rehabilitation. Crisafulli et al.[69] conducted a retrospective study of 2962 individuals with chronic obstructive pulmonary disease admitted for pulmonary rehabilitation. The presence of "metabolic" comorbidities, including hypertension, diabetes, or dyslipidemia, was associated with 43% reduced odds of obtaining a clinically important improvement in exercise tolerance as measured by the Six Minute Walk (OR 0.57; 95% CI: 0.49–0.67). The presence of heart disease was associated with 33% reduced odds of obtaining a clinically important improvement in health-related quality of life as measured by the St. George's Respiratory Questionnaire (OR 0.67; 95% CI: 0.55–0.83).

Karatepe et al.[70] evaluated the impact of comorbidities on the effectiveness of an individualized home-based stroke rehabilitation program with 140 persons. Using a weighted comorbidity score, they found that higher scores were associated with less functional improvement based on the motor score of the Functional Independence Measure. These findings raise concern about particular comorbidities reducing the effectiveness of some MS rehabilitation programs, but relevant data are lacking. Further, we do not know if modifications to these programs could improve outcomes in these people. In the light of these issues, until further data are available, rehabilitation treatment goals should probably be more conservative for people with MS who have comorbidities, with the idea that exceeding goals is preferable to failing to meet unrealistic goals. The creation of more conservative treatment goals should not be construed, however, to mean that more conservative treatments should be used. Another issue that should be addressed in future research is whether people with MS and comorbidities would benefit with respect to improved symptom control and functional status if they were to participate in long-term rehabilitation programs or "maintenance" therapy.

So far we have considered the potential influence of comorbidity on MS and its treatments, but MS and its treatments may influence the risk and severity of comorbidities; for example, corticosteroids used for relapses may induce or worsen diabetes or aggravate mood disorders. Fampridine, used to improve walking ability and leg strength, is associated with an increased risk of seizures.[71] As people experience more chronic conditions they are more likely to experience polypharmacy. As the number of medications increases, the risk of adverse effects and drug interactions increases. Of particular concern is the under-recognized, but increased risk of falls associated with polypharmacy, which will respond to reduction of medications.[72] As noted above, complex drug regimens are perceived as a barrier to care. Thus, the addition of any new medication for MS requires careful consideration of the person's comorbidities and existing medications.

The physical medicine and rehabilitation clinician developing a treatment program to manage disease-related impairments may be challenged by the person with MS who has comorbidities. Such individuals have multiple clinical problems adversely affecting their daily function, independence, and quality of life while evidence-based guidelines for managing such individuals are lacking. However, the physical medicine and rehabilitation clinician can apply general principles of clinical reasoning to enhance the care of these people (see case study 2) and work with other members of the health care team (see Table 19.2). Given potential fluctuations in functional status due to MS and comorbidities, periodic reassessment of treatment goals and priorities may be required.

BOX 2 CASE STUDY 2

A 60-year-old man with MS presents with declining mobility and frequent falls. He has comorbid arthritis, which causes knee pain when walking. He has extensor spasms of his lower extremities, a right-foot drop, and marked weakness of his right hip flexors. He is obese with poorly controlled diabetes. Sensation in his feet is impaired.

- What are the clinical problems in this individual (client)?
- What are the individual's goals?

- Which of these problems are related to MS?
- Which of these problems are related to comorbid conditions?
- Which of these problems can you address?
- Which of these problems should be addressed by another health care provider?
- What clinical problems would you focus on in your immediate management of this individual? That is, which should take priority? Why? Consider which problems cause the most functional limitation and which are most amenable to therapy.
- What treatment options can be used to manage these problems? Consider medications, physical modalities, physical training (therapeutic exercise), movement and activities modification, adaptive equipment and assistive devices, orthotics (braces), environmental modifications, and so on.
- Will treatment of one clinical problem adversely influence another clinical problem?

Some individuals with MS and comorbidities will present with multiple clinical problems. In other situations, these people will present with one problem due to multiple factors. Pain, fatigue, and impaired mobility, for example, are common symptoms in MS that could be influenced by comorbidities, either recognized or unrecognized. Careful consideration must be given to the possible contributions of other conditions when these people are evaluated. Clinical practice guidelines for managing fatigue emphasize the importance of identifying "alternative, treatable causes that may not be directly related to the MS disease process" (p. 6).[73,74] Although this should be done in conjunction with the rest of the health care team, it may be helpful to use brief screening tools to inquire about the presence of other diagnosed health conditions, detect sleep disorders, and to identify common medications that augment fatigue (see Table 19.3).[78]

To ensure that the proposed treatment regimen is manageable for the individual and will be adhered to, it may be useful to review the other pharmacological and nonpharmacological treatments that have been prescribed. Although specific guidelines do not exist for evaluating and treating all symptomatic issues in MS, the approach used for fatigue can be generalized by incorporating queries about other health conditions and their impact on the initial assessment of the individual.

TABLE 19.3

Intake Data Regarding Medications and Comorbidities for Rehabilitation Programs

---

- History of medication or substance use
  - Particular attention should be paid to substances acting on the central nervous system, such as sedative hypnotics, antidepressants, anticonvulsants, etc.
- Comorbid diseases
  - Mood disorders
    - Possible instrument: Hospital Anxiety and Depression Scale[75]
  - Sleep disorders
    - Possible instrument: Cleveland Sleep Habits Questionnaire[76]
  - Other chronic conditions
    - Possible instrument: Self-report Comorbidity Questionnaire for Multiple Sclerosis[77]

19.4 Future Directions

Given the high frequency of comorbidity with MS, it is imperative that we develop a better understanding of the bidirectional interactions of comorbidity and all aspects of MS. This will require prospective population-based studies that comprehensively and accurately evaluate comorbidity. It will be necessary to evaluate whether people with comorbidities experience disparate access to symptomatic and rehabilitative therapies, and whether their outcomes following treatment differ from those in people without comorbidities. These data are a necessary precursor to determining whether specific interventions need to be modified and what modifications are necessary to optimize outcomes. The health care professions need to move toward a truly holistic approach to the care of the person living with MS.[79]

References

1. Gijsen R, Hoeymans N, Schellevis FG, Ruwaard D, Satariano WA, van den Bos GA. Causes and consequences of comorbidity: A review. Journal of Clinical Epidemiology 2001;54: 661–674.
2. Kinne S, Patrick DL, Doyle DL. Prevalence of secondary conditions among people with disabilities. American Journal of Public Health 2004;94:443–445.
3. Hillman LJ, Burns SP, Kraft GH. Neurological worsening due to infection from renal stones in a multiple sclerosis patient. Multiple Sclerosis 2000;6:403–406.
4. Metz LM, McGuinness SD, Harris C. Urinary tract infections may trigger relapse in multiple sclerosis. Axone 1998;19:67–70.
5. Redelings MD, Lee NE, Sorvillo F. Pressure ulcers: More lethal than we thought? Advances in Skin and Wound Care 2005;18:367–372.
6. Hoffman C, Rice D, Sung HY. Persons with chronic conditions: Their prevalence and costs. Journal of the American Medical Association 1996;276:1473–1479.
7. National Center for Chronic Disease Prevention and Health Promotion. Chronic diseases the power to prevent, the call to control. Hyattsville, MD: Centers for Disease Control; 2009.
8. Broemeling A, Watson DE, Prebtani F, on behalf of the Councilors of the Health Outcomes Steering Committee of the Health Council of Canada. Population patterns of chronic health conditions, co-morbidity and health care use in Canada: Implications for policy and practice. Healthcare Quarterly 2008;11(3):70–76.
9. Caughey G, Vitry A, Gilbert A, Roughead E. Prevalence of comorbidity of chronic diseases in Australia. BMC Public Health 2008;8(1):221.
10. Battafarano RJ, Piccirilo JF, Meyers BF, Hsu HS, Guthrie TJ, Cooper JD, Patterson GA. Impact of comorbidity on survival after surgical resection in patients with stage I non-small cell lung cancer. Journal of Thoracic and Cardiovascular Surgery 2006;123:280–287.
11. Braunstein JB, Anderson GF, Gerstenblith G, Weller W, Niefeld M, Herbert R, Wu AW. Noncardiac comorbidity increases preventable hospitalizations and mortality among medicare beneficiaries with chronic heart failure. Journal of the American College of Cardiology 2003;42:1226–1233.
12. Greenfield S, Apolone G, McNeil BJ, Cleary PD. The importance of co-existent disease in the occurrence of postoperative complications and one-year recovery in patients undergoing total hip replacement: Comorbidity and outcomes after hip replacement. Medical Care 1993;31:141–154.

13. Sprangers MAG, de Regt EB, Andries F, van Agt HME, Bijl RV, de Boer JB, Foets M, Hoeymans N, Jacobs AE, Kempen GI et al. Which chronic conditions are associated with better or poorer quality of life? Journal of Clinical Epidemiology 2000;53:895–407.

14. Edwards LJ, Constantinescu CS. A prospective study of conditions associated with multiple sclerosis in a cohort of 658 consecutive outpatients attending a multiple sclerosis clinic. Multiple Sclerosis 2004;10:575–581.

15. Midgard R, Gronning M, Riise T, Kvale G, Nyland H. Multiple sclerosis and chronic inflammatory diseases: A case–control study. Acta Neurologica Scandinavica 1996;93:322–328.

16. Seyfert S, Klapps P, Meisel C, Fischer T, Junghan U. Multiple sclerosis and other immunologic diseases. Acta Neurologica Scandinavica 1990;81:37–42.

17. Marrie RA, Horwitz R, Cutter G, Tyry T, Campagnolo D, Vollmer T. Comorbidity, socioeconomic status, and multiple sclerosis. Multiple Sclerosis 2008;14(8):1091–1098.

18. Barcellos LF, Kamdar BB, Ramsay PP, DeLoa C, Lincoln RR, Caillier S et al. Clustering of autoimmune diseases in families with a high-risk for multiple sclerosis: A descriptive study. Lancet Neurology 2006;5:924–931.

19. Ramagopalan SV, Dyment DA, Valdar W, Herrera BM, Criscuoli M, Yee IM, Sadovnick A, Ebers G. The occurrence of autoimmune disease in Canadian families with multiple sclerosis. Lancet Neurology 2007:604–610.

20. Bernstein CN, Wajda A, Blanchard JF. The clustering of other chronic inflammatory diseases in inflammatory bowel disease: A population-based study. Gastroenterology 2005;129:827–836.

21. Gupta G, Gelfand JM, Lewis JD. Increased risk for demyelinating diseases in patients with inflammatory bowel disease. Gastroenterology 2005;129:819–826.

22. Broadley SA, Deans J, Sawcer SJ, Clayton D, Compston DAS. Autoimmune disease in first-degree relatives of patients with multiple sclerosis. A UK survey. Brain 2000;123:1102–1111.

23. De Keyser J. Autoimmunity in multiple sclerosis. Neurology 1988;38:371–374.

24. Niederwieser G, Buchinger W, Bonelli RM, Berghold A, Reisecker F, Koltringer P, Archelos JJ. Prevalence of autoimmune thyroiditis and non-immune thyroid disease in multiple sclerosis. Journal of Neurology 2003;250:672–675.

25. Sloka JS, Pryse-Phillips WEM, Stefanelli M, Joyce C. Co-occurrence of autoimmune thyroid disease in a multiple sclerosis cohort. Journal of Autoimmune Diseases [online] 2005;2(9).

26. Wynn DR, Rodriguez M, O'Fallon WM, Kurland LT. A reappraisal of the epidemiology of multiple sclerosis in Olmsted County, Minnesota. Neurology 1990;40:780–786.

27. Centers for Disease Control and Prevention. Prevalence of disability and associated health conditions. Morbidity and Mortality Weekly Report 1994;40:730–739.

28. Marrie RA. The influence of comorbid diseases and health behaviors on clinical characteristics, disability at diagnosis, and disability progression in multiple sclerosis. Unpublished dissertation. Cleveland, OH: Case Western Reserve University; 2007.

29. Lamberg L. Sleep disorders, often unrecognized, complicate many physical illnesses. Journal of the American Medical Association 2000;284(17):2173–2175.

30. Brass SD, Duquette P, Proulx-Therrien J, Auerbach S. Sleep disorders in patients with multiple sclerosis. Sleep Medicine Reviews 2010;14(2):121–129.

31. Tachibana N, Howard RS, Hirsch NP, Miller DH, Moseley IF, Fish D. Sleep problems in multiple sclerosis. European Neurology 1994;34(6):320–323.

32. Ajayi OF, Chang-McDowell T, Culpepper WJN, Royal W, Bever C. High prevalence of sleep disorders in veterans with multiple sclerosis [abstract]. Neurology 2008;70(S1):A333.

33. Nishino S, Kanbayashi T. Symptomatic narcolepsy, cataplexy and hypersomnia, and their implications in the hypothalamic hypocretin/orexin system. Sleep Medicine Reviews 2005;9(4):269–310.

34. Auger C, Montplaisir J, Duquette P. Increased frequency of restless legs syndrome in a French-Canadian population with multiple sclerosis. Neurology 2005;65(10):1652–1653.

35. Gomez-Choco M, Iranzo A, Blanco Y, Graus F, Santamaria J, Saiz A. Prevalence of restless legs syndrome and REM sleep behavior disorder in multiple sclerosis. Multiple Sclerosis 2007;13:805–808.

36. Kessler RC, Berglund P, Demler O, Jin R, Koretz D, Merikangas KR, Rush AJ, Walters EE, Wang PS. The epidemiology of major depressive disorder: Results from the national comorbidity survey replication (NCS-R). Journal of the American Medical Association 2003;289:3095–3105.

37. Goldman Consensus Group. The Goldman consensus statement on depression in multiple sclerosis. Multiple Sclerosis 2005;11:328–337.

38. Patten SB, Svenson LW, Metz LM. Psychotic disorders in MS: Population-based evidence of an association. Neurology 2005;65(7):1123–1125.

39. Rodgers J, Bland R. Psychiatric manifestations of multiple sclerosis: A review. Canadian Journal of Psychiatry 1996;41:441–445.

40. Fisk J, Morehouse SA, Brown MG, Skedgel C, Murray TJ. Hospital-based psychiatric service utilization and morbidity in multiple sclerosis. Canadian Journal of Neurological Sciences 1998;25:230–235.

41. Joffe RT, Lippert GP, Gray TA, Sawa G, Horvath Z. Mood disorder and multiple sclerosis. Archives of Neurology 1987;44:376–378.

42. Korostil M, Feinstein A. Anxiety disorders and their clinical correlates in multiple sclerosis patients. Multiple Sclerosis 2007;13:67–72.

43. Poder K, Ghatavi K, Fisk J, Campbell T, Kisely S, Sarty I, Stadnyk K, Bhan V. Social anxiety in a multiple sclerosis clinic population. Multiple Sclerosis 2009;15(3):393–398.

44. Jongbloed L. Factors influencing employment status of women with multiple sclerosis. Canadian Journal of Rehabilitation 1996;9:213–222.

45. Attarian HP, Brown KM, Duntley SP, Carter JD, Cross AH. The relationship of sleep disturbances and fatigue in multiple sclerosis. Archives of Neurology 2004;61:525–528.

46. Lobentanz IS, Asenbaum S, Vass K, Sauter C, Klosch G, Kollegger H, Kristoferitsch W, Zeitlhofer J. Factors influencing quality of life in multiple sclerosis patients: Disability, depressive mood, fatigue and sleep quality. Acta Neurologica Scandinavica 2004;110:6–13.

47. Marrie RA, Horwitz R, Cutter G, Tyry T, Campagnolo D, Vollmer T. Comorbidity delays diagnosis and increases disability at diagnosis in MS. Neurology 2009;72:117–124.

48. Dallmeijer AJ, Beckerman H, de Groot VD, van de Port IG, Lankhorst GJ, Dekker J. Long-term effect of comorbidity on the course of physical functioning in patients after stroke and with multiple sclerosis. Journal of Rehabilitation Medicine 2009;41(5):322–326.

49. Marrie RA, Rudick R, Horwitz R, Cutter G, Tyry T, Campagnolo D, Vollmer T. Vascular comorbidity is associated with more rapid disability progression in multiple sclerosis. Neurology 2010;74:1041–1047.

50. Fortin M, Bravo G, Hudon C, Lapointe L, Almirall J, Dubois MF, Vanasse A. Relationship between multimorbidity and health-related quality of life in patients in primary care. Quality of Life Research 2006;15:83–91.

51. Hermann B, Vickrey B, Hays RD, Cramer J, Devinsky O, Meador K, Perrine K, Myers LW, Ellison GW. A comparison of health-related quality of life in patients with epilepsy, diabetes and multiple sclerosis. Epilepsy Research 1996;25:113–118.

52. Lankhorst GJ, Jelles F, Smits RC, Polman CH, Kuik DJ, Pfennings LE, Cohen L, van der Ploeg HM, Ketelaer P, Vleugels L. Quality of life in multiple sclerosis: The disability and impact profile (DIP). Journal of Neurology 1996;243:469–474.

53. Nortvedt MW, Riise T, Myhr KM, Nyland HI. Quality of life in multiple sclerosis: Measuring the disease effects more broadly. Neurology 1999;53:1098–1103.

54. Rudick RA, Miller DM, Clough JD, Gragg LA, Farmer RG. Quality of life in multiple sclerosis: Comparison with inflammatory bowel disease and rheumatoid arthritis. Archives of Neurology 1992;49:1237–1242.

55. Wu N, Minden SL, Hoaglin DC, Hadden L, Frankel D. Quality of life in people with multiple sclerosis: Data from the Sonya Slifka longitudinal multiple sclerosis study. Journal of Health and Human Services Administration 2007;30(3):233–267.

56. Turpin KV, Carroll LJ, Cassidy JD, Hader WJ. Deterioration in the health-related quality of life of persons with multiple sclerosis: The possible warning signs. Multiple Sclerosis 2007;13:1038–1045.

57. Warren SA, Turpin KV, Pohar SL, Jones CA, Warren KG. Comorbidity and health-related quality of life in people with multiple sclerosis. Internal Journal of MS Care 2009;11:6–16.
58. Bayliss EA, Steiner JF, Fernald DH, Crane LA, Main DS. Descriptions of barriers to self-care by persons with comorbid chronic diseases. Annals of Family Medicine 2003;1:15–21.
59. Kalsekar ID, Madhavan SS, Amonkar MM, Douglas SM, Makela E, Elswick BLM, Scott V. Impact of depression on utilization patterns of oral hypoglycemic agents in patients newly diagnosed with type 2 diabetes mellitus: A retrospective cohort analysis. Clinical Therapeutics 2006;28:306–318.
60. Redelmeier DA, Tan SH, Booth GL. The treatment of unrelated disorders in patients with chronic medical diseases. New England Journal of Medicine 1998;338(21):1516–1520.
61. Turner BJ, Hollenbeak CS, Weiner M, Ten Have T, Tang SSK. Effect of unrelated comorbid conditions on hypertension management. Annals of Internal Medicine 2008;148(8):578–586.
62. Nuyen J, Spreeuwenberg PM, Van Dijk L, den Bos, GAMV, Groenewegen PP, Schellevis FG. The influence of specific chronic somatic conditions on the care for co-morbid depression in general practice. Psychological Medicine 2007;38(2):265–277.
63. Bosworth HB, Oddone EZ, Weinberger M. Patient treatment adherence: Concepts, interventions, and measurement. Mahwah, NJ: Lawrence Erlbaum Associates, Publishers; 2006.
64. Mohr DC, Goodkin DE, Likosky W, Gatto N, Baumann KA, Rudick RA. Treatment of depression improves adherence to interferon beta-1b therapy for multiple sclerosis. Archives of Neurology 1997;54:531–533.
65. Adherence to long-term therapies: Evidence for action [Internet]; c2003 [cited 2009 March, 2]. Available from: http://www.who.int/medicinedocs/collect/medicinedocs/pdf/s4883e/s4883e.pdf.
66. Oken BS, Kishiyama S, Zajdel D, Bourdette D, Carlsen J, Haas M, Hugos C, Kraemer DF, Lawrence J, Mass M. Randomized controlled trial of yoga and exercise in multiple sclerosis. Neurology 2004;62:2058–2064.
67. Romberg A, Virtanen A, Ruutiainen J, Aunola S, Karppi SL, Vaara M, Surakka J, Pohjolainen T, Seppanen A. Effects of a 6-month exercise program on patients with multiple sclerosis: A randomized study. Neurology 2004;63:2034–2038.
68. Wiles CM, Newcombe RG, Fuller KJ, Shaw S, Furnival-Doran J, Pickersgill TP, Morgan A. Controlled randomised crossover trial of the effects of physiotherapy on mobility in chronic multiple sclerosis. Journal of Neurology, Neurosurgery and Psychiatry 2001;70:174–179.
69. Crisafulli E, Costi S, Luppi F, Cirelli G, Cilione C, Coletti O, Fabbri LM, Clini EM. Role of comorbidities in a cohort of patients with COPD undergoing pulmonary rehabilitation. Thorax 2008;63(6):487–492.
70. Karatepe AG, Gunaydin R, Kaya T, Turkmen G. Comorbidity in patients after stroke: Impact on functional outcome. Journal of Rehabilitation Medicine 2008;40:831–835.
71. Goodman AD, Brown TR, Krupp LB, Schapiro RT, Schwid SR, Cohen R, Marinucci LN, Blight AR. Sustained-release oral fampridine in multiple sclerosis: A randomised, double-blind, controlled trial. The Lancet 2009;373(9665):732–738.
72. Weiner DK, Hanlon JT, Studenski SA. Effects of central nervous system polypharmacy on falls liability in community-dwelling elderly. Gerontology 1998;44(4):217–221.
73. Multiple Sclerosis Council for Clinical Practice Guidelines. Fatigue and multiple sclerosis: Evidence-based management strategies for fatigue in multiple sclerosis. Washington, DC: Paralyzed Veterans of America; 1998.
74. Netzer NC, Stoohs RA, Netzer CM, Clark K, Strohl KP. Using the Berlin questionnaire to identify patients at risk for the sleep apnea syndrome. Annals of Internal Medicine 1999;131(7): 485–491.
75. Honarmand K, Feinstein A. Validation of the hospital anxiety and depression scale for use with multiple sclerosis patients. Multiple Sclerosis 2009;15(12):1518–1524.
76. Mustafa M, Erokwu N, Ebose I, Strohl K. Sleep problems and the risk for sleep disorders in an outpatient veteran population. Sleep and Breathing 2005;9(2):57–63.

77. Horton M, Rudick RA, Hara-Cleaver C, Marrie RA. Validation of a self-report comorbidity questionnaire for multiple sclerosis. Neuroepidemiology 2010;35:83–90.
78. Sangha O, Stucki G, Liang MH, Fossel AH, Katz JN. The self-administered comorbidity questionnaire: A new method to assess comorbidity for clinical health services research. Arthritis & Rheumatism 2003;49:156–163.
79. Grumbach K. Chronic illness, comorbidities, and the need for medical generalism. Annals of Family Medicine 2003;1:4–7.

20

Coping

Kenneth I. Pakenham

CONTENTS

Although multiple sclerosis (MS) is a challenging illness, people with MS vary greatly in their levels of well-being and in how they adapt to it regardless of their level of disability and disease progression. In fact, illness and disease factors are typically not strong predictors of well-being in people with MS.[1] What seems to be a potent determinant of a person's well-being in the context of MS is how the individual copes with his or her illness, disability, and related stressors. Hence, facilitating people's ability to cope with their MS is an important focus of rehabilitation efforts. "Coping" is a term that can have many different

449

meanings and encompass a variety of characteristics of a person. However, most of the research into the role of coping in adapting to MS and other chronic illnesses has been guided by Lazarus and Folkman's[2] stress and coping theory. Hence, a framework derived from this theory will be used in this chapter to organize and present the many different variables and processes that are often implied or referred to under the rubric "coping." After reading this chapter you will be able to:

1. Present a framework for understanding the processes involved in coping with MS,

2. Examine the role of each coping process in adapting to MS and the relevance of this knowledge to members of the rehabilitation team,

3. Identify assessment instruments for each of the coping processes that can be used in rehabilitation settings, and

4. Review published interventions to foster adaptive coping processes in people with MS.

20.1 Stress and Coping Framework

Lazarus and Folkman's[2] stress and coping theory proposes that stress emerges when the relationship between the person and the environment is appraised as exceeding his or her resources and as threatening well-being. According to this theory, beyond the effects of the person's biographics, illness, and treatment, adjustment to MS is determined by three broad coping processes: cognitive appraisal, coping strategies, and coping resources. In brief, stress and coping theory proposes that when a person is confronted with a stressor, he or she appraises its potential for harm, threat, loss, controllability, and challenge and considers coping options. This appraisal process largely determines the extent to which the person perceives the event as stressful. The person then behaviorally responds to the stressor and these responses are referred to as coping strategies. The person may also draw on available resources to manage the event and his or her responses to it. Together, these processes shape a person's adaptation to the event. People may adjust their appraisals, coping strategies, or accessing of resources according to how well they are adjusting to the event. Hence, coping is a dynamic, transactional process responsive to internal and external feedback loops.

The utility of this framework in explaining adjustment to MS has been supported in cross-sectional[3–6] and longitudinal[1,7–9] studies. A similar model has been applied to caregiving in MS.[10] The particular relevance of this framework to MS is underscored by a recent review of research on the potential link between stress and MS disease status that concluded that there is evidence to support the association between stressful life events and relapses in MS.[11] According to stress and coping theory, appraisals, coping strategies, and coping resources have the potential to mitigate or exacerbate the negative physiological and psychological effects of stressful events.

A working diagrammatic summary of the framework is presented in Figure 20.1. This framework has been adapted from stress and coping theory and recent developments in positive psychology and is designed to be a tool for rehabilitation practitioners. The framework is not diagrammatically depicted as a tight theoretical model but as a working "map" of coping with MS; however, the framework is consistent with stress and coping theory.

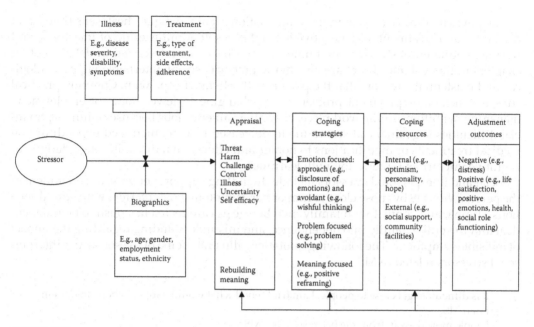

FIGURE 20.1

A working diagrammatic summary of a stress and coping framework for assessing and intervening in the coping processes that shape adjustment to multiple sclerosis.

The framework provides practitioners with a working map that can be used to build formulations of how their clients are coping with MS. It can be used to identify coping factors that need to be assessed and those that should be targeted by intervention. As is evident from Figure 20.1, each of the coping processes subsumes a potentially wide range of factors. The framework also includes consideration of the biographical characteristics of the person, the MS-related stressors faced by the individual, and the status of the person's illness and disease progression and treatment. These factors define the context within which the person is coping and provide important parameters for their coping efforts. Given that the focus of this chapter is on coping, only the stressor and coping process components of the framework will be discussed in detail. Other chapters deal with the other components of the framework: biographical characteristics (e.g., gender, culture, education, employment, and family), illness factors (e.g., disease severity and disability), and treatment (e.g., side effects and treatment adherence).

20.1.1 Stressors

Persons with MS face multiple physical and psychosocial problems or adaptive demands related to their illness that they have to cope with, and these are likely to change over time as the course of the illness unfolds and the person moves through different life stages.[6,12] It is interesting that, despite MS being a physical illness, one study found that psychosocial problems were reported as frequently as physical problems.[6] In particular, the ability to perform functional activities, distressing emotions, and relationship difficulties were frequently experienced. Furthermore, this study found that the number of physical stressors was related to poorer adjustment in the domains of depression and subjective health, whereas the number of psychosocial stressors was related to poorer adjustment in the areas of global distress and social role functioning.

The physical illness (e.g., symptom fluctuations, pain, fatigue, heat sensitivity, and disability) and treatment (e.g., negative side effects, self-administering of injections, and treatment adherence) stressors are numerous and have been described in detail in other chapters in this volume. There are also many practical, social, interpersonal, psychological, and existential stressors that the person with MS must cope with. Common practical stressors include employment problems (e.g., changing to fewer hours of employment, managing symptoms in the workplace, changing career or jobs because of illness), financial difficulties (e.g., financial strain due to decreased income, increased expenditure on medical treatments or modifications to home), and educational difficulties (e.g., fatigue or cognitive impairment interfering with learning).

Social and interpersonal stressors include dwindling support network, abandonment of the person with MS by those close to him or her, symptoms that interfere with sexual relations, stigma, and conflict with family members (e.g., care recipient's push to be independent and the pull of family into dependence, and misunderstanding regarding the impact of invisible symptoms). The following quotations illustrate some of the social and interpersonal stressors related to MS:

> It is difficult to get close to people, I am frightened what would happen when I told them I had MS.*
> People moved away from me because of my MS.*
> The family is too frightened to want to know what MS is and how it affects people.*

Regarding stigma, many people with MS perceive stigma related to their visible disabilities, or if their symptoms are less visible (e.g., pain, weakness, fatigue) they may become concerned that people will think they are faking.

There are numerous psychological stressors or issues that ebb and flow over the course of MS and several are described here. Uncertainty and fear of losing control are two related and common issues that people with MS face. MS is characterized by an uncertain course and unpredictable symptom relapses and remissions, or fluctuations in chronic deterioration. Loss of control over one's destiny and a sense of uncertainty about one's future can lead to fear, panic, and anxiety. Not being able to predict one's future makes planning difficult. Even during periods of stability there can be an underlying fear about when the next deterioration will occur. The following quotations illustrate how uncertainty can lead to feeling anxious about the future.

> I am scared of the future ... worried about exacerbations.*
> I don't know what's going to happen in the long term.*
> I am worried about pregnancy exacerbating symptoms.*

Another common stressor is the conflict between the dependency imposed by increasing disability and the person's struggle to retain his or her independence. The frustration and anger resulting from not being independent in basic activities of daily living can lead to self-harm or pushing oneself beyond endurance levels. Loss is a stressor that fluctuates over the course of the illness. The person with MS faces multiple losses, including loss of employment, career, income, mobility, engagement in valued activities, energy, physical

---

* Coping with MS Project, 1994–1996; Chief Investigator K. I. Pakenham; funded by a University of Queensland New Staff Grant.

strength, and contact with some friends. The following quotation illustrates the loss of family activities due to MS-related fatigue:

> I find I am very tired on the weekend and don't have the energy that my son wants me to have to spend time with him. My husband also gets frustrated with me that I am not doing anything "fun" with him if I don't do really physical things with him. I find that I just don't have the energy and prefer to do more quieter [sic] pursuits with my son, which my husband dislikes. This causes a lot of stress for me.*

Given the disruptive effects of MS in so many areas of a person's life, many people with MS confront existential issues around the meaning of their lives; mortality; physical limitations; the role of illness in their lives; and the revision of their self-definition, life goals, and values. Given that the onset of MS typically occurs in young adulthood, many people with MS are forced to face these existential issues prematurely.

Rehabilitation practitioners need to systematically identify and track the changes in the stressors that their clients are dealing with using some of the assessment strategies described below. Some of the internal stressors (e.g., uncertainty, existential conflict, and loss) may be more difficult to identify but will become apparent through careful probing during interview. Simply helping clients to identify their stressors is an important first step that often serves to place things in perspective for them.

20.1.2 Cognitive Appraisal

Cognitive appraisal has been defined as the evaluative process that reflects the person's subjective interpretation of an event.[2] Events are generally appraised in terms of threat, challenge, and controllability.[2] The appraisal of an illness-related event as threatening to one's well-being, limiting opportunities for personal growth (i.e., not challenging), and/or uncontrollable is likely to negatively influence adjustment to MS, given that these appraisals generate stress that may exceed the person's available coping skills and resources.[2] Regarding global appraisals of stress, higher perceived stress associated with MS-related difficulties, daily hassles, and psychosocial and financial stressors have been shown to be related to poorer adjustment across many domains, including distress, mental health problems, life satisfaction, and social adjustment regardless of level of disease severity (see reference 13 for review). Fewer studies have examined the specific threat, challenge, and controllability dimensions of appraisal. However, findings from the few studies that have examined threat or harm appraisals suggest that higher levels of these are related to concurrent worse adjustment.[1,3,6] The few studies that have examined challenge and controllability appraisals have yielded mixed findings.[13]

Other cognitive processes that have been shown to play a role in the adaptation to MS include illness uncertainty and self-efficacy. Illness uncertainty refers to perceptions of ambiguity; complexity; deficiencies in information; and unpredictability regarding symptoms, diagnosis, treatment, relationships, and future plans.[14] Higher illness uncertainty has been consistently linked with worse adjustment to MS.[13] Self-efficacy refers to a person's confidence in his or her ability to behave in ways that will lead to desired outcomes.[15] Higher self-efficacy specific to managing MS-related challenges seems to be related to better adjustment, whereas generalized self-efficacy is not consistently related to adjustment.[13] A small number

---

* Coping with MS Project, 1994–1996; Chief Investigator K. I. Pakenham; funded by a University of Queensland New Staff Grant.

of studies show that other problematic cognitive processes are related to poorer adjustment, including helplessness,[16] catastrophic pain beliefs,[17] and hypochondriacal beliefs.[18]

The assessment of appraisals is described in more detail below; however, simply probing for how the person perceives his or her stressor during interview is likely to reveal negatively biased and unhelpful appraisals. Practitioners need to listen carefully for comments that reflect the person's harm, threat, challenge, and controllability appraisals. The following quotation illustrates how one person appraises his MS-related physical limitations.

> I believe that if you think that you have a disability, then you will live like you're disabled!
> I accept that I now have limitations but I challenge myself to go as far as I can every day.*

20.1.2.1 Rebuilding Meaning

Adverse events, such as the diagnosis of MS or a sudden deterioration in health, have the potential to disrupt or shatter fundamental assumptions that bear on the benevolence and meaningfulness of the world and the worthiness of self. To the extent that an event undermines these assumptions, a sense of meaninglessness ensues that can cause existential distress which, in turn, is likely to trigger a rebuilding of meaning.[19] Two ways to restore meaning are (1) finding reasons or an explanation for what has happened and (2) looking for the positive aspects of the event, referred to as sense-making and benefit-finding, respectively.[19] Although the processes of meaning making are likely to involve both appraisals and coping strategies[20] for ease of understanding they are depicted in Figure 20.1 within the appraisal domain because they involve cognitive processes.

Making sense of adversity is achieved through developing new worldviews or via modifying existing assumptive worldviews that rebuild a sense of purpose, order, and self-worth. At the time of diagnosis, people with MS have described struggling with trying to make sense of their symptoms and their diagnosis.[12] Sense-making may also be triggered when a marked deterioration in MS symptoms occurs.[21] Pakenham[22] collected qualitative data from people with MS on how they made sense of their illness. Half the sample generated sense-making explanations for their MS, yielding 16 sense-making themes with the most frequently reported being causal explanations, acceptance (it is a fact of life), experienced growth, spiritual/religious explanations, MS a "wake-up call" for a lifestyle change, and MS a catalyst for relationship growth and change. These qualitative data were used to develop a multi-item Sense Making Scale[21] that is described below. Over a third of those people with MS who could not make sense of their situation were able to anticipate comprehending it, and the strength of this anticipation was related to greater life satisfaction.[22] Several studies show that sense-making predicts adjustment to MS concurrently[22] and over 12 months.[21] Notably, those sense-making dimensions that were characterized by perceptions of self-worth, controllability, and predictability (i.e., redefined life purpose, acceptance, and spiritual perspective dimensions) were related to better adjustment. In contrast, sense-making dimensions that entailed perceptions that magnified the negative aspect of MS (the "luck" factor) or forced personal challenges (changed values and priorities) were related to poorer adjustment.[21] These findings suggest that practitioners need to listen carefully to how people with MS explain their MS and help foster sense-making that protects self-worth and provides a realistic perspective on the predictability and controllability of events using some of the intervention strategies described below.

---

* Coping with Parental MS Project, 2007–2009; Chief Investigator K. I. Pakenham; funded by MS Research Australia.

Regarding benefit-finding, although MS is a potentially devastating illness that affects most areas of a person's life,[6] recent studies show that people with MS report benefits associated with their illness. An example of how positives can be perceived even in symptoms that can cause severe impairment, such as memory loss, is illustrated in the following quotation: "bad memory means enjoying the same books several times."* Caregivers have also been shown to report benefits derived from caring for a person with MS.[23,24] Pakenham[25] collected qualitative data from people with MS about the benefits related to their illness. Seven benefit-finding themes emerged from this data: personal growth, strengthening of relationships, appreciation of life, new opportunities, health gains, change in life priorities/goals, and spiritual growth. In a subsequent study, Pakenham and Cox[26] developed a multi-item benefit-finding scale (Benefit Finding in Multiple Sclerosis [BFiMS]) that yielded seven factors (see description below). Benefit-finding has been shown to be related to better adjustment to MS in positive domains (life satisfaction, positive affect, and dyadic adjustment) concurrently[27] and over 12 months.[26]

There is some evidence to show that the sense-making and benefit-finding of persons with MS is correlated with the respective sense-making and benefit-finding of their caregivers, and that the sense-making and benefit-finding of one partner are positively related to better adjustment of the other partner.[23,28] These findings suggest that people with MS and their caregivers engage in a process of shared meaning making. Rehabilitation practitioners need to be aware of these meaning-making processes in all those affected by MS and support these where possible using some of the intervention approaches described below.

20.1.3 Coping Strategies

Coping is defined as "constantly changing cognitive and behavioural efforts to manage specific external and/or internal demands that are appraised as taxing or exceeding the resources of the person."[2] Coping strategies examined in research on MS have typically been categorized as problem-focused, emotion-focused, or meaning-focused strategies. Problem-focused strategies include behaviors that deal with the problem that is causing the distress. Examples of problem-focused strategies are taking problem-solving action, developing a plan to resolve or change the stressor, analyzing the problem, and seeking information. Regarding the latter, research has documented the ongoing need for current and relevant information about MS and treatment.[29] Rehabilitation practitioners can assist by providing relevant information as changes in the person's health or treatment occur. This information can also be made available to family members and caregivers.

Emotion-focused strategies are directed at dealing with the distress associated with the stressor. Emotion-focused strategies can be further divided into those that regulate emotion by avoidance or approach.[2] Avoidance strategies include wishful thinking, escaping the situation, and using alcohol or drugs. Emotional-approach coping strategies include humor, identifying and expressing feelings, accepting the situation, disclosure of emotions to others, and relaxation strategies.

Meaning-focused strategies involve creating, reinstating, or reinforcing meaning and include positive reframing, prayer, rearranging life priorities, finding the positives, and making sense of the event.[20] One way to reframe one's hardship is to consider oneself relative to others who may be worse off as illustrated by the following quotation, "Who am I to complain when I see other people so much worse off."* These strategies are likely to be

---

* Coping with MS Project, 1994–1996; Chief Investigator K. I. Pakenham; funded by a University of Queensland New Staff Grant.

particularly helpful when people with MS are faced with stressors (e.g., sudden increase in disability) that challenge their sense of the meaning of their lives and that cannot be changed but must be accepted and integrated into their lives. Meaning-focused strategies have only recently been examined in MS but it appears that they may play an important role in sustaining positive adjustment outcomes, especially positive affect.[9] The following quotation illustrates the use of meaning-based coping by a person with MS, "MS makes you think and reflect about life and get in touch with self."*

Coping strategies that appear to be more specific to coping with MS and other chronic illnesses include (a) seeking physical assistance to manage symptoms and disability (e.g., I use assistive equipment such as a wheelchair, computer, crutches, or incontinence aids†); (b) modifying the external environment (e.g., I modify my living environments to meet my needs, for example, make home modifications, install air conditioners†); (c) managing energy by resting, pacing oneself, prioritizing activities, or avoiding situations that may aggravate symptoms; and (d) taking control of one's health in as many ways as possible, including using relaxation and meditation strategies, complementary treatments, and exercise programs (e.g., hydrotherapy, physiotherapy exercises, gym programs).[5] Rehabilitation practitioners can facilitate these coping strategies by ensuring that relevant practical aids, physical assistance, and exercise programs are made available where necessary.

Consistent with research into other chronic illnesses, both cross-sectional[10,30,31] and longitudinal[1,7] MS studies have shown that poorer adjustment is related to reliance on emotion-focused coping strategies that involve avoidance (e.g., wishful thinking, self-blame). Although a somewhat weaker finding, positive reappraisal coping and acceptance coping (which include elements of emotional-approach coping and meaning-focused coping), and seeking social support (often assessed as a problem-focused strategy), have been found to be related to better adjustment.[3,5,9] Recent research suggests that negative (e.g., anxiety and depression) and positive (e.g., life satisfaction and positive affect) adjustment outcomes have both common and unique coping antecedents.[9] The unique coping predictors of positive outcomes over a 3-month interval were emotional release, personal health control, and physical assistance coping, whereas the unique coping predictor of distress was avoidance.[9] The only coping strategy related to both positive outcomes and distress was acceptance coping. There is also evidence that coping strategies may buffer the negative effects of stress on the MS disease process. For example, Mohr et al.[32] demonstrated the role of coping strategies in moderating the relationship between stress (assessed as conflict and disruption in routine) and MS disease activity (i.e., subsequent development of brain lesions).

The research on coping strategies in MS is hampered by numerous methodological weaknesses and, therefore, it is difficult to draw robust conclusions that lead to practice guidelines. First, a wide range of coping measures has been employed by researchers and most have used generic coping scales that do not tap the coping strategies specific to managing MS or chronic illness that have emerged in a recently developed measure of coping with MS.[10]

Second, most studies have not examined coping with specific MS problems. As mentioned above, MS is associated with a range of stressors, and the types of coping strategies that are likely to be effective in dealing with one type of stressor (e.g., exacerbation of symptoms) may not be effective in managing a different type of stressor (e.g., financial strain). Indeed, Pakenham et al.[6] found evidence suggesting that the effects of some coping strategies on adjustment vary according to the type of problem on which the person with MS is focusing his or her coping effort. Furthermore, Warren et al.[33] showed that reliance

---

* Pakenham (2007).
† Items from the Coping with MS Scale (Pakenham, 2001).

on various coping strategies varies according to illness phase. They found that there was a tendency for more patients in exacerbation than remission to favor emotion-focused coping techniques over problem solving or social support. However, more research is necessary to determine whether coping strategies systematically differ between the different courses of MS. There is evidence in the wider stress and coping literature that suggests that emotion-focused coping strategies are more beneficial when dealing with a stressor that is uncontrollable (e.g., a mobility symptom), whereas relying on problem-focused strategies and attempting to change a stressor that cannot be changed and must be accepted is likely to lead to frustration and a fruitless struggle.[34] In contrast, relying on problem-focused coping to resolve a controllable stressor (e.g., heat in the living environment) is likely to be more effective than focusing on dealing with distress associated with the stressor. However, in most situations, the flexible use of a range of strategies is optimal.

Third, most studies have used cross-sectional designs. The few studies that have used longitudinal designs show that many of the concurrent associations between coping strategies and adjustment do not hold when using coping strategies to predict change in adjustment over time. However, despite these limitations, one robust finding that does seem to hold in longitudinal designs and that is consistent with findings in other chronic illness research is that reliance on avoidance coping is associated with worse adjustment in the short and medium term. Hence, it would seem that avoidance in the context of dealing with a chronic illness is not adaptive, although when faced with an acute health threat avoidance does appear to be more effective than problem-focused coping strategies.[35]

20.1.4 Coping Resources

Coping resources are relatively stable characteristics of an individual's disposition and environment and refer to what is available when an individual evaluates a situation and develops coping strategies.[36] Optimism (an internal coping resource) and the availability of social support (an external coping resource) have received the most attention in the literature on coping resources and adjustment to MS and will be discussed in more detail below. However, there is a wide range of potential dispositional (e.g., hope and hardiness) and external (e.g., financial and community assets) resources that could be accessed and that have not been adequately examined in the MS literature. Regarding the latter, Aronson et al.[37] undertook a large population-based survey of people with MS and their caregivers in Ontario and identified a wide range of community services, including purpose-built housing, respite care, information in the form of newsletters and pamphlets, financial assistance, and equipment aids. The following quotations from people with MS who were asked to identify the resources they found helpful illustrate the wide variety of potential helpful external coping resources:

> I have an ironing lady and a cleaning lady as these tire me. I would prefer to spend my time and energy with my child.*
> We moved to a home so I could access taxis … Mobile phone is a huge help, whenever/wherever I can get help. Cane alert system—home delivery services/meals [help].*
> MS society also gave us plenty of useful information—books, etc., for kids.*
> Community nurse visits my home for children's checkups. Financial help would be great so I don't have to work at paid work. Counseling/cognitive behavior therapy would be useful if not expensive. Some equipment has helped in the past with looking after a newborn—for example, hospital bassinette hire for stability when moving baby. Assistance with caring for newborn in first six weeks [has helped] as previously had attack which made this hard.*

---

* Coping with Parental MS Project, 2007–2009; Chief Investigator K. I. Pakenham; funded by MS Research Australia.

20.1.4.1 Optimism

Construed as dispositional rather than situation specific, optimism has been defined as the generalized expectation of favorable outcomes.[38] Research has shown that higher optimism is associated with better psychological functioning via the mitigation of the negative effects of stressors.[39,40] In their review of research on optimism in MS, Dennison et al.[13] found that most studies showed that higher dispositional optimism was related to better adjustment across many domains. The following quotation illustrates an attempt to retain an optimistic attitude to help another family member cope with MS.

> I hate that I have MS, but I have to stay positive for my son and believe that it won't beat me and I will beat it!*

20.1.4.2 Social Support

There are different types of social support that people with MS are likely to need, including practical, emotional, and material support. A frequently reported social support need of people with MS is being listened to, particularly with respect to validating their reports of their symptoms.[12] There are also many potential sources of support for people with MS, although the designated "primary caregiver" is probably one of the most important sources of support. Chapter 22 discusses the types of support caregivers provide and the benefits and difficulties associated with the role. The following quotations illustrate the many different sources and types of social support that people with MS find extremely helpful.

> Friends help when weather is bad and I can't get my son to the bus by scooter for school. Friends ... understand that some invitations I just can't offer when I don't drive or I don't have the energy to go but can still be asked as I am the best judge of what I am capable of at the time.*
> I receive an incredible amount of assistance from my mother with shopping, collecting my child from school, meals at times, and chores in general. I would be in dire straits without her!*
> I have a helpful, understanding husband and mother.*
> Sometimes, especially when the boys were younger, we would have help from family and friends to take them to and from sport or music lessons, etc. Family and friends also take me clothes shopping or gift shopping—I rely on them because they push me in a wheelchair. I rely on my husband a great deal—he sometimes has to do some parenting on his own as I might be in hospital or unwell in bed.*

However, people with MS have reported barriers to obtaining social support that are associated with the illness. For example, some have described difficulties in communicating because of cognitive impairment and interpreting nonverbal cues due to visual changes.[12] Stigma and a reluctance to disclose about one's MS are also barriers to obtaining social support. The following quotations illustrate how stigma-related concerns can be a barrier to accessing social support:

> I don't want to be seen when I've got my symptoms.†
> Hate sitting in wheelchair with people looking at me. People wonder, "What's the matter with her?"†
> They see me in a wheelchair and they don't see me as a person. They believe there is something mentally wrong with you.†

* Coping with Parental MS Project, 2007–2009; Chief Investigator K. I. Pakenham; funded by MS Research Australia.
† Coping with MS Project, 1994–1996; Chief Investigator K. I. Pakenham; funded by a University of Queensland New Staff Grant.

A review of research into relations between social support and adjustment to MS found that most studies demonstrated cross-sectionally that higher perceived support (availability or satisfaction) was associated with better adjustment across numerous domains.[13] However, in addition to the research that has demonstrated the propitious effects of social support, there is a small body of evidence that has shown an apparent harmful impact of supportive efforts perceived as unsupportive,[41] overprotective,[42] and conflicting.[43] Providing social support that matches the specific needs of the person with MS is a challenging task. Practitioners need to undertake a careful assessment of the social support needs of clients using some of the measures described below. The likely causes of social support deficits need to be determined (e.g., social skills deficiencies, an inadequate social support network, interpersonal conflict, refusal to accept support, and mismatches in support that is provided).

20.1.5 Implications for Practitioners

What are the implications of the empirical and theoretical data on coping with MS for rehabilitation practitioners? Regarding the implications of the empirical findings, several important cautionary statements should be made. First, many of the weaknesses of the research identified for coping strategies summarized above also characterize the research on the other coping processes. Second, the quantitative research has not mapped out how coping processes may vary over the long haul of managing MS despite evidence that suggests that there are likely to be specific crisis points for many people with MS that tax coping skills and resources. These include the time of diagnosis and points at which there is a marked increase in disability or deterioration in health. The time of diagnosis can be a relief for some, but it can also be traumatic for others. Indicative of the potential trauma associated with diagnosis is the 16% incidence of posttraumatic stress disorder following a diagnosis of MS identified in one study.[44] However, there are many other points along the illness trajectory that may place strain on coping processes, including a marked change in symptoms resulting in increased disability. For example, experiencing the loss of mobility and the need to move to a wheelchair can be extremely stressful because it marks the loss of considerable independence, introduces a new barrier and limitation in negotiating activities of daily living, and raises fear and uncertainty about the future. For those with a relapse–remitting course of MS, each relapse can evoke similar challenges and issues.

Despite these limitations to the research on coping with MS, findings do provide a basis for offering some broad concluding guidelines. Coping processes likely to foster adjustment seem to be those that are characterized by confidence in dealing with specific illness-related stressors (self-efficacy), retaining a realistic optimistic outlook, seeking and utilizing social support, positively reappraising stressors, accepting things that cannot be changed, and actively controlling and changing things that can be changed. Coping processes that appear to hinder adjustment include those that are characterized by avoidance of internal (e.g., unwanted feelings and thoughts) and external (e.g., unwanted symptoms and disability) stressors, and focus on the negative aspects (e.g., threat) of stressors.

According to stress and coping theory, there is no absolute "good" or "bad" coping skill or process. The extent to which a coping skill is adaptive for a given individual depends on whether that skill or process is producing the desired outcomes given the particular characteristics of the person and his or her situation. Hence, the guidelines mentioned above provide only a broad direction; a given individual's coping must be understood in the context of his or her unique characteristics and situation. A signal suggesting that a person is engaging in problematic coping processes is the manifestation of adjustment difficulties (e.g., ongoing distress, interpersonal conflict, withdrawal). When such difficulties become apparent, the practitioner should conduct an individualized, systematic assessment of the person's adjustment status and coping possesses.

The framework (see Figure 20.1) described above suggests key sets of variables that need to be taken into account when a practitioner endeavors to understand how a person is coping with his or her MS. The first step is to assess the individual with respect to his or her adjustment, current stressors, the broader person (biographics), illness and treatment context, and each of the coping processes (appraisal, coping strategies, and coping resources). The extent to which a person with MS is having problems adapting to MS is likely to manifest by way of mental health difficulties or emotional distress (e.g., depression, anxiety, stress). Such difficulties signal the need to assess how the person is coping with current MS-related stressors. Although the biographics, illness, and treatment contexts have not been elaborated on in this chapter, it is important that the practitioner have a clear understanding of these factors and how they may influence the coping processes for a given individual. The assessment process and the instruments for assessing the coping processes are described below. The assessment process should identify both strengths and weaknesses in the coping processes. Strengths need to be highlighted and built on, and weaknesses need to be targeted with appropriate interventions. A range of intervention strategies for enhancing coping processes are discussed below.

20.2 Assessment

The assessment of how a person is coping with his or her MS should be tailored to the individual. For example, the extent to which specific coping processes require detailed assessment will vary from person to person. First, an initial interview should be conducted to inquire about each of the areas summarized in Figure 20.1. At this stage, it is helpful to interview the person's caregiver and other people closely connected to the person with MS. Obtaining information from significant others can help to corroborate information already gathered, provide new information, and give different perspectives on the person's difficulties. The initial interview will highlight potential coping weaknesses that can be assessed in more detail using some of the instruments described below. Assessment should be an ongoing process throughout rehabilitation. For example, subsequent assessments can provide important information on the effects of intervention and changes in coping processes. Conducting a comprehensive assessment of all key factors summarized in the framework depicted in Figure 20.1 is important to inform the clinical reasoning and intervention planning of rehabilitation practitioners throughout all phases of the rehabilitation process.

Only measures relevant to assessing stressors and coping processes will be described below. Instruments for assessing adjustment outcomes (e.g., depression and quality of life) and illness and disability factors are discussed in other chapters in this volume. There are many generic published instruments that assess stress and coping constructs; hence, the focus in this section is on those measures that have been specifically developed for people with MS. Many of the instruments below have not been developed and used for clinical purposes. Hence, they simply provide systematic and detailed information about specific areas that may need to be followed up with further interviewing and or observation.

20.2.1 Appraisal Measures

Pakenham[1] has developed measures of threat, challenge, and controllability appraisals used in research on people with MS. Regarding the threat appraisal measure, respondents rate seven threat appraisals on seven 7-point scales (1—no harm to 7—extremely harmful) indicating the extent to which their MS problems have the potential for harm to such areas as important life goals and financial security. Three items measure challenge appraisals; respondents rate the extent to which their MS problems provide the potential for personal growth, a personal challenge, or the strengthening of a relationship on a 7-point scale (1—nil potential to 7—high potential). Controllability appraisals are assessed with two items. Respondents rate on a 7-point scale the extent to which their MS problems could be changed and must be accepted or gotten used to. The scales have been shown to have adequate internal reliabilities and validity.[1,6]

An 18-item measure of self-efficacy in MS has been developed by Schwartz et al.[45] The MS Self-Efficacy Scale has two subscales (each with nine items): self-efficacy expectation of function and of control. The self-efficacy function subscale measures confidence to perform behavior that enables engagement in daily living activities, whereas the self-efficacy control subscale measures confidence in managing disease symptoms, reaction to disease-related limitations, and the impact of disease on life activities. Internal reliabilities were good (0.86–0.90), as were test–retest reliabilities (0.62–0.81).[45] Both subscales demonstrated convergent (i.e., related to measures of theoretically related constructs) and divergent (absence of correlation with unrelated constructs) validity.

A generic 30-item Uncertainty of Illness Scale[14] assesses uncertainty in symptomatology, diagnosis, treatment, relationship with caregivers, and planning for the future. Items are rated on a 5-point scale (1—strongly agree to 5—strongly disagree), and item ratings are summed. The scale has been shown to have adequate internal reliability and validity[14] and has been validated on samples of people with MS.[46]

20.2.2 Benefit-Finding Measures

Generic measures of benefit-finding-related constructs include the Stress-Related Growth Scale—Revised (SRGS-R)[47] and the Posttraumatic Growth Inventory (PTGI).[48] A measure used to assess benefit-finding in people with MS is the Benefit Finding Scale (BFS) developed by Mohr et al.[49] However, the BFS was not developed from a targeted systematic investigation of benefit-finding in people with MS, and Pakenham[25] showed that it failed to provide a comprehensive assessment of benefit-finding in MS. Pakenham[25] found that over one third of participants who had completed the BFS reported additional benefits that they considered were not adequately reflected by the 19 items of Mohr et al.'s BFS. Consequently, Pakenham and Cox[26] developed a more comprehensive measure of benefit-finding in MS called the Benefit Finding in Multiple Sclerosis Scale (BFiMSS).

The BFiMSS was developed from qualitative benefit-finding data collected from people with MS.[25] The scale consists of 43 potential benefits. Respondents indicate the extent to which they have experienced each of these on a 3-point scale (1—not at all to 3—a great deal). Factor analysis of the BFiMSS revealed seven psychometrically sound factors: Compassion/Empathy, Spiritual Growth, Mindfulness, Family Relations Growth, Life Style Gains, Personal Growth, and New Opportunities. The BFiMSS items also loaded onto one factor; hence, factor scores or a total score may be used. All internal reliability coefficients were >0.75. A social desirability scale was only weakly correlated with the total BFiMSSCare score, and five of the BFiMSS factors suggesting that the BFiMSS is not strongly affected by a socially desirable response bias. Retest correlations showed that all BFiMSS factors were stable over a 12-month interval. Regarding convergent validity, all of the BFiMSS factors except for Family Relations Growth and the total BFiMSS score were significantly positively correlated with a benefit-finding-related construct. With respect to external validity, caregiver ratings of positive changes that the person with MS had made as a result of MS were significantly positively correlated with all BFiMSS factors. Criterion validity analyses indicated that one or more of the BFiMSS factors predicted positive adjustment outcomes 12 months later (anxiety, depression, positive affect, and positive states of mind). The BFiMSS has potential research and clinical applications. Regarding the latter, the BFiMSS may be used to monitor changes in benefit-finding in response to rehabilitation interventions. The scale may be obtained from the authors of the scale.

20.2.3 Sense-Making Measures

The first published multi-item scale of sense-making, called the Sense Making Scale (SMS), was developed by Pakenham.[21] The SMS measures sense-making related to MS and was developed from qualitative data collected from people with MS.[22] The scale consists of 38 items that are statements that people with MS have used to make sense of their illness. Respondents are asked to indicate the extent to which each statement reflects how they have made sense of their MS by rating the degree to which they agree with each statement on a 5-point rating scale (1—strongly disagree to 5—strongly agree). Factor analyses revealed six factors: Redefined Life Purpose, Acceptance, Spiritual Perspective, Luck, Changed Values and Priorities, and Causal Attribution. All factors were psychometrically sound with internal reliability coefficients greater than 0.70 (except for Luck because it has only two items). A social desirability scale was weakly related to only one factor (Acceptance), suggesting that the SMS is not greatly influenced by social desirability response bias. Regarding convergent validity, all of the SMS factors were correlated with one or more related generic meaning constructs. The SMS factors predicted change in both positive (life satisfaction and positive states of mind) and negative (anxiety and depression) outcomes and in caregiver-rated adjustment of the person with MS over a 12-month interval after controlling for the effects of demographics and illness variables. The scale may be obtained from the authors of the scale.

20.2.4 Coping Strategies Measures

Two widely used generic measures of coping strategies are the COPE[50] and the Ways of Coping Questionnaire[51] and their respective abbreviated versions: Brief COPE[52] and revised Ways of Coping Questionnaire (WOCQ).[53] Pakenham[5] developed the Coping with MS Scale (CMSS) specifically to assess coping with MS. The CMSS is a self-report measure of 29 coping strategies specific to coping with MS. The CMSS was developed from qualitative data obtained from people with MS. The scale incorporates an open-ended question that asks respondents to describe their main MS-related problem experienced in the last month. In order to obtain a global stress appraisal of the main problem, respondents are then asked to rate how stressful this problem has been in the past month on a 7-point scale (1—not at all stressful to 7—extremely stressful). Respondents then indicate on a 4-point scale (0—does not apply/never to 4—very often) how often they have used each of the 29 coping strategies in dealing with their main MS problem in the past month. Factor analyses revealed seven factors: Problem Solving, Physical Assistance, Acceptance, Avoidance, Personal Health Control, Energy Conservation, and Emotional Release. In the derivation study, subscales had internal reliabilities comparable to similar scales ranging from 0.74–0.57[5] and in a subsequent study ranged from 0.63–0.75.[9] Similar internal reliabilities have been found in other studies.[3] Test–retest reliabilities over a 3-month interval ranged from 0.63 to 0.75 (all coefficients $p < 0.001$), demonstrating high stability over the short term. The CMSS has high face validity, and predictive validity data show that one or more of the CMSS factors predicted a wide range of adjustment domains (e.g., life satisfaction, distress, and positive affect) concurrently and over a 3-month interval.[5,9] Convergent validation data suggest that although the CMSS differs from the widely used Ways of Coping Checklist—Revised,[53] it does share certain conceptual similarities with this scale. The CMSS was shown to be a stronger predictor of adjustment than the Ways of Coping Checklist—Revised.[5] The scale may be obtained from the author of the scale.

20.2.5 Coping Resources Measures

There are a wide range of generic social support scales, and two that have been used in MS research. One is a scale developed by Zich and Temoshok,[54] which assesses the availability, use, and usefulness of emotional, problem-solving, physical, and indirect personal support. For each type of support, respondents are asked to indicate on a 5-point scale how available it is, how often it is used, and how useful it is. The scores for each of these three dimensions (availability, utilization, and usefulness of social support) are summed to form three subscale scores. The other scale is a brief measure of satisfaction with social support and is the six-item Social Support Questionnaire.[55] Participants are asked to rate their level of satisfaction with various types of support on a 6-point scale (1—very dissatisfied to 6—very satisfied).

The most widely used measure of optimism is the 10-item Life Orientation Test—Revised (LOT-R).[56] Respondents indicate how strongly they agree or disagree with each item on a 5-point scale (0—strongly disagree to 4—strongly agree). The scale consists of four filler items, and the six optimism items include three items that are reverse scored.

20.2.6 Stressor Measures

It is important to pinpoint specific stressors that the person is struggling with. This may become evident through interview. The practitioner may simply ask the person to identify his or her main MS-related problems, which can then be grouped according to themes.[6] However, the following measures may also be useful in tapping a wider range of potential stressors that the person is facing but does not mention in interview. Two widely used generic measures of stressors are the Holmes and Rahe Readjustment Rating Scale,[57] which has a checklist of 43 life events, and the Hassles Scale,[58] which contains 117 common stressful events.

20.2.7 Application of Measures

The assessment of coping processes is necessary to identify those that need to be targeted by interventions. For example, an assessment of a person's coping strategies that indicates an overreliance on avoidant coping strategies may suggest the use of intervention strategies to enhance alternative approach strategies for some identified stressors. The assessment of sense-making may highlight possible difficulties the person has in comprehending his or her illness predicament. The assessment of social support may reveal deficits in specific types of needed support, a diminished support network, or support efforts that fail to match the person's needs. The practitioner should clarify with the person any responses on the abovementioned measures that seem inconsistent with information obtained by interview or from other measures. Once the assessment is completed, the practitioner should repeat a summary of the assessment to the person highlighting both strengths and weaknesses. An intervention approach should then be devised and discussed that addresses the weaknesses and strengths identified through the assessment process.

20.3 Treatment

Based on the coping framework presented above, interventions that enhance coping with MS should target the three coping processes (cognitive appraisal, coping strategies, and coping resources), which may not only help clients to regulate distress but also enhance positive adjustment outcomes such as positive affect and life satisfaction. The most widely used intervention approach to enhance these processes in people with MS is cognitive and behavioral therapy (CBT). However, most of the CBT intervention studies have used measures of mental health (e.g., depression and anxiety) and quality of life (QoL) as the main outcome measures, and few have evaluated the extent to which the coping processes changed as a result of the intervention. It should be noted that depression is a common mental health problem among people with MS, and the assessment and treatment of depression is dealt with in detail in Chapter 10.

20.3.1 Intervening at the Cognitive Appraisal Level

Regarding cognitive appraisal, interventions should enhance realistic appraisals of stressors that minimize threat and harm and that promote challenge, controllability, self-efficacy, and positive reappraisal of MS-related problems (without diminishing the associated hardship). Cognitive restructuring techniques could be used to achieve these outcomes when intervening within a CBT framework. Cognitive restructuring techniques can also be used to foster more optimistic attitudes and thinking. Several CBT intervention studies have shown improvements in specific cognitions in people with MS (e.g., self-efficacy).[59] If using the more recent variant of CBT, Acceptance and Commitment Therapy (ACT), cognitive-distancing techniques would be used to manage unhelpful thoughts.[60]

BOX 1   CBT

CBT identifies problematic behaviors and dysfunctional thinking associated with a person's problems and attempts to replace them with more adaptive ones using the learning principles of classical conditioning, operant conditioning, or modeling. CBT techniques include systematic desensitization, exposure, contingency management (systematic modification of rewards and punishments for target behaviors), self-monitoring, behavioral rehearsal, and cognitive restructuring (identifying and challenging negative or inaccurate thoughts and replacing them with more adaptive thinking).

> **BOX 2 ACT**
>
> ACT is a "third-generation CBT" intervention that emphasizes cultivating mindfulness, acceptance, and behavior change through six core processes: acceptance, cognitive diffusion (creating distance from unhelpful thoughts), contact with the present moment, self-as-context (adopting a broader more fluid perspective on self), values, and committed action. The ultimate goal of ACT is living a valued life, which entails clarifying values and developing willingness to behave in ways that are more in harmony with these values.

20.3.2 Intervening at the Meaning-Making Level

With respect to promoting the rebuilding of meaning via the sense-making and benefit-finding processes, a range of intervention techniques is potentially helpful. As mentioned above, rebuilding meaning in the context of illness often encompasses fundamental existential life issues; hence, interventions designed to address the existential crises associated with illness may facilitate sense-making and benefit-finding. LeMay and Wilson[61] reviewed eight manualized interventions that explicitly addressed existential themes in the context of life-threatening illness. Interventions varied greatly but most used a combination of both existential and CBT techniques. Tedeschi and Calhoun[62] have proposed a "clinician as expert companion" approach with associated guidelines for fostering posttraumatic growth (which is similar to benefit-finding). CBT stress management interventions for breast cancer patients have also been shown to be associated with improvements in benefit-finding.[63] In addition, strategies such as relaxation, alternative therapies, and prayer may assist people with MS in finding meaning via reappraising benefits associated with their illness.[5] Refer to Pakenham[64] for a more detailed review of interventions for promoting benefit-finding and sense-making in the context of chronic illness.

20.3.3 Intervening at the Coping Strategies Level

With respect to intervening at the level of coping strategies, interventions have largely been conducted from a CBT approach. A range of coping strategies has been fostered, including cognitive reappraisal using cognitive-restructuring techniques, problem-solving skills using standard problem-solving techniques, goal setting, and seeking social support. One study showed that relative to a control group, people with MS who received a CBT group intervention demonstrated greater use of problem-focused coping strategies.[65] Benedict et al.[66] used CBT strategies to enhance self-control and behavior regulation to reduce socially aggressive behavior in people with MS with marked cognitive impairment. CBT has also been successfully used to help promote skills in managing fatigue in MS.[67] Exercise training may be conceptualized as a specific type of coping-skills training. A meta-analysis of the effects of exercise training on QoL in MS concluded that exercise training is associated with a small improvement in QoL.[68]

Given the recent evidence indicating the beneficial effects of acceptance coping and acceptance sense-making in people with MS,[3,5,9,21] it would seem appropriate that interventions should assist people with MS to accept the realities of their illness and integrate them into their lifestyles. It appears to be counterproductive for a person to hold on to his or her life as it was prior to MS and to try to avoid chronic illness-related stressors.

Avoidance appears to represent a failure to move forward and adapt constructively to the situation. These practice implications are consistent with the recent upsurge in the application of ACT to chronic illnesses.[69] A cornerstone of ACT is acceptance, which is defined as "the active and aware embracing of private events ... without unnecessary attempts to change their frequency or form."[69] Pakenham and Fleming[70] found that after controlling for the effects of initial adjustment and relevant demographic and illness variables, greater (ACT-defined) acceptance was related to better adjustment to MS (lower distress, higher life satisfaction and positive affect, and better health) over a 12-month interval. In addition, mindfulness (which is also used in ACT and encompasses acceptance) has been successfully applied to helping people with MS cope with mobility difficulties[71]; however, there are no published trials of the application of ACT to people with MS.

20.3.4 Intervening at the Coping Resources Level

Regarding the promotion of coping resources, most interventions have focused on enhancing social skills. Gordon et al.[72] conducted a group social skills training intervention that focused on assertiveness training, presentation of self, acknowledgment and disclosure of disability, and interpersonal communication skills. Benedict et al.[66] used social skills training to improve the capacity of people with MS-related cognitive impairment to appreciate the perspective of others.

20.3.5 Cognitive Behavior Therapy Interventions

As mentioned above, the most frequently used framework for promoting a person's capacity to cope with MS is CBT. A recent Cochrane review[73] of the effectiveness of psychological interventions for people with MS concluded that there is reasonable evidence that CBT interventions help to improve depression, coping, and adjustment to MS. The majority of CBT interventions reviewed were delivered by psychologists, only two studies used interventions delivered by nurses, and none were led by laypersons. Most of these interventions targeted problematic cognitions and coping skills and deficiencies in coping resources. CBT interventions have been administered individually[65] and in group[74] formats, and by telephone.[75] A recent qualitative study of the applicability of generic computerized CBT interventions for people with MS found that these programs needed to be adapted for people with chronic physical conditions.[76] Intervention techniques used in the published CBT treatment trials have varied widely and include relaxation training (e.g., guided imagery and progressive deep muscle relaxation); self-monitoring of stressors and responses to these; rehearsal of coping strategies in response to anticipated stressors using role plays; behavioral activation using techniques such as pleasant events scheduling; cognitive structuring to identify and change dysfunctional negative thoughts; positive self-statements; problem-solving skills training; goal setting; and coping skills training for dealing with specific symptoms and problems (e.g., fatigue, pain, and sexual dysfunction). Most of the published CBT treatment studies refer to therapist manuals and participant workbooks that can be requested from authors.

20.3.6 Other Eclectic Interventions

There are several intervention approaches developed to enhance coping with MS that are more eclectic. A structured eclectic coping skills training program that was evaluated in a randomized trial is the intervention developed by Schwartz and Rogers[77] to teach coping flexibility for people with chronic illness. The intervention was compared to a peer telephone support program.[78] The intervention consisted of eight weekly, 2-h group educational and supportive sessions that included relaxation training, group discussion, home assignments, goal setting, expression of feelings, communication with caregivers, and strategies for dealing with cognitive impairment and for coping with distressing emotions, including loss. Caregivers attended three sessions on improving communication with others, and they were able to attend other sessions if desired. At the end of the intervention, participants were matched with coping partners from the group and they were invited to telephone each other monthly for 10 months. The peer telephone-support group program offered nondirective support by laypeople with MS who were trained in active listening. Compared to the peer telephone-support group, the coping skills intervention was more directive, required more personal and family commitment, and involved a health professional. The coping skills group participants reported increased functioning in psychosocial roles, less reliance on avoidant coping strategies, and greater reliance on more active and approach-oriented coping strategies (e.g., reframing), and enhanced well-being. These gains were achieved despite objective functional decline in many participants across the course of the study. Although overall the coping skills intervention helped participants more, the peer telephone-support intervention seemed most effective for those people with affective (depression and anxiety) problems.

A generic lay-led, community-based intervention called the Chronic Disease Self-Management Course for people with a range of long-term health conditions (including MS) has been delivered in a number of countries (e.g., United Kingdom, United States, Canada, Australia).[79] The course is manualized and comprises six weekly, 2-h sessions delivered in community settings by pairs of lay tutors trained in course delivery. The program draws on self-efficacy theory,[15] providing mastery experience, role modeling, persuasion, and reinterpretation of psychological and affective states. Generic topics are covered, including self-management principles, exercise, pain-management relaxation techniques, dealing with depression, nutrition, communication with family and professionals, and goal setting. Specific CBT techniques are also presented, including distraction, relabeling pain-associated sensations, and positive self-talk. The program is designed to promote generic self-management skills and, therefore, does not provide disease-specific management techniques or information. The program is primarily an educational and self-management intervention rather than a therapy-based intervention. Reviews of the effectiveness of the Chronic Disease Self-Management Course in the United States[79] and the United Kingdom[80] showed that the intervention was fairly consistently associated with improvements in self-management techniques, physical and mental health status, and self-efficacy. Qualitative studies have shown that, overall, participants report satisfaction with the program and report a range of benefits.[81] However, some disadvantages have also been reported, including course format too rigid, too rushed, and lacking in time for discussion of topics of interest to participants.[81] Barlow, Edwards, and Turner[82] conducted a qualitative study of the experience of 10 people with MS who participated in a Chronic Disease Self-Management Course. Participants were interviewed before attending the program and again at 4-month follow-up. While some participants learned new self-management techniques, others were reminded of techniques they had already learned, particularly with respect to managing fatigue. Valued components were goal setting and the mutual support gained from meeting others with similar health problems. However, some noted that the lack of MS-specific self-management techniques and information was a shortcoming of the program.

20.3.7 Involvement of Caregivers

The involvement of caregivers or family members in interventions for enhancing coping with MS has not been well investigated, although some of the abovementioned interventions have included caregivers. The inclusion of family or caregivers is likely to be particularly important when the person with MS presents with interpersonal stressors. For example, independence versus dependence conflicts when the person with MS aggressively holds on to independence and perhaps puts self at risk such that the caregiver attempts to help or, less often, when the person with MS readily accepts dependency. Both entail role changes among family members, which is likely to engender tension and conflict. Furthermore, given the evidence that suggests that differences and similarities on some coping strategies between the person with MS and their caregivers influences the adjustment of both partners, it would seem important to include caregivers in coping skills interventions when possible. It is, of course, preferable to include both the person with MS and his or her caregiver when possible in the rehabilitation assessment and intervention stages (refer to Chapter 22 for more details on how caregivers may need to be considered and included).

20.3.8 Summary

In general, it appears that interventions that are CBT-based are beneficial in improving mental health and QoL in people with MS and in enhancing the three coping processes (appraisal, coping strategies, and coping resources). However, it is not clear how interventions should be modified according to stage or course of MS or other characteristics of the illness. For example, intervention components may need to be modified to suit the needs of individuals with cognitive impairment. Please refer to Chapter 8 for a detailed discussion on the impacts, assessment, and treatment of cognitive impairment. A hallmark of CBT and the framework depicted in Figure 20.1 is individualized assessment and intervention. This is easier to achieve in one-on-one interventions. Some of the CBT group-based programs mentioned above incorporate this by having participants self-monitor and identify stressors and then match these with appropriate coping strategies, appraisals, and coping resources. Practitioners are encouraged to attend not only to the reduction of distress using the intervention techniques reviewed above, but also to the enhancement of the affective, cognitive, and meaning dimensions of well-being.

BOX 3   CASE STUDY

Bill is a 41-year-old, single male who lives alone in an apartment. He was diagnosed with MS seven years ago, and for the first few years he experienced alternating relapses and remissions with slightly more health decline occurring after each relapse. However, over the past three years he has experienced a consistent and progressive increase in disability. Bill experienced a crisis point about 18 months ago when his employment as a senior accountant was terminated. Bill believes that this was due to his increasing disability, although this was never formally acknowledged by his employers. Bill said that work was "my life." Bill tried to find work as an accountant or in related areas and became very despondent when these efforts failed to yield employment.

Bill currently uses a motorized scooter and has recently lost function in his right hand and his left hand is becoming progressively weaker. Bill has had to have modifications to his bathroom and kitchen. He has sleep problems because sometimes he experiences "cramps" and can't move for a period of time. When he goes to bed, he worries about not being able to move. He also has problems getting himself out of the bed and transferring to the bathroom. He says his sleep is broken, and he only gets about five hours sleep each night.

He has difficulty in visiting friends because most have homes that he can't easily access, and the visits he receives from friends have dwindled. He experiences frustration with the lack of access to many public buildings. He often feels overlooked when socializing because being in a scooter he is waist height to those standing, who often speak over him or about him rather than to him. His partner of four years left him about six months after he was diagnosed with MS. He yearns for intimacy with a person but feels overwhelmed by his changing and deteriorating symptoms and believes that no one would want to be with him.

Bill has "care staff" who visit daily to assist him with showering and meal preparation. He feels as if he has lost his privacy and his home, which has been a refuge for him in the past. His routine is now determined by the needs of the "care staff" rather than his own needs.

His elderly mother provides Bill with some assistance, mainly by driving him to places. He does not get along with his father so they have very little contact. There are periodic tensions and disagreements between Bill and his parents. Bill says that his parents insist on helping at times and in ways that he does not like. He also feels that when they visit him they do things in his home without asking him first. When disagreements occur, Bill withdraws and silently "fumes."

At times Bill becomes so despondent and "down" that he wonders about the point of his life, and this line of musing sometimes leads to fleeting thoughts about ending his life. He worries about losing all his independence and not being able to end his life even if he wanted to. Bill tends to spend much of his time distracting himself from his concerns with various activities that don't get finished and don't really lead to any personal satisfaction. Much of the time he says he tries to "push my worries to the back of my mind." He doesn't voice his concerns to anyone, but instead keeps them to himself. He sees his MS as subtracting from his life. He feels his life has shrunken to the scope of his scooter and apartment. He feels overwhelmed by what he describes as an ongoing "avalanche of losses." Yet at times he attempts to "look on the bright side" and tries to find positive things, like how since his MS he has time to devote to interests that he had neglected in the past. At these times he says he is more like his "old optimistic self." He sometimes reflects on his life and remembers how thinking about the positives got him through difficult situations in the past. At these times he feels "a little upbeat."

A case study is included as a means of bringing together the stress and coping, assessment, and intervention sections of this chapter. The stress and coping framework depicted in Figure 20.1 is applied to a hypothetical case (Bill) to illustrate the clinical relevance of the coping processes, how they might be assessed, and the various intervention options. As is evident from the case vignette, it is not possible to provide a prescriptive "one size fits all" approach to fostering coping in a person with MS. The evidence-based and theoretically

TABLE 20.1

Summary of Bill's Behaviors and Experiences Relevant to the Coping Processes

Stressors	Loss of friends, independence, intimacy, privacy, and employment; interpersonal conflict with parents; sleep difficulties (fear of cramps and difficulty transferring from bed to bathroom); and physical access barriers
Appraisals	Perceptions of little control over most areas of his life and functioning; worry about the uncertainty of his future and increasing disability; perceptions that he would not be desirable as a partner; perceptions of hopelessness and thoughts of ending his life
Meaning	Sense-making (e.g., sees MS as subtraction from his life and bringing an avalanche of losses); benefit-finding (e.g., more time to pursue interests)
Coping strategies	Avoidance strategies such as distraction and "pushing thoughts to the back of my mind"; interpersonal withdrawal; thinking about suicide as a solution; thinking about the positives; problem solving (evidenced in his attempts to find employment)
Coping resources	External (e.g., home "care staff" and parental assistance); internal/dispositional (e.g., optimism and self-reflection)
Adjustment outcomes	Despondency; depressed mood; frustration; role-functioning difficulties in many life areas; fleeting experiences of positive emotions

coherent framework provided in this chapter offers a broad practice guide for practitioners that enables a systematic consideration of key coping processes and that informs assessment, clinical reasoning, and intervention to maximize coping effectiveness.

In applying the framework depicted in Figure 20.1 to Bill, the first step is to assess Bill's behavior and experiences with respect to each of the model components. The focus of the following discussion will be the coping processes. This assessment could occur informally through observations and discussions with Bill (and his parents, if possible), or it might occur more formally through a more structured interview and/or the use of questionnaires (mentioned above). Based on the information presented in the case vignette, a summary of Bill's behavior and experiences relevant to each of the coping processes is presented in Table 20.1.

The predominance of negative adjustment outcomes (e.g., emotional distress and limited engagement in various areas of living, such as socializing) suggests deficiencies in some of the coping processes. Weakness at the appraisal level included negatively biased cognitive processes. Regarding coping strategies, there appears to be a heavy reliance on avoidant and escape coping strategies. With respect to coping resources, although external coping resources are utilized, they are also associated with costs and Bill's dispositional resources appear to be underutilized, possibly because of his being overwhelmed by the multiple, sometimes chronic, stressors.

20.3.8.1 Issues to Be Addressed in Fostering Bill's Coping Capacities

20.3.8.1.1 Assessment Approaches

Each of the coping processes has been examined at a macro level (i.e., in the context of multiple stressors); however, all of the processes could be examined with respect to a single stressor. For example, coping processes in response to the unemployment stressor could be examined in detail. Individual practitioners may want to do further fine-grained assessment of a particular stressor, coping process, or adjustment outcome more in line with their professional training and expertise. For example, a psychologist may want to conduct a detailed assessment of Bill's coping strategies using a coping questionnaire and interview and further assess his depression using a standardized measure of mood.

Another area of investigation is how Bill has coped with past crises and how effective his coping efforts were.

20.3.8.1.2 Which Stressors to Address First?

Negotiations in prioritizing stressors to be addressed must include Bill and may include several practitioners and family members. If several stressors are targeted by different practitioners, it will be important to ensure that efforts are coordinated by way of team meetings or other communication avenues.

20.3.8.1.3 Who Will Address Which Stressors?

Considerations will include availability of practitioners and their training. Some practitioners, because of their training, will be better suited to dealing with certain stressors. For example, an occupational therapist may be best placed to deal with the transfer from bed to bathroom stressor, whereas a psychologist may be best suited to dealing with the loss stressors.

20.3.8.1.4 Interventions

The intervention approach will depend on the specific stressors targeted and the training of the practitioner delivering the intervention. For example, a psychologist might focus on the loss issues and related depressed mood by using ACT cognitive distancing techniques to manage negatively biased thoughts associated with the losses and acceptance and mindfulness strategies for managing unwanted emotions such as sadness. A social worker may focus on the conflict between Bill and his parents. This may involve some mediation work and/or family therapy intervention. The vocational rehabilitation counselor may do formal career review and work on clarification of values, and problem solving and goal setting regarding employment directions. For example, the practitioner may help Bill to explore "employment" options such as short-term consultancy work and volunteer work and then set goals for exploring these.

20.3.8.1.5 Informal Fostering of Coping

Practitioners may intervene at an informal level as illustrated by some of the following examples. If, in the course of a discussion, Bill begins to talk about the positive aspects of MS (i.e., benefit-finding), the practitioner could simply reinforce this view of MS by engaging in discussion around these positives. Practitioners could informally explore other coping strategies with Bill. For example, Bill and a practitioner could discuss other options for responding to disagreements between him and his parents. A practitioner could engage Bill in a discussion about coping resources that he has not utilized, such as support groups and online facilities.

References

1. Pakenham KI. Adjustment to multiple sclerosis: Application of a stress and coping model. Health Psychology 1999;18(4):383–392.
2. Lazarus RS, Folkman S. Stress, appraisal, and coping. New York, NY: Springer Publishing; 1984.
3. Chalk HM. Mind over matter: Cognitive-behavioral determinants of emotional distress in multiple sclerosis patients. Psychology, Health and Medicine 2007;12(5):556–566.

4. McCabe MP, McKern S, McDonald E. Coping and psychological adjustment among people with multiple sclerosis. Journal of Psychosomatic Research 2004;56:355–361.

5. Pakenham KI. Coping with multiple sclerosis: Development of a measure. Psychology, Health and Medicine 2001;6(4):411–428.

6. Pakenham KI, Stewart CA, Rogers A. The role of coping in adjustment to multiple sclerosis-related adaptive demands. Psychology, Health and Medicine 1997;2(3):197–211.

7. Aikens JE, Fischer JS, Namey M, Rudick RA. A replicated prospective investigation of life stress, coping, and depressive symptoms in multiple sclerosis. Journal of Behavioral Medicine 1997;20(5):433–445.

8. McCabe MP. A longitudinal study of coping strategies and quality of life among people with multiple sclerosis. Journal of Clinical Psychology in Medical Settings 2006;13:369–379.

9. Pakenham KI. Investigation of the coping antecedents to positive outcomes and distress in multiple sclerosis (MS). Psychology and Health 2006;21(3):633–649.

10. Pakenham KI. Application of a stress and coping model to caregiving in multiple sclerosis. Psychology, Health and Medicine 2001;6(1):13–27.

11. Heesen C, Mohr DC, Huitinga I, Bergh FT, Gaab J, Otte C, Gold SM. Stress regulation in multiple sclerosis: Current issues and concepts. Multiple Sclerosis 2007;13:143–148.

12. Courts NF, Buchanan EM, Werstlein PO. Focus groups: The lived experience of participants with multiple sclerosis. Journal of Neuroscience Nursing 2004;36(1):42–47.

13. Dennison L, Moss-Morris R, Chalder T. A review of psychological correlates of adjustment in patients with multiple sclerosis. Clinical Psychology Review 2009;29:141–153.

14. Mishel MH. The measurement of uncertainty of illness. Nursing Research 1981;30:258–263.

15. Bandura A. Self-efficacy: Toward a unifying theory of behavioral change. Psychological Review 1977;84(2):191–215.

16. Shnek ZM, Foley FW, LaRocca NG, Gordon WA, DeLuca J, Schwartzman HG, Halper J, Lennox S, Irvine J. Helplessness, self-efficacy, cognitive distortions and depression in multiple sclerosis and spinal cord injury. Annals of Behavioral Medicine 1997;19:287–294.

17. Osborne TL, Jensen MP, Ehde DM, Hanley MA, Kraft G. Psychosocial factors associated with pain intensity, pain-related interference, and psychological functioning in persons with multiple sclerosis and pain. Pain 2007;127:52–62.

18. Taillefer SS, Kirmayer LJ, Robbins JM, Lasry JC. Psychological correlates of functional status in chronic fatigue syndrome. Journal of Psychosomatic Research 2002;53:1097–1106.

19. Janoff Bulman R, Yopyk DJ. Random outcomes and valued commitments: Existential dilemmas and the paradox of meaning. In: J. Greenberg, S. L. Koole, T. Pyszczynski, editors. Handbook of experimental existential psychology. New York, NY: Guilford Press; 2004.

20. Park CL, Folkman S. Meaning in the context of stress and coping. Review of General Psychology 1997;1(2):115–144.

21. Pakenham KI. Making sense of multiple sclerosis. Rehabilitation Psychology 2007;52:380–389.

22. Pakenham KI. Making sense of illness or disability: The nature of sense making in multiple sclerosis (MS). Journal of Health Psychology 2008;13:93–105.

23. Pakenham KI. The positive impact of multiple sclerosis (MS) on carers: Associations between carer benefit finding and positive and negative adjustment domains. Disability and Rehabilitation 2005;27(17):985–997.

24. Pakenham KI, Cox S. Development of the benefit finding in multiple sclerosis (MS) caregiving scale: A longitudinal study of relations between benefit finding and adjustment. British Journal of Health Psychology 2008;13:583–602.

25. Pakenham KI. The nature of benefit finding in multiple sclerosis. Psychology Health and Medicine 2007;12:190–196.

26. Pakenham KI, Cox S. The dimensional structure of benefit finding in multiple sclerosis (MS) and relations with positive and negative adjustment: A longitudinal study. Psychology and Health 2009;24:373–393.

27. Pakenham KI. Benefit finding in multiple sclerosis and associations with positive and negative outcomes. Health Psychology 2005;24(2):123–132.

28. Pakenham KI. The nature of sense making in caregiving for persons with multiple sclerosis (MS). Disability and Rehabilitation 2008;30(17):1263–1273.

29. Baker LM. Sense making in multiple sclerosis: The information needs of people during an acute exacerbation. Qualitative Health Research 1998;8(1):106–120.

30. Mohr DC, Goodkin DE, Gatto N, Van der Wende J. Depression, coping and level of neurological impairment in multiple sclerosis. Multiple Sclerosis 1997;3:254–258.

31. O'Brien MT. Multiple sclerosis: The relationship among self-esteem, social support, and coping behaviour. Applied Nursing Research 1993;6:54–63.

32. Mohr DC, Goodkin DE, Nelson S, Cox D, Weiner M. Moderating effects of coping on the relationship between stress and the development of new brain lesions in multiple sclerosis. Psychosomatic Medicine 2002;64:803–809.

33. Warren S, Warren KG, Cockerill R. Emotional stress and coping in multiple sclerosis (MS) exacerbations. Journal of Psychosomatic Research 1991;35(1):37–47.

34. Park CL, Folkman S, Bostrom A. Appraisals of controllability and coping in caregivers and HIV+ men: Testing the goodness-of-fit hypothesis. Journal of Consulting and Clinical Psychology 2001;69:481–488.

35. Clutton S, Pakenham KI, Buckley B. Predictors of well-being following a "false positive" breast screening result. Psychology and Health 1999;14:263–275.

36. Moos RH, Billings AG. Conceptualizing and measuring coping resources and processes. In: L. Goldberger, S. Breznitz, editors. Handbook of stress: Theoretical and clinical aspects. New York: Free Press; 1982.

37. Aronson KJ, Cleghorn G, Goldenberg E. Assistance arrangements and use of services among persons with multiple sclerosis and their caregivers. Disability and Rehabilitation 1996;18(7): 354–361.

38. Scheier MF, Carver CS. Optimism, coping, and health: Assessment and implications of generalized outcome expectancies. Health Psychology 1985;4:219–247.

39. Scheier MF, Carver CS. Effects of optimism on psychological and physical well-being: Theoretical overview and empirical update. Cognitive Therapy and Research 1992;16:201–228.

40. Segerstrom SC, Taylor SE, Kemeny ME, Fahey JL. Optimism is associated with mood, coping, and immune change in response to stress. Journal of Personality and Social Psychology 1998;74:1646–1655.

41. Wineman NM. Adaptation to multiple sclerosis: The role of social support, functional disability, and perceived uncertainty. Nursing Research 1990;39(5):294–299.

42. Schwartz C, Frohner R. Contribution of demographic, medical and social support variables in predicting the mental health dimension of quality of life among people with multiple sclerosis. Health and Social Work 2005;30:203–212.

43. Stuifbergen A. Health-promoting behaviors and quality of life among individuals with multiple sclerosis. Scholarly Inquiry for the Nursing Practice 1995;9:31–55.

44. Chalfant AM, Bryant RA, Fulcher G. Posttraumatic stress disorder following diagnosis of multiple sclerosis. Journal of Traumatic Stress 2004;17(5):423–428.

45. Schwartz CE, Coulthard-Morris L, Zeng Q, Retzlaff P. Measuring self-efficacy in people with multiple sclerosis: A validation study. Archives of Physical Medicine and Rehabilitation 1996;77:394–398.

46. Lynch SG, Kroencke DC, Denney DR. The relationship between disability and depression in multiple sclerosis: The role of uncertainty, coping, and hope. Multiple Sclerosis 2001;7:411–416.

47. Armeli S, Gunthert KC, Cohen LH. Stressor-appraisals, coping, and post-event outcomes: The dimensionality and antecedents of stress-related growth. Journal of Social and Clinical Psychology 2001;20:366–395.

48. Tedeschi RG, Calhoun LG. The posttraumatic growth inventory: Measuring the positive legacy of trauma. Journal of Traumatic Stress 1996;9(3):455–472.

49. Mohr DC, Dick LP, Russo D, Pinn J, Boudewyn AC, Likosky W, Goodkin DE. The psychosocial impact of multiple sclerosis: Exploring the patient's perspective. Health Psychology 1999;18(4):376–382.

50. Carver CS, Scheier MF, Weintraub JK. Assessing coping strategies: A theoretically based approach. Journal of Personality and Social Psychology 1989;56(2):267–283.
51. Folkman S, Lazarus RS. Manual for the ways of coping questionnaire. Palto Alto, CA: Consulting Psychological Press; 1988.
52. Carver CS. You want to measure coping but your protocol's too long: Consider the brief COPE. International Journal of Behavioral Medicine 1997;4(1):92–100.
53. Vitaliano PP, Russo J, Carr JE, Maiuro RD, Becker J. The ways of coping checklist: Revision and psychometric properties. Multivariate Behavioral Research 1985;20(1):3–26.
54. Zich J, Temoshok L. Perceptions of social support, distress, and hopelessness in men with AIDS and ARC: Clinical implications. In: L. Temoshok, A. Baum, editors. Psychological perspectives on AIDS. New Jersey: Lawrence Erlbaum; 1987.
55. Sarason IG, Sarason BR, Shearin EN, Pierce GR. A brief measure of social support: Practical and theoretical implications. Journal of Social and Personal Relationships 1987;4:497–510.
56. Scheier MF, Carver CS, Bridges MW. Distinguishing optimism from neuroticism (and trait anxiety, self-mastery, and self-esteem): A re-evaluation of the life orientation test. Journal of Personality and Social Psychology 1994;67:1063–1078.
57. Holmes TH, Rahe RH. The social readjustment rating scale. Journal of Psychosomatic Research 1967;11(2):213–218.
58. Kanner AD, Coyne JC, Schaefer C, Lazarus RS. Comparison of two modes of stress measurement: Daily hassles and uplifts versus major life events. Journal of Behavioral Medicine 1981; 4:1–39.
59. Rigby S. Coping with MS. Unpublished PhD Thesis. Liverpool, UK: University of Liverpool; 2003.
60. Hayes SC, Strosahl K, Wilson KG. Acceptance and commitment therapy: An experiential approach to behavior change. New York, NY: Guilford Press; 1999.
61. LeMay K, Wilson KG. Treatment of existential distress in life threatening illness: A review of manualized interventions. Clinical Psychology Review 2007;28(3):472–493.
62. Tedeschi RG, Calhoun LG. The clinician as expert companion. In: C. L. Park, S. C. Lechner, M. H. Antoni, A. L. Stanton, editors. Medical illness and positive life change: Can crisis lead to personal transformation? Washington, DC: APA; 2009.
63. Antoni MH, Lehman JM, Kilbourn KM, Boyers AE, Culver JL, Alferi SM, Yount SE, McGregor BA, Arena PL, Harris SD et al. Cognitive-behavioral stress management intervention decreases the prevalence of depression and enhances benefit finding among women under treatment from early-stage breast cancer. Health Psychology 2001;20(1):20–32.
64. Pakenham KI. Benefit finding and sense making in chronic illness. In: S. Folkman, editor. Oxford handbook on stress, coping, and health. New York: Oxford University Press; 2011.
65. Foley FW, Bedell JR, LaRocca NG, Scheinberg LC, Reznikoff M. Efficacy of stress-inoculation training in coping with multiple sclerosis. Journal of Consulting and Clinical Psychology 1987;55(6):919–922.
66. Benedict RHB, Shapiro A, Priore R, Miller C, Munschauer F, Jocobs L. Neuropsychological counseling improves social behavior in cognitively impaired multiple sclerosis patients. Multiple Sclerosis 2000;6(6):391–396.
67. van Kessel K, Moss-Morris R, Willoughby E, Chalder T, Johnson M, Robinson E. A randomized controlled trial of cognitive behavior therapy for multiple sclerosis fatigue. Psychosomatic Medicine 2008;70:205–213.
68. Motl RW, Gosney JL. Effect of exercise training on quality of life in multiple sclerosis: A meta-analysis. Multiple Sclerosis 2008;14:129–135.
69. Hayes SC, Luoma JB, Bond FW, Masuda A, Lillis J. Acceptance and commitment therapy: Model, processes and outcomes. Behavior Research and Therapy 2006;44:1–25.
70. Pakenham KI, Fleming M. Relations between acceptance of multiple sclerosis and positive and negative adjustment. Psychology and Health 2011; 26:1292–1309.
71. Mills N, Allen J. Mindfulness of movement as a coping strategy in multiple sclerosis: A pilot study. General Hospital Psychiatry 2000;22:425–431.

72. Gordon PA, Lam CS, Winter R. Interaction strain and persons with multiple sclerosis: Effectiveness of a social skills program. Journal of Applied Rehabilitation Counseling 1977;28: 5–11.

73. Thomas PW, Thomas S, Hillier C, Galvin K, Baker R. Psychological interventions for multiple sclerosis (review). Cochrane Database of Systematic Reviews 2006(1): CD004431.pub2.

74. Larcomgbe NA, Wilson PH. An evaluation of cognitive-behaviour therapy for depression in patients with multiple sclerosis. British Journal of Psychiatry 1984;145:366–371.

75. Mohr DC, Hart SL, Julian L, Catledge C, Honos-Webb L, Vella L, Tasch ET. Telephone-administered psychotherapy for depression. Archives of General Psychiatry 2005;62: 1007–1014.

76. Hind D, O'Cathain A, Cooper CL, Parry GD, Isaac CL, Rose A, Martin L, Sharrack B. The acceptability of computerized cognitive behavioral therapy for the treatment of depression in people with chronic physical disease: A qualitative study of people with multiple sclerosis. Psychology and Health 2010;25(6):699–712.

77. Schwartz CE, Rogers M. Designing a psychosocial intervention to teach coping flexibility. Rehabilitation Psychology 1994;39(1):57–72.

78. Schwartz CE. Teaching coping skills enhances quality of life more than peer support: Results of a randomized trial with multiple sclerosis patients. Health Psychology 1999;18(3):211–220.

79. Lorig KR, Sobel DS, Stewart AL, Brown BW, Bandura A, Ritter P, Gonzalez VM, Laurent DD, Holman HR. Evidence suggesting that a chronic disease self-management program can improve health status while reducing hospitalization: A randomized trial. Medical Care 1999;37(1):5–14.

80. Griffiths C, Foster G, Ramsay J, Eldridge S, Taylor S. How effective are expert patient (lay-led) education programmes for disease? BMJ 2007;334(1254–1256).

81. Barlow J, Bancroft GV, Turner A. Self-management training for people with chronic disease: A shared learning experience. Journal of Health Psychology 2005;10(6):863–872.

82. Barlow J, Edwards R, Turner A. The experience of attending a lay-led, chronic disease self-management programme from the perspective of participants with multiple sclerosis. Psychology and Health 2009;24(10):1167–1180.

83. Pakenham KI. Couple coping and adjustment to multiple sclerosis in care receiver-carer dyads. Family Relations 1998;47(3):269–277.

21

Physical Environment

Dory Sabata and Nancy Lowenstein

CONTENTS

A person's performance in meaningful life activities can be affected by the presence of a health condition, such as multiple sclerosis (MS), as well as by the demands posed by the environment while engaging in the selected activity. This chapter focuses specifically on the demands of the physical environment as they pertain to a person with MS, in particular, the home environment. After reading this chapter, you will be able to:

1. Explain the demands of physical environment,
2. Select assessment tools for understanding how the physical environment affects the health and functioning of people with MS, and
3. Identify home modification options to support the health and functioning of people with MS.

21.1 Conceptualizing the Physical Environment

Aspects of the physical environment can be described using the *International Classification of Functioning, Disability and Health* (ICF). The ICF categorizes physical environmental factors into (a) products and technology, (b) human-made changes, and (c) the natural environment.[1]

Products and technology are objects that can be added to the environment or can be modi-fied from their original form. Human-made changes to the environment are structural changes built into the environment, such as steps inside a home or landscaping surround-ing the home. In this chapter, products and technology, and human-made changes will all be considered as the built environment. The natural environment takes into account aspects of the environment that are beyond human control, such as weather and time of day. Each type of environmental feature can create barriers or help facilitate participation in a par-ticular space.

Another model that can be used to conceptualize and describe how the environment contributes to disability is the social model of disability. This model proposes that disabil-ity results when the demands of the environment exceed a person's abilities.[2] This per-spective considers the dynamic interactions of the environment and the effects of MS. For instance, a walking path can have smooth sidewalks and good lighting, while another path has potholes, cracks, and obstructions. These varying environmental factors can affect disability. Similarly, people are dynamic and their own abilities change from day to day as well as over time. One day a person can have a "good day," feeling energetic, strong, and able to accomplish goals for the day. Another day, the same person can feel exhausted and sluggish. A person's abilities need to fit with the demands of the environment for the person to successfully minimize disability and maximize participation.

Consider the variability in both the environment and the person's abilities. The person having a "good day" can manage walking on the smooth path and may or may not be able to manage the obstacles on the less optimal path. Therefore, the interaction between the person and the environment needs to be considered dynamic. At times, the environment can exceed a person's capacity. Not only do features of the physical environment need to be understood, but they can also be changed to better fit with a person's skills and maxi-mize his or her participation.

21.2 Relevance of the Physical Environment during MS Rehabilitation

Multiple sclerosis affects individuals in the prime of life. During prime work and child-bearing years, losing independence can be devastating. Tasks can be successfully accom-plished, though, when the demands of the environment fit with the capacity of the person. The built environment can often be modified to decrease demands that limit a person's participation. When taking this perspective, the environment is really responsible for whether a person is considered "disabled."

Features of both the natural environment and the built environment can be identified. However, the built environment is more readily modifiable to support participation. Clients often report that the physical barriers they encounter present the greatest chal-lenges to engaging in meaningful daily activities in their home environments.[3,4]

Think about how a hot summer day or a cold winter evening might affect whether a person goes out of the house. In these examples, the natural environment is posing increased demands as compared with a fair weather day. The extreme temperatures may contribute to fatigue. Also, both these conditions interact with a person's vision. The sunny day may result in glare, while the night time limits visual acuity. For someone who already is at risk of fatigue or visual impairment, these natural environmental conditions can fur-ther increase the demands imposed upon the person who is trying to complete an activity within these environments.

The built environment can similarly be supportive of participation or present barriers. One common physical barrier in homes is difficulty managing steps. Navigating steps and moving around spaces can be particularly challenging for people with impaired balance, strength, vision, or tactile sensation, all of which can occur with MS. Stairs or steps can vary in number, height of rise, depth of each step, direction, available lighting, available upper-body support (e.g., handrail), surface type (e.g., slick, nonslip), and alternative way to access elevation change (e.g., ramp or lift). For someone who can stand for short periods of time and who needs to access only a few steps when entering or exiting the home, a handrail may be supportive enough to achieve that short distance. However, a flight of steps may exceed a person's capacity to change floor levels, particularly if he or she is regularly using a mobility device. A lift or elevator in the built environment would be needed to access the second floor. This example illustrates not only the features currently present in the built environment but also identifies how some modifications reduce the demands of the built environment.

Multiple sclerosis affects individual lives in different ways. A variety of symptoms can influence function and participation in everyday life. Each individual's symptoms manifest differently. Therefore, rehabilitation professionals need to listen to each client and assess the client's strengths and barriers pertaining to their physical, cognitive, and social-emotional status. In some instances, the abilities of a person with MS can be preserved or even improved. However, for many clients, the progression of the disease may require some changes to the environment in order to minimize disability.

21.3 Symptoms That Influence the Need for Home Modifications

Some of the common symptoms of MS include fatigue, cognitive changes, impaired motor skills, and sensory deficits. Each of these symptoms can limit a person's participation. However, changes to the environment can sometimes minimize or compensate for the negative effects of these symptoms.

Fatigue is a symptom that has a major impact on daily functioning for individuals with MS (see Chapter 5). Physical exhaustion, as well as cognitive challenges, can contribute to fatigue, which is the primary reason that people have to leave work and go on disability.[5] Strategies to simplify tasks can help reduce fatigue and give the person more control over the environment. The environment can be arranged to reduce the amount and frequency of movement required. Simplified tasks can reduce muscle fatigue as well as cognitive fatigue.

Changes in vision and cognition are other symptoms of MS (see Chapters 11 and 8). Blurred vision can result from diplopia, nystagmus, and optic neuritis. Vision is commonly used to orient our bodies to balance, for instance, when navigating stairs. People with MS can have difficulty in problem solving, initiation, and organizing materials and space. These are examples of cognitive executive-functioning deficits. The physical environment can be organized and managed through environmental cues in order to reduce cognitive or visual demands required for completing an activity. For instance, visual cues may be used to orient a person with cognitive deficits. A person who forgets to take his or her medicine may need a pill box placed in a location that the person finds easily visible. If the medicine is taken in the morning, perhaps the person sees it on the bathroom counter each morning and is reminded to take it. If the medicine needs to be taken with food, it could be placed in the kitchen near a food item that is commonly consumed. When the medicine is in a place that visually stimulates the person, it serves as a cue to take the medicine.

Tactile or auditory cues can be helpful to people with visual deficits. For instance, raised letters or marks can be felt to communicate information to a person who has difficulty seeing a particular word or mark. Sounds can be used to indicate danger or to alert a person to additional information. For people with MS, environmental features can support performance despite cognitive and sensory changes or physical changes in motor functioning.

Many motor skills that are needed for mobility can be affected by MS (balance, decreased muscle strength, spasticity, and muscle fatigue, see Chapters 6 and 7). These symptoms affect transfer skills, climbing stairs, fall risks, activities of daily living (ADL), household management, and caregiving activities, to name a few. Månsson and Lexell[6] looked at ADL performance of individuals with moderate to severe MS and found that mobility, transfers, and self-care were the areas that had the greatest limitations. These researchers stated that "the quality of their instrumental activities of daily living (IADL) was reduced to such a degree that their ability to live independently in the society could be compromised."[6]

Not only loss of independence but also the general safety of individuals with MS is of concern. People with MS have a greater risk for falls, partly due to changes in balance and mobility and use of mobility devices.[7] Poor concentration and forgetfulness are cognitive risk factors for falls, while fear of falling is a psychosocial risk.[7] Falls can have severe consequences, such as fractures, high costs for injury recuperation, or even death. Home safety is part of home modifications and can address prevention of falls.

21.4 Home Modification Process

Home modification is both a product and a process.[8,9] As a product, home modifications can be changes or adaptations in the home to support performance, accessibility, and participation. As a process, home modification includes identifying needs and possible solutions as well as implementing the changes and reviewing them.[9]

A team approach can be an effective way of determining the demands of the environment on the person, to consider long-term needs, and to identify best options for home modifications. Several professions have valuable contributions to the home modification process. Table 21.1 outlines some of the most common professionals involved and the typical roles each one contributes.

Each professional brings his or her own disciplinary perspective and contributes different strengths to a home modification assessment, recommendations, and implementation. Some general guiding theories and models also cut across disciplines and shape how professionals think about, solve problems concerning, and make decisions about home modifications.

21.4.1 Theories That Guide Home Modification Process and Practice

Ecological models suggest that the environment plays a major role in facilitating positive outcomes. These models suggest that even if the person is not able to adapt or change, the environment can, and as a result of a good person–environment match, the person's participation is not compromised. The ecological model of aging describes the relationship between a person's capacities and the demands, or "press," from the environment and identifies that a "just right fit" needs to be achieved in order for a person to be able to adapt.[10]

TABLE 21.1

Roles of Interdisciplinary Team Members in Environmental Modifications

Team Member	Potential Tasks	Examples
Neurologist	Recognizes needs for modifications; makes referrals.	During a routine appointment with patient, doctor and client discuss difficulty managing the stairs to the upstairs bathroom (the only bathroom in the house). Doctor explains that environmental modifications might be possible.
Physiatrist	Recognizes needs for modifications; makes referrals; works with occupational therapy and physical therapy to help identify solutions.	Identifies mobility issues, and patient talks about having difficulty managing stairs and the laundry is in the basement. Doctor explains that environmental modifications might be possible.
Physical therapist	Identifies types of mobility aids required by patient; identifies interior and exterior barriers and support to mobility issues; makes recommendations to patient and family; and works with contractor; may work with family to identify payment sources for renovations, equipment.	Patient requires a rollator walker to use indoors and is unbalanced on stairs. PT feels that a stair lift would be beneficial to the patient.
Occupational therapist	Identifies tasks that patient wants and needs to complete and where they are usually completed; evaluates the barriers and supports; makes recommendations to patient and family; works with contractor; may work with family to identify payment sources for renovations, equipment. Understands ADA requirements and minimum requirements for doorways, ramps, etc.	Patient uses a wheelchair in the home and has an inaccessible bathroom. Patient has expressed a desire to make the bathroom fully accessible by wheelchair.
Social worker	Identifies family's financial resources and other resources to pay for renovations and equipment.	Patient has identified a need for home modifications/renovations and is looking for outside resources other than a bank loan. Social worker puts client and family in touch with not-for-profit organizations to request financial assistance for renovations. Patient may also qualify for vocational rehabilitation funds.
Contractor	Provides designs and works with patient, family, and occupational therapy and physical therapy to ensure proper renovations are done with future needs of the client and family in mind.	Family requests that contractor meet with OT and PT to discuss the renovations to ensure they will meet the future needs of the client and his or her family.
Durable medical equipment provider	Works with occupational therapy and physical therapy to make sure mobility devices can be maneuvered around the house and through necessary doorways.	Durable medical equipment provider meets with occupational therapy and physical therapy to discuss mobility devices and to ensure they will meet the future needs of the client and that they are going to work in the client's different environments.

The consumer decision model helps to guide thinking about a client's readiness for change.[11] According to this model, a person will make changes when he or she anticipates severe consequences from a recognized risk and also feels that the recommended intervention will work and is affordable.[11] Practitioners must consider their client's readiness for change to the home environment. Research indicates that individuals feel more isolated when the accessibility of their homes is a challenge.[3] Yet, participants in the same study felt that some changes to their homes (such as removing furniture and objects, installing grab bars and ramps) made their homes feel more "institution-like."[3] Providers must be mindful of the client's readiness to accept even simple modifications, such as tub seats, extra stair railings, and removal of carpets. Home should not be the place where a person feels most disabled. Clients' values and concerns are the foremost considerations for professionals who make recommendations for home modification. Professionals make suggestions and provide education about home modifications, but the final decision to accept or reject these suggestions resides with the client and often times the family. Families and clients can feel that an environmental modification is "taking away" an important function or, that by climbing the stairs, the family member with MS is "getting exercise," and removing this activity will "be giving in to the MS" or hastening the functional decline. These myths are difficult to debunk and it can take many months of patiently educating on the part of the therapy team to help the client and the family understand the advantages of environmental modifications.

Each of the theoretical models presented emphasizes the importance of this person–environment interaction. However, often our assessment approaches and tools attempt to consider the environment in isolation from the person. When using assessment tools, therapists often need to supplement the data collected from instruments with professional reasoning to determine home modification recommendations and to guide implementation.

21.4.2 Assessment Strategies

Several assessment approaches can be used to evaluate not only the features in the physical environment but also how the client interacts with the physical environment. Each client will present with a unique set of deficits and a unique set of supports and barriers. Common assessment approaches include self-assessment checklists, interviews, observation, and performance measures. Checklists serve as a quick screening. Therapists can interview clients and families to gain their perspectives. Observation offers information about the environment and a person's performance in his or her natural context of the home. Performance measures allow clients to demonstrate how they are currently doing an activity. Each of these assessment methods has strengths and limits to their application.

21.4.2.1 Checklists/Self-Assessment

Several checklist-type tools for assessing the physical environment are available to ascertain the presence or absence of features. Checklists can be useful tools early in the process to engage clients and families, to empower clients to identify hazards and barriers in the environment, and to serve as discussion starters for professionals to help families determine options. Home safety checklists typically consider the presence or absence of physical features, including hazards and safety concerns. Many include general recommendations for hazard removal or reduction. The information gathered from these tools is limited. The tools themselves are based on expert opinion and lack empirical evidence to support their claims of hazards and barriers.

TABLE 21.2

Online Resources for Home Safety Checklists and Home-Modification Funding

AARP	www.aarp.org
Centers for Disease Control and Prevention	www.cdc.gov
Independent Living Research Utilization	www.ilru.org
Home Safety Council	www.homesafetycouncil.org
MS Society	www.nationalmssociety.org
Medicaid Waiver Programs	www.cms.gov/MedicaidStWaivProgDemoPGI/01_Overview.asp
Rebuilding Together	www.rebuildingtogether.org

Hazard identification alone may not be the problem. Unfortunately, this assessment approach often does not take into account how the environment is used or not used. For instance, the checklist may consider the presence of grab bars but not whether they are in a useful location or how frequently the person uses them. These types of checklists are readily available online from numerous places (see Table 21.2).

Self-assessments are often good screening tools to help focus both the client and the therapist to key factors of concern in the environment as well as spaces in the home that are most relevant to the client. Other evaluation methods can be used in conjunction with self-assessment.

21.4.2.2 Interview and Observation

Everyone uses observation. Different professionals focus their observations on different considerations. Building and design professionals consider structural features, space plans, and functional use of rooms. Health professionals focus more on the person with the particular health condition and the family. Occupational therapists consider how the person and the environment affect performance in everyday life activities. Professionals use a critical-reasoning process to analyze their observations and relate them to their disciplinary perspectives. Ideally, the observations take place in the spaces where the everyday life activities occur for that individual. Each context presents its own unique characteristics that must be considered.

For instance, eating is an everyday activity. Eating at home alone while sitting on the couch has different demands than eating at the dining room table with family at a holiday gathering. Different expectations and different physical and social demands are expected at each of those two eating experiences. One is more casual and solitary; the other is more interactive.

In this example, the person at home alone has fewer environmental demands. Perhaps this individual is simply having cereal in a bowl while wearing loungewear. In this situation, the person does not have to worry about appearance or self-care-type activities, such as bathing or dressing prior to eating. The person is not expected to interact during the meal. The food is contained in a single bowl and can be eaten while sitting comfortably. The meal consists of a single course that required little preparation and no cooking. Eating in this environment has low social demands, low physical demands, and low cognitive demands.

On the other hand, eating a meal at a family gathering will likely have greater environmental demands. Appearance may be very important when seeing family or friends that are not commonly seen. The family may expect to see everyone dressed and eating a full meal with multiple dishes. The preparation for getting ready for this eating experience will require more energy than lounging at home. Also, the complexity of the holiday meal is much greater than a simple bowl of cereal. The group meal may take a longer time to eat. Table manners can be an expectation when dining with others. The family gathering will likely involve engaging in conversations when dining.

The complexity of eating with others at a big event involves more than social demands. A person with cognitive changes due to MS may have difficulty keeping up with the multiple demands of remembering unfamiliar faces, engaging in conversations, eating, remembering manners, and doing all of these things simultaneously. Someone with fatigue may need to prioritize activities in order to participate in the activities that are most important. For instance, the person may decide to forgo preparing the dinner. If the social aspect of a family dinner is most important to the person, then that may be where energy is focused.

The comparison of eating at home alone and eating at a family gathering illustrates the importance of context and environment. Both these situations pose demands from the environment as well as different types of social participation. Each individual must weigh his or her options and manage the environment in order to successfully participate in the activities that are important and meaningful. Theses priorities may also fluctuate on a daily basis.

The cognitive and physical demands presented in these environments are just some of the contextual issues that need to be considered. These examples highlight the importance of observing and discussing the actual environment where living occurs. Assessment in the natural context provides a greater sense of the demands that affect performance.

For the most part, observations are informal and constant. Observations start even before entering a home. When one meets a client in his or her home, observations start with the neighborhood, sidewalks, transportation access, and the entrance of the home before ever meeting the client. All of these observations inform planning for home modifications. Observations also lead to more questions.

Interviews are one way to ask the questions that develop from our observations. Informal interviews allow therapists to ask questions that are of interest. A semistructured assessment tool is used as a guide in more formal interviews. Whether using formal tools or informal skills to observe and interview, one commonly uses these assessment techniques in home evaluations.

21.4.2.3 Performance Measures

Observations can provide even more detail and definition when a person is asked to perform a particular task or activity. Performance-based measures inform how well a person can complete a specific skill, task, activity, or occupation. While many performance-based measures are available to assess movement, vision, cognition, and other performance skills, few of these performance measures take into account the role of the context or environment.

Measures used to determine home modification needs and options require considering more than just the environment. The tools used in assessment need to provide a clear picture of how the person performs within the home, what activities may be more demanding than necessary, and what can change to decrease those demands on the individual.

Tools with standard procedures provide consistency across clients and reduce tester bias. Measures with demonstrated reliability and validity help to ensure that the data collected are an accurate assessment of the person's performance. Table 21.3 outlines some

TABLE 21.3

Sample Assessment Tools

Name of Measure	Author/ Source	Constructs Measured	Who Should Administer	Examples of Evidence Where Tool Has Been Used with MS Population
Canadian Occupational Performance Measure (COPM)	Law et al.[12]	Importance, satisfaction, performance of desired occupations	Occupational therapist or skilled interviewer with understanding of self-care, work, and leisure activities	Bodiam[13]
Fatigue Impact Scale	Fisk et al.[14]	Fatigue and activity	Self-assessment rated by the person with MS	Téllez[15]; Kos et al.[16]; Mathiowetz[17]
Functional Status Questionnaire	Cleary and Jette[18]	ADL performance	Health professional skilled in assessing functional abilities	Paltamaa et al.[19]
Housing Enabler	Iwarsson and Isacsson[20]; www.enabler.nu[21]	Home environment intersecting with functional limitations	Health professional skilled in assessing functional abilities and able to follow protocol for environmental assessment	Fange and Iwarsson[22]

specific assessment tools that may be used to guide intervention, observation, and capturing a formal measure of performance in order to inform the home modification process.

The environment also needs to be assessed. Some environmental features will support the person in participation while other aspects of the environment may place unnecessary demands on the individual. Interventions can target the built environment and reduce demands upon the person. The process of making changes to the built environment where a person lives is called home modification. Not only is home modification a process but it also includes products in the built environment that are used to support participation at home and in the community.

21.5 Home Modification Implementation

In the literature, three types of intervention approaches to home modification are commonly discussed: visitability, universal design, and home adaptations/renovations.

Visitability is a concept to promote three specific features in new construction of homes. People who use a wheelchair or other mobility devices want to be able to visit family, neighbors, and friends whose homes are "visitable."[23] The three features include a zero-step entrance, widened doorways, and a first-floor bathroom.[23] A zero-step entrance means that at least one entrance to the home does not require the use of steps. Widened doorways, specifically 36 inches, allow additional clearance for someone using a wheelchair or other mobility device. A first-floor bathroom would allow someone to use the toilet without having to navigate steps. While this seems like a simple idea, most homes today do not have these three features. People with MS use mobility devices, such as scooters or power wheelchairs, to minimize fatigue. Visitablility features would create opportunities for this population to visit others, particularly when they need to use a mobility device.

Changes to the environment can help everyone in the family, not just the individual with MS. This is the basic concept of *universal design*. "Universal design is the design of products and environments to be usable by all people, to the greatest extent possible, without the need for adaptation or specialized design."[24] For instance, a shower can be designed for various people within a family. A tall adult who stands to shower needs the water to be overhead. Someone seated while showering needs space to sit and access to the water at a lower height. The water controls, as well as the soap and shampoo, need to be easily accessible. Each of these features can be considered in the original design of the space to meet the needs of a range of users of the space. Seven specific principles of universal design have been established to guide the design of products and spaces; these are summarized in Table 21.4.[25]

Universal design can be used to design and develop a space or product from the beginning. Thereby, a person may avoid the need to make major modifications later. However, most people are not designing a new space and may have to work with the existing space to meet their needs. These principles can still be considered during modifications to existing spaces or products.

Home adaptations and renovations are the most common type of home modification. Today's housing options have not typically been designed to respond to changing needs of people with chronic, progressive conditions such as MS. Therefore, the home needs to change to support the person with MS. Assistive devices and technology can also be part of the physical environment to support participation.

TABLE 21.4

Summary of the Principles of Universal Design[25]

Principles	Implications
Equitable use	Means of use should be the same for all users, and users should not be segregated or stigmatized. Opportunities for privacy, security, and safety should be provided to all users. Design should be appealing to all users.
Flexibility in use	Design should offer choice in methods of use, accommodate both right- and left-handed users, allow for variability in the user's accuracy and precision, and adapt to the user's pace.
Simple and intuitive use	Design should eliminate unnecessary complexity, be consistent with user expectations and intuition, accommodate variability in literacy and language skills, and provide prompting and feedback during and after task completion. Information should be arranged consistent with its importance.
Perceptible information	Design should allow essential information to be differentiated from its surroundings, presented in multiple redundant modes (pictorial, verbal, tactile), and be maximally legible. Design features should make it easy to give instructions or directions and should be compatible with strategies or devices used by people with sensory limitations.
Tolerance for error	Design elements should be arranged to provide warnings and minimize hazards and errors, provide fail-safe features, and discourage unconscious action in tasks that require vigilance.
Low physical effort	Design should allow the user to maintain a neutral body position, use reasonable operating forces, minimize repetitive actions, and minimize sustained physical effort.
Size and space for approach and use	Design should allow important elements to be visible and comfortable for both seated and standing users, accommodate variations in hand and grip size, and provide enough space to allow the use of assistive devices or personal assistance.

21.5.1 Home Modification Products

A range of everyday activities performed at home can be challenging for people with common symptoms related to MS. For the purposes of this chapter, home modification strategies have been organized according to task modification, assistive technology, and structural changes. Table 21.5 identifies some of the most common challenging activities in the home and provides examples of modifications to the task, assistive technology options, and structural changes to the home.

Tasks can be simplified to reduce physical and cognitive demands. For instance, items needed for a specific task can be organized together. A toothbrush, toothpaste, and floss can be kept together. Visually, the items can cue a person in the bathroom to use these

TABLE 21.5

Home Modification Interventions for Commonly Challenging Activities at Home

Commonly Challenging Activities	Task Modifications	Assistive Technology	Structural Changes
Moving around the house	Relocate rooms to main floor, create "control center," provide places to rest, use upper-body supports for stability	Cane, walker, wheelchair, scooter, chair for resting, smart home sensors that notify others of need for assistance, threshold ramp	Elevator or lift for managing steps to second level, door width at least 2″ clearance beyond size of person and mobility device, less than 0.5″ threshold, improved lighting
Bathing	Decrease frequency or duration of bathing, have hair washed at salon	Shower seat or bath bench, bath lift, smart home sensors that notify others of need for assistance	Shower or curbless shower, handheld showerhead, grab bars
Toileting	Reduce distance to get to toilet/commode	Raised seat, grab bars, commode chair in bedroom/ living room, smart home sensors that notify others of need for assistance	Bathroom size/space for assisted transfer, toilet paper location, space between sink and toilet
Dressing	Provide place to sit, select clothes with fasteners that are easily manipulated	Dressing stick, shoe horn, reacher, labeling	Chair, grab bar, height of closet rod
Grooming	Provide place to sit, group supplies for a specific task	Built-up handles; use packaging that is easily accessed, such as pumps and sprays; labeling	Roll-under sink, chair, lever-handled faucet, nearby storage for supplies
Meal preparation	Seated rather than standing, grouping supplies for a specific task, recipe/checklist	Timers and reminders, labeling	Roll-under counter or workspace, raised appliances, storage drawers, pull-out cabinets, lazy Susan, nonslip surfaces
Sleeping	Establish pattern/ habit of going to bed at the same time	Height of bed to support safe transfers, temperature controls, window coverings	Lighting, sound barrier
Getting in and out of the house	Leave the house at time of day when energy level is highest	Cane, walker, wheelchair, scooter, threshold ramp	Handrails, ramp, lift, lighting, nonslip surfaces, door width at least 2″ clearance beyond side of person and mobility device (for standard size wheelchair approx 32″ clear space) >0.5″ threshold

FIGURE 21.1
Different types of bath chairs are available. (Courtesy of Shutterstock.)

products. Minimal movement is needed to gather supplies when they are kept in a collective convenient location. Each of the activities in Table 21.5 has examples for the three identified modification strategies.

A shower chair is an assistive device that can be added to the physical environment to support and facilitate the strengths of a person with MS. Sitting instead of standing can reduce fatigue and reduce the risk of falling in the shower. A seat in the shower can help someone to bathe independently (see Figure 21.1). Figures 21.2 and 21.3 illustrate

FIGURE 21.2
Curbless shower with built-in shelves, seat, hand-held shower, and grab bar. (Courtesy of Nancy Lowenstein.)

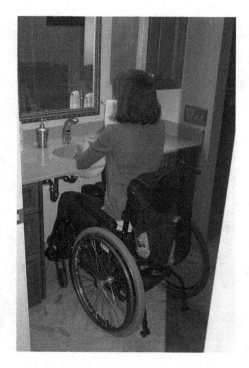

FIGURE 21.3
Sink with roll-under top. (Courtesey of Teresa Morreale.)

additional bathroom modifications that can be made to facilitate the completion of self-care tasks.

Home modification may require construction and remodeling to the physical structure, such as building a ramp or changing a tub to a shower, which is a structural change. One study identified some of the most common home modifications made by unemployed women with MS. Dyck[4] found that the top modifications were ramps, widened door-frames, grab bars, bath lifts, and other bathroom modifications. Additional modifications included moving laundry facilities to the main living floor, moving furniture, clearing out rooms to manage mobility devices, and changing locks to create a keyless entry.[4] This study also found that individuals with MS think about their future needs, and as the disease progresses, the environmental barriers can change.

Changes in mobility were previously identified as one of the physical symptoms commonly affecting people with MS. Home modifications designed to address mobility issues must build on the remaining capacities of the individual and consider his or her safety when moving around the home or in the community. Figures 21.4 and 21.5 illustrate kitchen modifications that enable a person to cook and do dishes while sitting, on a chair, wheelchair, or the seat of a walker. Figure 21.6 illustrates an exterior ramp that facilitates mobility. Stair lifts are another option for indoor and outdoor accessibility.

21.5.2 Funding Home Modifications

Home modifications can potentially benefit everyone who lives in or visits that home, but the cost of modification can be a barrier in making changes. Moving a washer and dryer upstairs or making a bathroom more accessible can be costly. The National Multiple

FIGURE 21.4
Modified stove top (with roll-under counter and lowered height). (Courtesy of Nancy Lowenstein.)

Sclerosis Society is a resource for finding contractors, low-interest loans, and, sometimes, providing some financial support for the modifications. Other national organizations also provide funding or services pertaining to home modifications. Clients may be more likely to make needed changes if they are aware of the available funding resources in their communities.

21.6 Home Modification in Various Home Settings

Home modifications are dependent upon understanding the person–environment interaction. Therefore, the assessment process must consider the various places that people call home. "Home," or the place where people live, can change over time. Among adults aged 45 and older, the most common residential places in the United States are single-family homes, rental property, and mobile homes.[26] As people age and need additional supports, they increasingly live in assisted-living facilities and long-term care facilities. Also, many policy issues influence the implementation of home modification interventions. Home modification assessment and intervention will vary across different residential settings.

Congregate-living facilities may temporarily or permanently become home to some. For people requiring more medical monitoring, assisted living is an option that often supports IADLs, offering amenities such as meals, housekeeping services, transportation, and laundry assistance. For more medically frail individuals, skilled nursing facilities

FIGURE 21.5
Sink with roll-under clearance and covered plumbing. (Courtesy of Nancy Lowenstein.)

provide even more support for activities of daily living, such as dressing, bathing, toileting, and feeding.

Public facilities with public spaces are subject to the American with Disabilities Act Accessibility Guidelines (ADAAG).[27] These Guidelines provide design specificity intended to address accessibility (to entrances, level changes, and bathrooms, to name a few). As a

FIGURE 21.6
An exterior ramp added to a home entrance. (Courtesy of Shutterstock.)

result, residential facilities (e.g., assisted living, skilled-nursing facilities) often have an elevator or are located on one level for easy access to congregate dinning. While the ADAAG provides specific design features, these guidelines were not originally developed to meet the needs of people living in institutional housing. Therefore, not all of these features required by law actually meet the individual needs of the people who live there. Features such as bathrooms, walkways, and entrances have specifications that need to be met regardless of whether those are useful to the people living in the facility. In order to go beyond what is specified in ADAAG, the provider must apply for waivers for alternative design features.

People who live in community settings may have rental, home-owner association, or property-association policies that can challenge the implementation of home modification. In neighborhoods where uniformity in housing is desired and supported by policies, people have to appeal to their association in order to have a ramp or lift on their home.

In rental properties in the United States, the Fair Housing Act[28] applies. This piece of human rights legislation protects people with disabilities (among other protected classes) from discrimination when renting housing. This policy allows for renters to make needed modifications with the condition that the property is returned to its original configuration when they move out. Any associated costs of home modifications while living there or costs of returning the unit back to its original state are the responsibility of the renter. In addition, common areas may be subject to ADAAG.

Mobile homes are challenging due to the lack of structural supports in the housing unit. Mobile homes do not have walls supported by studs and drywall. Reconfiguring the walls or the spatial layout of rooms within the mobile home is extremely limited. Even installing simple supports such as grab bars becomes complex for mobile home owners. Due to this complexity, finding contractors willing to make changes to mobile homes is limited. As a result, finding safe and feasible solutions to support mobile home owners can be very challenging.

People who live in single-family homes must meet structural building codes when making major modifications. Typically, accessibility is not a part of those building code ordinances. The lack of rigid housing policies allows for freedom in choosing a range of home modification interventions. Yet, the challenge is finding a builder knowledgeable about home accessibility features beyond ADAAG.

People in a single-family home may also have more options in terms of programs available to assist with the home modification assessment and interventions. Individuals receiving inpatient or outpatient care can be eligible for a home visit prior to discharge that may be covered by insurance. The home assessment can be conducted to prepare the person for returning home. People receiving home health can also qualify for a home assessment.

Community-based programs may have their own providers who conduct assessments. Many of these community programs vary in the range and type of services they provide. Some offer assessment and intervention. Some will pay for products or modifications but may or may not have an assessment as part of their process. In some cases, the individual may receive funding for the modifications but will need to find his or her own contractor to do the work.

The home environment is as variable as the effects of MS on each client. Regardless of what type of home a person has, each person with MS can benefit from assessment and interventions directed at reducing environmental demands and increasing participation. The obstacles to home modifications in each of these places may be different, yet they all have potential for improvement. The practices to determine effective interventions must be based on available evidence.

21.7 Available Evidence and Future Directions

The literature on home modification includes summaries, review papers, and various research studies. In one article, authors outline various types of assistive technology that may benefit people with MS, ranging from low-tech solutions such as nonglare paper to minimize the effects of visual impairment to high-tech computer adaptations for communication and cognitive cuing.[29] While this article provides a clear summary of technology available, more information is needed about the results when people with MS use these assistive technologies.

Researchers have conducted a systematic review of studies with people with MS and the use of assistive devices to support mobility. Results indicated that changes in gait and balance can affect work, leisure, and social participation for people with MS.[30] This evidence points to the importance of individuals finding mobility devices that fit their needs and will support their engagement in desired activities.[30]

Another study highlights some of the concerns raised by people with MS and their families in obtaining power mobility devices. This study found that as people with MS experience declines in mobility, both the person and the family recognize the need for additional support to improve mobility.[31] A power mobility device such as a scooter or power wheelchair can greatly minimize physical demands needed for moving from place to place. Once the need is recognized, finding a good fit can be challenging. Participants in this study indicated difficulty finding sufficient resources to help with selecting a power mobility device.[31] In addition, the space requirements, vehicle accessibility, and community accessibility may not be adequate to accommodate the mobility device. One person felt participation with the family actually decreased, since relatives did not have a home that was accessible with a power mobility device.[31]

Two articles about mobility devices help outline some of the benefits and concerns of such interventions.[30,31] People readily recognize the need for an assistive device to support home and community mobility. Professionals can help people with MS and their families benefit most from the addition of the assistive technology product when the human-made environment also supports the use of the device. For instance, when someone's home has adequate space and accessibility for using a power mobility device, then a person can move about freely. However, when the home has barriers such as high thresholds, level changes with steps, and narrow spaces, a power mobility device can limit a person's participation.

Sometimes people use their own knowledge and resources to choose products and technologies; however, what they choose may not be the best fit. As a result, technology can often be abandoned. Another study suggests that an interdisciplinary approach to identifying and implementing assistive technology interventions can reduce abandonment of technology.[32] Professionals from different disciplines offer varying perspectives that can help a family recognize a range of considerations. A person may recognize the need for a mobility device. A sales person can describe all the features of the device and funding options. A rehabilitation professional can provide valuable information about the capacities of the person to operate the mobility device. Perhaps the family recognizes the need to build a ramp to enter the home in place of steps. A building professional can help determine the local building codes and safety features that need to be addressed when building a ramp. In this example, various disciplines contribute to understanding the whole picture and to providing a comprehensive plan. Alternatively, a person can make a purchase without fully knowing about various mobility devices and about ways to get it funded. A person may not realize what skills and abilities he or she has, or will need, in order to operate the device. Family may recognize the need for a ramp and attempt to build it on their own. An inexperienced ramp builder may create an unsafe surface which is too steep, too narrow, or lacks side rail supports. As described in Table 21.1, various team members can contribute to this process.

Mobility is not the only change with MS that can be addressed by home modifications, including assistive technology. Other functional declines may occur with MS, yet people may not be knowledgeable about other environmental supports to maximize their participation. In particular, people may not know of resources available for those additional changing abilities. People with MS are filled with "fears of the future" and realize that with aging they are likely to face further decline and increased dependency on others.[33] Professionals can help families identify ways to decrease environmental demands and provide supports in the home environment in efforts to minimize the effects of changing abilities as the person with MS ages.

Many studies regarding home modifications have not specifically focused on the needs of people with MS, but rather focused on older adults. A review of the literature on home modifications and older adults suggests many challenges to generalizing the current body of knowledge regarding home modifications.[34] Wahl et al.,[34] who conducted this review, cautiously suggest that home modification interventions can help minimize the effects of disability. However, the evidence in support of home modifications is specific to functional outcomes. Other studies in this review that used fall-risk reduction outcomes were not as supportive of home modification interventions. Researchers suggest that valid, reliable assessment tools need to be consistently used in home modification research in order to accurately capture client outcomes. An additional finding from this review article is that studies that focused on a distinct modification and with a functionally related outcome resulted in stronger evidence in support of home modifications. These researchers also suggest that outcomes are dependent on the person–environment interaction more than simply the environment. Overall, this review resulted in a call for stronger evidence, for more consistency in how terms are defined and assessed, and for greater theoretical grounding of research accounting for the person–environment interactions.[34]

Research on home modifications is limited, and few studies specifically address the impact of the environment on MS and daily functioning. Therefore, findings have limited application and generalizability to a population with MS. With changing technologies and greater access to supportive products off the shelf, current studies are needed to better capture the current options and how those may impact people with MS.

References

1. World Health Organization (WHO). International classification of functioning, disability and health. Geneva, Switzerland: World Health; 2001.
2. Hughes B, Paterson K. The social model of disability and the disappearing body: Towards a sociology of impairment. Disability & Society 1997;12(3):325–340.
3. Lund ML, Nygård L. Occupational life in the home environment: The experiences of people with disabilities. Canadian Journal of Occupational Therapy 2004;71(4):243–251.
4. Dyck I. Hidden geographies: The changing lifeworlds of women with multiple sclerosis. Social Science Medicine 1995;40(3):307–320.
5. National Multiple Sclerosis Society. Social security disability benefits: A guide for people living with multiple sclerosis. National MS Society; 2010.

6. Månsson E, Lexell J. Performance of activities of daily living in multiple sclerosis. Disability and Rehabilitation 2004;26(10):576–585.
7. Finlayson M, Peterson E, Cho C. Risk factors for falling among people aged 45 to 90 years with multiple sclerosis. Archives of Physical Medicine and Rehabilitation 2006;87(9):1274–1279.
8. Siebert C. Occupational therapy practice guidelines for home modifications. Bethesda, MD: AOTA Press; 2005.
9. Sanford JA, Pynoos J, Tejral A, Browne A. Development of a comprehensive assessment for delivery of home modifications. Physical and Occupational Therapy in Geriatrics 2001;20(2):43–45.
10. Lawton MP, Nahemow L. Ecology and the aging process. In: C. Eisdorfer, M. P. Lawton, editors. Psychology of adult development and aging. Washington, D.C.: American Psychological Association; 1973.
11. Ohta RJ, Ohta BM. The elderly consumer's decision to accept or reject home adaptions: Issues and perspectives. In: S. Lanspery, J. Hyde, editors. Staying put: Adapting the places instead of the people. Amityville, NY: Baywood Publishing Company Inc.; 1997.
12. Law M, Baptiste S, McColl M, Opzoomer A, Polatajko H, Pollock N. The Canadian Occupational Performance Measure: An outcome measure for occupational therapy. Canadian Journal of Occupational Therapy 1990;57(2):82–87.
13. Bodiam C. The use of the Canadian Occupational Performance Measure for the assessment of outcome on a neurorehabilitation unit. British Journal of Occupational Therapy 1999;62(3): 123–126.
14. Fisk JD, Ritvo PG, Ross L, Haase DA, Marrie TJ, Schlech WF. Measuring the functional impact of fatigue: Initial validation of the Fatigue Impact Scale. Clinical Infectious Diseases 1994;18:S79–S83.
15. Tellez N. Does the modified fatigue impact scale offer a more comprehensive assessment of fatigue in MS? Multiple Sclerosis 2006;11(2):198–202.
16. Kos D, Nagels D, D'Hooghe MB, Duportail M, Kerckhofs E. A rapid screening tool for fatigue impact in multiple sclerosis. BMC Neurology 2006;6:27–34.
17. Mathiowetz V. Test-retest reliability and convergent validity of the Fatigue Impact Scale for persons with multiple sclerosis. The American Journal of Occupational Therapy 2003;57(4):389–395.
18. Cleary PD, Jette AM. Reliability and validity of the functional status questionnaire. Quality of Life Research 2000;9:747–753.
19. Paltamaa J, Sarasoja T, Leskinen E, Wikström J, Mälkiä E. Measuring deterioration in international classification of functioning domains of people with multiple sclerosis who are ambulatory. Physical Therapy 2008;88(2):176–190.
20. Iwarsson S, Isacsson A. Development of a novel instrument of occupational therapy assessment of the physical environment in the home—A methodologic study on "the enabler." Occupational Therapy Journal of Research 1996;16(4):227–244.
21. The enabler web site: Providing tools for professional assessments of accessibility problems in the environment [Internet] Lund, Sweden: Susanne Iwarsson; c2011 [cited 2011 November 21]. Available from: http://www.enabler.nu.
22. Fänge A, Iwarsson S. Changes in accessibility and usability in housing: An exploration of the housing adaptation process. Occupational Therapy International 2005;12(1):44–59.
23. Visitable Homes, Visitable Communities. [Internet]; c2008 [cited 2010 January 25]. Available from: http://www.concretechange.org/resources_handout1.pdf.
24. What is universal design? [Internet] Raleigh, NC: Center for Universal Design; c1997 [cited 2011 February 3]. Available from: http://www.ncsu.edu/www/ncsu/design/sod5/cud/about_ud/about_ud.htm.
25. The Principles of Universal Design, Version 2.0 [Internet] Raleigh, NC: Center for Universal Design; c1997 [cited 2011 August 8]. Available from: http://www.ncsu.edu/www/ncsu/design/sod5/cud/about_ud/udprinciples.htm.

26. Mathew Greenwald & Associates, Inc. These four walls... Americans 45+ talk about home and community. Washington DC: AARP; 2003.

27. Americans with Disabilities Amendments Act of 2008. PL 110-325. ADA Title III Regulation 28 CFR Part 36; 2008.

28. Fair Housing Act. PL 100-430, 102 Stat. 1619, 42 U.S.C. 3601; 1988.

29. Blake DJ, Bodine C. An overview of assistive technology for persons with multiple sclerosis. Journal of Rehabilitation Research & Development 2002;39(2):299–312.

30. Souza A, Kelleher A, Cooper R, Cooper RA, Iezzoni LI, Collins DM. Multiple sclerosis and mobility related assistive technology: Systematic review of literature. Journal of Rehabilitation Research and Development 2010;47(3):213–223.

31. Boss TM, Finlayson M. Responses to the acquisition and use of power mobility by individuals who have multiple sclerosis and their families. American Journal of Occupational Therapy 2006;60(3):348–358.

32. Verza R, Carvalho ML, Battaglia MA, Uccelli MM. An interdisciplinary approach to evaluating the need for assistive technology reduces equipment abandonment. Multiple Sclerosis 2006;12(1):88–93.

33. Finlayson M. Concerns about the future among older adults with multiple sclerosis. American Journal of Occupational Therapy 2004;58(1):54–63.

34. Wahl H, Fänge A, Oswald F, Gitlin LN, Iwarsson S. The home environment and disability-related outcomes in aging individuals: What is the empirical evidence? Gerontologist 2009;49(3):355–367.

22

Caregiving

Kenneth I. Pakenham and Marcia Finlayson

CONTENTS

Everyday, people with multiple sclerosis (MS) face a wide range of complex symptoms, unpredictable fluctuations in function, and compromised quality of life (QoL).[1,2] As previous chapters in this book have indicated, the symptoms of MS can contribute to significant restrictions in activity and participation in self-care tasks, domestic life responsibilities, education and employment, and social and community engagement. In order to effectively manage these consequences, people with MS often seek physical, emotional, and instrumental support from informal caregivers.[3,4] An informal caregiver is someone who does not receive financial compensation for providing assistance and is usually a family member or friend.[5] These individuals are the focus in this chapter.

Informal caregivers are essential throughout the MS rehabilitation process since the disease affects both the caregiver and care recipient, and they appear to respond as a social unit.[6] During the assessment process, caregivers can be a vital source of information about the daily functioning of the person living with MS, the physical setup of the home, and the ability of individuals within the social network to support or carry out rehabilitation plans.[7,8] Together with data collected by other means (e.g., direct observation, objective testing, chart review), the information provided by caregiver(s) can help ensure that intervention goals are relevant and identified outcomes are achievable. After reading this chapter, you will be able to

1. Describe the nature of MS caregiving,

2. Examine the impact of the caregiving role on the well-being of the caregiver and consider the relevance of this knowledge to members of the rehabilitation team,

3. Examine the factors likely to shape caregiver well-being and explore how rehabilitation professionals can use this knowledge to guide and develop their intervention plans,

4. Identify assessment instruments relevant to MS caregiving that can be used in rehabilitation settings, and

5. Review published MS caregiver interventions and commonly available community services to support caregivers in their role.

As we address these objectives, we not only consider the role of adult caregivers but also the care and support that children provide to family members with MS.

22.1 The Nature of MS Caregiving

Being a caregiver can be very challenging as the role requires the performance of a wide range of time-consuming and physically and emotionally demanding tasks, most of which are learned through trial and error rather than through any formal education and training.[9] Caregivers of persons with MS spend considerable time on these tasks, all of which are of immense economic value and an essential contribution to the rehabilitation of persons living with MS.[4,10] Although the caregivers of people with mild MS-related disability spend less than 1 hour per day providing assistance, this number can be greater than 12 hours per day when the person with MS experiences greater disability.[4,11]

For some people with MS, caregivers provide only periodic assistance with activities like transportation and shopping. For others, the need for support is greater and may include high levels of intimate care with regular activities such as toileting, bathing, and dressing.[11–13] A recent study found that, overall, MS caregivers ($N = 302$) provided assistance for an average of 7.6 activities (sd = 5, range = 1–24).[14] There are several articles in the literature that describe the proportion of caregivers who report assisting a person with MS with specific tasks (see Table 22.1). While direct comparison of these various studies is

TABLE 22.1

Proportion of MS Caregivers Reporting the Provision of *Any Level* of Assistance to a Person with MS for Various Everyday Tasks, by Study

	Buchanan et al.[15]	Finlayson and Cho[14]	Cockerill and Warren[16]	Forbes et al.[13]	O'Brien[17]
Basic sample caregiver information	N = 530 Mean age = 60 47% female	N = 302 caregivers Mean age = 59 54% male	N = unclear Mean age = 48 52% female	N = 257 Mean age = 52 40% female	N = 20 Mean age: 52 45% female
Indication of care recipients' MS disability level	PDDS ≥ 5	Very minimal: 23.1% Minimal: 29.0% Moderate: 18.9% Severe: 29.0%	Walk unaided: 22% Walk aided: 16% Wheelchair: 55% Nonambulatory: 7%	Minimal: 12.0% Mild: 18.0% Moderate: 32.0% Severe: 38.0%	Not specified
Mobility	Transportation: 90.2% Walking inside: 71.7% Walking outside: 90.6% Out of bed: 52.8%	Transportation: 69.2% Walking inside: 42.7% Up/Down stairs: 36.1% In/out of bed: 33.4%	Transfers: 30% Accompany on outings 48%	Lifting and moving: 74%	Transfers: 55%
Bathing/ Washing	Bathing: 54.2% Personal hygiene: 49.8%	Shower/Tub transfers: 23.5%	Bathing: 22% Hair care: 28% Skin care: 17%	Bathing/ Washing: 55%	Bathing/ Washing: 75%
Dressing	56%	41.1%	26%	69%	90%
Toileting	47.4%	Toilet transfers: 8.9% Bladder care: 24.2% Bowel care: 21.2%	Bowel care: 22% Bladder care: 26%	48%	Toileting: 55% Continence 75%
Eating/Feeding	Eating/Feeding: 27.9%	Assisting with eating: 9.0%	Eating/Feeding: 13%	Eating and drinking: 51%	Feeding: 70%
Meal prep	82.6%	Hot meal: 72.3%	38%	Not addressed	Not addressed
Health management	Giving medications or pills: 66% Giving injections: 49.5% Arranging or supervising other care services: 28.9%	Dealing with insurance claims: 37.4% Coordinating appointments: 38.7% Managing medications: 38.1%	Managing medications: 18%	Not addressed	Not addressed
Household management	Indoor housework: 91.8% Outdoor work: 78.4% Groceries: 93.9% Managing finances: 68.3% Shopping (general): 88.1%	Heavy housework: 43.7% Light housework: 42.7%	Housecleaning: 60% Laundry: 52% Groceries: 55% Personal shopping: 45% Banking: 46% Handling money: 24% Locking/Unlocking doors: 24%	Not addressed	Not addressed

complicated by differences in the average age of caregivers and care recipients, their relationship to each other (spouse, other family members, friends), the level of disability of the person with MS, and the way that the data were collected, several general observations can be made.

Observation #1: A large proportion of caregivers of people with MS report helping with mobility-related tasks (transportation, walking, stair climbing, transfers) (see Table 22.1). Depending on the level of disability of the person with MS and the relative height and weight of the caregiver and care recipient, providing this assistance has the potential to be physically demanding for caregivers. Caregivers do not consistently receive training about how to safely assist a person using a mobility device, how to protect themselves from injuries during physical caregiving, or what technologies may be available to assist them during these tasks (e.g., transfer boards, gait belt, Hoyer lift).[11] If training does occur, it might not be done in the home where the tasks are actually carried out or be updated as MS disability progresses. Because of the potential risks to both the caregiver and care recipient, at least one member of the rehabilitation team should be assigned the task of inquiring about the nature and extent of a caregiver's physical caregiving responsibilities. Once involvement in these tasks is disclosed, the caregiver–care recipient dyad should be referred to a physical or occupational therapist to ensure that the pair receive the necessary education and support to enable the caregiver to fulfill his or her responsibilities safely (e.g., transfer training, proper lifting techniques).

Observation #2: Caregivers are commonly involved in the completion of household management tasks such as cooking, shopping, cleaning, and laundry (see Table 22.1). For many caregivers, these are tasks that are taken on insidiously as roles within the household shift and change with MS progression.[14] For example, a woman with MS may have always done the cooking and laundry for her family, but, with her increasing fatigue and mobility limitations, her partner and children begin to share these responsibilities. Over time, they gradually take over these tasks. Rehabilitation providers must be sensitive to and inquire about shifts in household routines and responsibilities. Caregivers who become overwhelmed with their responsibilities are often unable to maintain their physical and mental health; this risk applies to the entire family unit.[18] Occupational therapists, social workers, and psychologists can all play an important role in enabling caregivers to examine work–life balance, and assess and modify their daily routines, and assess their use of external supports. At times, caregivers will require assistance to find, arrange, and pay for additional assistance to ensure that their health and that of the care recipient can be maintained over the long term. The use of services to give caregivers a break is discussed later in this chapter.

Observation #3: While fewer caregivers are involved in providing assistance with basic activities of daily living (ADL), such as dressing, toileting, eating, and medication management, it is important to note that some of these tasks are completed several times a day (e.g., bladder care). Some basic ADL tasks are highly intimate and the completion of them can alter the nature of the relationship between the person with MS and his or her caregiver. Managing incontinence is one area that has been identified by MS caregivers as particularly challenging emotionally[14] and one that often influences the sexual relationship of a couple.[19] Therefore, rehabilitation providers who recommend that a family member provide assistance with toileting must consider the broader consequences of their recommendations carefully. Open communication with the person with MS and the caregiver provides an opportunity for them to make a decision about what type of assistance is best for their own relationship. External assistance provided by a home-care aide may be a better choice for some families.

Medication management can be a complex task, depending on the number and nature of medications that the person with MS is taking and their scheduling. Education and training of caregivers to fulfill this role is paramount to minimize risk of serious side effects and drug interactions. Nurses and pharmacists play a primary role in this regard. One study found that 49% of MS caregivers were providing assistance with injectable medications.[15] Although injection anxiety among people with MS is related to lower medication adherence,[20] no research to date has focused on MS caregivers in exploring this issue or its implications.

In summary, there is tremendous variability in the nature and number of tasks with which caregivers provide assistance. As MS disability increases, the number of hours spent in caregiving increases and the nature of the caregiving tasks usually becomes more intimate. Inquiring about the routines and responsibilities of the caregiver is an important part of the MS rehabilitation process to ensure both the safety of the person with MS post-discharge and the feasibility of the discharge plan for the caregiver. Further details regarding assessment of and interventions for MS caregivers are provided later in this chapter.

22.2 The Impact of Caregiving

There are a number of distinctive characteristics of MS that present considerable challenges for caregivers, including onset in young adulthood; the usually degenerative nature of the illness, the absence of a cure; the variations in disease course; the unpredictability of exacerbations; the wide variations in clinical symptoms; and the profound disruptions in employment, sexual functioning, family life, and activities of daily living caused by these symptoms. It is therefore understandable that caring for a person with MS has a wide range of adverse impacts on many areas of a caregiver's life. There have been two reviews of MS caregivers that have reported on the impacts of caregiving.[5,9] McKeown et al.[9] found that caregivers experienced adverse impacts on their QoL, physical health, psychological well-being, social life, and financial situation. Corry and While[5] reviewed the literature subsequent to McKeown et al.'s[9] review and confirmed this pattern of adverse impacts.

Given the negative impacts of MS caregiving across so many life domains, it is not surprising that MS caregivers have been found to be less satisfied with life as a whole compared to the general population.[21] Interestingly, several studies show that the negative impacts emerge soon after diagnosis, before the emergence of marked physical disability, and before the consequent physical demands of caregiving are apparent.[22,23] Hence, given that people are typically diagnosed with MS in their 20s and 30s, the caregiving journey is lengthy. Below we discuss the negative, as well as the positive, impacts of MS caregiving.

22.2.1 The Negative Impacts of Caregiving

22.2.1.1 Psychological and Emotional

Qualitative studies have documented the nature of the psychological and emotional impacts of caring for a person with MS. Caregivers are confronted with many psychological issues related to their caregiving roles, including grief, confrontations with their own mortality as the family member struggles with the degenerative illness, uncertainty about the future, disruptions to established patterns of living, challenges to their definition of self, and changes to their life goals. All of these can lead to psychological distress. In the early stages of dealing with the diagnosis, caregivers report shock, relief, helplessness, frustration, and confusion.[23] The ongoing emotional impact of caregiving in MS has been described as "chronic sorrow."[24] Commonly reported distressing emotions include sadness and helplessness.[12] In one study, emotional distress was the second most commonly reported caregiving problem,[25] and, in other research, psychological and emotional difficulties together were one of the most commonly reported caregiving problems.[14,26]

Quantitative studies show that, on average, MS caregivers report higher levels of psychological distress than the general population.[18] Other studies have found that between 24%[22] and 28%[25] of MS caregivers have clinically significant levels of global distress. Janssens et al.[22] found that 48% of people with MS and 46% of their partners reported clinically significant levels of anxiety, depression, or distress in the early postdiagnosis phase (0–24 months; mean 8 months). Dewis and Niskala[27] found that MS caregivers reported four times as many stress symptoms as the general population and one-third more than a heterogeneous group of caregivers. Benbow and Koopman's[28] survey of MS caregivers showed that psychological needs (e.g., for being listened to and for information) constituted the highest number of needs. It appears that caregivers report similar levels of distress to their care recipients,[22] particularly when the effects of gender are controlled for.[29]

Regarding depressive symptoms specifically, Pakenham[25] found that 9% and 17% of caregivers fell in the moderate–severe and mild ranges on the Beck Depression Inventory, respectively. Solari et al.[30] found that 19% of caregivers reported depressed mood, which was twice as high as healthy controls. Anxiety also appears to be elevated in many MS caregivers. Janssens et al.[22] found that 40% of spouse caregivers reported clinically significant levels of anxiety and, on average, they reported higher levels of anxiety than a comparison group. There is some evidence suggesting that spouse caregivers continue to report elevated anxiety and distress levels for 2–3 years after the diagnosis.[31]

In view of the relatively high-prevalence rates of clinically significant levels of distress among caregivers, it is important that rehabilitation practitioners screen for mental health problems in caregivers using some of the assessment instruments described later in this chapter. In addition, practitioners need to be aware of the psychological issues that caregivers may struggle with and provide, or refer them for, assistance where necessary. Significant psychological or emotional problems in the caregiver may interfere with the rehabilitation process in that the caregiver may have difficulties in providing care and in supporting the care recipient with rehabilitation interventions. In addition, given the data that suggest distress levels of caregivers and care recipients are correlated,[6] elevated distress in the caregiver indicates the likely presence of distress in the care recipient. Hence, a critical element of the rehabilitation process is attending to psychological and emotional distress in the patient's social system.

22.2.1.2 Relationships

Caregiving in MS also appears to affect the relationship between the caregiver and the care recipient. Several studies show that at times the caregiver–care recipient relationship is characterized by disagreement and detachment, and the caregiver's loyalty often turns into a sense of duty.[32] Spouse caregivers often report loneliness and a sense of loss of their partners.[32] The illness introduces many changes to the relationship, including less time spent with each other, and the loss of activities in which they both had engaged. The quotations below illustrate the impact of MS on the marital and family relationships.

I just feel more like a burden than a wife. I'm not a wife anymore (Woman with MS).*

It has been hard coping with the changes in behavior and ability to do things we once enjoyed. Mood swings are very hard to get used to, and also there has been a big drop in intimacy in our relationship (Spouse of a person with MS).†

The impact of my husband with MS has been significant, particularly on his physical limitations to enable him to interact with our children. Now they are older they are more understanding, but at his first diagnosis our children were confused and upset by his changing moods and physical abilities. We all try to be understanding and tolerant of his changing requirements [relapsing-remitting MS] and try to keep the family unit as harmonious as possible and continue to be as supportive of him as we can to avoid him being depressed by the huge impact MS has had on his life [and our marriage] (Spouse of a person with MS).†

Perrone et al.[33] compared MS caregivers with normative data and found that caregivers reported lower overall marital satisfaction, communication, and satisfaction with physical intimacy. A common complication of caregiver–care recipient dyads is the tendency for one or both partners to protect the other partner from stress and burden by concealing his or her distress.[23] This potential complication should be considered by rehabilitation providers during the intervention process. Individual interviews with both the caregiver and care recipient may be warranted in order to obtain a clearer understanding of the situation and the likelihood of communication issues. This information may provide important information in terms of whom to involve in different aspects of the rehabilitation process, how to customize educational interventions, and at what stage rehabilitation interventions need to be focused on the individual with MS rather than on the dyad.

22.2.1.3 Social, Recreational, and Leisure Activities

Several studies document the negative impacts of MS caregiving on holidays and social, recreational, and leisure activities. The caregiver's social circle diminishes because friends may not fully understand the illness or because the person with MS wants to conceal his or her symptoms from others, particularly while the symptoms are still invisible.[23] The following quotation from a spouse of a person with MS highlights the impact of MS on the patient's social system with respect to restricting social and leisure activities.

[We] cannot function to 100% as a family, i.e., planning leisure activities—limitations on where/what we can do—a "happy" family gathering can be quickly marred by onset of my husband's illness.†

Occupational and recreational therapists can play an important role in helping caregivers and family members find leisure alternatives that will allow them to spend quality leisure time alone or together, despite the disease and its demands.

22.2.1.4 Financial

Many studies report adverse impacts on finances.[25] Aronson[21] found that compared to the general population, MS caregivers were less satisfied with their finances. As well as the

---

* From Finlayson and Van Denend (2003, p. 1177).
† Coping with Parental MS Project, 2007–2009; Chief Investigator K. I. Pakenham; funded by MS Research Australia.

costs associated with medical treatments, rehabilitation, complementary therapies, physical aids, and home modifications, there is also the loss of the earnings of the person with MS.[34] For some caregivers there is the additional loss or reduction of employment and associated decrements in income. As the duration of caregiving increases so does the financial burden.[35,36] This is likely to be due to the progression of the disease and resulting disability and associated costs. Rehabilitation practitioners must be especially sensitive to these realities when making recommendations for home modifications and other assistive technologies. Even the smallest cost may be more than a family can bear. Social workers are often able to help families affected by MS negotiate the maze of rehabilitation-equipment funding options.

22.2.1.5 Career and Employment

Caregivers may have to turn down job opportunities, change from full-time to part-time employment, take early retirement, or not seek promotions because of the demands of caregiving.[17,37] In a sample of 232 MS caregivers, Pakenham[38] found that a third reported that caregiving prevented them from taking on outside employment. In another study, Pakenham[12] showed that greater caregiving demands were related to not taking on outside employment, and Hakim et al.[37] found that caregivers of people with MS who had more severe disability reported more adverse effects of caregiving on their careers. Rehabilitation practitioners can inquire about whether caregiving is hindering the caregiver's functioning in the domain of employment and career and, if it is, assist with finding ways to reduce the caregiving workload.

22.2.1.6 Physical Health

Both reviews[5,9] summarized evidence showing that many MS caregivers report health problems, with arthritis being the most commonly reported condition. The most common reason cited for being unable to continue caring for a person with MS is failing health of the caregiver.[35] Although there is no clear causal link between poorer caregiver health and greater caregiving demands, it is noteworthy that O'Brien[17] found an association between higher care-recipient dependency and lower spouse-caregiver health-promoting behavior. Before any education and training for physical caregiving is provided to a caregiver, the rehabilitation practitioner must obtain a basic health history to ensure that there are no contraindications or need to modify the performance of particular physical tasks (e.g., lifts, transfers) to minimize any risks to the caregiver's health.

22.2.1.7 Caregiver Burden

Caregiver burden refers to the strain that arises from the physical, psychological, emotional, social, and financial stressors associated with caring for an ill or disabled family member[39] and consequently subsumes the negative impacts mentioned above. However, the caregiver burden construct has been extensively criticized because there is no widely agreed-upon definition of caregiver burden; the concept is broad and vague, and the measurement of burden is problematic.[40,-42] Common themes in studies of caregiver burden in MS include the caregiver's reported lack of knowledge and skills necessary to provide adequate care[43,44] and the disruption of schedules.[45,46] Both MS caregiver reviews[5,9] found that higher caregiver burden was consistently related to poorer psychological health in MS caregivers.

Rehabilitation practitioners need to screen for the presence of a wide range of potential negative impacts of caring for a person with MS using some of the instruments described below. The identification of specific problems will then flag appropriate interventions. Many of these negative impacts are experienced by both the caregiver and the person with MS, and sometimes it may only be the caregiver that gives voice to these problems. One potential adverse consequence of the strain and stress of caregiving is the intentional or unintentional neglect and/or abuse of a highly dependent and vulnerable care recipient. This issue has not been investigated in people affected by MS in the peer-reviewed literature; however, it has been researched among the frail elderly. Cooper et al.[47] reviewed elder abuse and neglect screening instruments that may be completed by caregivers, care recipients, health professionals, or administrators. Some of these instruments could be modified for the purpose of screening for abuse and neglect of highly dependent persons with MS. Although such instruments will help to identify indicators of caregiver abuse or neglect of a person with MS, practitioners should be alert to signs that might suggest a need for investigation of possible abuse or neglect. Examples of signs that may need further investigation are evidence of malnutrition and poor hygiene, repeated missed appointments, and a caregiver who will not leave the room when the person with MS is being examined. Such information should be reported to other members of the team or the primary care provider for further investigation.

22.2.2 The Positive Impacts

There are both costs and rewards in caregiving, and unfortunately most of the research has focused on the negative outcomes of caregiving in MS. Evidence suggesting that MS caregivers may experience rewards associated with their caregiving comes from Aronson's[21] study that compared the QoL of MS caregivers ($N = 345$) to that of the general population. Although the MS caregivers reported lower satisfaction in some areas (the costs), they reported similar levels of satisfaction regarding health, job, family relations, and friendships despite the added burden of caregiving. Similarly, Perrone et al.[33] compared MS caregivers with normative data and found that they reported greater love for their spouses than the normative sample. Hakim et al.[37] found that 11% of spouses reported that their marital relationships had improved since the diagnosis of MS. Recently, researchers have investigated the nature of the potential positive outcomes or rewards of caring for a person with MS.

Pakenham[48] collected qualitative data on the benefits that MS caregivers perceived were associated with their caregiving role. Seven benefit-finding themes emerged from these data: greater insights into illness and hardship, caregiving gains, personal growth, the strengthening of relationships, health gains, increased appreciation of life, and a change in life priorities and personal goals. In a subsequent study, Pakenham and Cox[49] developed a multi-item benefit-finding scale (Benefit Finding in Multiple Sclerosis Caregiving [BFiMSCare]) and found that these themes were reflected by six factors. Both studies showed that caregiver benefit finding was related to positive caregiver adjustment outcomes (positive affect and dyadic adjustment) concurrently and over 12 months.

People with MS also report benefits from living with their illness,[50-52] and some of the benefits reported by both the person with MS and the caregiver are qualitatively similar. In addition, their benefit-finding reports regarding personal growth have been shown to be correlated.[48] However, care recipients have been shown to report higher levels of benefit finding than caregivers.[48] There is some evidence suggesting that caregiver benefit finding is linked to satisfaction with caregiver–care recipient relationships.[48]

Identifying the presence of positive outcomes (in addition to the negative impacts) associated with caregiving is important in determining the cost and reward balance. The following quotation from a parent caring for her daughter with MS illustrates the costs and benefits of caregiving.

> Apart from dealing with her physiological problems, which I find very demanding, I think coping with her character change is the hardest thing to cope with. I feel I have lost my daughter to this disease and feel I am now looking after a stranger. I love my daughter no matter what. Caring for my daughter has taught me more patience, compassion, and understanding. Caring for her has made me grow as a human being.*

The positive aspects of caregiving help to maintain caregivers over the long haul of caregiving.[53] For example, positive emotions may provide relief and serve to refresh, broaden, or restore the individual's psychological state.[53,54] Practitioners can help caregivers build and strengthen the benefits of caregiving, which, in turn, may ameliorate the negative impacts of caregiving. Highlighting the positive aspects associated with caregiving and the caregiver's resources is consistent with an empowerment framework whereby the strengths, assets, and resources of caregivers are identified to ensure that these are built upon, along with identifying potential deficits, needs, and barriers. A caution in refocusing caregiver attention to the benefits and rewards of caregiving is the possible minimizing by the practitioner of the distress and hardships associated with caregiving. It is important that the practitioner empathize with the caregiver's distress while at the same time reinforcing the caregiver's identification of the positives associated with his or her caregiving. Caregivers will inevitably report positive aspects of their caregiving experience. Practitioners need to listen carefully for such comments and explore these positive perceptions with the caregiver using open-ended probing. Of course, fundamental to any effective rehabilitation intervention is the relationship between practitioner and client, and Tedeschi and Calhoun[55] propose a "clinician as expert companion" (p. 215) approach with associated guidelines for fostering benefit finding.

22.2.3 Children Caring for a Parent with MS

Young people who have a parent with MS often assume some responsibility for the parent's care and have been referred to as young caregivers. Caring for a parent with MS is embedded in the privacy and reciprocity of family interrelations as illustrated by the following quotation from a 16-year-old daughter of a mother with MS.

> My mum means the world to me; of course I want to help her; I wouldn't want anyone else to help her. MS has changed her in a way, but she is still my mum. Wouldn't change her for the world.†

The caregiving roles of these young people and the potential adverse impact from taking on these adult roles are widely acknowledged.[56] Children of parents with MS often take on a range of caregiving responsibilities. Pakenham and Cox[57] identified four dimensions of youth caregiving in the context of parental MS: instrumental care (e.g., managing finances and arranging and attending appointments and meetings), social–emotional care

---

* Coping with MS Project, 1994–1996; Chief Investigator K. I. Pakenham; funded by a University of Queensland New Staff Grant.
† Coping with Parental MS Project, 2007–2009; Chief Investigator K. I. Pakenham; funded by MS Research Australia.

(e.g., comforting the parent and keeping the parent safe), personal–intimate care (e.g., toileting and bathing), and domestic–household care (e.g., cooking and shopping). These authors also found fairly high agreement between parent and child reports of child caregiving activities. Some of these caregiving responsibilities necessitate the child's assuming an adult caregiving role, and the provision of personal intimate care is particularly challenging for children as illustrated by the following quotation from an 18-year-old daughter of a mother with MS.

> Showering and toilet help makes me very uncomfortable.*

The following are examples of other difficulties reported by children caring for a parent with MS.

> I wish that I had more emotional support/help in dealing with day-to-day life with MS (18-year-old daughter).*

> Sometimes when she falls down or suddenly blanks out and I don't know what to do (15-year-old daughter).*

One study showed that compared to children of "healthy" parents, children of a parent with MS undertook significantly higher levels of family responsibilities.[58] The following quotation from the 10-year-old son of a parent with MS illustrates this finding:

> I am very different to other kids [child's perception]. I have bigger responsibilities.*

Exacerbating these caregiving challenges are the difficulties many children report in understanding MS. Cross and Rintell[59] interviewed 21 children aged 7–14 years of age who had a parent with MS and found that most children lacked adequate understanding and knowledge of MS. Furthermore, sometimes children are asked to conceal their parent's MS as illustrated below.

> My mother asks that we don't tell anyone she has MS because of the stigma involved.*

Not surprisingly, as summarized below, there is evidence to indicate that caring for a parent with MS can have a range of negative impacts on the child. Compared to children who have "healthy" parents, children of a parent with MS have been shown to have higher levels of somatization, anxiety, dysphoria, hostility, and interpersonal difficulties and lower life satisfaction and less positive affect.[58–62] Other studies that have not employed comparison groups have also found somatic complaints[63] and emotional distress[64,65] to be prominent in children of parents with MS. For a more detailed discussion on issues relevant to children caring for an ill or disabled family member, refer to Pakenham.[51]

The difficulties experienced by children who have a parent with MS are further supported by studies that show that the parents with MS report concerns about the effect of their illness on their children.[66,67] This concern is understandable given that parenting can be disrupted by many of the symptoms characteristic of MS, particularly mood disturbance, cognitive impairment, fatigue, and mobility problems. Findings from several studies provide support for the disruptive impact of MS on family functioning, including parenting tasks and roles.[68,69] These findings are underscored by the following quotations from parents with MS who were asked to discuss problems related to parenting.

---

* Coping with Parental MS Project, 2007–2009; Chief Investigator K. I. Pakenham; funded by MS Research Australia.

Being able to help teach fine motor skills (e.g., tying shoelaces, writing), participate in physical activities, participate in school activities, lack of patience, low tolerance when fatigued.*

Fatigue. Difficulty looking after children and house when having an attack/flare-up—especially when hands or mobility affected. Shielding children from my increased stress levels and depression can be tough.*

Given the difficulties in fulfilling parenting tasks and roles, often the parent without MS has to take on a greater parenting role as illustrated by the quotation below.

Spreading myself across roles as mother and father for the last decade. Having to be the disciplinarian, mentor, driver, bread winner, etc. All usual teen issues provide me with challenges and now feel quite burnt out.*

Practitioners do not routinely collect information regarding the parenting status of patients; hence, there is a need for a shift away from the focus on the individual to include the family. A "whole family approach" to meeting the needs of families affected by illness or disability has been recommended.[70] Flexible alternative care and supports for parents with MS across a number of areas depending on need (e.g., physical, medical, psychological, practical, and financial) are necessary to reduce the caregiving demands placed on children. In addition to assessing the parenting needs of parents affected by MS, rehabilitation practitioners need to identify the needs of children caring for a parent with MS using some of the assessment instruments described later in this chapter, and then harness appropriate assistance, which may include the provision of age-appropriate information about MS and support services.

22.3 Factors Associated with Impacts of Caregiving

There is marked variation in the outcomes for MS caregivers that has led researchers to consider factors that might predict caregiver adjustment. Such variations may be due to a range of factors that define the caregiving context, including patient illness and caregiver characteristics. In addition, well-established psychological theories, such as stress and coping theory, propose various constructs that explain the process of adaptation to caregiving. Having a basic understanding of the factors that may influence caregiver adjustment is important to inform the clinical reasoning of rehabilitation practitioners throughout all phases of the rehabilitation process.

22.3.1 Patient Illness

Psychiatric symptoms and cognitive impairment appear to be two important correlates of the adverse impacts of MS caregiving.[46,71,72] as illustrated by the following quotations from partners of a person with MS.

More information/warning, etc., should be given to families re "cognitive issues" associated with MS. I was not aware how bad this could be, always believed more "physical issues." So it has been [and it is] difficult to adjust to these changes.*

[I've had] difficulties dealing with constant mood swings.*

* Coping with Parental MS Project, 2007–2009; Chief Investigator K. I. Pakenham; funded by MS Research Australia.

Figved et al.[72] found that higher levels of psychiatric symptoms and cognitive impairment in people with MS were associated with poorer QoL and higher personal distress in caregivers over and above the effects of mobility problems. Depression is the most common psychiatric symptom in MS,[73] and Figved et al.[72] found depressive symptoms in the patient to be related to depression in the caregiver. Other psychiatric symptoms associated with caregiver distress included delusions, disinhibition, agitation/aggression, and irritability. In general, greater care-recipient functional disability and more severe disease progression have been associated with more adverse caregiving impacts for adults[5,25,71] and young caregivers.[58] Greater care-recipient disability and illness progression are associated with higher care activity, which, in turn, is related to poorer caregiver well-being.[12] There are a range of other patient illness-related factors that may adversely influence adult caregiver well-being, including the patient's lack of awareness of his or her functional deficits[18] and low patient life satisfaction.[74] An additional care-recipient characteristic associated with caregiver burden is the advancing age of the person with MS.[5,9]

22.3.2 Caregiver Characteristics

A variety of caregiver-related factors have also been found to be associated with negative caregiver-adjustment outcomes, including being female,[25,75] being the spouse of the person with MS,[21,72] longer caregiving duration,[21] and high levels of activities of daily living (ADL) and psychological and emotional care activities (e.g., managing the care recipient's mood swings, emotional distress, personality changes, and cognitive problems).[12] Regarding the latter, psychosocial and emotional care tasks seem to be related to higher caregiver depression over time, whereas ADL care activities are related to lower positive outcomes.[12]

22.3.3 Caregiver Stress and Coping Factors

Lazarus and Folkman's[76] stress and coping theory has been widely used to guide research into how caregivers adjust to the demands of their caregiving role. This theory proposes that stress emerges when an event or situation is appraised by the person as exceeding his or her resources and as threatening his or her well-being. Although various stress and coping models of caregiving have been derived from Lazarus and Folkman's[76] framework,[25] most are based on the premise that beyond the effects of care-recipient and caregiver characteristics, adjustment to caregiving is determined by three processes, which are cognitive, behavioral, and interpersonal: cognitive appraisal, coping resources, and coping strategies. The stress and coping framework has applied utility because it provides practitioners with a theoretically coherent and empirically supported way of understanding and identifying key factors that are likely to shape a caregiver's adjustment to his or her caregiving role. The framework points to factors that influence the coping process and that need to be assessed and modified. A similar model has been applied to people with MS[77] and is described in Chapter 20. Hence, another benefit of this framework is that it can be used to understand how both the caregiver and care recipient cope with the demands of MS.

The framework is briefly described here; however, for a more detailed discussion see Chapter 20. The *appraisal* component of the framework refers to an evaluative process involving the person's perception and interpretation of an event. The appraisal of an event as stressful will generate stress, which may exceed the coping skills and resources available to the person. The second component, *coping resources*, includes the relatively stable external (e.g., social support) and internal (e.g., optimism) resources that a person can draw on when dealing with the event. Lazarus and Folkman[76] refer to the third component, *coping strategies,* as "constantly changing cognitive and behavioral efforts to manage specific external and/or internal demands that are appraised as taxing or exceeding the resources of the person" (p. 141). Coping strategies have been categorized as problem-focused strategies (that deal with the problem that is causing the distress, such as problem solving), emotion-focused strategies (that regulate emotion either by avoidance or emotional approach),[76] meaning-focused strategies (that involve creating, reinstating, or reinforcing meaning, such as positive reframing),[78] and relationship-focused strategies (such as supportive engagement or criticism).[79]

In general, findings from MS caregiver studies suggest that better caregiver well-being is associated with lower stress appraisals, higher social support, less reliance on avoidant coping and relationship-focused coercion and criticism, and greater reliance on meaning-focused coping and relationship-focused supportive engagement.[25,26,45,74,75] Pakenham and Bursnall[61] found a similar pattern in a study of young MS caregivers. Notably, stress and coping predictors have been shown to be antecedents to both negative and positive caregiving impacts.[74] With respect to coping strategies, evidence from one study suggests that whether caregivers and their care recipients rely on similar coping strategies may influence the adjustment of both partners.[6] There also appear to be gender differences in the types of coping strategies caregivers rely on and the resources they access.[5] Rehabilitation practitioners can screen for deficits in the coping processes specified by the stress and coping framework by using some of the measures described below. Deficits in a particular area can then be targeted by appropriate intervention. For example, a caregiver who has a limited and unsatisfactory social support network may benefit from assistance with developing ways to expand social supports or become more effective in mobilizing support.

Sense making is another potentially important predictor of caregiver adjustment. Sense making refers to the development of explanations for adversity or the making sense of it within existing "assumptive schemas" or worldviews.[80,81] The many stressful demands and losses associated with the caregiving role can threaten a caregiver's self-definition, fundamental life goals, and sense of meaningfulness, resulting in psychological distress. Restoration of meaning, life purpose, and one's self-worth are important elements when adjusting to adversity.[82] One frequently cited meaning reconstruction process is sense making, which involves making sense of adversity through developing new worldviews or via modifying existing worldviews such that self-worth is protected and randomness and uncontrollability are minimized.[81] It is not clear where sense making fits within the stress and coping framework, but it has been proposed that it helps to reduce any mismatch between appraised meaning of a stressful event and global meaning structures.[78]

Pakenham[38] collected qualitative data from MS caregivers on how they made sense of their caregiving situations. Half the caregivers generated sense-making explanations for their caregiving situations, yielding 11 sense-making themes: catalyst for relationship changes, relationship ties, acceptance, duty, experienced growth, gift of giving, spiritual/religious explanations, insight into suffering, everything has a purpose, "wake up call" for a life-style change, new life goals, and negative explanations. Over a third of those caregivers who could not make sense of their situation were able to anticipate comprehending it, and the strength of this anticipation was related to greater life satisfaction. Caregiver and care-recipient sense making were positively correlated, and the sense making of one partner was positively related to life satisfaction of the other partner. These findings suggest that both caregiver and care recipient become engaged in a process of shared sense making. Several studies show that sense making predicts caregiver adjustment concurrently[38] and over 12 months.[83] It is interesting that those sense-making factors that were characterized by perceptions of self-worth, controllability, and predictability (i.e., the acceptance, relationship ties, and spiritual perspectives factors) were related to better adjustment. In contrast, sense-making factors that entailed perceptions that magnified the negatives associated with the caregiving situation (the "incomprehensible" factor) or forced personal challenges (the catalyst for change factor) were related to poorer adjustment.[83] In view of the role of sense making in the process of adjusting to caregiving and illness, it is important for rehabilitation practitioners to listen carefully for their clients' sense-making explanations and perhaps, where appropriate, engage in dialogue about the meaning that caregiving or MS have in the context of the individual's living. For example, a caregiver may mention "in passing" how that, since taking on caregiving for the person with MS, they have re-prioritized what is important in life. The practitioner may explore this experience of re-prioritizing using open-ended probes. People develop their own personalized sense-making explanations for their adversity, and it is important for practitioners not to impose their own explanations or be judgmental of those that a carer derives. Refer to Pakenham[84] for a more detailed review of interventions for promoting sense making in the context of chronic illness.

22.4 Assessment

Given the evidence that both partners of caregiver–care-recipient dyads help to shape each other's adaptation to the illness predicament, it is important that both caregiver and care recipient are routinely assessed in rehabilitation settings.[13] This section on assessment instruments describes those that have been developed specifically for MS caregivers (see Table 22.2). These tools can and should be used in conjunction with a high-quality intake interview, observation of task performance, and physical assessment, when appropriate. All of the instruments mentioned below have been used effectively in both research and clinical settings. Most of the instruments are relatively brief self-report questionnaires that can be easily administered. For example, questionnaires may be completed by caregivers at home at a time most convenient to them without imposing undue demands on caregiver or clinician time constraints.

22.4.1 MS Caregiving Tasks

Researchers have tended to rely on purpose-built measures of MS-related caregiving tasks. For example, Carton et al.[4] used a prospective diary and Aronson et al.[10] used a questionnaire that included modified items from scales developed by others. One measure of MS caregiving tasks that has undergone considerable scale development is the Caregiving Tasks in MS Scale (CTiMSS). This questionnaire was developed by Pakenham[83] from qualitative data obtained from MS caregivers.[38] The scale consists of 24 caregiving tasks. Respondents indicate how much help they give their care recipient for each of the tasks using a 5-point scale (0 = No help to 4 = Lots of help). Factor analyses revealed four caregiving factors: instrumental care, activities of daily living care, psychological–emotional care, and social–practical care. However, the CTiMSS items also loaded onto one factor; hence, factor scores or a total score may be used. The CTiMSS factors were shown to be psychometrically sound (for further psychometric details, see Pakenham[83]). The scale may be obtained from the author of the scale.

TABLE 22.2

Overview of Potential Assessment Tools Related to MS Caregiving

Areas for Assessment	Potential Tools	Recommended Timing	Useful for ...	How Information Can be Used
Understanding what caregivers are actually doing	Caregiving diary Care Tasks in MS Scale	Initial assessment Repeat when MS status changes	Nurse Occupational therapist Physical therapist Psychologist Social worker	OT/PT/Nurse: Target education & training SW/Psych: Help to identify and arrange potential services & additional supports
General caregiver outcomes	Brief Symptom Inventory–18 Depression, Anxiety and Stress Scale Zarit Caregiving Burden Interview Caregiver Reaction Assessment Satisfaction with Life Scale Abbreviated Spanier Dyadic Adjustment Scale Positive Affect scale	Pre- and post-intervention	Nurse Occupational therapist Psychologist Social worker	Identify mental health needs that require intervention Outcome measure for caregiver interventions focused on reducing stress, improving coping abilities, and supporting overall caregiver adjustment and mental health
MS-specific caregiver outcomes and coping processes	Benefit Finding in MS Caregiving Scale Coping with MS Caregiver Inventory Carer Sense Making Scale	Pre- and post-intervention	Nurse Occupational therapist Psychologist Social worker	Identify coping difficulties that require intervention Outcome measure for caregiver interventions focused on reducing stress, improving coping abilities, and supporting overall caregiver adjustment and mental health
Involvement of children in caregiving	Youth Activities of Caregiving Scale Young Caregiver of Parents Inventory	Initial assessment Repeat when MS status changes Pre- and post-intervention	Developmental therapist Nurse Occupational therapist Psychologist School counselor Social worker	Guide goal setting and intervention development by providing information about what children are doing and how they are coping Outcome measure for child caregiver interventions

Note: OT = occupational therapist, PT = physiotherapist, SW = social worker, Psych = psychologist.

Rehabilitation practitioners could use the CTiMSS to develop a care-task profile of individual caregivers that would show areas of elevated caregiving demands. Such information could be used as a basis for planning individualized education and support or service resources for both caregiver and care recipient. Discussing the CTiMSS care-task profile with a caregiver would provide an important first step in developing an individual care plan. Readministering the CTiMSS at appropriate points could reveal changes in caregiving activities overtime, which could be used to adjust care plans.

22.4.2 Caregiving Outcomes

There are a range of published generic measures that may be used to assess the negative impacts of MS caregiving, including the Brief Symptom Inventory-18 (BSI-18) for assessing depression, anxiety, and somatization[85] and the Depression, Anxiety, and Stress Scale (DASS),[86] which assesses stress in addition to depression and anxiety. Buhse[87] highlights the importance of assessing burden in MS caregivers and recommends the 21-item Burden Interview,[88] which takes about 10 min to complete. It measures caregivers' perceptions of the impact of their caregiving role on their emotional, physical, social, and financial status. She recommends early recognition of burden so appropriate interventions can be put in place. Another measure of burden that has been used with MS caregivers[26] is the Caregiver Reaction Assessment.[89] This 24-item scale assesses caregivers' reactions to caring for a physically ill person in five domains, including how caring affects caregivers' health, daily schedule, finance, their sense of self-worth (self-esteem), and their families. Development of the scale was based on a derivation sample of Alzheimer's and cancer caregivers. The instrument has been shown to have satisfactory reliability and validity across various groups of caregivers.[89] Participants rate each item on a 5-point scale from 1 = Strongly disagree to 5 = Strongly agree. After reverse scoring some items, all items are summed to form a total caregiving impact score. There are several reviews of instruments for assessing caregiver burden that rehabilitation providers may find useful as they explore options that could fit within the constraints of their own practice environments.[39,41]

There are also many published generic measures that may be used to assess the positive outcomes of MS caregiving, including Satisfaction with Life Scale,[90] Positive States of Mind Scale,[91] Abbreviated Spanier Dyadic Adjustment Scale,[92] and a Positive Affect Scale (see Pakenham and Cox[49] for a modified version of the Bradburn Affect Balance Scale[93]).

Pakenham and Cox[49] developed a measure of benefit finding specifically related to caring for a person with MS called the Benefit Finding in Multiple Sclerosis Caregiving (BFiMSCare) Scale. The BFiMSCare was developed from qualitative benefit-finding data collected from MS caregivers.[48] The scale consists of 27 potential benefits of caregiving. Respondents indicate the extent to which they have experienced each of these on a 3-point scale (1 = Not at all to 3 = A great deal). Factor analysis of the BFiMSCare scale revealed six psychometrically sound factors: enriched relationship, spiritual growth, family-relations growth, life-style gains, inspiration, and relationship opportunities. The BFiMSCare items also loaded onto one factor; hence, factor scores or a total score may be used. The BFiMSCare factors were shown to be psychometrically sound (for further psychometric details, see Pakenham and Cox[49]). Regarding clinical applications, the BFiMSCare may be used to monitor changes in benefit finding in response to rehabilitation interventions. The scale may be obtained from the authors of the scale.

Rehabilitation practitioners may use standardized measures of psychological distress (e.g., BSI-18 or DASS) to screen for mental health problems. Scores can be compared to norms to determine whether they fall within a "normal" range or above clinical cut-offs. Caregivers who score above clinical cut-offs can be referred for further specialized mental health assessment, monitoring, and intervention. Caregiver-burden measures are useful for determining areas of caregiving strain, which may then be used to determine appropriate supportive interventions. Often the positive aspects of caregiving are neglected in the rehabilitation process. Routinely assessing both the positive and negative impacts of caregiving helps to identify whether there is an imbalance of high costs and low rewards. Identifying low levels of positive outcomes provides a starting point for rehabilitation practitioners to develop a plan with the caregiver for enhancing potential benefits of caregiving, infusing daily activity with positive meaning, and engaging in activities that are self-restorative. The caregiving outcome measures can also be used for monitoring the impact of interventions.

22.4.3 Stress and Coping Factors

There are many generic published instruments that assess stress and coping constructs. Pakenham[25,94] has developed self-report scales for assessing stress, challenge, threat, and control appraisals related to caregiving. There are numerous generic scales for assessing coping resources such as social support and optimism. Regarding coping strategies, Pakenham[26] developed the Coping with MS Caregiving Inventory (CMSCI). The CMSCI is a self-report measure of 34 coping strategies specific to MS caregiving. The CMSCI was developed from qualitative data obtained from MS caregivers. The scale incorporates an open-ended question that asks respondents to describe their main caregiving problem experienced in the last month. In order to obtain a global stress appraisal of the main problem, respondents are then asked to rate how stressful this problem has been in the past month on a 7-point scale (1 = Not at all stressful to 7 = Extremely stressful). Respondents then indicate on a 4-point scale (0 = Does not apply/Never to 4 = Very often) how often they have used each of the 34 coping strategies in dealing with their main caregiving problem in the past month. Factor analyses revealed five factors: supportive engagement, criticism and coercion, practical assistance, avoidance, and positive reframing. The CMSCI has been demonstrated to be psychometrically sound (for further psychometric details, see Pakenham[26,74]). Several studies have shown that one or more of the CMSCI factors have predicted caregiver adjustment concurrently and over a three-month interval.[26,74] The CMSCI has also been shown to be sensitive to change within an intervention trial.[95] The CMSCI was shown to be a stronger predictor of caregiver adjustment than the Ways of Coping Checklist–Revised.[26] The scale may be obtained from the author of the scale.

Pakenham[83] developed a questionnaire that measures sense making related to caregiving called the Carer Sense Making Scale (CSMS). The scale was developed from qualitative data collected from MS caregivers.[38] The scale consists of 57 item statements that caregivers of people with MS have used to make sense of their caregiving situations. Respondents are asked to indicate the extent to which each statement reflects how they have made sense of their care recipients' having MS and their caring for them by rating the degree to which they agree with each statement on a 5-point rating scale (1 = Strongly disagree to 5 = Strongly agree). Factor analyses revealed six factors: catalyst for change, acceptance, spiritual perspective, incomprehensible, relationship ties, and causal attribution. The CSMS has been shown to be psychometrically sound.[83] The CSMS factors have been shown to predict change in both positive (life satisfaction and positive affect) and negative (anxiety and depression) caregiver outcomes over a 12-month interval after controlling for the effects of caregiver demographics and care-recipient illness variables. The scale may be obtained from the authors of the scale.

Assessing coping processes may be useful in identifying those that need to be targeted by interventions. For example, an assessment of a caregiver's coping strategies that indicates a heavy reliance on avoidant coping strategies may suggest the use of intervention strategies to enhance alternative approach strategies for some identified stressors. The assessment of sense making may alert the rehabilitation practitioner to possible difficulties the caregiver may have in comprehending his or her caregiving predicament and the care-recipient's illness.

22.4.4 Young Caregiving

While there are several published measures of caregiving tasks undertaken by children who care for a family member with an illness or disability, these instruments suffer from many limitations (see review by Ireland and Pakenham[96]). Addressing the weaknesses of these earlier measures, Ireland and Pakenham[96] developed a questionnaire called the Youth Activities of Caregiving Scale (YACS) to assess the care tasks performed by young people in the context of family illness/disability. The derivation sample was composed of children who had a family member with an illness or disability. The scale consists of 28 care tasks. Respondents are asked to indicate how much help they provide their ill or disabled family member with each care task on a 5-point scale (0 = No help at all to 4 = Lots of help). Factor analyses yielded four factors: instrumental care, social/emotional care, personal/intimate care, and domestic/household care. The items also loaded onto one factor; hence, factor scores or a total score may be used. The YACS has been applied to children of a parent with MS.[57] The factor structure was replicated with this sample, although some items loaded onto different factors. The YACS has been shown to be psychometrically sound (for details, see Ireland and Pakenham[96] and Pakenham and Cox[57]).

Pakenham et al.[61] developed a questionnaire called the Young Caregiver of Parents Inventory (YCOPI) to measure young caregivers' diverse positive and negative caregiving experiences (as distinct from actual care tasks). The measure was developed from qualitative data obtained from young caregivers. The YCOPI was designed for use with youth who perform care tasks for a parent; however, the word "parent" can be changed to "family member" if the ill or disabled family member is not restricted to parents. Young caregiving experiences are assessed across eight subscales: caregiving responsibilities, perceived maturity, worry about parents, activity restrictions, isolation, caregiving compulsion, caregiving discomfort, and caregiving confidence. The 48 YCOPI items consist of statements to which participants rate the strength of their agreement on a 5-point scale (0 = Strongly disagree to 4 = Strongly agree). For psychometric details on the YCOPI, see Pakenham et al.[61] The YCOPI has shown sensitivity to change in an intervention study for children caring for a parent with MS.[97] Pakenham et al.[61] also developed other measures relevant to young caregiving, including self-report scales that assess illness unpredictability of the care recipient, choice in caregiving, and functional impairment of the care recipient. These measures have been validated on a sample of children caring for a parent with MS.[58]

Rehabilitation practitioners working with families in which a child or children are involved in caring for a family member with MS can use any of the abovementioned young caregiver measures to collect further specific information to assist in identifying potential difficulties a child may be having in providing care and support or in adjusting to their parent's disability. Children may be more willing to express their difficulties via a questionnaire than in face-to-face interactions with an adult practitioner.

22.5 Interventions

As this chapter has demonstrated thus far, MS caregiving can be a complex, long-term process with many potential implications (positive and negative) for the MS caregiver and the care recipient. Typically, MS rehabilitation targets the person with the disease and focuses on reducing impairments and enhancing activity and participation. While caregivers can contribute to the rehabilitation process, they are not often considered the primary recipient of these efforts. Given the critical role that caregivers play in the everyday lives of people with MS, it is necessary to examine opportunities to offer specific interventions that target the needs, issues, and concerns of caregivers and focus on maintaining their health, well-being, and overall ability to meet the demands associated with caregiving.

22.5.1 Interventions Targeting Adults Who Provide Care

Interventions to support the adult caregivers of people with MS can be divided into three broad categories: (a) interventions that give caregivers a break by providing practical and instrumental support and assistance, (b) interventions that offer emotional support to caregivers through various forms of counseling, and (c) interventions that focus specifically on developing knowledge, skills, and strategies to enable caregivers to effectively manage their caregiving responsibilities.

22.5.1.1 Interventions That Give Caregivers a Break

Like the caregivers of older adults or other individuals with chronic or disabling conditions, the caregivers of people with MS juggle many responsibilities and often struggle to complete all of their tasks in the time that they have available. To support caregivers as they attempt to manage these multiple demands, many organizations and communities offer programs and services that allow caregivers to take a break from their responsibilities (e.g., respite care, day programs, in-home health care aides). The literature suggests that MS caregivers avail themselves of these programs and services at very low rates.[10,98] In addition, the extent to which utilization effectively reduces stress and burden or improves MS caregiver health and QoL has not been established in the peer-reviewed literature.

Looking more broadly, a recent systematic review of the effectiveness of these services that provided a break in care to caregivers of frail or disabled older adults living in the community concluded that there may be some positive effects, but the evidence is limited and weak in quality.[99] This same review also examined the barriers to using these programs and services. Researchers found that the actual use of respite care was influenced by caregiver attitudes and sense of obligation toward the care recipient, their knowledge of and the availability of the services, the acceptability and impact of the services on the care recipient, the hassles associated with using the services, and the extent to which the services matched needs (e.g., appropriateness, flexibility). These issues are consistent with those studies that have addressed the low rates of utilization of supportive services among MS caregivers.[10,98]

Ultimately, the available evidence does not provide clear direction about the appropriateness or effectiveness of recommending interventions that are intended to support caregivers by giving them a break from their responsibilities. What is known is that utilization is low, and there are many complex psychological, social, and logistical barriers to use. Therefore, rehabilitation practitioners must use their clinical judgment and knowledge of the caregiver and care recipient to guide decision making about recommendations or referrals on a case-by-case basis. What is very important during this process is to realize that finding and using resources is a complex process, both logistically and emotionally. Providers must resist labeling caregivers as "resistant" if they do not seem open to using external assistance, but rather they should take the time to explore potential barriers to use and ways to overcome them. The materials from Chapter 20 on coping with MS may be useful in this regard.

22.5.1.2 Interventions That offer Emotional Support to Caregivers through Various Forms of Counseling

This group of interventions includes individual or group-based cognitive-behavioral or psychodynamic therapies, usually focused on enabling caregivers to manage distress, depression, and anxiety associated with their responsibilities.[100] These interventions are most often delivered by psychologists and social workers, although other rehabilitation professionals may be involved, depending on the setting and their specific training. While these types of intervention are recognized as having potential to reduce the negative impacts of caregiving and help caregivers reframe their experiences, no MS-specific caregiver intervention studies using these approaches are published in the current peer-reviewed literature. A recent review of caregiver interventions for frail older adults found three cognitive-behavioral intervention studies and findings that were promising.[100] Research is needed to determine if these types of interventions could also produce positive results for MS caregivers.

22.5.1.3 Interventions That Focus Specifically on Developing Caregiver Knowledge, Skills, and Strategies

Despite calls for MS caregiver interventions,[101] there is only one small group-based, face-to-face intervention for MS caregivers published in the peer-reviewed literature. This program, which is titled "Meeting the Challenges of MS," was developed using the findings of interviews from 302 MS caregivers together with an extensive review of existing interventions targeting caregivers of older adults and people with other chronic diseases.[11] "Meeting the Challenges of MS" focuses on building the knowledge, skills, and self-efficacy of MS caregivers in order to support them in their caregiving roles.

Using a self-management framework[102] and the person–environment–occupation model[103] to select and sequence content, "Meeting the Challenges of MS" involves five sessions (see Table 22.3). Self-management is a specific form of health education that focuses on developing skills for problem solving, decision making, goal setting, finding and using resources, developing partnerships with health care providers, and self-tailoring of information to address individual-specific problems.[102] These skills are critical for managing health-related challenges that are long term, variable, and unpredictable in outcome.[104] The person–environment–occupation model recognizes that the ability of a person to perform a given task is dependent on the dynamic interactions among the person's knowledge and abilities, the demands of the environment in which the task is being performed (e.g., social, physical, cultural), and the specific requirements of the task itself (e.g., cognitive, physical, emotional).[103] By applying the model to the work of caregiving, it is possible to identify potential mismatches across the caregiver, the environment, and the demands of the caregiving tasks that may compromise the caregiving effort and contribute to caregiver distress.

A before and after pilot study of the "Meeting the Challenges of MS" program included 19 MS caregivers. The Coping with MS Caregiving Inventory was the primary outcome tool used for the study.[26] After completing the program, participants felt more prepared for their caregiving roles ($p = 0.02$), were able to reframe their experiences positively ($p = 0.01$), and sought more practical assistance ($p = 0.02$).[95] Feedback from the participants emphasized the value of learning a problem-solving process and how to apply it to different caregiving challenges. They reported having greater knowledge of MS and greater knowledge of community resources and how to access them. They also reported that the program helped them appreciate the value of delegating, asking for help, and planning for future

TABLE 22.3

Highlights of the Sessions of "Meeting the Challenges of MS"

Session Number and Focus	Key Objectives
#1: Understanding MS, caregiving, and the importance of communication and problem solving to manage challenges	• Discuss common caregiving challenges • Describe basic features of MS • Introduce strategies to promote good communication • Introduce a problem-solving method
#2: Skills to manage the unpredictability of MS (i.e., managing emotional challenges)	• Recognize feelings and responses to MS unpredictability • Recognize that some caregiving challenges may require external assistance and support • Introduce constructive methods of managing feelings and responses to MS unpredictability
#3: Skills to manage changes in disability over time (i.e., managing physical challenges)	• Identify tasks that can be completed safely and those for which assistance is needed • Understand role of proper body mechanics during provision of physical assistance • Identify equipment options for now or in the future
#4: Getting and managing the help you need now and in the future	• Describe steps for planning for the future • Articulate strategies to involve the care recipient and other family members in planning for the use of formal services • Apply problem-solving method to address challenges with finding and using services
#5: Setting goals to apply program contents in everyday life	• Reinforce learning from program • Support efficacy for caregiving • Set at least one short-term and one long-term goal related to meeting the challenges of MS

Source: Adapted from Finlayson M et al. British Journal of Occupational Therapy, 2009; 72(1): 11–19.

needs. However, despite these benefits, participants reported the need to have more flexibility in the timing and delivery of the program, which suggests that other formats, such as teleconference and online, should be examined.[95] Given the promising findings of this pilot study, further research on this program using more rigorous designs is warranted. Efforts are underway to extend this work further.

22.5.2 Interventions Targeting Children Who Provide Care

In a study of parents with a disability who had children aged 11–17 years, Olkin et al.[105] concluded that in comparison with other disability groups, families with MS had a particular need for psychosocial support. The need for more proactive family-centered approaches has been echoed by others.[70] Although they did not carry out a "true" family intervention, Coles et al.[97] conducted the only published evaluation of the efficacy of an intensive intervention for children of a parent with MS in which both parents and children were included in the evaluation phases and received intervention resources. Twenty children (aged 9–14 years) of a parent with MS attended a six-day camp intervention. The intervention, called the Fun in the Sun Camp, involved both recreational activities and eight group sessions providing education about MS, opportunities to share experiences within a supportive environment, and training in various coping strategies and life skills. Each day, one to two structured group sessions were held in the morning, and recreational activities were offered in the afternoon. Group sessions lasted one to two hours and were mostly interactive. A primarily cognitive-behavioral, psychoeducational approach was used. All group sessions were coordinated by two facilitators. Materials included the "Fun in the Sun Facilitator Manual," the "Fun in the Sun Participant Workbook," and "Fun in the Sun Parent Manual" (manuals are available on request from the authors). The parent manual provided an outline of the children's program and tips on how parents could reinforce at home the strategies children learned at camp.

Children completed questionnaires (many of which are discussed above) at pre- and post-intervention and at a three-month follow-up, and parents completed questionnaires at pre-intervention and follow-up. Both quantitative and qualitative data were used to examine intervention efficacy. Results showed that after the intervention, children reported statistically significant decreases in distress, stress appraisals, caregiving compulsion, and activity restrictions and increases in social support and knowledge of MS. Parental data confirmed the significant increase in the children's knowledge of MS, and, overall, the qualitative data supported the quantitative findings. Qualitative data showed that the intervention had indirect effects on some parents and families.

22.5.3 Intervention Summary and Observations

Throughout this chapter, references have been made to ways in which rehabilitation practitioners may intervene to assist caregivers in informal ways. Examples of informal ways to assist caregivers mentioned above are listening for caregivers' benefit-finding and sense-making comments and reinforcing them by engaging in discussion on these issues. Sections 22.5 and the following discuss specific formal intervention resources and programs.

Intervention is always based on an assessed need or problem. Sometimes the caregiver will self-assess as having a problem or need and access or request a resource or intervention independently, and this should be encouraged. Other times a practitioner will identify a caregiver problem or need via direct contact with the caregiver or indirectly via the care recipient or another practitioner. The practitioner should then adequately assess the nature of the need or problem to determine an appropriate intervention that matches the need.

Once the most appropriate intervention has been identified, it should be decided who will facilitate the intervention. The "who" will depend on many factors, including whether the practitioner is in a solo practice or part of a rehabilitation team and what the mix, training, and competencies of the individual practitioners on that team are. It is not possible to be prescriptive about who can or cannot facilitate particular interventions. Such decisions depend on available resources and personal training and expertise; professional standards; agency protocols and policies; staff position descriptions; and individual ethical considerations, such as only providing interventions that one has been trained to competently administer.

Each of the various practitioners involved in rehabilitation in MS is likely to have an opportunity to intervene and assist caregivers. For example, occupational therapists may assist a caregiver in developing skills in activity analysis and modification to reduce the physical demands of providing assistance with everyday functional activities. Together, occupational therapists and physical therapists may be involved in training caregivers to use mobility equipment such as wheelchairs, stair climbers, or mechanical lifts that can aid in transfers. A psychologist or psychiatrist may assist with respect to a caregiver's mental health issues, and a social worker may link a caregiver with needed social welfare resources. Nurses and pharmacists can provide education and training on medication management, giving injections, and watching for medication side effects.

Another consideration is the format of the intervention. Many of the more formal interventions mentioned above are in group format, and it is not clear how these may be modified and delivered in other ways, such as in individual, couple, or family consultations or via the telephone or online, and whether they would be effective if delivered in these formats. In the absence of such data, practitioners must rely on their professional training, professional and ethical standards, and consultations with peers regarding their use of the informal and formal caregiver intervention strategies mentioned above.

22.6 Conclusions and Future Research and Practice Directions

Caregivers are clearly a valuable member of the MS rehabilitation team and may, at times, be the target of intervention. Caregivers engage in a wide range of important caregiving activities that change over the unpredictable course of a complex illness. Caregivers and their care recipients both go through parallel and related processes of adjusting to the changing realities of MS. For the caregiver, costs and rewards ebb and flow over the course of the caregiving journey. Caregiving occurs in a context and can only be fully understood when that context is considered, including characteristics of the person with MS and his or her illness, the caregiver's biographics, the caregiver's perceptions of his or her predicament, meaning making, coping strategies, and coping resources. Practitioners need workable frameworks, like those presented in this chapter, that inform understanding of how the caregiver copes with the caregiving role, and that can guide a systematic and responsive approach to assessment and intervention.

In drawing on available research findings to provide direction for rehabilitation practice, we acknowledge that the research into MS caregiving is relatively small compared to caregiving research in other domains (e.g., dementia and frail elderly) and is particularly limited in some areas (e.g., MS caregiver interventions). There are also methodological limitations that pervade many of the studies reviewed, including (1) convenience sampling with caregivers often recruited from MS societies, (2) a lack of control or comparison groups, (3) few truly longitudinal investigations to examine changes over time, (4) few studies that obtain data from both caregiver and care recipient, and (5) an over-representation of research from North America, the United Kingdom, and Australia. For more in-depth critical analysis of the research into MS caregiving, refer to the recent reviews that have already been referenced in this chapter.[5,9]

In addition to the methodological limitations of existing MS caregiving research, there are also many issues on the MS caregiving research agenda that have not been investigated to date. Below is a list of research questions that need to be investigated. Pursuing this work would go far to strengthen and support rehabilitation outcomes for people with MS.

- To what extent do the experiences of caregivers of care recipients receiving care in the community differ from those of caregivers of a person with MS who is in acute care, receiving inpatient rehabilitation, or is attending outpatient rehabilitation?

- What specific strategies are rehabilitation practitioners using to engage caregivers in the rehabilitation process, how regularly does this engagement occur, and how effective are these engagement strategies in improving rehabilitation outcomes and supporting the caregiver in continuing in his or her role?

- How effective are respite care, counseling, and coping skills interventions for MS caregivers?

- What factors facilitate or impede decision making concerning the transition to residential care?

- What are the effects on the caregiver of the care recipient's transfer to residential care?

- How are the risk of and occurrence of caregiver abuse or neglect of care recipients best assessed, what is the prevalence of this phenomena among MS caregivers, and how is it most effectively managed?

- Are there definable stages in the MS caregiving trajectory when caregivers are likely to experience elevated strain and require intensified support?

With respect to future practice directions, community services and individual practitioners need to develop protocols that more clearly and prominently articulate a focus on caregivers. For example, it should be clear to care recipients and their caregivers when they access a service how caregivers are routinely consulted and actively included in the rehabilitation process. A "whole family" approach that includes consideration of all family members is necessary, particularly when the person with MS is a parent. Community organizations need to develop an array of services for children, adolescents, and young adults affected by parental MS. Such services may include counseling, online information on MS, and "chat forums." Given the inextricable, potent, reciprocal influences between the person with MS and the caregiver, the recognition, inclusion, and support of caregivers should be integral to the rehabilitation process.

References

1. Stuifbergen AK, Blozis SA, Harrison TC, Becker HA. Exercise, functional limitations and quality of life: A longitudinal study of persons with multiple sclerosis. Archives of Physical Medicine & Rehabilitation 2006;87(7):935–943.
2. Burks JS, Johnson KP. Multiple sclerosis: Diagnosis, medical management, and rehabilitation. New York, NY: Demos Medical Publishing; 2000.
3. O'Hara L, De Souza L, Ide L. The nature of care giving in a community sample of people with multiple sclerosis. Disability and Rehabilitation 2004;26(4):1401–1410.
4. Carton H, Loos R, Pacolet J, Versieck K, Vlietinck R. A quantitative study of unpaid caregiving in multiple sclerosis. Multiple Sclerosis 2000;6:274–279.
5. Corry M, While A. The needs of carers of people with multiple sclerosis: A literature review. Scandinavian Journal of Caring Sciences 2009;23(3):569–588.
6. Pakenham KI. Couple coping and adjustment to MS in care receiver–carer dyads. Family Relations 1998;47:269–277.
7. Edwards D, Baum C. Occupational performance: Measuring the perspectives of others. In: M. Law, C. Baum, W. Dunn, editors. Measuring occupational performance: Supporting best practice in occupational therapy. 2nd ed. Thorofare, NJ: Slack Incorporated; 2005.
8. Hinojosa J, Kramer P, Christ P. Evaluation: Obtaining and interpreting data. 3rd ed. Bethesda, MD: AOTA Press; 2010.
9. McKeown LP, Porter-Armstrong AP, Baxter GD. The needs and experiences of caregivers of individuals with multiple sclerosis: A systematic review. Clinical Rehabilitation 2003;17:234–248.
10. Aronson KJ, Cleghorn G, Goldenberg E. Assistance arrangements and use of services among persons with multiple sclerosis and their caregivers. Disability and Rehabilitation 1996;18(7):354–361.

11. Finlayson M, Garcia J, Preissner K. Development of an education program for caregivers of people aging with multiple sclerosis. Occupational Therapy International 2008;15(1):4–17.
12. Pakenham KI. The nature of caregiving in multiple sclerosis: Development of the caregiving tasks in multiple sclerosis scale. Multiple Sclerosis 2007;13:929–938.
13. Forbes A, While A, Mathes L. Informal carer activities, care burden and health status in multiple sclerosis. Clinical Rehabilitation 2007;21(6):563–575.
14. Finlayson M, Cho C. A descriptive profile of caregivers of older adults with MS and the assistance they provide. Disability and Rehabilitation 2008;30(24):1848–1857.
15. Buchanan RJ, Radin D, Chakravorty BJ, Tyry T. Informal caregiving to more disabled people with multiple sclerosis. Disability & Rehabilitation 2009;31:1244–1256.
16. Cockerill R, Warren S. Care for caregivers: The needs of family members of MS patients. Journal of Rehabilitation 1990;56(1):41–44.
17. O'Brien MT. Multiple sclerosis: Stressors and coping strategies in spousal caregivers. Journal of Community Health Nursing 1993;10:123–135.
18. Sherman TE, Rapport LJ, Hanks RA, Ryan KA, Keenan PA, Khan O, Lisak RP. Predictors of well-being among significant others of persons with multiple sclerosis. Multiple Sclerosis 2007;13:238–249.
19. Koch T, Kelly S. Identifying strategies for managing urinary incontinence with women who have multiple sclerosis. Journal of Clinical Nursing 1999;8(5):550–559.
20. Turner AP, Williams RM, Sloan AP, Haselkorn JK. Injection anxiety remains a long-term barrier to medication adherence in multiple sclerosis. Rehabilitation Psychology 2009;54(1):116–121.
21. Aronson KJ. Quality of life among persons with multiple sclerosis and their caregivers. Neurology 1997;48(1):74–80.
22. Janssens, ACJW, van Doorn PA, de Boer JB, van de Meche, FGA, Passchier J, Hintzen RQ. Impact of recently diagnosed multiple sclerosis on quality of life, anxiety, depression and distress of patients and partners. Acta Neurologica Scandinavica 2003;108:389–395.
23. Bogosian A, Moss-Morris R, Yardley L, Dennison L. Experiences of partners of people in the early stages of multiple sclerosis. Multiple Sclerosis 2009;15:876–884.
24. Hainsworth MA. Helping spouses with chronic sorrow related to multiple sclerosis. Journal of Psychosocial Nursing 1996;34:36–40.
25. Pakenham KI. Application of a stress and coping model to caregiving in multiple sclerosis. Psychology, Health & Medicine 2001;6:13–27.
26. Pakenham KI. Development of a measure of coping with multiple sclerosis caregiving. Psychology & Health 2002;17:97–118.
27. Dewis MME, Niskala H. Nurturing a valuable resource: Family caregivers in multiple sclerosis. Axon 1992;13:87–94.
28. Benbow C, Koopman W. Clinic-based needs assessment of individuals with multiple sclerosis and significant others: Implications for program planning—Psychological needs. Rehabilitation Nursing 2003;28:109–116.
29. Pakenham KI, Cox S. Manuscript in preparation.
30. Solari A, Ferrari G, Radice D. A longitudinal survey of self-assessed health trends in a community cohort of people with multiple sclerosis and their significant others. Journal of Neurological Sciences 2006;243:13–20.
31. Janssens, ACJW, Buljevac D, Van Doorn PA, Van de Meche, FGA, Polman CH, Passchier J, Hintzen RQ. Predictors of anxiety and distress following diagnosis of multiple sclerosis: A two year longitudinal study. Multiple Sclerosis 2006;12:794–801.
32. Eriksson M, Svedlund M. "The intruder": Spouses' narratives about life with a chronically ill partner. Journal of Clinical Nursing 2006;15:324–333.
33. Perrone KM, Gordon PA, Tschopp MK. Caregiver marital satisfaction when a spouse has multiple sclerosis. Journal of Applied Rehabilitation Counseling 2006;37:26–32.
34. De Judicibus MA, McCabe MP. Economic deprivation and its effect on subjective wellbeing in families of people with multiple sclerosis. Journal of Mental Health 2005;14:49–59.

35. Woollin J, Reiher C, Spencer N, Madl R, Nutter H. Caregiver burden: Meeting the needs of people who support the person with multiple sclerosis. International Journal of Multiple Sclerosis Care 1999;1:6–15.

36. Sato A, Ricks K, Watkins S. Needs of caregivers of clients with multiple sclerosis. Journal of Community Health Nursing 1996;13(1):31–42.

37. Hakim EA, Bakheit A, Bryant T, Roberts MWH, McIntosh-Michaellis SA, Spackman AJ, Martin JP, McLellan DL. The social impact of multiple sclerosis—A study of 305 patients and their relatives. Disability and Rehabilitation 2000;22(6):288–293.

38. Pakenham KI. The nature of sense making in caregiving for persons with multiple sclerosis (MS). Disability and Rehabilitation 2008;30(17):1263–1273.

39. Pearlin LI, Mullan JT, Semple SJ, Skaff MM. Caregiving and the stress process: An overview of concepts and their measures. Gerontologist 1990;30:583–594.

40. Braithwaite V. Caregiver burden: Making the concept scientifically useful and policy relevant. Research on Aging 1992;14(1):3–27.

41. Visser-Meily JMA, Post MWM, Riphagen I, Lindeman E. Measures used to assess burden among caregivers of stroke patients: A review. Clinical Rehabilitation 2004;18:601–623.

42. George LK. Caregiver burden and well-being: An elusive distinction. Gerontologist 1994; 34(1):6–7.

43. Courts NF, Newton AN, McNeal LJ. Husbands and wives living with multiple sclerosis. Journal of Neuroscience Nursing 2005;37(1):20–27.

44. Gulick EE. Coping among spouses or significant others of persons with multiple sclerosis. Nursing Research 1996;44:220–225.

45. O'Brien MT, Wineman NM, Nealon NR. Correlates of the caregiving process in multiple sclerosis. Inquiry for the Nursing Practice: An International Journal 1995;9:323–338.

46. Khan F, McPhail T, Brand C, Turner-Stokes L, Kilpatrick T. Multiple sclerosis: Disability profile and quality of life in Australian community report. International Journal of Rehabilitation Research 2006;29:87–96.

47. Cooper C, Selwood A, Livingston G. The prevalence of elder abuse and neglect: A systematic review. Age and Aging 2008;37:151–160.

48. Pakenham KI. The positive impact of multiple sclerosis (MS) on carers: Associations between carer benefit finding and positive and negative adjustment domains. Disability and Rehabilitation 2005;27(17):985–997.

49. Pakenham KI, Cox S. Development of the benefit finding in multiple sclerosis (MS) caregiving scale: A longitudinal study of relations between benefit finding and adjustment. British Journal of Health Psychology 2008;13:583–602.

50. Pakenham KI. The nature of benefit finding in multiple sclerosis. Psychology, Health & Medicine 2007;12:190–196.

51. Pakenham KI. Children who care for their parents: The impact of disability on young lives. In: C. A. Marshall, E. Kendall, M. Banks, R. M. S. Gover, editors. Disability: Insights from across fields and around the world. Westport, CT: Praeger Press; 2009.

52. Pakenham KI, Cox S. The dimensional structure of benefit finding in multiple sclerosis (MS) and relations with positive and negative adjustment: A longitudinal study. Psychology and Health 2009;24:373–393.

53. Folkman S. The case for positive emotions in the stress process. Anxiety, Stress, & Coping 2008; 21(1):3–14.

54. Fredrickson BL. Positive emotions. In: C. R. Snyder, S. J. Lopez, editors. Handbook of positive psychology. New York: Oxford University Press; 2002.

55. Tedeschi RG, Calhoun LG. The clinician as expert companion. In: C. L. Park, S. C. Lechner, M. H. Antoni, A. L. Stanton, editors. Medical illness and positive life change: Can crisis lead to personal transformation? Washington: APA; 2009.

56. Antoun MZ, Frank AO. Caregivers of people with multiple sclerosis. Clinical Rehabilitation 2003;17(7):804–805.

57. Pakenham KI, Cox S. The nature of caregiving in children of a parent with multiple sclerosis from multiple sources and the associations between caregiving activities and youth adjustment over time. Psychology and Health 2012;27(3):324–346.

58. Pakenham KI, Bursnall S. Relations between social support, appraisal and coping and both positive and negative outcomes for children of a parent with MS and comparisons with children of healthy parents. Clinical Rehabilitation 2006;20:709–723.

59. Cross T, Rintell D. Children's perceptions of parental multiple sclerosis. Psychology, Health and Medicine 1999;4(4):355–359.

60. Arnaud SH. Some psychological characteristics of children of multiple sclerotics. Psychosomatic Medicine 1959;21:8–22.

61. Pakenham KI, Bursnall S, Chiu J, Cannon T, Okochi M. The psychosocial impact of caregiving on young people who have a parent with an illness or disability: Comparisons between young caregivers and non-caregivers. Rehabilitation Psychology 2006;51:113–126.

62. Yahav R, Vosburgh J, Miller A. Emotional responses of children and adolescents to parents with multiple sclerosis. Multiple Sclerosis 2005;11:464–468.

63. Friedemann ML, Tubergen P. Multiple sclerosis and the family. Archives of Psychiatric Nursing 1987;1(1):47–54.

64. Kikuchi JF. The reported quality of life of children and adolescents of parents with multiple sclerosis. Recent Advances in Nursing 1987;16:163–191.

65. Turpin M, Leech C, Hakenberg L. Living with parental multiple sclerosis: Children's experiences and clinical implication. Canadian Journal of Occupational Therapy 2008;75:149–156.

66. Braham S, Houser HB, Cline A, Posner M. Evaluation of the social needs of non-hospitalized chronically ill persons: 1. Study of 47 patients with multiple sclerosis. Journal of Chronic Diseases 1975;28:401–419.

67. De Judicibus MA, McCabe MP. The impact of parental multiple sclerosis on the adjustment of children and adolescents. Adolescence 2004;39(155):551–569.

68. Deatrick JA, Brennan D, Cameron ME. Mothers with multiple sclerosis and their children: Effects of fatigue and exacerbations on maternal support. Nursing Research 1998; 47(4):205–210.

69. Peters LC, Esses LM. Family environment as perceived by children with a chronically ill parent. Journal of Chronic Diseases 1985;38(4):301–308.

70. Aldridge J, Becker J. Punishing children for caring: The hidden cost of young carers. Children and Society 1993;7:376–387.

71. Chipchase SY, Lincoln NB. Factors associated with carer strain in carers of people with multiple sclerosis. Disability and Rehabilitation 2001;23:768–776.

72. Figved N, Myhr KM, Larsen JP, Aarsland D. Caregiver burden in multiple sclerosis: The impact of neuropsychiatric symptoms. Journal of Neurology, Neurosurgery, & Psychiatry 2009;78:1097–1102.

73. Feinstein A. The neuropsychiatry of multiple sclerosis. Canadian Journal of Psychiatry 2004; 49:157–163.

74. Pakenham KI. Relations between coping and positive and negative outcomes in carers of persons with multiple sclerosis. Journal of Clinical Psychology in Medical Settings 2005; 12(1):25–38.

75. Knight RG, Devereux RC, Godfrey HPD. Psychosocial consequences of caring for a spouse with multiple sclerosis. Journal of Clinical and Experimental Neuropsychology 1997; 19(1):7–19.

76. Lazarus RS, Folkman S. Stress, appraisal, and coping. New York: Springer Publishing Inc.; 1984.

77. Pakenham KI. Adjustment to multiple sclerosis: Application of a stress and coping model. Health Psychology 1999;18:383–392.

78. Park CL, Folkman S. Meaning in the context of stress and coping. Review of General Psychology 1997;1(2):115–144.

79. Kramer BJ. Expanding the conceptualization of caregiver coping: The importance of relationship-focused coping strategies. Family Relations 1993;42:383–391.

80. Davis CG, Nolen-Hoeksema S, Larson J. Making sense of loss and benefiting from the experience: Two construals of meaning. Journal of Personality and Social Psychology 1998; 75(2):561–574.

81. Janoff Bulman R, Yopyk DJ. Random outcomes and valued commitments: Existential dilemmas and the paradox of meaning. In: J. Greenberg, S. L. Koole, T. Pyszczynski, editors. Handbook of experimental existential psychology. New York: Guilford Press; 2004.

82. Taylor SE. Adjustment to threatening events: A theory of cognitive adaption. American Psychologist 1983;38(11):1161–1173.

83. Pakenham KI. Making sense of caregiving for persons with multiple sclerosis (MS): The dimensional structure of sense making and relations with positive and negative adjustment. International Journal of Behavioral Medicine 2008;15:241–252.

84. Pakenham KI. Benefit finding and sense making in chronic illness. In: S. Folkman, editor. Oxford handbook on stress, coping, and health. New York, NY: Oxford University Press; 2011.

85. Derogatis LR. Administration and procedures manual: BSI–18. Minneapolis, MN: National Computer Systems Inc; 2000.

86. Lovibond SH, Lovibond PF. Manual for the depression anxiety stress scales, 2nd ed. Sydney: Psychology Foundation of Australia; 1995.

87. Buhse M. Assessment of caregiver burden in families of persons with multiple sclerosis. Journal of Neuroscience Nursing 2008;40(1):25–31.

88. Zarit S, Reever K, Bach-Peterson J. Relatives of the impaired elderly: Correlates of feelings of burden. Gerontologist 1980;20:649–655.

89. Given CW, Given B, Stommel M, Collins C, King S, Franklin S. The caregiver reaction assessment (CRA) for caregivers to persons with chronic physical and mental impairments. Research in Nursing and Health 1992;15(4):271–283.

90. Pavot W, Diener E. Review of the satisfaction with life scale. Psychological Assessment 1993;5(2):164–172.

91. Horowitz M, Adler NE, Kegeles S. A scale for measuring the occurrence of positive states of mind: A preliminary report. Psychosomatic Medicine 1988;50:477–483.

92. Spanier GB. Measuring dyadic adjustment: New scales for assessing the quality of marriage and similar dyads. Journal of Marriage and the Family 1976;38(1):15–28.

93. Bradburn NM. The structure of psychological well-being. Oxford, England: Aldine; 1969.

94. Pakenham KI, Stewart CA, Rogers A. The role of coping in adjustment to multiple sclerosis-related adaptive demands. Psychology, Health and Medicine 1997;2:197–211.

95. Finlayson M, Preissner K, Garcia J. Pilot study of an educational programme for caregivers of people ageing with multiple sclerosis. British Journal of Occupational Therapy 2009; 72(1):11–19.

96. Ireland M, Pakenham KI. The nature of young caregiving in families experiencing chronic illness/disability: Development of the youth activities of caregiving scale (YACS). Psychology and Health 2010;25:713–731.

97. Coles AR, Pakenham KI, Leech C. Evaluation of an intensive psychosocial intervention for children of parents with multiple sclerosis. Rehabilitation Psychology 2007;52(2):133–142.

98. Cheung J, Hocking P. The experience of spousal carers of people with multiple sclerosis. Qualitative Health Research 2004;14:153–166.

99. Shaw C, McNamara R, Abrams K, Cannings-John R, Hood K, Longo M, Myles S, OMahony S, Roe B, Williams K. Systematic review of respite care in the frail elderly. Health Technology Assessment 2009;13(20):1–224.

100. Coon DW, Evans B. Empirically based treatments for family caregiver distress: What works and where do we go from here? Geriatric Nursing 2009;30:426–436.

101. Khan F, Pallant JF. Use of International Classification of Functioning, Disability and Health (ICF) to describe patient-reported disability in multiple sclerosis and identification of relevant environmental factors. Journal of Rehabilitation Medicine 2007;39(1):63–70.

102. Lorig KR, Holman H. Self-management education: History, definition, outcomes, and mechanisms. Annals of Behavioral Medicine 2003;26(1):1–7.

103. Law M, Cooper BA, Strong S, Stewart D, Rigby P, Letts L. The person–environment–occupational model: A transactive approach to occupational performance. Canadian Journal of Occupational Therapy 1996;63(1):9–23.
104. Bodenheimer T, Lorig K, Holman H, Grumbach K. Patient self-management of chronic disease in primary care. Journal of the American Medical Association 2002;288(19):2469–2475.
105. Olkin R, Abrams K, Preston P, Kirshbaum M. Comparison of parents with and without disabilities raising teens: Information from the NHIS and two national surveys. Rehabilitation Psychology 2006;51(1):43–49.

23

Cultural Considerations

Yolanda Suarez-Balcazar, Fabricio E. Balcazar, and Celestine Willis

CONTENTS

Loretta, an African American woman in her early 30s, shares with her husband doubts about going back to see her physician for further testing. He thinks that the symptoms she has been experiencing—numbness, dizziness, tingling sensations, and muscle weakness—are the result of stress and nervousness, while she worries that this might be something more. She believes it might be a punishment from God for not lending money to her mother, who needed it to take care of some personal business. Loretta decides not to go back to see her physician or any other professional and to wait to see what happens. In her daily prayers, Loretta implores God to make her symptoms go away.

This is not an uncommon scenario for many African Americans experiencing the early signs of multiple sclerosis (MS). Despite the fact that MS is more common among populations of European descent, the incidence of MS is rising among other ethnic groups.

As asserted by Maghzi et al.,[1] "There is an overall increase in the worldwide prevalence and incidence of MS" (p. 359). Cordova et al.[2] and Corona and Román[3] reported increased MS in Mexico, and although there is less MS across Asian countries, Kira[4] reported a small but statistically significant increase. In addition, research documenting the impact of MS and the rehabilitation process in ethnically/racially diverse groups is scarce.[5]

The purpose of this chapter is to provide an overview of cultural considerations that will assist rehabilitation professionals in the United States when working with individuals from diverse ethnic/racial backgrounds. After reading this chapter, you will be able to

1. Define cultural competence,
2. Recognize the importance of cultural competence during the MS rehabilitation process,
3. Implement strategies to enhance cultural competence in the MS rehabilitation team,
4. Identify issues to be considered across treatment settings, and
5. Identify directions for future research.

In 1999, Loveland[6] reported on the scarcity of information about people from diverse ethnic/racial backgrounds with MS. Although 12 years later new studies have been conducted, researchers continue to assert that little information is available about the experience of ethnically diverse populations with MS and the outcomes of their rehabilitation processes and treatment.[7–10] One such study, conducted by Buchanan et al.,[11] suggested that African Americans with MS have a greater level of disability and worse treatment outcomes when compared with Whites. Binetti[9] pointed out that "one of the biggest challenges facing health care providers when treating African-Americans with MS is to identify how race plays a role in the development and course of the disease and on the impact of rehabilitation treatment plans" (p. 4). In an effort to examine clinical characteristics of the disease among African Americans, Cree et al.[5] found that African Americans required the use of a cane for ambulation and became wheelchair dependent more rapidly than their Caucasian counterparts. In the United States, some of the few research studies available on ethnically/racially diverse populations with MS have been conducted with African Americans, but there is little information about Latinos, Asian Americans, or Native Americans with MS.

Although the rates of MS are lower among people of color, researchers assert that when comparing individuals from European descent and those from ethnically/racially diverse backgrounds, the disparities in rehabilitation and health outcomes are widening.[12] The Office of Minority Health and Health Disparities;[13] Braithwaite et al.;[14] Hill-Briggs et al.;[15] and Smedley et al.[16] reported growing disparities in health outcomes while Wilson[17] and Wilson and Senices[18] alluded to large disparities in vocational rehabilitation outcomes between Whites and ethnically and racially diverse individuals. Several complex factors contribute to the disparities in health outcomes; among them are patient-level factors, health care system factors, and care-process factors[16] in addition to complex sociopolitical and contextual factors.[14] Examples of these factors include lack of outreach to diverse populations as well as the lack of health care insurance and the experiences of prejudice and stereotyping these populations often experience when seeking care. Shabas and Heffner[8] reported that 32% of individuals with MS receiving Medicaid had never seen an MS specialist and were most likely to be from diverse ethnic/racial backgrounds.

To address the gap in health and rehabilitation outcomes, researchers and practitioners alike argue that culturally competent interventions hold promise for improving the delivery of successful rehabilitation interventions for ethnically/racially diverse populations.[19,20] In fact, Zalaquett et al.[21] recognized that disability and rehabilitation service providers need to develop a broad range of multicultural competencies in order to work more effectively with clients from diverse ethnic and racial backgrounds with MS. We acknowledge that multicultural competencies can apply to a variety of diverse populations experiencing MS, including, but not limited to, racial/ethnic groups, low-income individuals, people who are geographically isolated, and gays and lesbians. However, in this chapter we will focus on those who are from ethnically/racially diverse backgrounds, particularly African Americans, Latinos, Asians, Pacific Islanders, and Native Americans.

There has been considerable debate in the literature about how to refer to the ethnic and racial groups listed above. Some refer to them as "minorities"; however, the current trend is to move away from using this term because of its negative connotation (e.g., subordinate group, of less power and resources than the dominant group[16]). We acknowledge that race and ethnicity are socially constructed categories, but, despite the current controversy surrounding their use, we will consistently use "racial" to refer to African American, Native American, and Asian-American populations, and "ethnic" will refer to Latinos.

23.1 Definitions of Cultural Competence

A variety of fields like psychology, counseling, nursing, rehabilitation, and education are contributing to the vast literature available on the topic of cultural competency. Noteworthy is the lack of a universally accepted definition and an ongoing debate among professionals and researchers about how best to define cultural competence.[22] The word *culture*, which "has been widely debated and broadly defined,"[16] refers to an integrated pattern of behaviors, norms, and rules that are shared by a group of individuals and involves their beliefs, values, expectations, worldviews, communication, common history, and institutions.[23] Fiske et al.[24] refer to culture as the "belief systems and value orientations that influence customs, norms, practices and social institutions, including psychological processes (e.g., language, caretaking practices) and organizations (e.g., media, educational systems)" (p. 380). *Cultural competence*, a very broad term,[25] denotes having the capacity to function effectively as an individual and as an organization within the context of the cultural beliefs, behaviors, and needs presented by consumers and their communities.[26,27]

Suarez-Balcazar and Rodakowski[28] suggested that "becoming culturally competent is an ongoing contextual, developmental, and experiential process of personal growth that results in a greater ability to adequately serve individuals who look, think, and behave in ways that are different from us" (p. 15). The authors have asserted that the *us* in this context refers to health professionals whose practice might be guided by middle-class, Western values of service delivery.

In an attempt to underscore the importance of cultural competence, several professions have adopted specific guidelines within their practices and professional frameworks. For instance, the National Rehabilitation Association's[29] document titled *Multicultural Rehabilitation Concerns* includes a mission statement and specific goals that speak of cultural competence among rehabilitation professionals. Likewise, the *American Occupational*

Therapy Association Practice Framework[30] speaks to the need to meet the concerns of multi-cultural populations and the importance of eliminating disparities.

Suarez-Balcazar et al.[31] validated a conceptual model of cultural competence with occupational therapists (see Figure 23.1) proposed by Suarez-Balcazar et al.[32] and found that the process of gaining cultural competence among health professionals can be explained with three interrelated factors: cognitive (i.e., seeking critical awareness and knowledge), behavioral (i.e., developing appropriate skills), and organizational (i.e., support for cultural competence).

The cognitive factor, seeking critical awareness and knowledge, involves a complex process of self-reflection and information seeking on the part of the rehabilitation team. The treatment of individuals with MS is often conducted by a team of service providers and specialists, including physicians, physical therapists, occupational therapists, and rehabilitation counselors, referred to as the rehabilitation team. The process of increasing cultural awareness begins with a willingness to question one's own beliefs and potential biases, a desire to change, and willingness to learn about the client's cultural background. It also involves understanding and learning about the client's experiences of oppression, marginalization, and discrimination. The rehabilitation team is encouraged to engage in self-reflection and dialogue among themselves to examine their personal attitudes and readiness to serve a client with MS who is from a different ethnic/racial background, might not speak English, and might not share the same views about Western medicine as the rehabilitation team. Professionals are encouraged to discuss issues related to White

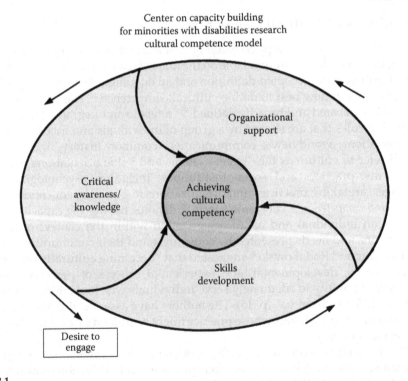

FIGURE 23.1
Cultural competence conceptual model. (Adapted from Balcazar FE, Suarez Balcazar Y, Taylor-Ritzler T. Cultural competence: Development of a contextual framework. Disability Rehabilitation 2009;31(14):1153–1160; Courtesy of *Journal of Comparative Neurology* by Wistar Institute of Anatomy and Biology. Reproduced with permission of John/Wiley & Sons, Inc. in the format Journal via Copyright Clearance Center.)

privilege. This refers to perceived advantages that White people may experience as a result of their race.[33–35]

Part of the process of self-reflection involves putting aside personal biases and reactions to provide the best care in situations in which the client may communicate to the therapist his or her perceived beliefs or negative emotions against the ethnic group to which the therapist belongs. Although the percentage of African American and Latino health and rehabilitation providers is very small compared to White providers, the percentage of Asian health providers is increasing.[16]

Along with seeking cultural awareness is gaining cultural knowledge that leads to familiarization with others' cultural characteristics, history, values, belief systems, and behaviors. Several researchers[36,37] have conducted detailed reviews of the complex cultural characteristics of multiple ethnic groups, particularly regarding health-related practices, rituals associated with birth and death, nutritional practices, and the role of spirituality in health, among other factors. Seeking critical awareness and knowledge implies a cognitive function that is facilitated by rehabilitation professionals asking themselves questions for self-reflection. See Table 23.1 for examples of questions for self-reflection.

The behavioral factor, developing culturally competent skills, includes professional practices and behaviors designed to improve service delivery to diverse populations, including effective verbal and nonverbal communication skills; utilization of translation services when the clients with MS speak a language other than that spoken by the rehabilitation treatment team; and understanding the clients' concept of personal space, physical and eye contact, and other forms of verbal and nonverbal expression. Service providers can develop cultural competence skills through several engagement processes, including critical self-reflection about their skills and repeated encounters with individuals from a variety of ethnic groups. Capacity to empathize with the client with MS and overcome mistrust includes the ability to understand the client's experiences of oppression and marginalization and to place the client's culture at the center of the relationship.[38]

The organizational factor of the model is seeking organizational support for implementing culturally competent rehabilitation services. This factor includes the context in which rehabilitation treatment is implemented. The organization in which the MS rehabilitation team practices needs to articulate its commitment to cultural competence in their mission and vision statements and provide opportunities for health care professionals to attend training, participate in discussions, and be open to practicing in ways that respect the culture of the

TABLE 23.1

Sample Questions for the Rehabilitation Team to Consider

Critical awareness and knowledge

- What has been my experience with people who look, think, and behave differently from me?
- Do I hold any biases against people who are different from me?
- Do I engage in honest dialogue with others regarding my feelings about and personal biases toward people who are different from me?
- What have been experiences of oppression in my life?
- Do I consider the experiences of oppression of my clients?
- Am I willing to change my personal attitude toward people who are different from me?
- Do I know about the culture, traditions, customs, language, values, and beliefs of the clients I serve?
- Am I willing to examine the institutional biases of traditional practices and services?
- Am I willing to try nontraditional interventions?
- Do I consider my position of privilege in society?

client base.[32,39] Practitioners may attempt to implement culturally competent adaptations in their work, but often they need the support of their supervisors/administrators in order to implement changes in the delivery of rehabilitation services that are necessary to meet the needs of diverse clients.[32,34,40] For example, rehabilitation work settings may need to allocate resources for training, foster a learning culture related to valuing diversity, support and allow rehabilitation professionals to adopt culturally competent models of practice, allow physical changes to their work place (e.g., placing posters of ethnically/ racially diverse people with MS on the walls), hire practitioners from racially and ethnically diverse backgrounds, have access to translation services as needed, and support changes in work schedules that allow employees to conduct treatments or outreach activities in the community on evenings and weekends or to attend community events. To underscore the importance of reaching out to communities, the 2010 report from the US Surgeon General states that one of the most important ways to begin to address health disparities is community engagement. This involves the rehabilitation team working together with relevant partners in the community—such as faith organizations, recreation settings—who share common goals and interests to better reach out, serve, and work with racially and ethnically diverse individuals.[14] The rehabilitation specialist needs to build personal connections with the gate keepers in the communities in which they work. Rehabilitation settings are responsible for establishing a culture and environment conducive to culturally competent care in which the MS rehabilitation team can engage in innovation and is supported in their efforts to reach out and treat individuals from diverse ethnic backgrounds.

The development of cultural competence means the willingness of professionals to engage in a series of practices that examine the institutional biases of traditional practices and services, which have typically been created by and for middle-class Whites. Professionals have to consider nontraditional interventions or procedural changes in order to address the individual needs of their racially diverse clients while challenging racist practices, discrimination, and oppression, when observed.

23.2 Importance of Cultural Competence during the MS Rehabilitation Process

In this section we discuss the importance of cultural competence across the different phases of the MS rehabilitation process, including screening, assessment, goal planning, treatment implementation, and follow-up. A synthesis of the reasons why cultural competence is important in MS rehabilitation is presented in Table 23.2.

23.2.1 Screening for MS

Many ethnically and racially diverse individuals are simply not screened for MS or are more likely to experience misdiagnosis. Loveland[6] interviewed 100 individuals with MS and found that 30% of African-American women, compared to 11% of Caucasian women, were told by their medical care professionals that their symptoms were emotional or psychological and not MS. They were referred to psychological support or counseling services. Patients who go through that experience are less likely to seek screening and assessment or develop treatment goals.[41] Further, Loveland[6] found that a common perception among African

TABLE 23.2

Reasons Why Cultural Competence is Important in the Rehabilitation Process

Why is cultural competence important during the MS rehabilitation process?
- Individuals of ethnically/racially diverse backgrounds are not being screened for MS.
- Low-income and diverse populations with disabilities experience challenges in obtaining optimal health care.
- The disparities between White and non-White clients continue to increase.
- Linguistically and culturally appropriate assessments are often unavailable.
- Culturally competent interventions can improve rehabilitation service outcomes for individuals with disabilities from diverse backgrounds.
- Diverse individuals face a number of challenges that make it harder to attain desired rehabilitation goals.

Americans is that "MS is a White person's disease" (p. 23), making it harder for them to accept being diagnosed with MS and seek treatment.

Latinos in the United States are hardly screened for MS, in part, because they have limited access to health care services, and there is a lack of culturally adapted, translated information and educational materials about MS. Cook et al.[42] found that among some Asians and Latinos, the onset of a physical disability or a chronic debilitating disease such as MS is often perceived as a temporary transitional state and not as a disability.

In general, there are many factors that contribute to the poor screening for MS among African-American and Latino populations. These include lack of access to health insurance and preventive care, instances of discrimination and fear of being misunderstood or misdiagnosed, lack of knowledge about MS, and overall mistrust of health care professionals.[43,44] These authors also argue that building trust and investing time in developing an empathic relationship are likely to increase collaboration and compliance in all stages of the rehabilitation process. The process of building trust from the beginning of the relationship with the client with MS and his or her family may help deconstruct the client's mistrust of health professionals and possible history of discrimination or inferior care.[16]

23.2.2 The Use of Assessments in MS

Disparities among Whites and African Americans and other diverse populations with MS in the United States are greatest in terms of screening, assessment, and access to rehabilitation professionals.[7,9,12] Assessments need to reflect the fact that African Americans and Latinos tend to have a more advanced stage of the disease and are less educated about MS when they are first diagnosed.[7,11,41,45] These differences affect rehabilitation outcomes and reflect limited access to preventive care or early intervention.

A common problem for non-English-speaking clients with MS is that assessments often used in rehabilitation are not necessarily available in other languages, nor have they been validated with ethnically diverse populations. Assessments used in rehabilitation practice most likely reflect an ethnocentric view of culture, mostly representing White, middle-class values.

23.2.3 Setting Goals and Implementing Treatment in MS Rehabilitation

When setting goals and implementing MS treatment, it is of the utmost importance to take into consideration the cultural values and beliefs of the client (positive outcomes is a given). Hasnain et al.[46] conducted a meta-analysis of 22 rehabilitation and health intervention experimental studies that utilized culturally competent interventions and found that the interventions had positive effects on several rehabilitation outcomes for individuals of diverse ethnic and cultural backgrounds who had disabilities. This systematic review found benefits in four outcome measures: disability symptom-related measures; client knowledge of disability measures; behavioral measures (e.g., disability self-management and treatment compliance); and psychosocial measures of well-being, self-efficacy, and quality of life. The authors concluded that culturally competent interventions can improve rehabilitation outcomes for ethnically and culturally diverse persons with disabilities who reside in Western cultural contexts.

The US Department of Health and Human Services, Office of Minority Health[47] developed the *National Standards on Culturally and Linguistically Appropriate Services* (CLAS), directed at health care organizations. These standards have been implemented nationally across health care settings. One of their focuses is *language access services*, which includes collecting and disseminating information on model programs for implementing language assistance services; supporting the financing of language assistance services at all levels of health services delivery; supporting the development of national standards for interpreter training, skills assessment, certification, and codes of ethics; developing national standards for the translation of health-related materials; developing standard language for key documents like consent forms, health information, and medication information; and developing an Internet clearinghouse of downloadable samples of translated documents (see US Department of Health and Human Services [2001] for a complete description of the standards). The US Department of Health and Human Services, Office of Minority Health[47] also argues that the principles and activities of culturally and linguistically appropriate services and interventions should be integrated throughout an organization (any health care setting) and undertaken in partnership with the communities being served. Table 23.3 provides a listing of the CLAS standards for health care settings.

People with MS from ethnically/racially diverse backgrounds are more likely to experience discrimination and the double jeopardy of race and disability or the triple jeopardy of race, disability, and language (limited English proficiency, as in the case of Latino immigrants in the United States or African and Arab immigrants in European countries).[48] Shabas and Heffner[8] asserted that the challenges that low-income, racially and ethnically diverse populations with physical disabilities experience in obtaining optimal health care, such as discrimination on the part of health care providers coupled with contextual and logistic barriers (e.g., lack of transportation) contribute to exacerbating poor health outcomes, in part, because care is rapidly curtailed and the rate of treatment noncompliance is high. Hahn[49] calls this the *social production of illness*, meaning that poverty and social organization promote illness and predict poor rehabilitation outcomes.

In a study of a large database obtained from the Registry of the North American Research Committee on Multiple Sclerosis, Buchanan et al.[41] found significant racial/ethnic differences in care received by individuals with MS. The authors reported that African Americans were less likely to be treated or evaluated for MS than Whites and Latinos, and less likely to have seen a neurologist specializing in MS care. The authors also reported that a larger proportion of Latinos never received occupational therapy rehabilitation services and had never used the services of a home aide when compared with African Americans and Whites. Shabas and Heffner[8] found that Latinos often do not understand the course of the MS disease nor do they have sufficient knowledge about management of medications. Unfortunately, when populations lack understanding of the disease, they are less likely to

TABLE 23.3

National Standards on Culturally and Linguistically Appropriate Services (CLAS) in Health Care
(US Department of Health and Human Services, Office of Minority Health)

Standard 1	Health care organizations should ensure that consumers receive understandable and respectful care that is provided in a manner compatible with their cultural health beliefs and practices and preferred language.
Standard 2	Health care organizations should implement strategies to recruit, retain, and promote at all levels of the organization a diverse staff and leadership that are representative of the demographic characteristics of the service area.
Standard 3	Health care organizations should ensure that staff at all levels and across all disciplines receive ongoing education and training in culturally and linguistically appropriate service delivery.
Standard 4	Health care organizations must offer and provide language assistance services, including bilingual staff and interpreter services, at no cost to each patient/consumer with limited English proficiency at all points of contact, in a timely manner during all hours of operation.
Standard 5	Health care organizations must provide to patients/consumers in their preferred language both verbal offers and written notices informing them of their right to receive language assistance services.
Standard 6	Health care organizations must assure the competence of language assistance provided to limited English proficient patients/consumers by interpreters and bilingual staff. Family and friends should not be used to provide interpretation services (except on request by the consumer).
Standard 7	Health care organizations must make available easily understood patient-related materials and post signage in the languages of the commonly encountered groups and/or groups represented in the service area.
Standard 8	Health care organizations should develop, implement, and promote a written strategic plan that outlines clear goals, policies, operational plans, and management accountability/ oversight mechanisms to provide culturally and linguistically appropriate services.
Standard 9	Health care organizations should conduct initial and ongoing organizational self-assessments of CLAS-related activities and are encouraged to integrate cultural and linguistic competence-related measures into their internal audits, performance improvement programs, patient satisfaction assessments, and outcomes-based evaluations.
Standard 10	Health care organizations should ensure that data on the individual patient's/consumer's race, ethnicity, and spoken and written language are collected in health records, integrated into the organization's management information systems, and periodically updated.
Standard 11	Health care organizations should maintain a current demographic, cultural, and epidemiological profile of the community as well as a needs assessment to accurately plan for and implement services that respond to the cultural and linguistic characteristics of the service area.
Standard 12	Health care organizations should develop participatory, collaborative partnerships with communities and utilize a variety of formal and informal mechanisms to facilitate community and patient/consumer involvement in designing and implementing CLAS-related activities.
Standard 13	Health care organizations should ensure that conflict and grievance resolution processes are culturally and linguistically sensitive and capable of identifying, preventing, and resolving cross-cultural conflicts or complaints by patients/consumers.
Standard 14	Health care organizations are encouraged to regularly make available to the public information about their progress and successful innovations in implementing the CLAS standards and to provide public notice in their communities about the availability of this information.

Note: For a full reading of the CLAS standards see US Department of Health and Human Services, Office of Minority Health (http://minorityhealth.hhs.gov/assets/pdf/checked/executive.pdf).

participate in setting goals and implementing rehabilitation plans, and they are more likely to stop or curtail medication, which can increase the frequency of relapses and speed up the progression of the disease.

23.2.4 Compliance with Treatment Implementation and Follow-Up

Another reason for promoting cultural competence is that when patients feel that the MS rehabilitation team understands their cultural values and beliefs and takes them into consideration, they demonstrate better compliance with the treatment implementation and follow-up, which, in turn, results in a better response to treatment.[6] Follow-up with low-income ethnically and racially diverse populations is a challenge because they move often and change phone numbers, making it harder for the rehabilitation team to locate them.[50] Long-term compliance with medications can also be complicated by changes in Medicaid policies, requirements to determine continued eligibility, and bureaucratic barriers for getting services reinstated if they are lost at any time for any reason.

23.3 Strategies for Acknowledging Cultural Differences during the MS Rehabilitation Process

Below are key recommendations for providing culturally competent rehabilitation services and a brief synthesis of the literature available on the topic. These recommendations are adapted to the context of the MS rehabilitation process with ethnically/racially diverse individuals, including African Americans, Latinos, Asians, Pacific Islanders, and Native Americans. The synthesis by no means exhausts all the recommendations already made by researchers and health care providers alike in the area of cultural competence.

23.3.1 Utilize Instruments and Tools That Have Been Validated with the Population of Interest during the MS Screening and Assessment Process

Should these instruments not be available, then carefully review the cultural appropriateness and fit of the available instruments with ethnically/racially diverse groups. Hernández et al.[51] assert that clinical and psychological tests that have not been standardized with different groups are often used to determine eligibility for vocational rehabilitation services. From a review of psychological assessment instruments used in rehabilitation, the authors concluded that "race and ethnicity are important variables to consider throughout the assessment process (i.e., test selection, test administration, data interpretation, and writing of reports)" (p. 73). Racial and ethnic variables might include the client's experience of discrimination and marginalization in obtaining services; cultural mistrust; cultural beliefs about authority and related practices; and culturally bound beliefs about MS, independence, self-management, and health, among others. All of these variables influence not only what clients disclose and how they behave during the MS assessment process, but also the entire intervention process, from selecting rehabilitation goals to implementing the MS rehabilitation plan.

23.3.2 Include Members from Diverse Ethnic/Racial Backgrounds on the MS Rehabilitation Treatment Team

Although ethnic identity does not guarantee empathy and positive client outcomes, it usually helps.[52] With regard to client–provider interactions, a direct way to support the cultural values of ethnically/racially diverse groups is to match the client with a provider of the same ethnicity, someone who speaks the client's native language, or someone who has experience working with that particular population.[53] Previous research[52] found that an ethnic/racial match has a positive effect on rehabilitation outcomes.

23.3.3 Translate Informational Materials into the Client's Language

In the United States for instance, the growing percentage of Latinos raises the importance of the availability of Spanish translators and of translating educational materials.[8] Conducting training and educational activities in the preferred language of the patient is also important. Many Latino immigrants are more likely to attend training workshops and events that are offered in Spanish. Team members should make sure that the clients who do not speak the dominant language understand the treatment, the types of medications, and the course of the disease. All translations should be checked for quality and accuracy.

23.3.4 Consider Cultural Conceptions and Ideas Assigned to Disability, Chronic Health Disease, and to MS

Clients' cultural values and beliefs have a profound influence on how they experience disability, how they approach it, and what they think should be done about their disabilities. A recent Hmong immigrant from Laos who does not understand the gradual debilitating effects of MS might interpret her difficulty in mobility as a result of evil spirits.

A common Western value coming from the rehabilitation team is to assume that independence is the "gold standard" of rehabilitation. This might, in fact, be considered threatening to an elderly male from the Philippines experiencing MS, whose family is expected to assist with all his activities of daily living.[54] As discussed by Cook et al.;[42] Garate et al.;[55] and Purnell and Paulanka,[36] the notion of "independence" is interpreted in the context of social interactions and it is culturally constructed. In that sense, for most Asian Americans and Latinos the desired goal is to maintain social interdependence and not "independence." The person with a disability is seen as part of a collective for whom the immediate kinship is responsible. However, the interpretation of independence proposed by advocates of the social model of disability refers to the right of the person with a disability to have control over relevant life decisions and to have choices.[56] This interpretation is not readily understood by many immigrants with disabilities and minority individuals unfamiliar with the principles and values advocated by the social model of disability. In fact, they may never have heard such ideas before.

In addition, many diverse racial/ethnic groups with disabilities are likely to experience poverty and experience challenges in acquiring the assistive devices they need, and in making their homes accessible by making necessary accommodations, such as adding a ramp or making a bathroom accessible. Paying for these modifications can be economically unfeasible for some families. The rehabilitation team needs to educate the family about possible resources in the community they can tap into, such as applying to vocational rehabilitation for accessibility accommodations in the home so the person with MS can get to work.

23.3.5 Consider the Cultural Beliefs about Authority and Decision Making

As Brach and Fraserirector[57] suggest, it is important to be sensitive to perceptions of authority and the role of the family and elderly relatives in making health-related decisions in some cultures. The MS treatment team should make an effort to include family members and persons of authority in the process of screening, assessment, treatment, and follow-up. For instance, in some Asian families the parents of the husband are the ones who make the health decisions for the family. In some African-American, Latino, and Native American families, the individual with MS might not make decisions without consulting with others in the family, in particular his or her elders.[36]

Furthermore, in some Latino cultures, it is disrespectful to say "no" to a person of authority, such as a doctor or other medical personnel. It is more respectful to "agree" and say "yes" even if there is no intention of following through, complying with treatment, or coming back for subsequent appointments. The MS rehabilitation team needs to gain knowledge about the person's cultural values as mediated by factors like class status and level of education. Team members should also consider the cultural norms related to decision making, self-control, self-determination, and self-management among patients with MS and how they might influence decisions about health. These concepts are often not part of the language repertoire of low-income, ethnically diverse groups.

23.3.6 Consider the Intersect between Race, Culture, and MS

Individuals from ethnically/racially diverse backgrounds are more likely to experience discrimination, oppression, and marginalization. They are more likely to feel cultural mistrust, meaning that they often doubt if the decisions that service providers make about the services they will receive are fair or equal to those received by other clients.[43] Also, the MS rehabilitation team needs to consider multiple differences within ethnic groups due to regional and social differences. For instance, it is important to consider that Asian includes more than half of the world's population and comprises a large number of countries and cultures, including different religions, languages, spiritual beliefs, and health practices. The same is true to a lesser degree of Africans and Latinos, who have a variety of subgroups and cultures unique to their geographical and sociopolitical contexts.

23.3.7 Consider the Practice of Using Non-Western Medicines and Healers

Some ethnic/racial groups may use herbs and natural home remedies before they seek Western-type health services. Even if they do seek Western medicine for MS, they might be hesitant to share their experience with home remedies with their MS rehabilitation team for fear of being criticized. In some cultures, individuals might believe that their home remedies are enough and that there is no need to pursue medical treatment. For instance, many individuals from African American and Latino backgrounds have a deep sense of spirituality, and praying for a cure (and lighting candles) is often seen as the best possible remedy for all illnesses.

The MS rehabilitation team might inquire about the individual's home practices and examine ways that they can be respected when appropriate. When ethnically/racially diverse groups are asked to stop practicing their home remedies, this can contribute to their decision to drop or not comply with the MS rehabilitation team's treatment plan.[36] Alternative health practices (e.g., taking herbs, using prayer, being examined by local healers, consuming certain foods) need to be understood before asking a client to change his or her practices. Most important is to become familiar with the client's folk medicine and alternative practices in order to prescribe treatments that are complementary, have no negative interaction, and yet are respectful of the client's wishes.[37,54]

23.3.8 Offer Training in Cultural Competence to the MS Rehabilitation Team

Health care organizations and vocational rehabilitation offices in the United States are beginning to require health professionals and counselors to undertake training on how to provide culturally competent services to diverse clients, in particular when large disparities in outcomes are reported. We suggest that MS rehabilitation team members be provided with opportunities to receive training on how to work with ethnically/racially diverse clients. As discussed throughout this chapter, ignoring cultural competence and issues of ethnicity and race may accentuate the disparities in treatment outcomes among diverse groups.

There is a well of resources in the literature on multicultural education and training in cultural competency. Brach and Fraserirector[57] reviewed the literature on cultural competence and health disparities and concluded that there are several topics that need to be covered in the training of cultural competency, including providing appropriate interpreter services, having policies for recruiting and retaining minority staff and providing cultural competence training, using minority community health workers, utilizing culturally competent health promotion techniques, including family or community members in the rehabilitation treatment process, immersing staff into community and cultural events, and implementing a diversity-friendly workplace. Another essential component of training, which is consistent with the literature presented above, is the opportunity to engage therapists and rehabilitation professionals in self-reflection about their perceived level of competence, challenges they experience, and their effectiveness in reaching out and treating clients from diverse ethnic backgrounds. Part of this self-reflection should include collecting feedback from clients. Self-reflection will allow the therapist to provide the best intervention possible.

23.4 Cultural Competence Issues to Consider Across Settings

MS rehabilitation professionals are likely to work in a variety of settings, including, but not limited to, home care, institutional care (e.g., acute care, inpatient care, nursing homes), and community care (e.g., rehabilitation settings, community agencies). Some of these settings call for attention to specific cultural beliefs, rituals, and traditions that affect a client's ways of engaging in everyday activities and interacting with others and their environments. Home care provides practitioners an opportunity to observe more closely routines and habits, artifacts, and symbols while institutional and community care settings allow practitioners the opportunity to work with the setting in developing policies and procedures conducive to culturally competent care.

Purnell and Paulanka[36] suggest exploring several dimensions of culturally competent care when working with diverse clients. These dimensions include verbal and nonverbal communication, temporal and space relationships, family organization and roles, health care practices, and the role of spirituality. We also suggest observing the role of the arts; artifacts and symbols; and ideas about work, occupation, and leisure. A synthesis of the aspects to consider under each dimension is included in Table 23.4.

TABLE 23.4

Dimensions to Consider in Culturally Competent MS Rehabilitation Across Settings

1. Verbal and nonverbal communication (across all settings)
   - Identify dominant language and dialects
   - Identify contextual speech patterns (volume and tone)
   - Observe gestures and nonverbal behavior
   - Explore cultural patterns in sharing feelings and thoughts
   - Explore practice and meaning of touch
   - Explore eye contact (presence or absence)
   - Explore facial expressions
   - Identify ways of greeting others (including family members, health care professionals, and strangers)
2. Temporal and spatial relationships (across all settings)
   - Observe distance kept when communicating with others
   - Explore concept of time (e.g., being on time for appointments)
   - Explore concepts of past, present, and future
   - Explore concept and use of space
3. Family organization and roles (across all settings)
   - Observe decision making (who makes decisions in what contexts?)
   - Identify gender roles (e.g., who is present at rehab appointment besides the client? Who is the main caregiver of the person with MS?)
   - Family organization (role of elderly, role of children, role of extended family)
4. Family rituals and traditions (most relevant in family care)
   - Be aware of taboos and restrictions
   - Identify rituals and traditions observed by the family and individual
5. Health care practices (most relevant in rehab, acute care, but can apply across settings)
   - Understand beliefs about health and health behaviors
   - Note who has responsibility for health care
   - Respect folk practices (e.g., nontraditional practices such as use of herbs and healers)
   - Understand beliefs about illness and rehabilitation
   - Understand beliefs about blood transfusion and organ donation
   - Identify perceptions the client holds about health care practitioners
   - Explore beliefs about MS
   - Explore beliefs about health care settings
   - Explore experience with health care settings in home country when working with immigrants
6. Role of spirituality (across all settings)
   - Explore religious/spiritual practices
   - Explore the use of prayer and meditation
7. Arts and artifacts (most relevant in rehab and family care)
   - Explore the meaning of symbols
   - Recognize the role of the arts and music in the culture
8. Cultural conceptions of work, occupations, and leisure (across all settings)
   - Explore culturally bound concepts about work and occupation
   - Explore culturally bound concepts about leisure
   - Explore collective ideas about recreation versus individual ideas
   - Explore culturally bound concepts of self

It is important to note that although dimensions like verbal and nonverbal communication apply across treatment settings (home care, acute care, rehab, nursing homes), specific aspects might be more relevant in some settings. For instance, exploring family rituals and traditions might be important to consider when working in home care or when an individual is transitioning from acute care to home care. For example, Asian families expect people who come into the home to take off their shoes and place them by the entrance door to avoid bringing in germs from the outside. Most Latino families would offer food or drink to visitors in their homes, and it would be impolite to refuse.

Organizational contextual factors that support cultural competence play a critical role in different settings. Institutional and organizational environments can embrace mission statements that allow for health professionals to go beyond the prescribed practice and adopt policies that transmit a message that embraces diversity. One example is the use of dual language text for all signage. Organizations and institutions that provide care have a responsibility for setting the stage for enforcing policies, practices, and programs that are culturally competent. Some of such policies that foster cultural competency should include practices like not scheduling appointments for assessments and intervention around important religious holidays, allowing practitioners to reach out to communities outside of their clinic or health center, and allow flexibility to adjust to cultural practices of the target group to facilitate active engagement.

23.5 Future Research

In this chapter we argue that delivering culturally competent rehabilitation services to people of diverse backgrounds with MS is of the utmost importance. However, many questions remain unanswered and need to be further explored to advance both practice and research in this area. Binetti[9] suggests several questions that researchers need to examine, such as: How do different ethnic and racial groups experience and live with MS? How do minority individuals like African Americans, Latinos, Asian Americans, and Native Americans differ in their responses to treatment? What are the specific needs related to managing the disease and self-management among different ethnic groups? And how does poverty affect access to MS medications and treatment?

To provide culturally appropriate MS rehabilitation treatment, we need to have well-trained, culturally competent practitioners and rehabilitation team members who are ready to take on these challenges. We need to further investigate how different cultural beliefs affect the rehabilitation process of people with MS and how cultural differences affect the participation of individuals in the self-management of the disease. What is the impact of using the strategies highlighted in this chapter on MS outcomes in terms of progression of the disease among diverse groups? In what ways does providing culturally competent rehabilitation care affect MS rehabilitation outcomes for people from diverse backgrounds?

Future research also needs to consider the inclusion of ethnically diverse populations in standardizing assessments used for screening and diagnostic purposes. Most tests utilized in MS rehabilitation have been standardized with White populations. Future research needs to expand the availability of culturally adapted assessment instruments to evaluate individuals from minority groups. Finally, clinical trials research needs to include more representatives from African Americans and Latinos in their samples.

23.6 Conclusion

Cultural competence is a process of *becoming*, which represents a commitment toward addressing well-documented disparities in health and rehabilitation. Understanding the experience of ethnically and racially diverse populations with MS is a complex process and requires investment of resources and time.

One might argue that a client-centered approach, as described elsewhere in this book (see Chapters 4 and 24), might be sufficient to address the needs of ethnically/racially diverse clients with MS. The fact is that the disparities in rehabilitation outcomes between Whites and non-Whites continue to increase. Until we address issues of culture and ethnicity/race, diverse groups will continue to experience difficulties in seeking services, being screened for MS, receiving the appropriate assessment of their level of functioning, developing rehabilitation goals, meeting rehabilitation goals, and receiving appropriate follow-up support and medications.

Becoming culturally competent is an intentional endeavor, a journey, and a lifelong process (see Balcazar et al.[27] for a synthesis of the literature on the topic). The development of cultural competence means our willingness to engage in a series of practices, like examining the organizational biases of traditional practices and services; being open and willing to accept individuals from other cultures; trying nontraditional interventions or changing standard procedures to fit individual needs; and challenging racist practices, discrimination, and oppression. But it also means to think differently about rehabilitation practice. We need to continue to develop a deeper knowledge and understanding of ethnicity/race, poverty, and oppression in the life experiences of diverse and disadvantaged individuals, especially ethnically diverse individuals with MS.

References

1. Maghzi AH, Ghazavi H, Ahsan M, Etemadifar M, Mousavi S, Khorvash F, Minagar. An increasing female preponderance of multiple sclerosis in Isfahan, Iran: A population-based study. Multiple Sclerosis 2010;16(3):359–361.
2. Cordova J, Vargas S, Sotelo J. Western and Asian features of multiple sclerosis in Mexican mestizos. Clinical Neurology and Neurosurgery 2007;109:146–151.
3. Corona T, Román GC. Multiple sclerosis in Latin America. Neuroepidemiology 2005;26:1–3.
4. Kira J. Multiple sclerosis in the Japanese population. Neurology 2003;2:117–127.
5. Cree BAC, Reich DE, Khan O, De Jager PL, Nakashima I, Takahashi T, Bar-Or A, Tong C, Hauser SL, Oksenberg JR. Modification of multiple sclerosis phenotypes by African ancestry at HLA. Archives of Neurology 2009;66(2):226–233.
6. Loveland CA. The experiences of African Americans and Euro-Americans with multiple sclerosis. Sexuality and Disability 1999;17(1):19–35.
7. Johnson SK, Terrell D, Sargent C, Kaufman M. Examining the effects of stressors and resources on multiple sclerosis among African Americans and whites. Stress and Health 2007;23:207–213.
8. Shabas D, Heffner M. Multiple sclerosis management for low-income minorities. Multiple Sclerosis 2005;11:635–640.
9. Binetti K. Living with MS: An African American perspective. MS Exchange [Internet]. [revised 2004;8(4). Available from http://www.msexchange.org.

10. Marrie RA, Cutter G, Tyry T, Campagnolo D. Does multiple-sclerosis-associated disability differ between races? American Academy of Neurology 2009;66:1235–1240.

11. Buchanan RJ, Martin RA, Zuniga M, Wang S, Kim M. Nursing home residents with multiple sclerosis: Comparisons of African American residents to white residents at admission. Multiple Sclerosis 2004;10:660–667.

12. National Healthcare Disparities Report [Internet]; c2008. Available from: www.ahrq.gov/qual/qrdr08.htm.

13. About minority health. Centers for Disease Control [Internet]. Available from: http://www.cdc.gov/omhd/AMH/AMH.htm.

14. Braithwaite R, Taylor S. Health issues in the black community. San Francisco, CA: Jossey-Bass; 2009.

15. Hill-Briggs F, Kelly K, Ewing C. Challenges to providing culturally competent care in medical rehabilitation to African Americans. In: F. Balcazar, Y. Suarez-Balcazar, T. Taylor-Ritzler, C. Keys, editors. Race, culture and disability: Rehabilitation science and practice. Sudbury, MA: Jones and Bartlett; 2010.

16. Smedley BD. Unequal treatment: Confronting racial and ethnic disparities in health care. Washington, DC: The National Academies Press; 2003.

17. Wilson KB. Exploration of VR acceptance and ethnicity: A national investigation. Rehabilitation Counseling Bulletin 2002;45(3):168–176.

18. Wilson KB, Senices J. Cultural access to vocational rehabilitation services for black Latinos with disabilities: Colorism in the 21st century. In: F. Balcazar, Y. Suarez-Balcazar, T. Taylor-Ritzler, C. Keys, editors. Race, culture and disability rehabilitation science and practice. Sudbury, MA: Jones and Bartlett Publishers; 2010.

19. Balcazar FE, Suarez-Balcazar Y, Taylor-Ritzler T, Keys C. Race, culture, and disability: Rehabilitation science and practice. Sudbury, MA: Jones and Bartlett; 2010.

20. Kumas-Tan ZO, Beagan BL, Loppie C, MacLeod A, Frank B. Measuring cultural competence: Examining hidden assumptions in instruments. Academic Medicine 2007;82(6):548–557.

21. Zalaquett CP, Foley PM, Tillotson K, Dinsmore JA, Hof D. Multicultural and social justice training for counselor education programs and colleges of education: Rewards and challenges. Journal of Counseling and Development 2008;86:323–329.

22. Bonder BR, Martin L, Miracle AW. Culture emerging in occupation. American Journal of Occupational Therapy 2004;58:159–168.

23. Gladding ST. The counseling dictionary: Concise definitions of frequently used terms. Upper Saddle River, NJ: Prentice-Hall; 2001.

24. Fiske AP, Kitayama S, Markus HR, Nisbett RE. The cultural matrix of social psychology. In: D. T. Gilbert, S. T. Fiske, G. Lindzey, editors. The handbook of social psychology. 4th ed. Boston, MA: McGraw-Hill; 1998.

25. Black RM, Wells SA. Culture and occupation: A model of empowerment in occupational therapy. Bethesda, MD: AOTA Press; 2007.

26. Cross TL, Bazron BM, Dennis KW, Isaacs MR. Towards a culturally competent system of care: A monograph on effective services for minority children who are severely emotionally disturbed, Vol. 1. Washington, DC: Georgetown University, Child and Adolescent Service System Program; 1989.

27. Balcazar FE, Suarez-Balcazar Y, Willis C, Alvarado F. Cultural competence: A review of conceptual frameworks. In: F. Balcazar, Y. Suarez-Balcazar, T. Taylor-Ritzler, C. Keys, editors. Race, culture and disability: Rehabilitation science and practice. Sudbury, MA: Jones and Bartlett; 2010.

28. Suarez-Balcazar Y, Rodakowski J. Becoming a culturally competent occupational therapy practitioner. OT Practice 2007, September;12(17):14–17.

29. National Association of Multicultural Rehabilitation Concerns [Internet]; c2009. Available from: http://nationalrehabvaassoc.weblinkconnect.com/cwt/external/wcpages/divisions/namrc.aspx.

30. American Occupational Therapy Association. Occupational therapy practice framework: Domain and process. 2nd ed. American Journal of Occupational Therapy 2008;62:625–683.
31. Suarez-Balcazar Y, Balcazar F, Taylor-Ritzler T, Rodakowski JT, Garcia-Ramirez M, Willis C. Development and validation of the cultural competence assessment instrument: A factorial analysis. Journal of Rehabilitation 2011;77:4–13.
32. Suarez-Balcazar Y, Rodakowski J, Balcazar F, Taylor-Ritzler T, Portillo N, Barwacz D, Willis C. Perceived levels of cultural competence among occupational therapists. American Journal of Occupational Therapy 2009;63(4):496–503.
33. McIntosh P. White privilege: Unpacking the invisible knapsack. Peace and Freedom 1989;July/August:10–12.
34. Taylor-Ritzler T, Balcazar F, Dimpfl S, Suarez-Balcazar Y, Willis C, Schiff R. Cultural competence training with organizations serving people with disabilities from diverse cultural backgrounds. Journal of Vocational Rehabilitation 2008;29:77–91.
35. Taylor-Ritzler T, Balcazar F, Dumfl S, Willis C, Suarez-Balcazar. Empirical validation of McIntosh's observations of white privilege. Under review.
36. Purnell LD, Paulanka BJ. Transcultural health care: A culturally competent approach. Philadelphia, PA: FA Davis Company; 2003.
37. Stone JH. Culture and disability: Providing culturally competent services. Thousand Oaks, CA: Sage Publications; 2005.
38. Atdjian S, Vega WA. Disparities in mental health treatment in U.S. racial and ethnic minority groups: Implications for psychiatrists. Psychiatric Services (Washington, DC) 2005;56(12):1600–1602.
39. Balcazar FE, Suarez Balcazar Y, Taylor-Ritzler T. Cultural competence: Development of a contextual framework. Disability Rehabilitation 2009;31(14):1153–1160.
40. Alston RJ, Harley D, Middleton K. The role of rehabilitation in achieving social justice for minorities with disabilities. Journal of Vocational Rehabilitation 2006;24(3):129–136.
41. Buchanan RJ, Zuniga MA, Carrillo-Zungia G, Chakrovorty BJ, Tyty T, Moreau RL, Huang C. Comparison of Latinos, African Americans, and Caucasians with multiple sclerosis. Ethnicity & Disease 2010;20:451–457.
42. Cook JA, Razzano LA, Jonikas JA. Cultural diversity and how it may differ for programs and providers serving people with psychiatric disabilities. In: F. Balcazar, Y. Suarez-Balcazar, T. Taylor-Ritzler, C. Keys, editors. Race, culture and disability: Rehabilitation science and practice. Sudbury, MA: Jones and Bartlett; 2010.
43. Alston RJ, Bell TJ. Cultural mistrust and the rehabilitation enigma for African Americans. Journal of Rehabilitation 1996;62:16–20.
44. Adegbembo AO, Tomar SL, Logan HL. Perceptions of racism explain the difference between blacks' and whites level of healthcare trust. Ethnicity & Disease 2006;16(4):792–798.
45. Cree BAC, Khan O, Bourdette D, Goodin DS, Cohen JA, Marrie RA, Glidden D, Weinstock-Guttman B, Reich D, Patterson N, et al. Clinical characteristics of African Americans vs. Caucasian Americans with multiple sclerosis. Neurology 2004;63:2039–2045.
46. Hasnain R, Kondratowicz D, Portillo N, Balcazar F, Johnson T, Gould R, Borokhovski E, Bernard RM, Hanz K. The use of culturally adapted competency interventions to improve rehabilitation service outcomes for culturally diverse individuals with disabilities: A systematic review and meta-analysis. Campbell Collaboration. Under review.
47. National Standards for Culturally and Linguistically Appropriate Services in Health Care Executive Summary [Internet]; c2001. Available from: http://minorityhealth.hhs.gov/assets/pdf/checked/executive.pdf.
48. Block P, Balcazar F, Keys C. Race, poverty and disability: Three strikes and you're out! or are you? Social Policy 2002;33(1):34–38.
49. Hahn R. Sickness and healing: An anthropological perspective. New Haven, CT: Yale University Press; 1995.

50. Taylor-Ritzler T, Suarez-Balcazar Y, Balcazar F, Garcia-Iriarte E. Conducting disability research with people from diverse ethnic groups: Challenges and opportunities. Journal of Rehabilitation 2008;74(1):4–11.
51. Hernández B, Horin EV, Donoso OA, Saul A. Psychological testing and multicultural populations. In: F. Balcazar, Y. Suarez-Balcazar, T. Taylor-Ritzler, C. Keys, editors. Race, culture and disability: Rehabilitation science and practice. Sudbury, MA: Jones and Bartlett; 2010.
52. Maramba GG, Hall GCN. Meta-analysis of ethnic match as a predictor of dropout, utilization and level of functioning: Brief report. Cultural Diversity and Ethnic Minority Psychology 2002;8(30):290–297.
53. Sue S, Zane N, Hall GCN, Berger LK. The case for cultural competency in psychotherapeutic interventions. The Annual Review of Psychology 2009;60:525–548.
54. Royeen M, Crabtree JL. Culture in rehabilitation from competency to proficiency. Upper Saddle River, NJ: Pearson Prentice Hall; 2006.
55. Garate T, Charlton J, Luna R, Townsend O. Implications for practice in rehabilitation. In: F. Balcazar, Y. Suarez-Balcazar, T. Taylor-Rtizler, C. Keys, editors. Race, culture and disability rehabilitation science and practice. Sudbury, MA: Jones and Bartlett; 2010.
56. Charlton JI. Nothing about us without us: Disability oppression and empowerment. Berkeley, CA: University of California Press; 1998.
57. Brach C, Fraserirector I. Can competency reduce racial and ethnic health disparities? A review and conceptual model. Medical Care Research and Review 2000;57(1):181–217.

Section V

Moving into the Future

24

Optimizing Rehabilitation Experiences

Marcia Finlayson and Lisa I. Iezzoni

CONTENTS

Previous chapters in this book have described how multiple sclerosis (MS) typically strikes people in early adulthood, just as they are finding partners, starting families, entering the labor force, building careers, and planning their futures. Because MS is currently incurable, many people can anticipate living for decades with a condition that has an uncertain course and often becomes more disabling over time. Over the last 10–20 years, there has been increasing recognition that health care systems worldwide are not well designed to serve people with chronic conditions,[1-5] such as MS. Instead, these systems are designed to serve people with acute illnesses who can be diagnosed, treated, cured, and then sent home to continue on as before. For people with chronic conditions and disabilities, the return home means finding ways to perform daily tasks and pursue those activities that give meaning and pleasure to their lives. Their family members and close friends often join in these efforts.

While rehabilitation providers often work intensely with their patients for several weeks or months at a time, the reality is that—for the most part—we move into and out of people's lives at critical junctures only, for example, during an MS exacerbation or when a significant new functional problem has emerged. Rarely do we have opportunities to maintain longitudinal relationships with our patients across the full continuum of their health and functioning, over time, or across settings. As a result, it is imperative that we understand their preferences and values for their various activities in their homes and communities. We must provide the highest quality care possible to enable them to maximize their abilities and achieve their goals when we are no longer actively involved in their care.

Building off the foundational work of Donabedian,[6] de Jonge et al.[7] described three domains that must be considered when examining quality in health care: structural (e.g., availability of resources, provider skill), process (e.g., service access, provider–patient interactions), and clinical (e.g., changes in patient health status, patient satisfaction). Problems in any of these levels put safety, effectiveness, efficiency, equity, timeliness, and, most important—our relationships with our patients—at risk. Only through high-quality MS rehabilitation services can we effectively support people with MS in managing this chronic condition and its ongoing impact on their everyday lives.

Internationally, several important reports have been released since 2001 advocating for health care system changes to address the realities of living with chronic conditions.[1,3,5] While the reports and the recommendations they contain tend to focus primarily on medical care, they are certainly applicable to the design and delivery of MS rehabilitation services. Drawing on these broader dialogues can only strengthen our efforts to ensure high-quality rehabilitation experiences for people with MS. After reading this chapter, you will be able to:

1. Summarize key shifts in thinking about the delivery of health care to better serve people with chronic conditions;

2. Examine key recommendations emanating from these shifts that can be used to engage in continuous quality improvement in MS rehabilitation, particularly:

   a. Knowledge-based care,

   b. Patient- or client-centered care, and

   c. Systems-minded care;

3. Discuss strategies to monitor and constantly improve the experiences that people with MS have during rehabilitation.

24.1 Shifts in Thinking about Health Care

Defining, measuring, and working to constantly improve quality are priorities for many industries and form emerging foci in health care and rehabilitation.[7] Although quality of health care is embedded within structures, processes, and clinical outcomes, Berwick[8] explains that the patients' experiences in the system must be the "truth north" for determining the quality of service delivery. Were the resources necessary to meet their needs available, accessible, and affordable? Did they receive respectful care from everyone on the team from the moment they arrived in the clinic? Were they provided with the information they needed, when they needed it, so they could make informed choices about treatment options? Was that information provided in a manner that was accessible and understandable to them? Were their preferences, values, and context given priority during the course of goal setting and treatment planning? Were they satisfied with the services received? Were clinically meaningful changes observed as a consequence of intervention? Did individuals and systems respond in a timely manner when gaps in quality were identified?

While MS rehabilitation providers likely aim for "yes" answers to all of these questions, evidence suggests shortfalls in achieving these goals. Persons with MS often struggle to gain access to the rehabilitation services they want or need.[9-14] Even when access is available, authors throughout this book have made it clear that evidence is not always available to indicate which specific intervention approach is the best or most effective or what constitutes a clinically meaningful change in outcome. Evidence also indicates that people with MS do not always experience positive interactions with their health care providers.[15,16] For example, Thorne et al.[17] write:

> Participants [people with MS] recounted certain practices ... that were particularly unhelpful. For example, when health care professionals assumed that participants had a low level of knowledge and they needed only essential information presented to them in the most simplistic of terms, [their] fear and frustration tended to escalate ... These kinds of communication were considered "patronizing" in that they failed to take into account the participants' increasing familiarity with the meaning of their symptoms and their increased understanding of clinical terminology .... (p. 11)

Internationally, individuals and organizations are working diligently to limit these types of negative experiences and ensure that systems and their personnel are striving constantly to improve the quality of health care delivery.[1-5] For example, in 2001, the Institute of Medicine (IOM)[1] in the United States released a report titled *Crossing the Quality Chasm: A New Health System for the 21st Century*. The authors of the report called for a redesign of the US health care system in order to reduce the "chasm" between patients' expectations for quality health care and what is actually delivered. The recommendations stemming from the IOM report were organized using a four-level framework in which the experience of the patient is the central determinant of quality (level 1).[8] To improve patients' experiences, Berwick[8] summarized how changes must occur among service delivery teams (level 2, e.g., an MS rehabilitation team), health care organizations (level 3), and the overall health care environment (level 4, e.g., financing, accreditation, regulation). The proposed changes echo recommendations made in more recent reports released by the National Health Service in the United Kingdom,[18] the Picker Institute Europe,[19] the World Health Organization,[5] and with models for improving care for people with chronic conditions.[2,4]

While the IOM report is specific to the US health care system, it contains information and recommendations that are internationally relevant for ensuring that people receiving MS rehabilitation obtain the best quality of care possible. Berwick[8] summarized three core principles for redesigning service delivery teams based on the IOM document. The three principles are knowledge-based care, patient-centered care, and systems-minded care. These principles provide a solid foundation on which to develop comprehensive strategies to optimize the experience and quality of rehabilitation for people with MS as we move into the future.

24.1.1 Principle #1: Knowledge-Based Care

Knowledge-based care is more than just evidence-based practice; it also includes ensuring that clinicians and patients have access to information when and where they need it, in a format they can understand, so that they can make informed decisions.[8,20] As a result, knowledge-based care can be viewed as dependent on three related elements (see Figure 24.1).

24.1.1.1 Element #1: Documenting Needs

Pursuit of knowledge-based care requires a clear understanding of what information rehabilitation stakeholders want and need, and where and when they need it. Stakeholders refer primarily to clinicians and patients, but family members, managers, and third-party payers also have wants and needs related to rehabilitation service delivery. Comprehensive, holistic, and multi-method needs assessments are critical for obtaining information on

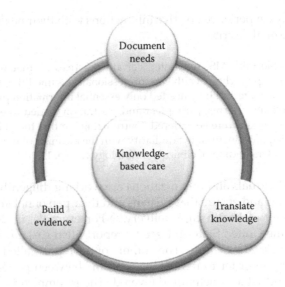

FIGURE 24.1
Elements of knowledge-based care.

stakeholder needs, and they are also necessary to inform knowledge-based care. Because needs assessments are inherently political and value laden, care must be given to the guiding definition of need (e.g., normative, felt, expressed, comparative[21]) and whose perspective is given precedence in the development of subsequent recommendations.[22] Attention must also be given to differences in information needs over time and across settings for all potential stakeholders.

24.1.1.2 Element #2: Building Evidence

Knowledge-based care also depends on building high-quality programs of research that focus on developing, implementing, and evaluating processes and interventions to address identified needs. These programs of research are necessary to generate new knowledge and evidence that support knowledge-based care and, ultimately, quality rehabilitation experiences for people with MS. Studies that have already identified gaps in service access[9–14] or particular needs in terms of patient–provider interactions[15–17] provide ample opportunities to build programs of research to address structural and process outcomes in MS rehabilitation. In addition, several promising research directions have been identified throughout this book in relation to specific clinical outcomes across all areas of functioning—impairment, activity, and participation. Finally, research opportunities also abound to address cost-effectiveness, cost–benefit, and comparative effectiveness in MS rehabilitation. Collectively, the pursuit of new knowledge is the only way to unpackage the "black box" of MS rehabilitation, identify its active ingredients, and determine whether or not it enables people to fully participate in everyday life.

24.1.1.3 Element #3: Constructing an Infrastructure for Knowledge Translation

Knowledge-based care also requires an effective infrastructure through which available evidence is systematically synthesized into relevant tools and then disseminated in a timely way. In response, there is a growing emphasis on knowledge translation in health care, including the development and dissemination of decision-support tools for both clinicians and patients. The tools for clinicians often take the form of clinical practice guidelines,[23] while the tools for patients are often referred to as "patient decision aids" and discussed in the context of shared decision making.[24] Regardless of who will use a particular decision tool, Kitson[25] describes how successful knowledge translation must address:

- How people access and understand new knowledge,
- The extent to which people can use new knowledge to make informed and autonomous decisions that improve outcomes, and
- Whether systems are in place to facilitate and support changes in resources, practices, and contexts that improve care.

The efforts of groups such as the Cochrane Collaboration,[26] the Campbell Collaboration,[27] the Centre for Reviews and Dissemination,[28] and the National Guideline Clearinghouse[29] have made significant contributions to knowledge translation efforts in health care. Yet, from the literature it is clear that there are multiple factors at the level of the individual (e.g., attitude towards research, available time), organization (e.g., productivity demands, funding constraints) and system (e.g., regulation, accreditation) that influence the process of translating research into action in rehabilitation and other areas of health care,[30–33] Nevertheless, the imperative of knowledge-based care is clear. Without it, patients will be subject to inconsistencies in service delivery that are a function of local service delivery habits and routines rather than available evidence.[8]

In MS rehabilitation, knowledge-based care requires that people with MS have consistent, high-quality experiences that are informed by rigorously developed clinical practice guidelines. Several organizations have developed guidelines for MS care, for example, the American Academy of Neurology, the National Institute of Clinical Excellence, and the European Multiple Sclerosis Platform.[34] Nevertheless, existing guidelines have focused primarily on disease-modifying therapies (pharmaceuticals), medical treatment of acute relapses, and use of MRI technology.[34] As a result, Trisolini[34] has called for more attention to symptom management and rehabilitation, and more focus on outcomes that matter to people with MS, such as quality of life.

While rehabilitation-specific clinical practice guidelines have the potential to strengthen the quality of MS rehabilitation service delivery, supports for people with MS are also needed as part of a comprehensive knowledge translation infrastructure. People with MS must have easy access to information about rehabilitation service options (evidence, benefits, harms) so that they can make informed choices about which providers and intervention (s) best fit with their concerns, goals, personal values, and context. Information format must account for differences in health literacy as well as cultural and economic variations in information-seeking behaviors.

Health literacy reflects "the degree to which individuals have the capacity to obtain, process, and understand basic health information and services needed to make appropriate health decisions" (p. 3).[35] Consequently, it depends not only on a person's literacy level but also on the clarity and presentation of content. Written materials must have a clear purpose, be behavior or action oriented (rather than presenting abstract concepts), ensure that the most important information is provided first, connect with and build onto the person's current knowledge and context, and use formatting and graphic elements to emphasize and clarify information.[36,37] Web-based materials have many of the same requirements, but issues of navigation, interactivity, and placement of different information on the page also come into play.[35]

In sum, achieving knowledge-based care in MS rehabilitation will be challenging. Currently, the knowledge about the wants and needs of rehabilitation stakeholders is limited, evidence to support many of the processes and interventions used during rehabilitation is patchy, and consistent and effective knowledge translation mechanisms are not in place. To achieve knowledge-based care in MS rehabilitation, we must:

- Examine the structural, process, and clinical needs of stakeholders (e.g., people with MS, families, providers, managers, third-party payers);
- Build more evidence in all areas of clinical rehabilitation (body structure and function, activity, and participation);
- Engage in economic evaluations and comparative effectiveness research;
- Develop easily accessible and useful syntheses of rehabilitation-specific evidence for application throughout the rehabilitation process (e.g., clinical reasoning algorithms, clinical practice guidelines); and—most importantly
- Develop decision-making tools for people with MS and their families so that they can make fully informed decisions about their rehabilitation based on expected outcomes, risks, and benefits.

24.1.2 Principle #2: Patient- or Client-Centered Care

Patient-centered care is a core principle for redesigning health care systems worldwide.[1,5,19,38,39] The core principles of patient-centered care (also called client-centered care or person-centered care) can be tracked back over 60 years to Carl Rogers, a counselor and psychotherapist, who coined the term client-centered therapy.[40]

In a paper in 1946, Rogers described key elements of client-centered therapy, including good communication, belief in the client's capacity, and a focus on creating an environment in which the client can work toward his or her own goals.[41] In particular, Rogers noted that clients possess "constructive forces whose strength and uniformity have been either entirely unrecognized or grossly underestimated" (p. 418). He went on to emphasize that these forces must be trusted if therapy is to be successful and that therapists must not impose their goals and preferences on the client. Instead, the therapist must look to the client for direction. These basic ideas continue to form the foundation of current approaches to patient-centered care, which is a fundamental component of best practices in MS rehabilitation.[42,43,44]

In an extensive review of person-centered health care across medicine, nursing, and occupational therapy, Dow et al.[38] found that its essence was best described as a collaborative, respectful partnership between the health care provider and the person. Developing these partnerships requires much more than just asking a person with MS what his or her rehabilitation goals are. True client-centered care requires sharing power and decision-making responsibility throughout the course of care.[38,40,45] This partnership is critical because people will return home and will use only those aspects of rehabilitation that enable them to engage in the activities that are most relevant and meaningful to them. In other words, they will make choices independent of their rehabilitation providers.

Shared decision-making is considered a global health care concern because health care providers often do not recognize, or are slow to recognize, that people want to understand their health conditions, know their options, and make their own decisions about their care.[46] According to Coulter and Collins,[47] "Shared decision-making is a process in which clinicians and patients work together to select tests, treatments, management or support packages, based on clinical evidence and the patient's informed preferences. It involves the provision of evidence-based information about options, outcomes and uncertainties, together with decision support counseling and a system for recording and implementing patients' informed preferences" (p. vii).

Shared decision making is only now gaining attention in MS care, particularly in relation to decisions about disease-modifying therapies.[24,48] Because shared decision making is a critical component of client-centered care,[38] it is necessary to pay greater attention to it in MS rehabilitation. Shared decision making acknowledges that people with MS have the right to ask questions about their rehabilitation options and seek the information and advice they need to make informed decisions. It also acknowledges that people have the right to follow, or not to follow, the advice of their rehabilitation providers. This presumes that providers have met their obligations and responsibilities,[19,38,39] which include:

- Ensuring that the person has been given accurate, up-to-date, and comprehensive information, including risks and benefits, in a form that he or she can understand;
- Being accessible for ongoing questions and clarifications;
- Involving family and friends; and
- Providing the time and emotional support necessary for the person to make an informed decision.

Clearly, patient-centered care and shared decision making require excellent communication skills, easily accessible evidence, and a supportive infrastructure.[38,40,49,50] All of these potential barriers raise the question, Is it worth it? Evidence suggests that it is. Compared to standard care, a client-centered approach is more effective for improving recipients' satisfaction with services;[51] facilitating clients' disclosure of problems;[52] strengthening overall quality of care, patient safety, and adherence to treatment regimes;[53] and producing better clinical outcomes and lower costs.[54-58]

Overcoming the barriers to client-centered care requires a multipronged approach. First, rehabilitation providers must work continually to develop and strengthen their communication skills (e.g., asking open-ended questions, probing, paraphrasing, summarizing, reflecting feelings, clarifying) in order to more effectively engage their patients throughout the care process.[59,60] Second, focused attention must be given to the broader lifestyle of the person so that processes can be tailored and culturally appropriate.[40,53] Third, rehabilitation providers must develop expertise to engage in collaborative goal setting, to build patients' self-management capacity, and to incorporate patient decision tools in their practices. Finally, greater emphasis must be placed on interprofessional education and collaboration,[20,38,39] and strategies must be developed to give providers regular feedback to promote practice change.[20,38,39] Without feedback, practice change is unlikely to occur.[61]

24.1.3 Principle #3: Systems-Minded Care

The final IOM principle for redesigning the health care system to improve quality and patient experiences is systems-minded care.[8] Providers who commit to systems-minded care take responsibility to coordinate, integrate, and work efficiently across settings, disciplines, and roles. Systems-minded care is particularly important for people who have chronic conditions such as MS because their needs shift and change over time, requiring input from different providers and settings at different points in their disease course.

Within systems-minded care, the small unit of people who actually provide care to the patient is referred to as a "microsystem" (e.g., a rehabilitation team). Berwick[8] summarized the 10 "simple rules" for redesigning health care microsystems that were presented in the IOM report for improving the quality of patient experiences. These rules can be used to envision systems-minded care in MS rehabilitation and are summarized in Table 24.1.

TABLE 24.1

Rules for Redesigning Health Care Microsystems Applied to MS Rehabilitation

Ten Simple Rules for Improving Quality of Patient Experiences during MS Rehabilitation	Possible Strategies for Implementation
Care is based on continuous relationships.	Create infrastructures that permit regular follow-up with rehabilitation patients (e.g., regular telephone contacts to "check-in"). Assign the same provider(s) to the person with MS at each visit.
Care is customized to address the needs and concerns of the person with MS.	Use tools such as the *Canadian Occupational Performance Measure*[62] to elicit client concerns to guide MS rehabilitation goals and interventions. Engage in collaborative goal setting and shared decision-making about interventions on an ongoing basis.
The person with MS is the source of control.	Develop patient decision aids and systems that are attentive to issues of health literacy, culture, and socioeconomic differences. Provide information people need, when they need it, in the format they prefer. Recognize and accept that people may not follow your advice.
Knowledge is shared freely with the person with MS.	Use patient decision-making tools. Have patient education information available in different formats. Ensure that people have contact information if they have questions between appointments. Have systems in place to permit timely responses to people's questions, all day and every day. Share risks, benefits, and potential consequences of intervention options.
Clinical decisions are based on evidence.	Use clinical practice guidelines when they are available. Operate journal clubs to stay up-to-date on available evidence. Partner with local researchers to generate evidence needed for your practice.
Safety is a built-in feature of the rehabilitation process.	Provide patients with information about the risks and consequences of treatment options.
Transparency is necessary.	Monitor the performance of rehabilitation teams (e.g., quality of care, outcomes, safety). Make performance information readily accessible to people (e.g., online report cards for rehabilitation facilities and providers). Make information about MS expertise of individual team members available to people.
Needs of the person with MS are anticipated.	Use knowledge of the disease and prognosis to raise issues for the person to consider (e.g., future home modification).
Waste is continuously decreased.	Examine the possibility of delivering interventions in groups. Find ways of ensuring that equipment that is not being used is recycled. Use rehabilitation support personnel effectively.
Cooperation among members of the rehabilitation team is a priority.	Take the time to learn and understand what each member of the team can do. Coordinate appointments to ease the burden on the person and family. Use electronic medical records (or shared records) so that information can be easily shared. Offer interdisciplinary care.

24.2 Stepping Forward to Optimize MS Rehabilitation

The authors contributing to this book used their expertise to summarize the best science available to support MS rehabilitation. Nevertheless, it is clear that high-quality rehabilitation requires more than effective interventions and good clinical outcomes. It also requires evidence of effective and equitable rehabilitation structures; efficient, safe, timely, and patient-centered care processes; and satisfied service users.[7,8] While much progress has been made since the first randomized trial of inpatient rehabilitation for people with MS in 1997,[63] many challenges lie before us.

If the "true north" of health care quality is the patient's experience,[8] rehabilitation researchers need to document how people with MS and their families perceive rehabilitation care. This documentation must include information about patients' experiences of safety, effectiveness, efficiency, timeliness, equity, and patient-centeredness.[8] This work must also examine the extent to which patients' experiences meet their expectations of rehabilitation across each dimension of quality (structure, process, clinical outcomes). The knowledge gained through these inquiries can then be used to define ways of reliably monitoring and continually improving how MS rehabilitation is accomplished.

It is also imperative that rehabilitation providers and researchers commit to all elements of knowledge-based care—identifying stakeholder needs, finding effective ways to address those needs, and translating evidence into action. This commitment necessitates partnership across many sectors—patients, families, providers, managers, third-party payers, and researchers. Diverse knowledge and skills will be required to achieve knowledge-based care, for example, qualitative methodologies, clinical trials, comparative effectiveness research, health services research, health economics, theories of behavior change (provider and patient), adult learning, and so on.

Embracing the full breadth and depth of person-centered care will require that rehabilitation providers move past simply asking people with MS about their rehabilitation goals. Instead, a fundamental shift in power must occur (from provider to patient) and recognition that people with MS are the true experts in their own health and functioning. Efforts to develop excellent communication skills must be ongoing so that collaborative, respectful partnerships can be developed between providers and patients. These partnerships must support informed decision making and self-management, well beyond discharge from rehabilitation.

Finally, rehabilitation providers must continuously seek ways to monitor and improve the quality of the services they deliver. People with MS have an often debilitating, incurable condition. Rehabilitation providers are in a unique position to provide information and support and to share their expertise in a way that can make it easier for people with MS to engage in the aspects of their lives that are important and meaningful to them. Rehabilitation providers have a responsibility to collaborate, educate, coach, and advise so that the decisions that people with MS make about health and functioning are fully informed and fit within the context of their everyday lives. Nevertheless, to determine if progress toward these outcomes is being made, rehabilitation professionals must monitor and measure the full range of outcomes that determine quality of care.

We must always be asking ourselves: How are we doing? Can we do better? Continuously seeking answers to these questions and acting on what we learn are necessary steps to optimize the experience of rehabilitation for people with MS.

References

1. Crossing the quality chasm: A new health system for the 21st century (free executive summary) [Internet]; c2001. Available from: http://www.nap.edu/catalog/10027.html.
2. Barr V, Robinson S, Marin-Link B, Underhill L, Dotts A, Ravensdale D, Salivaras S. The expanded chronic care model: The integration of concepts and strategies from population health promotion and the chronic care model. Hospital Quarterly 2003;7(1):73–81.
3. Department of Health. Creating a patient led NHS: Delivering the NHS improvement plan. London: National Health Service; 2005.
4. Wagner EH. Chronic disease management: What will it take to improve care for chronic illness? Effective Clinical Practice 1998;1(1):2–4.
5. World Health Organization. Innovative care for chronic conditions: Building blocks for action. A global report. Geneva, Switzerland: WHO; 2002.
6. Donabedian AE. Explorations in quality assessment and monitoring. Ann Arbor, MI: Health Administration Press; 1980.
7. de Jonge V, Nicolass JS, van Leerdam ME, Kuipers EJ. Overview of the quality assurance movement in health care. Best Practice & Research Clinical Gastroenterology 2011;25:337–347.
8. Berwick DM. A user's manual for the IOM's "quality chasm" report. Health Affairs 2002;21(3):80–90.
9. Ytterberg C, Johansson S, Andersson M, Widén Holmqvist L, von Koch L. Variations in functioning and disability in multiple sclerosis. A two-year prospective study. Journal of Neurology 2008;255:967–973.
10. Cup EHC, Pieterse AJ, Knuijt S, Hendricks HT, van Engelen BGM, van Oostendorp RAB. Referral of patients with neuromuscular disease to occupational therapy, physical therapy, and speech therapy: Usual practice versus multidisciplinary advice. Disability & Rehabilitation 2007;29(9):717–726.
11. Minden SL, Frankel D, Hadden L, Hoaglin DC. Access to health care for people with multiple sclerosis. Multiple Sclerosis 2007;13(4):547–558.
12. Buchanan RJ, Kaufman M, Zhu L, James W. Patient perceptions of multiple sclerosis-related care: Comparisons by practice specialty of principal care physician. NeuroRehabilitation 2008;23(3):267–272.
13. Forbes A, While A, Mathes L. Informal carer activities, care burden and health status in multiple sclerosis. Clinical Rehabilitation 2007;21(6):563–575.
14. Plow M, Cho C, Finlayson M. Utilization of health promotion and wellness services among middle-aged and older adults with multiple sclerosis in the mid-west US. Health Promotion International 2010;25(3):318–330.
15. Malcomson KS, Lowe-Strong AS, Dunwoody L. What can we learn from the personal insights of individuals living and coping with multiple sclerosis? Disability & Rehabilitation 2008;30(9):662–674.
16. Finlayson M, Van Denend T, Shevil E. Multiple perspectives on the health service need, use and variability among older adults with multiple sclerosis. Occupational Therapy in Health Care 2003;17(3/4):5–25.
17. Thorne S, Con A, McGuinness L, McPherson G, Harris SR. Health care communication issues in multiple sclerosis: An interpretive description. Qualitative Health Research 2004;14(1):5–22.
18. Ham C. Improving care for people with long-term conditions: A review of UK and international frameworks. Birmingham, UK: University of Birmingham Health Services Management and NHS Institute for Innovation and Improvement; 2006.
19. Picker Institute Europe. Using patient feedback. Oxford: Picker Institute Europe; 2009.
20. Washburn ER. Fast forward: A blueprint for the future from the institute of medicine. Physician Executive 2001;27(3):8–14.
21. Bradshaw JL. A taxonomy of social need. In G. McLachlan, editor. Problems and progress in medical care: Essays on current research. London: Oxford University Press; 1972.

22. Finlayson M. Assessing needs for services. In: G. Kielhofner, editor. Scholarship in occupational therapy: Methods of inquiry for enhancing practice. Philadelphia, PA: FA Davis; 2006.
23. Institute of medicine of the National Academies [Internet]; c2011 [cited 2011]. Available from: http://www.iom.edu/.
24. Heesen C, Kasper J, Kopke S, Richter T, Segal J, Muhlhauser I. Informed shared decision making in multiple sclerosis—Inevitable or impossible? Journal of Neurological Sciences 2007;259(1–2):109–117.
25. Kitson AL. The need for systems change: Reflections on knowledge translation and organizational change. Journal of Advanced Nursing 2009;65(1):217–228.
26. The Cochrane Collaboration [Internet]; c2011 [cited 2011]. Available from: http://www.cochrane.org/.
27. The Campbell Collaboration [Internet]; c2011 [cited 2011]. Available from: http://www.campbellcollaboration.org/.
28. Centre for Reviews and Dissemination [Internet]; c2011 [cited 2011]. Available from: http://www.york.ac.uk/inst/crd/.
29. National Guideline Clearinghouse [Internet]; c2011 [cited 2011]. Available from: http://www.guideline.gov/.
30. Graham ID, Logan J, Harrison MB, Straus SE, Tetroe J, Caswell W, Robinson N. Lost in knowledge translation: Time for a map? The Journal of Continuing Education in the Health Professions 2006;26(1):13–24.
31. Metzler MJ, Metz GA. Analyzing the barriers and supports of knowledge translation using the PEO model. Canadian Journal of Occupational Therapy 2010;77:151–158.
32. Menon A, Korner-Bitensky N, Kastner M, McKibbon KA, Straus S. Strategies for rehabilitation professionals to move evidence-based knowledge into practice: A systematic review. Journal of Rehabilitation Medicine 2009;41(13):151–158.
33. Girard A, Rochette A, Fillion B. Knowledge translation and improving practices in neurological rehabilitation: Managers' viewpoint. Journal of Evaluation in Clinical Practice 2011; doi: 10.1111/j.1365–2753.2011.01769.x.
34. Comparison of multiple sclerosis guidelines underscores need for collaboration [Internet]; c2008. Available from: http://www.guideline.gov/expert/expert-commentary.aspx?id=16443&search=multiple + sclerosis.
35. U.S. Department of Health and Human Services, Office of Disease Prevention and Health Promotion. Health literacy online: A guide to writing and designing easy-to-use health web sites. Washington, DC: Author; 2010.
36. Griffin J, McKenna K, Tooth L. Written health education materials: Making them more effective. Australian Occupational Therapy 2003;50:170–177.
37. Hoffman T, Worrall L. Designing effective written health education materials: Considerations for health professionals. Disability and Rehabilitation 2004;26:1166–1173.
38. Dow B, Haralambous B, Bremner F, Fearn M. What is person-centred health care? A literature review. Melbourne: Department of Human Services; 2006.
39. Lewis S. Patient-centered care: An introduction to what it is and how to achieve it. A discussion paper for the Saskatchewan Ministry of Health. Saskatoon, SK: Access Consulting Ltd.; 2009.
40. Law M, editor. Client-centered occupational therapy. Thorofore, NJ: SLACK Incorporated; 1998.
41. Rogers, C. Significant aspects of client-centered therapy. Paper given at a seminar of the staffs of the Menninger Clinic and the Topeka Veteran's Hospital on May 15 1946 and later published in American Psychologist, 1(1), 415–422.
42. European Multiple Sclerosis Platform. Recommendations for rehabilitation services for persons with multiple sclerosis in Europe. Genoa, Italy: Associazione Italiana Sclerosi Multipla; 2004.
43. National Institute of Clinical Excellence. Multiple sclerosis: National clinical guideline for diagnosis and management in primary and secondary care. London, UK: Royal College of Physicians of London; 2003.
44. Consortium of Multiple Sclerosis Centers. White paper: Comprehensive care in multiple sclerosis. Teaneck, NJ: Consortium of Multiple Sclerosis Centers; 2010.

45. Mead N, Bower P. Patient centeredness: A conceptual framework and review of the empirical literature. Social Science & Medicine 2000;51:1087–1100.
46. Salzberg Global Seminar [Internet]; c2011 [cited 2011]. Available from: http://www.salzburgglobal.org/current/Sessions.cfm?IDSPECIAL_EVENT=2754.
47. Coulter A, Collins A. Making shared decision-making a reality: No decision about me, without me. London: The King's Fund; 2011.
48. Heesen C, Kopke S, Richter T, Kasper J. Shared decision making and self-management in multiple sclerosis: A consequence of evidence. Journal of Neurology 2007;254(2):16–21.
49. Kensit DA. Rogerian theory: A critique of the effectiveness of pure client-centred therapy. Counseling Psychology Quarterly 2000;13(4):345–351.
50. Sumsion T. Facilitating client-centred practice: Insights from clients. Canadian Journal of Occupational Therapy 2005;72(1):13–20.
51. Lewin S, Skea Z, Entwistle VA, Zwarenstein M, Dick J. Interventions for providers to promote a patient-centred approach in clinical consultations. Cochrane Database of Systematic Reviews 2001;4: CD003267.
52. Barrier PA, Li JT, Jensen N,M. Two words to improve physician–patient communications: What else? Mayo Clinic Proceedings 2003;78:211–214.
53. Bauman AE, Fardy HJ, Harris PG. Getting it right: Why bother with patient-centred care? Medical Journal of Australia 2003;179(5):253–256.
54. Rocco N, Scher K, Basberg B, Yalamanchi S, Baker-Genaw K. Patient-centered plan-of-care tool for improving clinical outcomes. Quality Management in Health Care 2011;20(2):89–97.
55. Helitzer DL, Lanoue M, Wilson B, de Hernandez BU, Warner T, Roter D. A randomized controlled trial of communication training with primary care providers to improve patient-centeredness and health risk communication. Patient Education and Counseling 2011;82(1):21–29.
56. Cooper LA, Roter DL, Carson KA, Bone LR, Larson SM, Miller ER, Barr M, Levine DM. A randomized trial to improve patient-centered care and hypertension control in underserved primary care patients. Journal of General Internal Medicine 2011;26(11):1297–1304.
57. Redfern J, Briffa T, Ellis E, Freedman SB. Patient-centered modular secondary prevention following acute coronary syndrome: A randomized controlled trial. Journal of Cardiopulmonary Rehabilitation and Prevention 2008;28(2):107–115.
58. Harth A, Germann G, Jester A. Evaluating the effectiveness of a patient-oriented hand rehabilitation programme. Journal of Hand Surgery (European Volume) 2008;33(6):771–778.
59. Haidet P, Pateriti DA. "Building" a history rather than "taking" one: A perspective on information sharing during the medical interview. Archives of Internal Medicine 2003;163(10) 1134–1140.
60. Taylor R. The intentional relationship model. Philadelphia, PA: FA Davis; 2008.
61. Shine KI. Health care quality and how to improve it. Academic Medicine 2002;77(1):91–99.
62. Law M, Baptiste S, Carswell A, McColl MA, Polatajko H, Pollock N. The Canadian Occupational Performance Measure. 3rd ed. Ottawa, Canada; Canadian Association of Occupational Therapists; 1998.
63. Freeman JA, Langdon DW, Hobart JC, Thompson AJ. The impact of inpatient rehabilitation on progressive multiple sclerosis. Annals of Neurology 1997;42:236–244.

Index

Printed in the United States
by Baker & Taylor Publisher Services